CHEMICALS AND LONG-TERM ECONOMIC GROWTH

CHEMICALS AND LONG-TERM ECONOMIC GROWTH
Insights from the Chemical Industry

Edited By

Ashish Arora
Heinz School of Public Policy and Management
Carnegie Mellon University
Pittsburgh, PA

Ralph Landau
The Center for Economic Policy Research
Stanford University
Stanford, California

Nathan Rosenberg
Department of Economics
Stanford University
Stanford, California

Published in Conjunction with
the Chemical Heritage Foundation

A WILEY-INTERSCIENCE PUBLICATION

JOHN WILEY & SONS, INC.

New York • Chichester • Weinheim • Brisbane • Singapore • Toronto

On the Cover

Aerial photograph of The Dow Chemical Company, Midland Plant, as it looks today. Courtesy Post Street Archives.

For ordering and customer service, call 1(800) CALL-WILEY.

Library of Congress Cataloging-in-Publication Data:

Chemicals and long term growth : insights from the chemical industry /
   edited by Ralph Landau, Ashish Arora, Nathan Rosenberg.
      p.   cm.
   "Publication co-sponsored by the Chemical Heritage Foundation."
   "A Wiley-Interscience publication."
   Includes bibliographical references and index.
   ISBN 0-471-d18247-8 (cloth : alk. paper); ISBN 0-471-39962-0 (paperback)
   1. Chemical industry — Case studies. 2. Chemical industry —
History. 3. Chemical industry — Technological innovations — History.
4. Chemical engineering — History. 5. Petroleum chemicals industry —
History. I. Landau, Ralph. II. Arora, Ashish. III. Rosenberg,
Nathan, 1927–   . IV. Chemical Heritage Foundation.
HD9650.5.C535 1998
338,4′766 — dc21   97-28674

Printed in the United States of America

10 9 8 7 6 5 4

Contents

Preface

My professional life has moved largely in two orbits. I spent the first, lasting from the mid-1940s until 1982, in the chemical industry, much of that time running Scientific Design-Halcon International, a specialized engineering and research firm that developed several innovative chemical processes which are still in widespread use. In 1982, I sold my business interests and entered into a second trajectory: the community of academic economics, first at Stanford University and then two years later at Harvard University's John F. Kennedy School of Government. The concept of this book has been heavily influenced by both my business and my academic experience.

Our book should be of interest to several different audiences:

1. Chemical industry participants who are too busy with current tasks to know about the evolutionary paths that this industry has taken since its beginnings in the 1850s will discover a very dramatic history. Having access to the experience of chemical firms in the past should be very useful in devising corporate strategies for the future.

2. Economists and economic historians who are interested in long-term growth will find in this book the most detailed case history yet of a major technology-based industry in four countries (Japan, the United States, Great Britain, and Germany). By tracing over the last 150 years how factors internal and external to firms create wealth, the book suggests important lessons for economic growth.

3. Business schools and their students and graduates will be able to read what is typically absent from their crowded curricula: a detailed historical study of all factors impinging on the competitive situation of firms. This book is possibly one of the most extensive studies of a single industry ever published and may well provide a template for analyzing other technology-based industries. Business managers can only profit from understanding better how much the external environment matters to their jobs and how their overseas competitors may be affected differently by industry developments because they reside in a different social environment.

4. Policy makers who are concerned with fiscal and monetary, tax, regulatory, educational, and many other government policies will find a long-term analysis of how these matters have affected the growth of a technology-based industry.

The book does not presuppose knowledge of chemistry and will be accessible to the general reader.

I had for some years been intrigued by the factors underlying the development of the chemical industry. As early as 1958 I had written on the role of the chemical engineering profession in fueling the American rise in the petrochemicals field. In that study I highlighted the vital role the American university system had played in creating the field of chemical engineering. Over the years I wrote a number of papers in reaction to the rapidly evolving technological and economic changes occurring in many countries. However, a real world macroeconomic experience that my company encountered in 1980 powerfully demonstrated to me that even great technology may not be enough in the face of adverse external forces.

Here is a brief summary of this experience that triggered my long interest in the intersection of micro- and macroeconomics. (A more detailed account appeared in Landau 1994, 1996a, b). My company (Halcon-Scientific Design) had developed nine major petrochemical processes since our founding in 1946 and had built over 300 plants in many countries. But with our last major discovery, a process for the manufacture of propylene oxide, a key ingredient in polyurethane plastics, foams, and rubbers, we felt strong enough to share in its exploitation with Arco Chemical Company as our 50 percent partner. This venture was extremely successful and had achieved a sales volume of $1 billion in just 11 years with plants in the United States, Spain, Japan, and the Netherlands. Our last plant was financed by a loan of $230 million at 11 percent interest. But then inflation roared in, fueled by an accommodating Federal Reserve chief appointed by President Jimmy Carter. By 1979, the President had to reverse course and appoint Paul Volcker, who immediately limited the money supply so that our interest rate rose almost overnight to 21 percent. Our cash flow disappeared and we realized we could not cope with the situation. We had to sell our half back to Arco, which had no difficulty raising cash. Economics had triumphed over great technology! The financial officer of Arco commented at the time, "I will teach you the value of money." He meant the value of possessing money versus technology. Technology, of course, had been our primary contribution to the 50 percent partnership with Arco.

Since joining the community of professional economists, I have been able to commit myself to attending conferences, workshops, and seminars; to writing papers; and to co-editing books. I became convinced that a book that would bring together academic economic inquiry and real world experience would be of great value for scholars and practitioners alike.

It quickly became clear to me, some years ago, that no one person had the expertise and background necessary to write the sort of book I had in mind, a volume that would combine an insider's knowledge of economics, the chemical industry, business history, social history, issues such as finance and the environment—and moreover address these topics in a way that would do justice to the experiences of the United States, Britain, Germany, and Japan over the last 150 years. But fortunately the editors were able to form a team with collective capabilities to take on this project. It has taken six full years to complete. My co-editors on this volume are Nathan Rosenberg of Stanford University, a pioneer in the study of the microeconomics of technical change, and Ashish Arora of Carnegie-Mellon University, also an expert in the economics of technical change (and once a student of Rosenberg's). A number of fine scholars and industry experts agreed to join us in this project and contribute individual chapters. Their backgrounds are described at the end of the book.

We have been helped by extensive advice and reviews given by our distinguished colleagues at Stanford Economics, especially Timothy Bresnahan, Lawrence Lau, Paul Romer, and Gavin Wright. This volume conveys the results of the extensive interactions among our authors over the last six years. We met in small and large groups, exchanged papers by mail, and talked on the phone.

Finally, we have had indispensable editorial supervision by Timothy Taylor, managing editor of *The Journal of Economic Perspectives*. He knitted together the different styles and emphases of our authors and minimized unnecessary overlap, while polishing the presentations and keeping them within our page allowance.

We have been much assisted by Deborah Carvalho and Donna Holm at the Center for Economic Policy Research at Stanford University, who helped manage the flow of messages, papers, and administrative matters. I also acknowledge gratefully the help of my two colleagues Lewis Gasorek and Judi Wind, who did extensive analysis of the data obtained for this book. I especially wish to thank my invaluable assistant, Angela Rey, who did the really hard job of keeping track of the flow over the years, and Johann Peter Murmann who assisted me in all sorts of tasks for this book.

We also wish to thank the Sloan Foundation, the Pine Tree Trust, and other donors for financial support of this project.

Now, as our book nears completion, we believe more strongly than ever in the importance of this kind of comprehensive cross-disciplinary industry study. We believe that building a greater understanding of the connections between the microeconomic world, exemplified in this book by the chemical industry, and the broader macroeconomic and social context of economic growth, competitiveness, and trade can lead to insights about how technological advantages are created and maintained across time and place. One of the great challenges will be to understand better the multitude of facts that connect technical change to longer-term economic growth and to the wealth of nations. After the efforts of the last six years, we can appreciate why such studies as ours are comparatively rare. To create a book that would weave together all levels in the matrix of comparative advantage, all participants committed phenomenal time and energy. It is not possible to eliminate all errors or omissions in such an effort. Nonetheless, we believe that the general picture assembled in the book is credible as well as illuminating.

I am confident that the reader will find novel insights, and perhaps inspiration, in the pages that follow.

RALPH LANDAU

REFERENCES

Landau, Ralph (1994) "Technology and the Matrix of Growth," *The Bridge*, National Academy of Engineering, Vol. 24, No. 4, Winter, pp. 16–19.

Landau, Ralph (1996a) "Strategy for Economic Growth: Lessons from the Chemical Industry," in Ralph Landau, Timothy Taylor, and Gavin Wright, *The Mosaic of Economic Growth*, Stanford University Press, Stanford, CA, pp. 398–420.

Landau, Ralph (1996b) "Managing Innovation," in *DÆDALUS*, Spring 1996, issued as Vol. 125, No. 2 of the *Proceedings of the American Academy of Arts and Sciences*, Cambridge, MA, pp. 19–37.

CHEMICALS AND LONG-TERM ECONOMIC GROWTH

PART I
Organization and Purpose

1 Introduction

ASHISH ARORA, RALPH LANDAU, and
NATHAN ROSENBERG

This book is a study of how commercial and technological leadership in the chemical industry has moved across countries and firms over the last century and a half, and the forces that powered this shifting leadership. It is aimed at those who are interested in how technology can be harnessed to create wealth, raise productivity, and sustain economic growth, and the constellation of factors that support this process. In other words, the book is aimed at economists, policy makers, and students of business and corporate strategy. We also hope that those working in the industry will find here a fresh perspective, because of the long historical perspective and broad sweep of the book.

The chemical industry is today the single largest manufacturing industry in the United States, and in Europe is the second largest after food and kindred products. In the mid-19th century, chemicals were the first science-based industry and, thus, the first industry legitimately to claim the distinction of being "high technology." Other high-tech industries of the post-World War II era, such as electronics and information technology, also depend on innovative science and the commercialization of technology. Thus, lessons gleaned from the development of the chemical industry may well be of more general value in enunciating policies that could nurture long-term economic health.

In seeking understanding and policy lessons, however, we have found that such questions cannot be answered by the study of only one country, even one as large as the United States. Fortunately, this science-based industry has at least a 150-year history, and it has existed for many of these years in a number of other countries, where conditions and circumstances have differed from one another. Only by such multi-country studies is it even possible to grasp the underlying organizational principles that must guide a study with many variables changing simultaneously, and often not independently of each other.

For the purposes of this book, therefore, we have selected four countries where enough data have accumulated, and whose chemical industries are large but of different strengths at various times. These are first, Britain and Germany, where the chemical industry definitively emerged. The United States' industry has since become the largest in the world. Japan is included partly because it is the world's second-largest industrial economy, with the second-largest national chemical industry, but also because the Japanese chemical industry has been quite different in its evolution from the others and

We acknowledge the numerous exchanges with colleagues, from which we have greatly benefited in the preparation of this introductory overview, as well as in subsequent chapters. On specifically economic matters, we wish particularly to thank Michael Boskin, Timothy Bresnahan, Dale Jorgenson, Larry Lau, and Paul Romer.

Japanese chemical industry has been quite different in its evolution from the others and less influential outside its borders. Of course, the industry is more widespread on a global basis, but these four countries offer the most extensive data for study.[1]

In delving into the chemical industry, or indeed any industry, it is necessary to move between different perspectives and levels of analysis. Much technology, of course, is created and commercialized within firms. The internal factors contributing to growth of firms have to do with management, the recruiting of trained personnel, the conduct of research and development, and investment and marketing. But the ability to draw upon science and technology also depends on the performance of institutions of learning, on the scientific and engineering research conducted by various public and other institutions, and on whether the broader social climate looks favorably upon science. Corporate investment and strategy decisions, especially in a capital-intensive industry like chemicals, depend on the cost and availability of capital, the demands of the owners of the companies for a share of the profits, and how labor markets function. Capital supply, in turn, depends on the functioning of the external (and now international) capital markets; the intermediating institutions, such as banks, that allocate capital from savers to investors; and the competition for capital by other firms and governments. Government also affects the conditions for business through tax policies, fiscal and monetary policies, antitrust, patent law, environmental regulation and trade policies, and a variety of other policies that set the ground rules of finance, disclosure, and competition in an economy. It is broadly agreed that certain constellations of these laws, practices, policies, and attitudes and their interactions are more conducive to long-term economic growth than others. If firms do not do well, the country's overall economy suffers, growth can slow, and trade will falter. But if the country does not do well, neither can the firms. Inquiries into the subject of what produces long-term growth have become the focal point of much current debate among politicians, economists, journalists, and scholars.

This book is not a history of the chemical industry per se, although the industry's history permeates the volume. Nor is it a professional historian's history of the nations studied; the numerous treatises on national history contain many details and controversies that are not germane to the pursuit of the fortunes of the chemical industry. Instead, only enough sociopolitical and historical background is presented to frame the environment in which the firms in the industry and the industry itself have had to function. Most obviously, this is an economics book with, we believe, considerable relevance for policy, and is intended for economists, economic historians and policy makers. But we also hope and believe that many participants in the chemical industry, as well as students of business and corporate strategy, will find the book accessible and pertinent. However pressing the problems of the day may be to participants in the chemical industry, a long-term perspective on the industry can also be helpful in assessing new situations as they arise and in gauging changes in the larger global environment that affect their competitors as well as themselves.

The next section offers an introduction to the chemical industry, aimed in particular at economists and policy makers who may know little of it. Conversely, the following section offers an introduction to the economic issues, aimed in particular at those hands-on participants in the chemical industry or in business schools who may know little of how their work lives are discussed and modeled by economists, and also at

---

[1]Despite the importance of France in European and world history, in our judgment its chemical industry *would not* add enough insight to justify lengthening the volume further.

policy-makers who may not be familiar with the economic approaches to these issues. Nevertheless, we hope that some in all camps will enjoy both sections. The introduction concludes with an overview of the structure of this volume.

A BRIEF INTRODUCTION TO THE CHEMICAL INDUSTRY

To the public, as well as to many economists and policy makers, and even to some industry participants, the chemical industry may seem almost invisible, except when it is implicated in an environmental problem. But the industry is important to consider for at least three reasons: its high-tech history, its size, and the fact that it touches so many other industries and products.

First, chemicals were the first science-based, high-technology industry, tracing back 150 years to the British dyestuff manufacturers of the 1850s. That chemicals are a high-tech industry often comes as a considerable surprise to those first considering the industry. The importance of knowledge-based industries has been so heavily discussed in the media that it sometimes sounds like a cliché, but it remains true that all of the plausible ways for America to continue increasing its standard of living, or for poor countries of the world to approach the living standards of wealthier countries, revolve around discovering, adopting, and commercializing new technology. Chemicals are also one of only two major high-tech industries in which the United States has maintained its competitive lead in international trade (aerospace is the other),[2] and the growth rate of the chemical industry has exceeded that of the economy for most of the years since World War II. The chemical industry can hardly be accused of excessive dependence on government support: For example, the industry's research and development has been financed almost solely by private investment. However, in policy areas including education, trade, and antitrust and environmental laws, as well as in various institutions of society, the history of the chemical industry offers clear illustrations of the interdependence between the government-created economic-policy environments, its institutions, and the actions of individual high-technology firms.

Second, the sheer size of the industry compels attention. Although it is large in all four countries, the data are available in greatest detail for the United States, so the material that follows is largely for this country. According to the Standard Industrial Classification (SIC) system used by the U.S. Department of Commerce, chemicals (SIC 28) are the largest manufacturing industry, followed by industrial machines and equipment, electrical and electronic equipment, and food and kindred products (Table 1-1). On a value-added basis, chemicals represent about 11.3 percent of U.S. manufacturing and produce about 1.9 percent of the U.S. gross domestic product. The chemical industry is America's number one exporter; in 1996, the exports amounted to $61.8 billion. From a global perspective, the U.S. chemical industry, with $372.4 billion in sales, had about 24 percent of the global market in 1996. Japan, at $216 billion in sales, had 14 percent. The two other nations discussed intensively in this book are Germany and Britain, with chemical industry sales of $117 billion and $56 billion, and a global market share of 8 percent and 4 percent, respectively. The rest of global chemical production is spread primarily across the rest of western Europe and Asia.

---

[2]In light of our subsequent discussion on the ambiguous relevance of trade to growth of a country, we cite this fact primarily to establish the strong performance of this industry relative to other high-tech industries and the economy more generally, as shown in Figure 1-1.

TABLE 1-1. GDP by Industry 1996 (Value Added)

U.S. Manufacturing Sector	$1,332 billion
(17.4% of total GDP)	$7,636 billion
Chemicals & allied products	$157.8
Industrial machines & Equipment	150.2
Electronic & electric equipment	143.8
Food & kindred products	122.6
Fabricated metal products	98.2
Printing & publishing	90.4
Motor vehicles & parts	85.1
Paper products	57.1
Instruments	52.3
Other transportation	49.7
Petroleum & coal	30.1
Other	294.9
Total	$1,332.2

Source: U.S. Bureau of Economic Analysis, Survey of Current Business, November 1997.

Third, the chemical industry has been responsible for the diffusion of new technologies and positive spillover effects to a wide range of other industries. The chemical industry is incredibly diverse. Its output includes more than 70,000 products: paints and coatings, pharmaceuticals, soaps and detergents, perfumes and cosmetics, fertilizers, pesticides, herbicides and other agricultural chemicals, solvents, packaging materials, composites, plastics, synthetic fibers and rubbers, dyestuffs, inks, photographic supplies, explosives, antifreeze, and many, many others. Few other industries approach this degree of complexity and diversity. Chemicals are building blocks at every level of production and consumption in agriculture, construction, manufacturing, and the service industries. The chemical industry has played a central role in generating technological innovations for use by other industries, such as autos, rubber, textiles, consumer products, agriculture, petroleum refining, pulp and paper, health services, construction, publishing, entertainment, and metals. In this regard, the chemical industry is evidence of the more general fact that the benefits of technology spill over to other industries. Only a quarter of chemical output, however, goes directly to the consumer, which may contribute to the relative invisibility of the industry.

A comparison of this industry's performance relative to other high-tech industries is given in Table 1-2 and Figures 1.1 to 1.4. Table 1-2 positions the major R&D industries to show that chemicals and pharmaceuticals remain first. Furthermore, although more recent data are not available, it nevertheless shows that chemicals are the most privately financed R&D performer and thus follow the market most closely. Trade has historically played an important role in economic growth, and Figure 1-1 illustrates that the chemical industry has a positive trade balance. Figure 1-2 gives profitability statistics, which show the cyclicality of the industry as it matures. Figure 1-3 shows that R&D and capital expenditures are strongly related — R&D is capital investment, which with the associated physical investment creates wealth in the industry. It is noteworthy that in Figure 1-4 the productivity rates for chemicals are below those for other high-tech industries. This may well relate to the cyclical nature of the industry in recent years (see Figure 1-2) as it approaches maturity and large plants must be built in expectation of

TABLE 1-2(a). The Major R&D Industries for 1997 (in bil. of current $)

Industry	Est. 1997 R&D
1. Chemicals and pharmaceuticals	31.4
2. Transportation	30.4
3. Telecommunications	29.0
4. Computers	22.5
5. Electronics	15.2
6. Software	9.9
7. Semiconductors	6.8
Total	145.2

Total R&D for 1997 is estimated by Battelle at $192 bil., of which 62.8% ($120.6 bil.) will be financed by industry, 32.4% ($62.2 bil.) by government, and 4.9% ($9.4 bil.) by others (such as non-profits, universities, research institutions).
Note: Numbers have been rounded and may not add to 100.
Source: "1997 R&D Funding Forecast" by Battelle.

TABLE 1-2(b). The Major R&D Industries 1991 & 1992 R&D Expenditures (in bil. of current $)

Industry	1991	1992	1992 Privately Financed	Percentage Privately Financed
1. Aerospace	16.63	16.12	6.25	39%
2. Electrical machinery & communications	13.42	13.55	9.69	72%
3. Machinery	14.78	15.14	14.07	93%
4. Chemicals	14.65	16.71	16.42	98%
5. Autos, trucks, transportation	10.80	10.37	9.48	91%
6. Professional & scientific instruments	8.71	9.65	7.43	77%
7. Computer software & services	5.77	6.66	3.89	58%
8. Petroleum	2.50	2.34	2.33	99%
Total	87.26	90.54	69.56	

Total R&D in 1992 was $154.5 bil., of which R&D performed by industry was $107.6 bil., so that the above are the bulk of R&D performers. 2. Electrical Machinery & Communications — Battelle estimates these figures for 1995 at $182 bil. and $130.6 bil. No breakdown after 1992 is available.
Source: National Science Foundation, Division of Science Resources Studies, "Selected Data on Research and Development in Industry: 1992" and "National Patterns of R&D Resources."

future markets, yet often have to operate below capacity. Although not illustrated, investment in this industry is a two-way street, with foreign investment in the U.S. slightly greater than U.S. investment abroad. For 1996, these figures were $75 billion versus $69 billion.

Given a recognition of how pivotal the chemical industry has been and is, the next barrier to serious consideration of the industry usually turns on vocabulary and

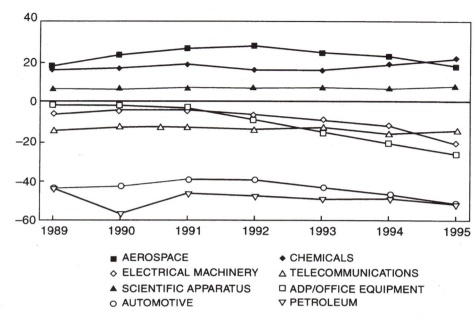

Figure 1-1. Trade Balance (in billions of $). Source: U.S. Bureau of Census, U.S. Industries with heavy R&D.

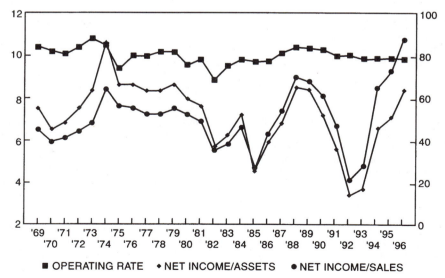

Figure 1-2. Return on Assets & Profit Margin (%) and Operating Rate (%). Source: Chemical Manufacturers Association.

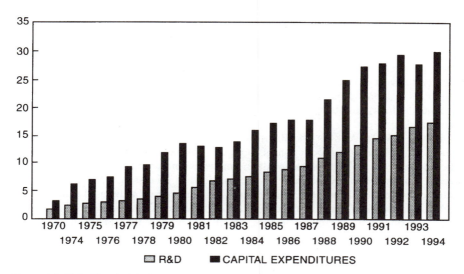

Figure 1-3. U.S. Chemical Industry R&D vs. Capital Expenditures. Source: Chemical Manufacturers Association.

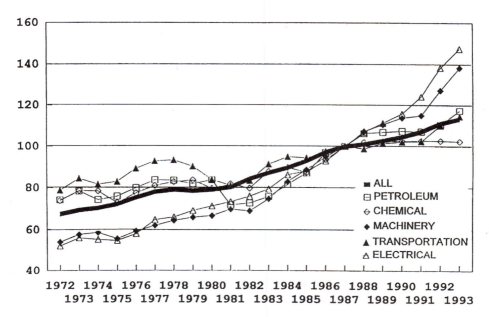

Figure 1-4. Labor Productivity Index (1987 = 100). Source: Bureau of Labor Statistics, U.S. Dept. of Labor.

historical background. While many economists and policy makers have dim memories of such terms as "organic" and "alkali" from a chemistry course in their distant past, they do not feel well-prepared to launch into a policy discussion on the subject! Nor do they feel prepared, based on a half-remembered course in 19th century European history, to discuss the evolution of the chemical industry in that time frame. Let us kill two birds with one stone by summarizing briefly the essential historical trajectory of the chemical industry and using that history as an opportunity to lay out some recurrent terms and concepts.

In the 1830s and 1840s, the British had the world's dominant chemical industry, which was focused on production of inorganic chemicals. Inorganic compounds are those taken from the earth, such as salt and minerals, and which are then processed into useful products employed directly or used in further processing. One leading set of products were the alkalis, such as lime, soda ash, and caustic soda, used extensively in textiles, glass making, fertilizers, etc.; another includes acids, such as sulfuric and nitric, which are often used in tanning, textiles, dyeing, and myriad other applications. Alkalis and sulfuric acid produced in large quantities are referred to in this book and elsewhere as heavy chemicals. Currently, organic chemicals such as ethylene, benzene, and propylene, which are produced in large quantities, are more typical examples of heavy chemicals.

The early version of the inorganic chemical industry was in some sense closer to mining than to the science-based chemistry of today. The date when the chemical industry moved to being science-based is usually given as 1856, when William Henry Perkin discovered the first synthetic dye (mauve) and launched the modern organic chemical industry. Since then, organic compounds have proved the most important class of chemicals because they are more varied and pervasive than the inorganic compounds. Organic chemistry begins with inputs that contain hydrocarbons (composed of hydrogen and carbon)—for instance, coal, oil, and natural gas—which form the backbone of final organic chemical outputs. In the first stage of processing, these raw materials are refined to produce primary outputs such as benzene and ethylene. In subsequent processing, chemicals such as chlorine and oxygen are added to the hydrocarbon backbones to give the compounds their desired characteristics. The final output may, for example, be nylon or polyester fiber, a plastic, or a pharmaceutical product. The hydrocarbon backbone for British dyestuffs, and for organic chemistry throughout almost all of the 19th century, was provided by coal.

Britain dominated the dyestuff industry until the 1870s. These were glory times for Great Britain. The nation was rich. Its organic chemical industry making dyestuffs had the technical know-how, the largest supply of basic raw material (coal), and the largest customer base (textiles). But those advantages slipped away, for reasons discussed at several places in this volume, and by the end of the 1880s the Germans dominated the organic chemical industry. By 1913, German companies produced about 140,000 tons of dyes, Switzerland produced 10,000 tons, and Britain only 4,400 tons. The American industry was a large producer of basic inorganic chemicals, but for its organic chemicals it depended mainly on German dyestuffs and other imports, except for use in the domestic production of explosives.

World War I changed the relative positions of nations in the chemical industry, at least for a few years. The United States was cut off from German dyestuffs and built its own organic chemical industry. The German industry, shattered by war, fell on hard times. Both Britain and Germany sought to create chemical companies that could be national standard bearers. In 1925, Germany formed IG Farben company, merging all dye firms into one stock company Britain created Imperial Chemical Industries (ICI) through a merger of smaller entities in 1926. In the U.S., a number of consolidations took place among private companies, forming such large and competitive entities as DuPont, Union Carbide, Allied Chemical, and American Cyanamid. IG Farben soon regained Germany's former dominance over the European chemical industry. At the same time, the U.S. chemical industry was gaining strength through the development of a large petroleum-refining base and also was building its skill in designing large-scale, continuous processing plants through the use of expert chemical engineering tools. This skill, largely in the hands of specialized engineering firms, was readily transferable to the burgeoning petrochemical industry, which was based on the cheap petroleum and nutural gas feed stocks with which the United States was rich. The European chemical industry, however, still used coal, rather than petroleum, as its main feedstock.

World War II resulted in the physical destruction of a significant portion of the German chemical industry. The U.S. industry was now using petrochemicals to produce fibers, plastics, and many other products, while dyestuffs shrank in importance. America's chemical industry grew enormously and dominated the market at least until the 1970s. However, as world prosperity returned in the decades after World War II, so did a successful chemical industry in Germany, and in Europe more generally. Petrochemical industries were soon well-established in Asia, in the oil-rich countries of the Middle East, and elsewhere. No longer did one country dominate; the industry's growth had made it a worldwide success story. Competitive advantages at the firm level truly came to the fore, with different companies in different countries excelling at particular skills or products and trading extensively with one another. Japan was an exception. Although the Japanese chemical industry grew to become the second largest in the world by providing inputs for Japan's home market, it has not yet become a major player in international markets for products or technology.

A BRIEF INTRODUCTION TO ECONOMIC PERSPECTIVES ON PRODUCTIVITY AND GROWTH

Many economists avoid using the term "competitiveness" when referring to national economies, because countries do not compete—firms and industries do. When economists do refer to the competiveness of a nation's overall economy, they typically mean the ability to sustain an acceptable rate of growth in the real standard of living of the population, while avoiding social costs such as high unemployment, environmental damage, or an excessively uneven distribution of income. Furthermore, current growth rates must be achieved without reducing the growth potential in the standard of living of future generations, which implies some constraint on borrowing from abroad. Over the long term, the most promising way to increase the standard of living is to raise the productivity of labor.

Notice that when a nation's competitiveness is defined in terms of its own productivity, the standard of competition is not so much the economic performance of other nations as how a nation manages to raise its own productivity over time. Although United States workers remain the most productive in the industrialized world, as they have been for most of the 20th century, their lead over workers in the other key countries has diminished in the last few decades. The convergence of productivity levels among OECD countries (the most industrialized) in recent years has given rise to a burgeoning literature (Abramovitz 1986).

What are the sources of long-run growth in productivity, and how might public policy and institutions affect the growth of productivity over time? Among economists, the thinking about issues of industrial productivity and a fertile climate for economic growth was for some time separated into two boxes whose sides did not touch.

One perspective on productivity and growth was essentially macroeconomic. The modern study of the macroeconomic sources of economic growth dates from the mid-1950s, when Moses Abramovitz (1956) and Robert Solow (1957) attempted to measure the contribution of labor and capital to long-term U.S. economic growth. Their studies demonstrated that only a small portion of per capita growth of the U.S. economy from the 1870s to the 1950s could be explained by increases in the quantities of conventionally measured inputs alone. Some other residual factor (sometimes called "technology" or, more accurately, multi-factor productivity) was responsible for one-half or more of economic growth. For several decades, a number of economists have worked within this framework, using various categories and subcategories of labor and capital, and attempting to capture research and development as a separate input (Kendrick 1961; Denison 1962, 1969). More recent studies continue to find that "technology," including that embodied in physical-plant and skilled workers, remains an important factor in U.S. economic growth (Jorgenson 1994; Boskin and Lau 1997). The findings, however, continue to be highly sensitive to the particular underlying assumptions and methodologies.

The economist's use of the word *technology* here is worth noting, because technology, as employed by growth economists in these studies, is by no means identical to technology as understood by an engineer or a voter. In these studies, improvements in technology — or technical progress — could not be measured directly, but instead could only be inferred as a residual factor. "Technology" was, in effect, all sources of productivity growth other than the measured inputs of capital and labor. Since this usage may include a large number of variables other than technological progress, Abramovitz astutely labeled the residual "a measure of our ignorance." Later, in a larger context, Abramovitz has also argued that this category of "technology" may include a large component of "social capability," a term embracing institutional and social factors that also may affect growth.

From a macroeconomic-growth perspective, the chief elements of a growth-oriented policy are to create, by using the levers of fiscal and monetary policy, a stable climate for savings and investment and for demand growth, as well as a climate for improving labor productivity by utilizing technology effectively, together perhaps with such steps as assuring a broad level of education and setting up an antitrust law to prevent the emergence of monopoly power. The efficiency of market forces is presumed to take care of most of the rest.

An alternative perspective on productivity and growth begins at the level of the firm or the industry, rather than at an aggregate level. It might be called the "worm's eye" view of productivity and technological growth. This microeconomic perspective focuses on how new technologies were developed, commercialized, and diffused. It is associated with students of technical change who have explored the complexity of the commercialization process that converts technology into wealth creation, and who have emphasized that this process is evolutionary, path-dependent, and incremental (see, for instance, Nelson and Winter 1982; Rosenberg 1986; Rosenberg, Landau, and Mowery 1992). This perspective brings in the processes of design and innovation; learning by doing and using; technology transfer from abroad; and the incentives for research and development; all of which can produce new and improved products, processes, and services. It asks how firms learn from experience and from one another, how good management differs from bad, how firms differ in the ways they gather and transmit information internally, why some firms compete successfully in international markets and others do not. In short, it does not assume that all firms are the same, as the earlier aggregate studies did. In the economics literature, this work often focused on a particular innovation such as the steam engine, the railroads, or the transistor. In the business literature, the focus would also include what management practices were used to revive a dormant company or to start up a firm that might drive an entire industry (Chandler 1990; Porter 1990).

The microeconomics perspective not only discusses how firms may differ but raises the possibility that some industries may be more important to long-term productivity growth than others. An industry such as computers (which might be termed a "general purpose" technology), which generates many spillover effects to a host of other industries, may well be more important to the nation's productivity growth than others, such as office construction, that have fewer spillovers.[3] It is at least plausible to argue that the breadth and diversity of products from the chemical industry give it the ability to contribute positive spillovers to many other industries as well. The new products created by an industry may stimulate the rise of another industry. Job formation thereby may be enhanced. An aggregate view of the economy cannot easily capture these sorts of structural differences.

Of course, these two views of growth — the macroeconomic and the micro-economic — are inextricably linked and complementary to each other. Firms create the growth and the wealth in a capitalist society, but governments and societies create the business climate, the conditions, the boundaries, within which firms must act. For new knowledge to contribute to economic well being, it must be complemented and realized commercially by investment. The investment and commercialization decisions of firms are, however, affected by macroeconomic and institutional conditions such as the availability of savings (investible funds), the rate of interest, and government legal and regulatory policies.

In the last few years, new mathematical modeling techniques and richer data have made it possible to build more realistic models that eventually could consider how events at the microeconomic level affect the long-term growth and behavior of the macroeconomic system. Scholarly interest in this area, commonly known as "endo-genous growth theory," has increased rapidly. Romer (1994) describes the eventual contributions of further development of endogenous growth in the following way:

> Ultimately, this endogenous view on growth will put us in position to offer policy makers something more insightful than the standard neoclassical prescription — more saving and more schooling. We will be able to rejoin the ongoing policy debates about tax subsidies for private research, antitrust exemptions for research joint ventures, the activities of multinational firms, the effects of government procurement, the feedback between trade policy and innovation, the scope of protection for intellectual property rights, the links between private firms and universities, the mechanisms for selecting the research areas that receive public support, and the costs and benefits of an explicit government-led technology policy...

To this list one might add such matters as tax and regulatory policies, subsidies to business, labor policies, and the like.

---

[3] But as *inter alia* Jorgenson (1996) notes, spillovers are difficult to model quantitatively at the aggregate level.

A glance around the globe certainly suggests that national, organizational, and firm efficiencies vary greatly. The collapse of the Soviet Union, despite high rates of capital investment and education (by world standards), shows that much more is involved in growth than pouring money into manufacturing plants run by unmotivated workers in the absence of market competition. The rapid growth of certain economies in East Asia — Taiwan, South Korea, Singapore, and others — suggests that the way nations handle economic organization can also have a profoundly positive influence on their economic outcomes (Stiglitz 1993). Britain's redistributionist policies after World War II had an impact on the trajectory of the British economy over those decades. Germany's dense set of institutional arrangements involving government employees, unions, and financial institutions, which enhanced the relative attractiveness of long-term investments, had real effects as well on her rapid rise from defeat after World War II.

While the study of growth in the aggregate offers a perspective on what happened, by its nature such a study fails to include the reason something happened (Crafts 1995, 1996). For example, an aggregate study may say that investment went up, but without explaining why investments were made at that time or why research and development may have lagged. By attempting to explain growth largely or solely in terms of aggregated inputs such as labor and capital, this work has tended to obscure the role of ideas, innovations, institutions, and policy in growth outcomes. These and other difficult-to-measure factors probably account for a significant part of the productivity growth of nations, and technological gain in the narrow sense used by engineers may well be only a fraction of the total. Our book cannot offer any direct estimate of the contribution of these other factors, except to show that they do matter and that their importance varies over time.[4]

Popular discussions of an industry's or an economy's productivity or competitiveness often seem to take a quick detour into international trade. All too often, the philosophy seems to be that an overall trade surplus, or a surplus in a particular industry, or a surplus with a particular country, should always be viewed in a positive light, while a deficit in any of these areas should be viewed as a negative. Several of these points can be dismissed quickly. For example, if U.S. workers and firms were to produce exports and ship them overseas, without receiving any imports in return, the U.S. would have a trade surplus; but by producing without receiving any goods in return, it would also experience a lower standard of living. In a world with nearly 200 nations, there is no reason at all why the United States should have equal imports and exports with every separate nation. After all, if the U.S. buys from Japan, Japan buys from Germany, and Germany buys from the U.S., the U.S. would have a trade deficit with Japan and a trade surplus with Germany, but an overall balance. One sometimes hears claims that we should balance our trade not only with individual countries but also within individual industries. This proposition is even more peculiar. Would anyone seriously uphold the

---

[4]An enlightening discussion and debate on these issues of history and growth is contained in The Brookings Institution's *Economic Activity* volume for 1995, with particularly interesting essays by Gregory Mankiw, Edmund Phelps, and Paul Romer. For another discussion that brings together the two perspectives of economic historians and growth economics, see the symposium on "New Growth Theory and Economic History: Match or Mismatch?" published in the May 1996 issue of the *American Economic Review*, with papers by Paul Romer, representing the new growth theorists, while the economic historians present were Nicholas Crafts and Martin Weitzman. Another discussion is contained in the *New England Economic Review* issue for Nov.-Dec. 1996, which summarizes an extensive conference organized by the Federal Reserve Bank of Boston, with contributions from Dale Jorgenson, Nathan Rosenberg, Alan Greenspan, and others. The subject is further elaborated in the *Conference Proceedings*, No. 40, entitled "Technology and Growth," edited by Jeffrey G. Fuhrer and Jane Sneddon Little, Federal Reserve Bank of Boston, Boston, MA, June 1996.

idea that the United States should export as much oil to nations of the Middle East as it imports from those countries? The guiding principle behind economic transactions, whether international or domestic, is to focus on producing in one's area of comparative advantage (see the following), not to try to duplicate all other producers.

The issue of an overall trade deficit is real and worth attention, but for a different reason than is usually given. A trade deficit indicates, in effect, that the nation is consuming (through its imports) more than it is producing in goods and services: it is importing capital. Conversely, a trade surplus occurs when a nation is producing more than it is consuming and investing the surplus abroad; thus, the nation is exporting capital. The high U.S. trade deficits since the mid-1980s meant, by definition, an inflow of foreign investment into the United States. Such flows of foreign investment can sometimes benefit a nation. For example, South Korea ran enormous trade deficits through the 1970s, which meant the country was borrowing from abroad. But that money was invested in productivity-enhancing ways, so that the country until recently has been able to benefit from its international borrowing and repay the money without incident. Several Latin American countries, for instance Brazil and Mexico, in the late 1970s and early 1980s borrowed heavily from abroad but did not invest the money productively, and so when it came time to repay, they were faced with economic austerity plans.

An overall trade deficit thus may have an effect on productivity growth, but the size and direction of such an effect is not as clear as might as first be imagined.[5] The mainstream economics view of the trade balance is that it is determined by a nation's overall levels of savings and investment. If our policy goal is long-term growth, then we do not wish to hinder investment. However, we may well wish to encourage a higher level of national savings, to reduce the nation's current consumption and focus it more toward investment. One method of increasing national savings available to private business would be to reduce the drain of funds going to the federal budget deficit, as is now already in sight. Another method would be proposals to call forth more private savings, via education or incentive plans.

Those who instead would call for reducing the flows of trade through protectionism tend to ignore or minimize the many ways in which competition from international trade benefits an economy. American consumers have a wider choice of less expensive products, which makes their dollars go farther. U.S. producers are forced to become more efficient by competing against the best in the world. Since competition holds down price increases, it allows monetary policy to be less concerned about inflation, and to permit lower interest rates. When other countries grew, this did not harm the United States. On the contrary, it gains from the greater demand for U.S. goods and services abroad, and the imports of valuable products (such as consumer electronic devices and automobiles) which American consumers desire.

A more vulgar error, most often committed by politicians, editorial writers, and occasionally businessmen, is to judge a nation's productivity and competitiveness based on whether sales are growing for a particular industry or even a particular leading

[5]Furthermore, the subject of international capital flows tends to overshadow trade flow patterns (see The Bank Credit Analyst for July 1995). For instance, a number of mainstream economists are concerned that the foreign investment received by America since the 1980s has been used more for consumption rather than invested in productivity-enhancing ways. If this is true, then the United States has been on a consumption binge, consuming imports and paying for them with dollars that foreign investors then had to invest or lend to countries that accept dollars. The persistently low American savings rates of the last 15 years or so have required sums from abroad to finance the deficit—which, again, are the mirror images of the trade flows. (Board on Science, Technology, and Economic Policy 1994.) Moreover, the large American deficit has reduced the flow of international capital to less-developed countries.

company. These groups are concerned mostly with stabilization of the economy over a short period of the few years between key elections, rather than with longer-term growth. Such stabilization policies tend to deal, for example, with balancing the costs of higher unemployment versus the risks of rising inflation. Some advocates of "industrial policy" even speak as if government can guarantee future prosperity by showering favors on private industry. This perspective is misguided in several ways. First, government subsidies, whether provided directly (say, through tax breaks) or indirectly (say, by trade protectionism) always come at a cost elsewhere in the economy. Perhaps there are very special circumstances when such subsidies make economic sense for a country, but that conclusion surely cannot be reached merely by looking at whether the recipient of the subsidy benefits from it! Second, the focus on a particular industry or company tends to minimize the critical importance of a propitious macroeconomic climate within which all business must work. For example, if an industrial policy benefits a few companies but leads to higher inflation, it is unlikely to represent an improvement in economic welfare.

Economists distinguish sharply between the competitive advantage of a firm or industry in global markets and the comparative advantage of a nation. Competitive advantage for a firm can be measured by the sales and profitability of that firm. But the principle of comparative advantage of a national economy in global markets holds that national economies will benefit if countries focus on areas in which their own economy has the greatest productivity advantage — or the smallest disadvantage — and then trade with each other. This principle has a sometimes-counterintuitive implication that all nations have comparative advantages, because any nation has an area in which its productivity disadvantage is smallest. To clarify the philosophy behind this principle, consider a camping trip with one Eagle Scout and four greenhorns. The Scout may be better at every aspect of camping — in our terminology, the Scout has a competitive advantage in every area — but that does not mean that everyone is better off if the Scout tries to do everything, while the greenhorns sit around and chat. Instead, the task is to identify the areas in which the productivity disadvantage of the greenhorns is the least — say, collecting firewood — and have them work in the areas of their comparative advantage. As a result of this division of labor, everyone can be better off. Thus, we can apply the terms of comparative advantage to industries, and even to institutions and policies, among different nations.

From a modern perspective, the original definition of comparative advantage was static, determined by nature or geography. It seems less relevant today, especially for high-technology products such as chemicals or electronics. The case of Japan is revealing: A country without much arable land and with few resources except the skill, determination, and hard work of its people has become, less than two generations after the heavy losses of World War II, the second-largest economy in the world. Thus, comparative advantage is dynamic and can shift over time, especially in a world of rapid technological change.

Let us be clear: Neither we nor any of the authors in this book believe that just because a policy helps the competitiveness of the chemical industry it is necessarily good for the nation. However, we presume to say all the authors in this book would agree that certain policies which benefit the chemical industry may also benefit certain other science-based industries, and the economy as a whole, by raising productivity growth. The challenge, of course, is to identify policies where the positive-spillover effects across other sectors of the economy are greatest, and where the costs are relatively low and distributed in ways that do the least harm to economic efficiency.

This perspective brings us back to the issue of what national policies would be most conducive to overall productivity growth and how that relates to growth for the chemical industry in particular. We believe that national economies do compete, but not in the absurd sense of playing a zero-sum game in which every nation tries to run a trade surplus with every other nation in every product. Instead, our vision is that industries and firms, aided by a government-sustained stable macroeconomic and business climate and supportive institutions favoring innovation and commercialization of new technology, search continually for new and improved products, processes, and services for customers. By favoring technology development and especially its commercialization, and by providing appropriate incentives for firms to take risks, government policy can help its firms to compete and thus raise the nation's standard of living relative to that of other nations, as well as its own past performance. That this is performed differently in different countries, even countries with a largely similar technology, is illustrated by the steady differences in the unemployment rates between Europe and the U.S.

THE MATRIX OF COMPARATIVE ADVANTAGE

Our point of departure is that a well functioning and growing economy depends on: (1) A complicated mix of institutions that goes well beyond the laissez faire view emphasizing the legal system, the budget authority, and the central bank: (2) Policies which, while not justifying naïve interventionism, favor all the institutions of a society working together with a market to generate higher economic growth but retaining a suitable macroeconomic environment and support for education and science, (3) The size of the market and the historical sociology of the country. As already noted, existing models and theories in economics are limited in their ability to accommodate many of these aspects.

We should not permit the limited models and theories of today to inhibit us. By studying the chemical industry in-depth, we hope to understand whether and how details of organization — of firms, the industry, financial institutions, and others — have larger, economy-wide implications. Our organizing framework, The Matrix of the Sources of Comparative Advantage, is illustrated in Table 1-3. This matrix has been used before in *The Mosaic of Economic Growth*, a more general study of economic growth (Landau, Taylor, and Wright (eds.) 1996), which should be referred to for a more in-depth approach to the economic issues we can only briefly describe in this book. However, that inquiry did not have the richness of detail and sharpness of focus that a study dedicated only to a single industry can provide. To use a metaphor, aggregate studies look only at the forest, not the trees, even though the trees collectively make the forest. Our book is an attempt to connect the trees as they grow with the growth that an aerial observer sees only as that of the forest. Of course, trees in a forest cannot take remedial action to correct their environment, but firms and nations can.

HOW THIS BOOK IS ORGANIZED

Country Overviews

The section following this Introduction consists of three chapters that cover the history of the chemical industry in the four countries. Inevitably, there is some overlap between this section of the book and the cross-cutting chapters that follow, but our hope is that the overlap helps cement the matrix together. The first chapter of this section, by Johann Peter Murmann and Ralph Landau, "On the Making of Competitive

TABLE 1-3. The Sources of Comparative Advantage

	U.S.	U.K.	Germany	Japan
National Governance				
Socio-Political Climate				
Macro Policies				
Fiscal				
Monetary				
Trade				
Tax				
Institutional Setting				
Financial				
Legal (including Torts and Intellectual Property)				
Corporate Governance				
Professional Bodies				
Intermediating Institutions (SEFs, Fraunhofer, etc.)				
Structural and Supportive Policies				
Education (including university-industry relations)				
Labor				
Science & Technology (including the role of Engineers and Scientists)				
Regulatory & Environmental				
The Industry				
Companies within the Industry				

Advantage: The Development of the Chemical Industries in Britain and Germany since 1850," describes the development of the industry in Britain and Germany. From a global perspective, the second half of the 19th century saw a contest between Britain and Germany for leadership in the chemical industry. Britain began with what appeared an almost insurmountable advantage—and lost it in a few decades. During the 20th century, much of the story involves how the chemical industries of the two nations were buffetted by two world wars, depression, and the rebuilding of Europe after World War II.

The next chapter, by Ashish Arora and Nathan Rosenberg, tells the tale of "Chemicals: A U.S. Success Story." The U.S. chemical industry has been a world leader at least since World War II, whether one measures crudely by sales or more tellingly by the disproportionate number of new products and processing innovations that have come from U.S. firms. This chapter analyzes the factors contributing to this success. It analyzes the importance of the size of the domestic market, the availability of key natural resources, the responsiveness of universities, investment in plant and equipment, investment in research and development, and institutional factors and government policies. This chapter sorts out which factors were most important in which time periods, and the role of public policy.

The final chapter of this section, by Takashi Hikino, Tsutomu Harada, Yoshio Tokuhisa, and James Yoshida, addresses "The Japanese Puzzle." The "puzzle" of the title refers to the fact that, while the Japanese chemical industry is the second largest in the world, many people — even many Japanese — cannot name a single Japanese chemical company. Such Japanese car companies as Toyota or Nissan, or electronics companies as Sony and Panasonic, span the globe. But when, if ever, have Western readers seen the names of Japanese chemical companies such as Showa Denko, Ube Industries, or Asahi Chemical? This chapter seeks to explain how Japan's chemical industry became so large while becoming even more invisible than its foreign competitors.

Innovation at the Heart of the Industry

The next section contains three chapters focusing on what is perhaps the key characteristic of the industry: the overwhelming importance to the industry of science-based technical changes, and the legal context in which they must be implemented.

The first chapter in this section, by Ralph Landau, discusses "The Process of Innovation in the Chemical Industry." Landau is well-situated from his long-term industry experience to describe how technologies appear and the long and complex process through which they are commercialized and subsequently improved. Indeed, Landau shows that when an innovation has occurred is often not especially clear-cut; often what is thought of as a significant innovation is really the accumulation of many gradual improvements. He highlights how the technology-led growth of the chemical industry has fed back and stimulated growth elsewhere in the economy. The discussion highlights the role that key individuals and small entrepreneurial companies can play at critical historical junctures. The chapter also illustrates how macroeconomic, institutional, and even chance historical events shape the complex route through which new knowledge is transformed into wealth.

The legal system plays a particularly important role in this process, and the second chapter, "The Impact of the Legal System on Innovation," gives a somewhat personal review of the legal system under which all innovation must function. Here Tom Campbell and Ralph Landau discuss the effects of America's extreme litigiousness and its effect on innovation. Other chapters in the book by Murmann and Landau, by Arora and Rosenberg, by Arora and Gambardella, and by Esteghamat, also take up aspects of the legal system.

The third chapter in this section, by Nathan Rosenberg, is on the subject of "Technological Change in Chemicals: The Role of University-Industry Relations." Rosenberg discusses the different paths along which the university systems in Britain, Germany, and the United States have evolved. In particular, the modern chemical industry has required both the science of chemistry and the expertise of engineering; indeed, early in the 20th century this interrelationship was already recognized as important and specialized enough that American universities created departments of chemical engineering. Even though the German chemical industry benefited greatly from German strengths in organic chemistry, Germany never fully embraced chemical engineering. Rosenberg takes on the question of why various countries took different paths, and how this mattered for the development of the industry.

Surveying the Levels of the Matrix

The chapters in this section gradually work their way down the Matrix of the Sources of Comparative Advantage, from the top to the bottom, although emphasizing only some of them. Thus, the first chapter in this section, "The Industry Evolves Within a Political, Social, and Public Policy Context" by Micheline Horstmeyer, discusses and compares national governance and socio-political systems across Britain, Germany, Japan, and the United States and places the chemical industry in the context of its time. The chapter is not intended to be a comprehensive scholarly history of these countries; rather, it is meant to sketch the background for the developments studied in this book.

The second chapter, by Barry Eichengreen, explores "Monetary, Fiscal, and Trade Policies in the Development of the Chemical Industry." Eichengreen begins with the fact that from 1913 to 1989, "the share of world chemical production accounted for by Japan rose from 2 to 14 percent, while that accounted for by Britain, Germany, and the United States declined from 12 to 4, 25 to 7, and 36 to 24 percent, respectively." Eichengreen then inquires into how much of this change can be accounted for by monetary, fiscal, and trade policies. His analysis focuses on impacts at particular times and places, which can be very important for extended periods. But, over the long time horizon, it appears difficult to argue that these particular policies were central in shaping the market share of the chemical industry, although they were vital to sustaining general economic growth rates that in turn influenced their chemical-industrial growth rates.

The third chapter, by Marco Da Rin, considers the subject of "Finance and the Chemical Industry." The chemical industry is capital-intensive, so the institutions that govern the flow of capital are potentially very important. Various institutions and practices have prevailed at different times: reinvestment of retained earnings in a family-owned firm, as in Britain during the 19th century; raising funds through joint-stock companies, as in German chemical firms after about 1873; ownership exercised through broadly distributed stockholders, as occurs in the United States today; or ownership exercised through more concentrated ownership, often through a bank that owns part of the company directly, as is common in Japan and Germany. Da Rin considers the strengths of these systems at various times — and how their weaknesses sometimes led to evolutionary institutional change.

Kian Estaghamet, in the fourth chapter of this section, addresses the "Structure and Performance of the Chemical Industry Under Regulation." Estaghamet focuses particularly on environmental regulations over the last half-century, since this is an area in which national regulations have changed and varied most, and it is also an area in which the chemical industry has been a particular target of policy. In the United States, such regulations have been largely enforced through lawsuits; in Germany, the more common enforcement mechanism is direct and strictly enforced government regulation. Among the four countries on which this volume focuses, the differences in the current extent of environmental regulation are not extreme — although Germany probably imposes the most onerous standards at present, and enforcement is the most inflexible in the U.S. But as the chemical industry grows worldwide, there is reason to wonder whether production of certain potentially hazardous products will be shifted to Latin America or Asia to avoid the strictest environmental regulation.

In the fifth chapter of this section, Ashish Arora and Alfonso Gambardella examine the "Evolution of Industry Structure in the Chemical Industry." Industry structure includes consideration of how many firms are active in the industry, the patterns of entry and exit of firms, and the degree of competition between chemical firms that are often making somewhat different products. The authors emphasize how the importance of technology in the chemical industry encouraged the market to bring forth a way to diffuse and sell technology: the so-called specialized engineering firms or SEFs. When an industry is young, it is driven by market opportunities and may require more entrepreneurial management; careful attention to cost minimization may be less important than exploiting the full potential of the emergent technologies. Their discussion shows how these forces have diminished in importance as the industry has matured and the implications of how the industry has had to restructure.

The final paper of this section, by Alfred D. Chandler, Jr., Takashi Hikino, and David C. Mowery, discusses the chemical industry at the level of individual firms. In "The Evolution of Corporate Capabilities and Corporate Strategy within the World's Largest Chemical Firms: The 20th Century in Perspective," these three authors survey the evolution of corporate strategy in the chemical industry, focusing on the leading chemical firms in the United States, the United Kingdom, Germany, and Japan. Of course, the authors also call on the insights accumulated earlier in the book to give a sense of when corporate strategy was heavily constrained by events, institutions, or national policy and when it was free, for better or worse, to make crucial decisions.

The Modern Chemical Industry and Corporate Governance

Perhaps the two great external influences on the chemical industry today are finance and public policy. In his chapter entitled, "Connecting Performance and Competitiveness with Finance — A Study of the Chemical Industry," Albert D. Richards discusses how chemical firms are evaluated in the increasingly intertwined global financial markets. The world has changed since 1980: two earlier oil shocks, with consequent deep macroeconomic disruptions; the globalization of the industry; the rapidly increasing flood of pension money into equity markets; the almost instantaneous flow of capital around the world; and the maturing of the industry and its technologies. This turmoil has resulted in significance changes in corporate governance, and Richards' chapter lays out the increased importance of financial analysts and their impact on management and shareholders. The chapter is a pioneering effort in many respects and therefore includes what might be called a "primer" that outlines the financial accounting methodology used. This is possibly the first published detailed account of the industry as seen by a chemical industry analyst. The empirical analysis is based on a rich dataset gathered from a variety of sources.

Conclusion

Finally, a concluding chapter attempts to integrate and review the findings of the individual chapters, and to draw some conclusions from them for policy and corporate management. Of course, conclusions must be drawn cautiously: One does not want to fall into the error of claiming that because a policy helped sales in the chemical industry at some particular time, it necessarily benefited the economy as a whole. However, after having considered the history of a key industry over 150 years in four countries, along a variety of dimensions, only excessive timidity should inhibit one from drawing at least some provisional conclusions. It is impossible to capture all of the elements of successful economic policy in a few lines or a few paragrahs or even a few volumes of text. But in either case, it is possible to give an impressionistic sense of distinguishing characteristics.

Economic growth is extraordinarily important. An average annual growth rate of 2.5 percent—close to the world average in recent years—will double the world's average standard of living in less then 30 years. When one considers the political and economic turmoil of the 20th century, combined with the rapid growth in population and dreadful wars, it seems remarkable that the standard of living during this century has increased so dramatically in the advanced industrial countries, and indeed increased to some extent almost everywhere, while providing millions of jobs for the growing workforce.

Policies for achieving success, be it for governments or for corporations, are not always obvious, especially when one looks back at 150 years of history that cover war, depression, changes in fundamental goals of government, the rise and fall of technologies, and much more. Our study suggests that simple (and simplistic) recipes for promoting economic growth are likely to be wrong. Indeed, sustained industrial growth seems to require a constellation of factors: sensible government policies, wise managements, and productive engineers and researchers. By combining the macroeconomic and microeconomic views of growth, and looking across the Matrix of the Sources of Comparative Advantage to consider key factors, and studying how the environment shapes the short-run and long-run incentives for firms, it is possible to develop an educated sense of which public policies are most likely to be successful in enhancing a nation's productivity and standard of living.

REFERENCES

Abramovitz, Moses (1956), "Resource and Output Trends in the United States since 1870," *American Economic Review*, May, 46(2), pp. 5–23.

Abramovitz, Moses (1986), "Catching Up, Forging Ahead, and Falling Behind," *Journal of Economic History*, June, pp. 386–406.

Board on Science, Technology and Economic Policy (1994), "Investing for Productivity and Prosperity," National Academy Press, Washington, DC.

Boskin, Michael and Lawrence J. Lau (1997), *The Sources of Long Term Economic Growth*, forthcoming. See also Lau (1996), same title, in Ralph Landau, Timothy Taylor, and Gavin Wright (1996), *The Mosaic of Economic Growth*, Stanford University Press, Stanford CA, pp. 63–91.

Chandler, Alfred D. (1990), *Scale and Scope: The Dynamics of Industrial Capitalism*, Harvard University Press, Cambridge, MA.

Crafts, Nicholas F. R. (1995), "Golden Age of Economic Growth in Western Europe, 1950–1973," *Economic History Review*, 48(3), pp. 429–47; see also manuscript entitled "The First Industrial Revolution: A Guided Tour for Growth Economists", London School of Economics.

Crafts, Nicholas F. R. (1996), "Deindustrialisation and Britain's Industrial Performance Since 1960," *The Economic Journal*, The Royal Economic Society, Blackwell Publishers, Oxford, UK, January, pp. 172–183; see also "Endogenous Growth: Lessons For and From Economic History," discussion paper, Centre for Economic Policy Research, London, January.

Denison, Edward F. (1962), "The Sources of Economic Growth in the United States and the Alternatives Before Us," Committee for Economic Development, New York.

Denison, Edward F. (1969), "Some Major Issues in Productivity Analysis: An Examination of Estimates by Jorgenson and Griliches," *Survey of Current Business*, May, Part II, 49(5), pp. 1–28.

Jorgenson, Dale W. (1994), "Investment and Economic Growth," Simon Kuznets Lecture, Yale University, November 9–11.

Jorgenson, Dale W. (1996), "Technology in Growth Theory," in Jeffrey G. Fuhrer and Jane Snedden Little, eds., *Technology and Growth*, Federal Reserve Bank of Boston Conference Series No. 40, pp. 45–77.

Kendrick, John W. (1961), *Productivity Trends in the United States*, Princeton University Press.

Landau, Ralph, Timothy Taylor, and Gavin Wright (1996), *The Mosaic of Economic Growth*, Stanford University Press, Stanford, CA.

Nelson, Richard R. and Sidney G. Winter (1982), *An Evolutionary Theory of Economic Change*, Harvard University Press, Cambridge, MA.

Porter, Michael (1990), *The Competitive Advantage of Nations*, Free Press, New York.

Romer, Paul (1994), "The Origins of Endogenous Growth," *Journal of Economic Perspectives*, Vol. 8, No. 1, Winter, p. 20.

Rosenberg, Nathan (1986), "An Overview of Innovation," in Ralph Landau and Nathan Rosenberg, *The Positive Sum Strategy*, National Academy Press, Washington, DC, pp. 275–305.

Rosenberg, Nathan, Ralph Landau, and David Mowery (1992), *Technology and the Wealth of Nations*, Stanford University Press, Stanford, CA.

Solow, Robert M. (1957), "Technical Change and the Aggregate Production Function," *Review of Economic Statistics*, August, 39(3), pp. 312–20.

Stiglitz, Joseph E. (1993), "Endogenous Growth and Cycles," National Bureau of Economic Research Working Paper No. 4286, Cambridge, MA.

PART II
Country Overviews

2 On the Making of Competitive Advantage: The Development of the Chemical Industries in Britain and Germany Since 1850

JOHANN PETER MURMANN and RALPH LANDAU

INTRODUCTION

Are competitive advantages of industries created by social action or are they inherited as immutable endowments of nations? This is a key question for entrepreneurs, corporate strategists, and public policy makers alike, as they must decide where to invest their personal, corporate, or national resources. The answer to this question holds enormous implications for how economic analysts should think about corporate strategy and public policy in relation to a particular branch of industry.

In this chapter, we examine the comparative performance of the British and German chemical industries since 1850. The history of these two national industries suggests that competitive advantages of industries are created by developing appropriate firm capabilities, market structures, educational systems, legal frameworks, and other public policies rather than inherited as unchangeable endowments of a particular national environment. As the first nation to experience the industrial revolution, Britain was the undisputed leader in the inorganic chemical industry in 1850. With the rise of new technologies in the second half of the nineteenth century, Germany was able to leap ahead in the new synthetic organic chemical sector and catch up with Britain in the inorganic sector. In response to the First World War (The Great War), Britain adopted policies that allowed an organic chemical industry to take hold on her soil in the interwar period. Later on, when oil replaced coal as the main chemical feedstock and when Germany lay in ruins at the end of World War II, Britain enjoyed a great opportunity to leap ahead of Germany once more and become the home of the leading European chemical industry. Yet it was Germany, rather than Britain, which was able to create an economic and social environment that allowed the German chemical

We would like to thank Ashish Arora, Marco Da Rin, Alfonso Gambardella, and Nathan Rosenberg for very helpful comments on earlier versions of this chapter.

industry to flourish and outperform the British one during the economic boom of the postwar period. In recent years, the change in British economic and social relations has improved the position of the British chemical industry vis-à-vis the German one, although Germany remains the largest chemical exporter in the world.

This shift of competitive advantages between these two national chemical industries over the last 150 years provides an excellent opportunity to investigate the factors that are responsible for the creation and destruction of competitive advantages on the level of a specific industry. A close examination of how Britain and Germany have gained and lost competitive advantages in chemicals over the last 15 decades strongly supports the idea that whether or not a national industry enjoys competitive advantages depends significantly on the extent to which the actions of governments, firms, and other stakeholders are mutually reinforcing in creating an environment in which a particular industry can flourish. When the diverse collection of actors who influence the different levels in the matrix of the sources of comparative advantages[1] interact and formulate a coherent set of policies, an industry is more likely to prosper than when these actors operate in isolation and do not take into consideration the impact of their policy choices on the development of the industry as a whole.

In the pages that follow, we will try to document this thesis by examining closely the causal factors that have shaped how competitive advantages in chemicals have developed over the last 150 years in the context of the two national environments. We begin with a close examination of the period from 1850 until the First World War, which saw the rise of synthetic organic chemical products, the growth motor of the modern chemical industry. Both world wars disrupted commercial trade between nations and forced a reorganization of traditional supply lines. As chemicals were of significant military importance for both wars, each national chemical industry was jolted because of the wars and catapulted to a different development trajectory. To understand the impact of these special events, we devote two shorter sections to tracing the implications of the wars on the competitive position of each national industry and the environment in which it operated. We examine again in more detail how competitive advantages developed in the interwar era. Finally, we investigate how the sources of competitive advantage that emerge from our investigation up to the Second World War can in turn explain the developments in the post-World War II period. The chapter closes by identifying some lessons for corporate strategists and public policy makers.

EARLY GROWTH IN THE CHEMICAL INDUSTRY, 1850–1914

In the middle of the nineteenth century, Britain was the home of the largest chemical industry in the world. The industrial revolution had created an enormous demand for chemical acids and alkalis to supply the voracious appetite of the textile, soap, glass, and steel industries. Germany had just recently begun her own industrialization with the building of a railroad network, but the German economy was at this point substantially smaller than the British one.

A few numbers will illustrate how much Britain was ahead of Germany. In 1850, raw cotton consumption per head of population was almost 20 times larger in Britain than in Germany. A total of 1,290,000 steam engine horse-power units were installed in

[1]A detailed description of the framework is provided in the introduction of this book.

British factories compared to only 260,000 in Germany (Landes 1969, p. 221). By 1870, British gross domestic product per head of population was twice as high as Germany's (Maddison 1991). Britain was the largest producer of heavy chemicals. Her output of 304,000 tons of soda in 1867 and 590,000 tons of sulfuric acid in 1870 was an order-of-magnitude larger than Germany's 33,000 tons of soda and 43,000 tons of sulfuric acid; in per capita terms, Britain's production of these two chemicals was 11 and 17 times, respectively, above that of Germany. Production of these two chemicals can be taken to represent a rough proxy for overall economic activity during the 19th century.

But from 1870 until the outbreak of the First World War, the competitive position of the German chemical industry improved dramatically.[2] By 1910, Germany was producing around 500,000 tons of soda, reducing the lead of Britain on a per capita basis to only 2.3 to 1. In 1913, Germany produced 1,600,000 tons of sulfuric acid, which surpassed British production by 350,000 tons and even pulled slightly ahead on a per capita basis (1.05 to 1). Although Britain was the birthplace of the synthetic dye industry, Germany's production of 137,000 tons of highly valued dyes in 1913 gave her about a 19-to-1 lead over Britain on a per capita basis. With a world market share of about 85% in dyes and a similar position in new pharmaceuticals, the German chemical industry had captured a dominant position in synthetic organic chemicals similar to the one Britain had enjoyed in soda half a century earlier. On the eve of World War I, the German chemical industry was more than twice as large in terms of value (2.4 versus 1.1 billion Reichsmark) and 1.5 times as large on a per capita basis as Britain's. British gross domestic product per capita, however, remained 1.5 times as large as that of Germany ($4024 versus $2606 in 1985 dollars).

How was the German chemical industry able to catch up with Britain in the inorganic sector and surpass Britain so dramatically in the synthetic organic sector at the eve of World War I? In setting out to identify the key factors, it is useful to recall that in the final analysis all competitive advantages are based on being able to offer a product at a lower cost or higher quality.[3] The various factors that have been highlighted in the matrix framework described in the introduction to this book differ in terms of how many steps they are removed in the causal chain from the processes directly responsible for price or quality advantages. To organize our analysis, we begin with examining the factors that were directly responsible for giving German chemical firms cost and quality advantages over their British counterparts. We will then move the causal analysis backward to more distant factors that allowed German firms to enjoy competitive advantages over their British counterparts. This analysis will take us from the behavior of British and German firms to the institutional setting in which the two chemicals industries were embedded.

The primary reason German chemical firms collectively outperformed their British rivals in the period from 1870 to 1913 was that the Germans made more frequent and

---

[2]For the classic accounts of how the German economy caught up with the British, see Veblen (1915) and Landes (1969).

[3]We focus in this chapter on competitive advantages and not on comparative advantages. To establish whether Germany or Britain had a comparative advantage in chemicals at a particular time, we would need to ascertain the relative productivity levels of all other national industries to fully express the basic Ricardian concept. Note that Ricardo's national-level concept of comparative advantages can also be applied to two companies that make several different products. Hence, it is possible to speak of how company A has a comparative advantage over company B. Today, chemical firms are clearly assessing their comparative advantages in the way they design alliances, swaps, and other restructurings vehicles. The recent swap of ICI's nylon for DuPont's acrylics business is a case in point.

earlier investments in what Chandler (1990) identified as the mutually reinforcing areas of manufacturing, marketing, and management. These investments allowed firms to reap cost advantages from large-scale or wide-scope production. In addition, German firms spent large sums on research and development that enabled them to create a continuous stream of new or superior products and processes. As we will show, German chemical firms were more likely than British ones to make these investments because they resided in a social context that placed much higher value on technical and scientific education than was the case in Britain, and they received more useful support from their nation's educational and political institutions.

The Creation of Firm Capabilities

The Case of the Dye Industry as an Illustration of the Organic Sector   The case of the dye sector in the chemical industry illustrates how differences in British and German business practices shaped the performance of the industry. The first synthetic dye was invented and manufactured by William H. Perkin in Great Britain. He was an 18-year-old student of August Wilhelm Hofmann, hired in 1845 as a professor at the newly founded Royal College of Chemistry in London to bring the teaching and laboratory techniques of the famous German chemist Justus von Liebig to Great Britain (Beer 1959). Hofmann had been working on the chemistry of coal tar dyes for over 15 years when his student Perkin in 1856 synthesized, by sheer coincidence, a purple substance as he tried to create quinine by oxidizing aniline (Suchanek 1992). Realizing the commercial value of his discovery, Perkin filed for a British patent and within a year set up a factory with his father and brother that would make aniline purple dye from coal tar via a number of intermediate steps.

The news of Perkin's discovery quickly crossed Europe. French chemists immediately tried to make colors by reacting aniline with every chemical on their shelves (Travis 1993), and Franois E. Verguin in Lyon found another way to make a purple color very much like Perkin's. In late 1858 or early 1859, Verguin made a discovery that was commercially even more important than Perkin's initial discovery: He synthesized a red dye which is known as fuchsine in France and magenta in Britain. Verguin swiftly filed for a French patent that became heavily contested in French courts as chemists discovered different pathways for making purple, red, and blue aniline dyes. Because high fashion developed a great demand for these brilliant colors, synthetic dye-making became a very lucrative business that attracted many new firms.

In this opening stage of the synthetic-dye industry, British firms performed extremely well. The German industry that started in the early 1860s was much smaller and relied on copying British and French inventions. Professor Hofmann, through whose laboratoryoratory most of the inventors of new synthetic dyes in Britain had passed, confidently predicted in 1862 that his adopted country was destined to develop the leading synthetic-dye industry in the world, since it was endowed with large amounts of coal and since it produced the greatest amount of coal tar as a by-product of making illuminating gas and iron (Hofmann 1863). Furthermore, Hofmann pointed out that Britain's enormous market for textiles gave British dye makers the advantage of a very large home market. In the first decade of the synthetic dye industry, British firms remained leaders in this new industry; between 1859 and 1861 alone there were 26 British patents on dyes from aniline and homologues (Hornix 1992, p. 163). But German firms increased their production steadily, and sometime between 1870 and 1880 they overtook the British industry.

How could Britain, as the most advanced industrial nation and the market with the largest demand, lose its competitive advantages in dyes to a much less-industrialized country like Germany? An analysis of this dramatic episode in industrial history suggests that British firms fell behind because they failed to make investments in crucial firm capabilities of manufacturing, marketing, management, and research.

Creating Competitive Manufacturing Capabilities   Many German firms sprang up in the early 1860s to jump on the synthetic-dye bandwagon. Initially, firms such as Hoechst (1863), Bayer (1863), BASF (1865), and AGFA (1867) copied the dyes that British and French manufacturers had put on the market. With many German firms trying to follow this imitative strategy, competition was keen (Schröter 1991). To survive, the successful firms had to improve the yield of chemical reactions and find processes that would use fewer chemical steps, so that they could undercut the price of the competitors and enlarge their market share. British firms, on the other hand, blessed with a larger and more sophisticated home market, focused more attention on product innovation than on process improvement. Since British dyers were very keen on getting synthetic colors other than purple and red, British firms experienced a particularly strong demand for new colors. German firms soon realized, to a greater extent than their British counterparts, that dye-making was susceptible to enormous economies of scope. Costs could be reduced, for example, by making a great many different dyes with the same production equipment. Since many dyes had similar organic intermediates, firms could achieve cost reduction by creating a higher internal demand for organic intermediates and thereby reap economies of scale. Every year the number of dyes offered by the successful German dye firms increased. By 1913 Hoechst, for example, made 10,000 different dye products and offered its customers a full spectrum of colors for every conceivable fabric (Hoechst 1990). No British firms offered a comparable line.

Economies of scale and scope could also be achieved by integrating backward into the production of basic chemicals and intermediates. BASF, for example, was formed by Friedrich Engelhorn, who had produced fuchsine as early as 1861 (BASF 1990), with the specific business plan of erecting a fully integrated plant that would make dyes and also the acids, alkalis, and intermediate organic chemicals needed for dye production. By 1890, BASF had created the largest chemical plant in the world in Ludwigshafen. Hoechst began to produce acids and basic chemicals around 1883. Bayer imitated this strategy when a contract with its supplier of sulfuric acid (the most important acid in dye production) expired in 1892. Bayer's management decided that it could reduce its cost by integrating backward into the three most important acids — sulfuric, hydrochloric, and nitric (Flechtner 1959, p. 138). The relocation to a larger site in Leverkusen enabled Bayer to replace its patchwork of production lines in Elberfeld (now called Wuppertal) with a rationally planned, integrated factory that was specifically designed to reap production economies (Duisberg 1933). British dye firms, in contrast, tended to buy intermediate organic chemicals, acids, and alkalis, from other firms.

In the early 1880s, scientists discovered that some of the chemical intermediates derived from coal tar also had therapeutical properties. A number of the German dyestuff manufacturers then leveraged their technological capabilities and diversified into pharmaceuticals. In 1883 Hoechst was the first dye firm to market a pain-relieving and fever-reducing pharmaceutical under the name of Antipyrin. Bayer also seized this opportunity. In 1888 Bayer started selling a fever-reducing drug called Phencetin; in 1899 it patented Aspirin, a pain-relieving, fever-reducing, anti-inflammatory wonder drug that made the firm famous all over the world. Together with German firms such as Schering and Merck, which confined their activities to pharmaceuticals, these dye firms dominated the world market for new chemical-based drugs until World War I. Another dye firm, AGFA, diversified into the production of photochemicals in 1887. To secure their ever-growing coal needs, the alliance of BASF, AGFA, and Bayer went so far as to buy the coal mine Auguste Victoria in Marl (Verg, Plumpe, and Schultheis 1988).[4]

British dye firms from the 1880s onward failed to offer a variety of dyes similar to that offered by the leading German dye firms. They also did not diversify forward into pharmaceuticals or photochemicals, or integrate backward into the production of acids or alkalis. When three of the leading German dye firms integrated backward into coal mining, German firms had captured over 80% of the world market.

Creating Competitive Marketing Capabilities   Large capital-intensive plants only pay off when a firm can ensure high-volume throughput and long production runs (Chandler 1990). The need to ensure high levels of plant utilization gives a firm that builds large plants a strong incentive to develop its marketing capabilities and find a steady stream of orders to keep the plants busy. Thus, simultaneously with making large investments in plants, German chemical firms devoted great resources toward building a network of sales representatives in textile markets throughout the world (Beer 1959; Haber 1958; Miall 1931).

These marketing capabilities evolved from the time when German firms were copying synthetic dyes from British and French producers. When many entrepreneurs sold the same product, marketing abilities were a key factor in winning German customers. As German firms found customers outside the boundaries of Germany, they translated their sales catalogs into foreign languages and built sales forces that were particularly extensive in such important markets as the United States, Britain, China, France, Russia, and India. By 1890, competition in the dye industry had become largely a rivalry among German firms, which collectively controlled over 80% of the world market as noted above. At the turn of the century, these sales networks had become so expensive that the largest dye firms combined into two alliances, the *Dreibund* (Union of Three) made up of Bayer, BASF, as well as AGFA; and the *Dreiverband* (Association of Three) made up of Hoechst, Cassella, and Kalle. The alliances were designed to reduce overlap in sales networks and thereby achieve further economies of scale and scope (Haber 1971). British firms did not make as extensive investments in sales forces and were increasingly unable to compete with German firms for market share in the international dye industry.

German firms employed other techniques to improve their marketing capabilities. A number of them very early on offered technical assistance to dyers who were steeped in the centuries-old craft tradition of attaching natural dyes to textiles and needed to learn how to fix these new dyes on different kinds of textiles. The process of distributing a dye evenly across a piece of cloth is a challenging activity that requires different procedures according to the fabric and the chemicals in the dye. German firms gained significant competitive advantages and improved the marketability of the new synthetic dyes by viewing the dye business as a technical service. These technical-assistance

---

[4]We do not have data on the cost of coal from the alliance's own mine and on prices for coal in the open market, making it impossible to judge whether the firms actually achieved a cost advantage by this move. Henry Ford found that it made no economic sense for an automobile manufacturer to integrate all the way back into steelmaking, which is a reminder that vertical integration has its limits. But note that British dye firms did not consume enough coal even to make the proposition of buying a coal mine worth considering.

programs comprised a variety of services: for example, the German firms dispatched experts to the plants of dyers and invited dyers to send personnel for a year-long training course at their factories in Germany (Haber 1958; Beer 1959). British firms did not match the quality and scope of these technical-assistance programs and thus were increasingly less popular with dyers around the world.

Creating Competitive Management Capabilities   One reason German firms quickly learned to sell technical assistance along with their dyes was that a majority of dye firms had among their founders, in addition to businessmen, academically trained chemists who could appreciate the intricacies of dying processes. Eugen Lucius and Adolf Brüning at Hoechst, August and Carl Clemm at BASF, and Carl Alexander von Martius at AGFA—just to name a few—were all formally educated in chemistry. When German dye firms initially lacked a connection to academic knowledge of organic chemistry, they quickly brought some chemical talent into the firm. For example, the founders of Bayer, Friedrich Bayer and Johann Weskott, who were trained as a trader of chemicals and a colorist respectively, hired their first academic chemist, August Siller, only a year after starting the firm (Verg, Plumpe, and Schultheis 1988). Successful British firms such as Perkin & Sons and Levinstein also had academically trained leaders, but in Britain this pattern was more the exception than the rule.

As the demand for synthetic dyes increased and existing firms grew to fill this increased demand, they became large complex organizations that could not be managed by a single owner or a small group of relatives. To take advantage of growth opportunities—that is, to make informed strategic decisions about product portfolios and vertical and horizontal integration opportunities (Chandler 1990)—it was necessary for dye firms to hire large staffs of salaried managers who would administer large plants and sales forces. Those firms that grew into large chemical enterprises all made the transition to professional leadership by salaried managers starting in the 1880s. German founding families often turned their firms into public corporations when the financial needs of the firms outstripped the founders' personal resources (for details see the Da Rin Chapter).

British firms of this time, however, remained personally managed and therefore small; it was not unusual for founders of British dye firms to sell their firms rather than build managerial hierarchies that would allow their firms to take advantage of growth opportunities. Perkin, for example, sold his firm in 1874 to Brooke, Simpson & Spiller and retired to devote himself to pure scientific work, as was characteristic for well-educated upper-class Britons.[5] It was not until the middle of the 1920s that managerial hierarchies comparable to those of the large German chemical firms were created in the British chemical industry.

When the founding families of German dye firms handed responsibility to the first generation of professional management, they selected chemists as chief executive officers because they thought that scientifically trained management would be in the best position to appraise the risks and opportunities faced by these expanding enterprises. In addition, the rational outlook of a technically competent management could infuse the entire organization with rational business practices whose aim was to ensure the growth and profitability of the enterprises. Karl König at Hoechst and Carl Duisberg at Bayer were among the first in a long line of chemists who have led large German

---

[5]For a masterful but not uncontroversial account of the relationship between British values and the decline of British economic peformance, see the famous essay by Wiener (1981). The chapter by Micheline Horstmeyer gives a more complete picture.

chemical firms. In fact, the majority of managers at the highest levels of the large German chemical firms remain scientists and technologists up to the present day. In general, the upper management of British firms in the late 19th century were less endowed with scientific and technological competency.

Creating Competitive R&D Capabilities   The most significant investments that determined which firms would maintain competitive advantages and grow occurred in research and development. Those firms that built large research and development organizations ensured themselves of a continuous stream of new products that created large profits. High profit margins could not be maintained with existing dyes, whose prices were constantly falling as new firms entered the market.

Initially, the search for new dyes was essentially a process of trial and error—more akin to skillful cooking than to the systematic synthesis of particular molecular structures. In the early period of synthetic dyes, British firms had more people engaged in searching for new dyes and thus were more successful than German firms. However, in 1865 Friedrich A. Kekulé determined that benzene, which is the building block for all coal-tar dyes, had a ring structure of six carbon atoms. This discovery opened up the possibility of searching for new dyes with the guide of chemical theory, and those advanced in organic chemistry were now much better equipped to create new dyes than those who were relying on empiricism and serendipity. While chemists had been employed from the very beginning of the dye industry, they initially were charged with a whole range of duties, from directing production and quality control to the search for new dyes (Homburg 1992). When dye firms had achieved a considerable size and chemistry was in the position to assist the process of finding new dyes, however, it became economically viable to create laboratories where academically trained chemists would look systematically for new chemical products. Between 1877 (BASF) and 1886 (Bayer), the seven largest dye firms in Germany set up research laboratories (Homburg 1992). By 1890, the big three German chemical firms employed about 350 academically trained chemists in a work force of around 14,000. By 1912 the number had risen to about 930 in a work force of about 30,000 (Marsch 1994).

The commitment of the large German dye firms to the pursuit of new products through research and development is illustrated by the race to develop a commercial process for man-made indigo, the "king" of all dyestuffs, and the last important natural dyestuff. After the university professor Adolph Baeyer synthesized the dye in the laboratory in 1880, the challenge was to find a process that could compete with natural indigo growing primarily in India. BASF worked for 17 years and spent 18 million marks—as much as the entire stock capital of the company—before it had a commercially viable process. Hoechst also invested heavily into indigo research and by 1904 had developed an even better process than that of BASF.[6] Between 1902 and 1904 alone, Hoechst invested 11 million marks in production facilities (Wetzel 1991, p. 64).

Those firms that committed large resources into R&D profited enormously from their innovations. This process, often called the industrialization of innovation, made it

---

[6]René Bohn developed in 1901 an even better blue synthetic dye based on anthracene. That dye grabbed much of the market of the first synthetic indigo dye developed by BASF, until the first synthetic indigo made a comeback after World War II because of the world-wide blue-jeans craze (Teltschik 1992, p. 10). The new dyestuff, Indanthrene (indigo + anthracene), became one of the most successful dyestuffs in the 20th century (Verg, Plumpe, and Schultheis 1988, p. 173). Fortunately for BASF, Indanthrene came out of its own laboratories. This is an example of how R&D efforts can be an insurance policy against new discoveries. For an account of the cultural, scientific, and technological significance of natural and synthetic indigo, see Seefelder (1994).

possible to achieve extraordinary returns on a large and diversified portfolio of research programs. From 1890 to the First World War, the largest German dyes firms were able to pay out dividends that averaged a 20% rate of return for stockholders, in addition to retaining large earnings for expansions. (The chapter by Da Rin later in this volume will deal with the difficulties of financing capital-intensive chemical plants. It will assess to what extent the German chemical firms possessed an advantage because of their large retained earnings in the early decades of the dye industry.) No British dye firm made similar investments in research and development. Ivan Levinstein, who was born and trained in Germany, also hired a sizable number of chemists, but his efforts remained on a much smaller scale than those of the largest German firms.

Patent statistics underline the differences between the British and German firms in research and development. While the six largest German firms for coal-tar products took out 948 patents between 1886 and 1900, the corresponding British firms took out only 86 (Landes 1969, pp. 352–53). The German research and development effort created a virtuous cycle: high profits made it possible to invest in research and development, and the innovations from the research and development efforts created the profits that could sustain research efforts even as it became more expensive to find the next blockbuster dye.

Soda and Sulfuric Acid as Examples from the Inorganic Sector: Advantages from Demand Spillovers   Dye production requires large quantities of sulfuric acid and alkalis; thus, German growth in dye production created an enormous demand for heavy chemicals (Hohenberg 1967). In turn, this demand spillover helped the German heavy-chemical industry catch up to the British.

The German soda industry basically did not exist in the 1860s, because British producers using the Leblanc process were so efficient. However, the German producers then gained competitive advantages and captured a large portion of the German home market by building soda plants based on the new and superior ammonia soda process developed by Ernst and Alfred Solvay. Ludwig Mond, who was born in Germany and educated there in chemistry as well as in a number of technological subjects, saw this opportunity and went to Britain, where he licensed the Solvay process to start an ammonia soda plant together with the businessman John Brunner. Because the ammonia soda process was cheaper to operate and also able to produce soda at higher purity, Brunner, Mond and Company became the fastest growing soda manufacturer in Britain from the 1880s onward. The company spent a great deal of money improving the process and scaling up the plants, making the firm almost as profitable as the successful German dye firms; it averaged a 15% return on investment from 1881 to 1926 (Reader 1970).

The success of Brunner-Mond was very detrimental to the profit margins of the large number of established British Leblanc soda producers. In 1890, 45 of the 50 existing British Leblanc soda manufacturers merged their firms into one company (Miall 1931, p. 20). The newly formed United Alkali Corporation even organized a central research laboratory, but no significant innovations were achieved that could have rejuvenated this technological path. Although United Alkali was the largest chemical firm in Britain when it was formed, it was technologically so behind that it averaged only about a 3% return on investment until it was absorbed into a larger corporate entity in 1926, as we describe later in the section on the interwar period.

The German sulfuric acid industry is also an example of considerable R&D investments in inorganic chemicals helping German producers to improve their competitive position. In the 1890s, BASF developed a commercial version of the new contact process for making sulfuric acid, partly because an economically viable synthetic indigo production required a stronger acid. The contact process became a widely adopted technology because it could make sulfuric acid at higher concentration and greater purity than that produced by the traditional lead chamber process. As technological capabilities were built within firms, they could often be leveraged into other product classes. The skills BASF cultivated, for example, in finding a suitable catalyst for the contact process added to the technological capabilities that were necessary to create the technologically daunting Haber-Bosch process in the first decade of the 20th century, which will be discussed in a later section. (See, however, the chapter by Arora and Gambardella for some illuminating comments on this contact process when adopted in the United States.)

The Institutional Environment as a Foundation of Competitive Advantages

British chemical firms lost their competitive advantages in the period before the First World War because they did not invest to the same degree as their German counterparts in manufacturing, marketing, management, and research and development. Those British firms that invested to some extent in these four firm capabilities — for example, Nobel in explosives, Brunner-Mond in soda, and Levinstein in dyes — performed significantly better than the other British firms. But why did not more British firms make these investments? Since British dye firms were at least as profitable as their German counterparts in the early stages of the industry, the explanation is not a lack of financial resources (see the Da Rin chapter for details on finance). What, then, led British entrepreneurs to make such a vastly different investment decision? The answer must be sought in the larger institutional context in which British and German firms were operating.

The Advantage of a Well-Developed Educational System   British firms did not face the same positive incentives that German firms faced in their social environment. The innovative capabilities of the German chemical industry rested, to a large extent, on the highly skilled organic chemists and engineers who were employed in corporate R&D laboratories. Both the German and the British chemical industries relied on universities, polytechnics, and trade schools to provide their scientific and technical personnel with fundamental education. Firms then hired these well-educated individuals and taught them the necessary industry- or company-specific skills.

But the German chemical industry enjoyed substantial competitive advantages over Britain's because of a superior educational system (see the chapter by Rosenberg for more details). After about 1830, Germany had the most advanced system of higher education in the world for a century (Ben-David 1977). Germany's institutions of higher learning, unlike their British counterparts, could rely on a strong primary and secondary school system. For instance, Prussia (by far the largest German state) made primary education compulsory by 1772. Britain had such laws only 130 years later, after it became apparent that a large, relatively uneducated population was a great disadvantage for a modern industrial society (Wrigley 1987). While Britain had a number of excellent secondary schools, the lack of broad primary education meant that they did not reach as widely into different classes of society as did Germany's highly regarded secondary schools. Germany thus created a larger number of students who could meet the challenges of university education.

The research universities, which were created in Germany starting with the University of Berlin in 1810, became the source of many important contributions to the natural sciences between 1830 and 1930 (Ben-David 1984). These institutions placed a new emphasis on original research as a fundamental part of a professor's job and a student's educational experience. This focus was enormously beneficial for the development of the natural sciences in Germany. In this context, Liebig opened his famous laboratory and became the leader in the relatively new field of organic chemistry (Holmes 1989). Students from all over the world flocked to Giessen in the 1830s and 1840s to receive training in organic chemistry. The large numbers of British students who studied under Liebig — they amounted to 83, or 12% of the entire student body of his laboratory between 1830 and 1850 (Fruton 1988, p. 16) — and who subsequently became the leaders of British academic organic chemistry are a testimony to how far ahead German laboratory teaching was compared to the education offered in Britain in the middle of the 19th century (Morrell 1972).[7]

To be sure, Britain had a number of scientists with international reputations in the nineteenth century. But such eminent figures as Michael Faraday, Humphry Davy, John Dalton, and William Wollaston did not create research schools as Justus von Liebig, Friedrich Wöhler, Hermann Kolbe, Robert Wilhelm Bunsen, and others did in Germany. Britain excelled in producing world-class "private" scientists, who were often associated with learned societies and academies, but did not develop the institutional framework for producing a large number of qualified students who could be employed by industry. German universities, in contrast, were often educating more chemists than society needed (Flechtner 1959; Titze 1987). A number of German chemistry students went to Britain — for instance, Heinrich Caro, Carl Alexander von Martius, Johann Peter Griess, and Otto Witt — because job opportunities for chemists in Germany were often scarcer than qualified applicants.

The overproduction of chemists in Germany was for a time a blessing for Britain. But when the expansion of the Germany chemical industry and academic chemistry created excellent opportunities for German chemists at home, it was increasingly difficult for British firms to hire German talent. The creation of the Royal College of Chemistry in 1845 was an attempt by British agriculturists and Prince Albert, the German husband of Queen Victoria, to bring Liebig's teaching and his famous laboratory instruction to Britain (Beer 1959). August Wilhelm Hofmann, whom we have already encountered in connection with the formation of the synthetic dye industry, became — on the recommendation of his teacher Liebig — the first professor of organic chemistry at the college. But the enthusiasm of the initial sponsors declined, since they were largely interested in training students in practical chemistry, not in supporting research scientists. Within a few years the college found itself in financial straits and was absorbed by the government-backed Royal School of Mines. This episode illustrates how difficult it was for a Liebig-like research institution to flourish in the context of British society. Nonetheless, until Hofmann's return to Berlin in 1865, the Royal College of Chemistry was the most important provider of home-trained organic chemists for the British chemical industry.

Traditional British universities, meaning especially Oxford and Cambridge, were slow to create serious research and teaching programs in the natural sciences. Eventually, between 1871 and 1880 the Cavendish Laboratory at Cambridge and the Clarendon Laboratory at Oxford were established. In addition, seven new colleges of

---

[7]Morrell (1972) provides a superb account of Liebig's research school in Germany and the much less successful research school of Thomson in Britain.

higher education, the so-called civic or red brick universities, were created in this period to compensate for the lack of science and technology education (Sanderson 1972). But despite the creation of the laboratories at the two elite institutions, their output of science students remained meager. Only 56 students received a bachelor of science honors degree at Oxford or Cambridge between 1880 and 1900 (Haber 1971). Meanwhile, the second industrial revolution of chemicals, electrical machines, and the internal combustion engine was rewarding those countries that could exploit technological opportunities. The German system of higher education was much more responsive to the new demands of industry than the British one.

For business enterprises, the polytechnic schools that sprang up throughout the German states in the 1830s were at least as significant as the universities in creating a broad base of technical talent. While laboratories like that of Liebig channeled a significant number of students into industrial firms — 17% of his students in Giessen went into industrial manufacture, whereas 30% went into pharmacy and 14% took academic jobs in chemistry (Fruton 1988, p. 17) — polytechnic schools formed the backbone of training for engineers and skilled craftsmen. Besides the need for scientifically-trained organic chemists, the chemical industry developed an ever-larger demand for science-based engineering. This demand was met by upgrading the polytechnics into technical universities from the 1870s onward and by expanding their enrollments. The number of full-time students rose from 2759 in 1870 to 10,591 in 1910 (Titze 1987). Professors at these technical universities fought a long battle against the traditional university faculties before they were finally granted the right to award doctorate degrees in the engineering sciences in 1902. Because these technical universities had both excellent engineering programs and strong chemistry curricula, large German chemical firms came to rely more heavily on talent from the new technical universities than from the traditional universities (Haber 1971).

While traditional forces resisted the expansion of science and technology teaching in Germany as well as in Britain, these forces were much stronger in Britain. Although the overall British student population increased 20% between 1900 and 1913, Germany's climbed 60% in the same period. Between 1891-96 and 1911 the numbers of students at Britain's civic universities (excluding Oxford and Cambridge) rose from around 6400 to 9000 or by 40%. Of those, only 2700 were engineering or science students (Haber 1971, p. 51). In contrast, Germany had about 11,000 full-time technology and science students at the technical universities, in addition to the sizable number of science students educated at the regular German universities. Germany's substantial advantage in the quantity of scientifically literate manpower was reflected at the highest echelons of industrial leadership. Recent research on successful British and German businessmen from 1870 to 1914 has shown that 13% of them were academically trained in Britain versus 24% in the case of Germany (Berghoff and Möller 1994).

Not surprisingly, a number of Britons were quite aware of the relative backwardness of British educational institutions in the 1860s. Lyon Playfair (a Liebig student) was among the most vocal critics of British science and technology education. He played a key role in the first of a series of special government committees that examined the state of British education. The creation of the civic universities in the 1870s and the science laboratoryoratories at Oxford and Cambridge was a response to these committee reports. However, British society as a whole and its government were not convinced until many decades later that matching Germany's emphasis on education was in the vital interest of the nation.

Interface between Technical Education and the Chemical Industry   When Hofmann was lured back to Germany in 1865 to head a large new university laboratory in Berlin, the British dye industry lost one of its most important intellectual resources. After his student Perkin had synthesized aniline purple, Hofmann used his laboratory in London to discover other dyes, while channeling students into British dye firms, and consulting widely. He continued this pattern when he moved to Berlin. Not only did he become the most important consultant to the AGFA (he may have held stock in the company), but he also became a tireless organizer of German academic chemistry and its relationship to industry.[8]

Hofmann was by no means an exception. Firms could draw on a large number of university researchers who were willing to interact in a variety of ways with industrial firms. The most famous episodes are the alliance Carl Graebe and Carl Lieberman formed with Heinrich Caro of BASF in 1869 to pool their discoveries and jointly file a patent for alizarin in Britain (Travis 1993). Hoechst collaborated extensively with leading medical professors. Hoechst's first drug, Antipyrin, was a joint venture with Ludwig Knorr at the University of Erlangen. Robert Koch, the discoverer of the tuberculosis virus, and Paul Ehrlich, the inventor of chemotherapy, are among the other distinguished scientists who participated closely in Hoechst's effort to bring new drugs on the market (Hoechst 1990). Even before Bayer had established its own central research laboratory, the firm made use of the superior research facilities at German universities. The first assignment Bayer handed to newly hired chemist Carl Duisberg was to work for six months in 1883 to 1884 in the organic chemistry laboratory at the University of Strasbourg and research basic problems of interest to the company (Duisberg 1933).[9] When Bayer had difficulty getting high yields in their newly set up acid plant in Leverkusen in the 1890s, the firm could call on the leading expert for inorganic chemical technology and famous textbook author Georg Lunge, from the Zurich Technical University (Flechtner 1959, p. 151). Perhaps the single most significant collaboration was the one between the physical chemist Fritz Haber and the BASF chemist-engineer Carl Bosch. Their Haber-Bosch process that created synthetic ammonia gave the world a cheap source of fertilizers and won both men the Nobel prize in chemistry.

In the first years of the 20th century, however, German chemical firms increasingly felt that German universities did not satisfy industry's need for basic research that could lead to industrial applications.[10] Thus, the German industry sought an alliance with German academics to form special research institutes. Following the tireless efforts of the industrialist Werner Siemens, the German government had already agreed in 1887 to finance the Imperial Institute for Physics and Technology. However, with social insurance schemes and the military buildup putting heavy demands on the Imperial

---

[8]Recently an edited volume was published to commemorate the 100th anniversary of Hofmann's death by focusing specifically on his role in facilitating the interaction between science and the chemical industry (Meinel and Scholz 1992). For other details on Hofmann's interaction with the industry, see Travis (1992).

[9]Duisberg later developed into a towering figure in the German chemical industry. He became CEO of Bayer; the principal architect of the German dye cartel; leader of the German Association of Chemists, the Chemical Industry Association; and, after his retirement from executive duty at I. G. Farbenindustrie Aktiengesellschaft, the representative of the Association of German Employers (Flechtner 1959).

[10]German industrial leaders had increasingly begun to see the United States, not Britain, as their true industrial rival. By all accounts, visiting German industrialists were awed by the enormous size of American plants and their productivity (Duisberg 1933). One of the motivations for creating more fundamental research capabilities was the fear that the newly formed Rockefeller and Mellon research institutes would jeopardize German leadership in science-based industries (Johnson 1990).

budget, the government was reluctant to fund a planned Imperial Research Institute for Chemistry. Leading researchers in both academic chemistry and industry therefore recruited the emperor to lend his name to a new scientific association, to be financed largely by private and corporate philanthropy (Johnson 1990). Under its wings, three chemical research institutes were formed between 1911 and 1914: the Kaiser Wilhelm Institutes for Chemistry and Physical Chemistry in Berlin Dahlem and a third Kaiser Wilhelm Institute dedicated to coal research, which opened in Mühlheim on the Ruhr river just before the war. (One of the most important discoveries made at this institute is touched on in the chapter "The Process of Innovation.") The giant coal corporations of the Ruhr shouldered large portions of the yearly costs of running the institute and gave high-caliber chemists the resources to work on projects that were too large to be financed by university research laboratories (Johnson 1990).

Undoubtedly, the creation of these chemical research institutes owed a great deal to the spirit of cooperation that prevailed among the firms in the German chemical industry, despite the fierce competition that often existed in the market place. Alfred Chandler (1990) found this behavioral trait so characteristic of the modern German economy that he contrasted the "cooperative capitalism" of Germany with the "personal capitalism" of Britain. When the larger interest of industry was at stake, German firms in the period before World War I found it easier to cooperate than did their British counterparts. For example, a very effective industry-wide trade association for chemicals was already formed in Germany by 1877, 39 years before the British one (Grant, Paterson, and Whitston 1988).

British chemical firms, of course, also interacted with institutions of science in addition to their traditional reliance on the consultants who traveled from one firm to the next. Levinstein, for example, was a keen supporter of the University of Manchester, and he relied on universities for talent just as the German firms did. Levinstein also acted as the president of the Society of Chemical Industry that was founded in 1881 by, among others, Professor Henry Roscoe, the country's leading academic chemist, and the industrialist Ludwig Mond to facilitate interaction between academia and industry (Miall 1931). Because the British academic system for chemical research and training was so much smaller than Germany's, the opportunities for British firms to draw on their local universities were limited. Thus, it is not surprising that German industry cultivated the interface with academia more extensively than did British industry: German entrepreneurs had more to gain from interacting with their local university system.

Government Policies   Germany offered greater financial support for higher technical and scientific education than did Britain. At the end of the 19th century, the British government paid £26,000 to universities for all purposes. Prussia, the largest German state, alone supported her universities with £476,000. The corresponding figures for the academic year 1911/12 amount to £123,000 and £700,000 (Haber 1971, pp. 45 and 51).

This dramatic variation in the public support of education can be understood in the context of the very different historical experiences of Britain and the German states. The British industrial revolution was more or less carried out by the hands of private individuals, but Germany's entry into the industrial era was orchestrated by government bureaucracies. German states often were eager to stimulate and sponsor industry in their territories to increase income for the crown and thus further the power of the ruler. Education of highly-qualified bureaucrats (who could mastermind industrialization efforts) and of engineers (who could transfer British technological achievements) was seen as a means to this end. Along with the goal of making the German states individually stronger, the experience of being run over by Napoleon probably also helped generate widespread backing across Germany for support of higher education (Schmookler 1965).[11]

In the first half of the 19th century, while German public support laid the foundation for what would become the leading system of higher education, British universities were financed largely by private grants and tuition fees (Sanderson 1972), with virtually no involvement of the British government. By the 1860s, a national myth of *Wissenschaft* (scholarship) had taken hold in German society; it interpreted her rapid industrialization as directly connected with advances achieved in the natural sciences since the 1830s (Turner 1989). But in Britain, where the industrial revolution was conducted by men who, in most cases, had no academic education, there was no perceived association between industrial success and a developed educational system. These differences help to explain why the German Imperial government and the individual states upgraded the polytechnics to technical universities in the 1870s, began to form schools of business administration in the 1890s, and expanded enrollments particularly in the traditional universities. The British government also increased its support for science education by giving grants to local institutions, expanding its support of universities, and merging the London technical colleges in 1907 to form the Imperial College, whose model was the Technical University of Berlin (Wrigley 1987). Because the British government started from a much lower level of involvement and less support for education in society, however, it responded more slowly, less systematically, and with fewer resources than did Germany.[12]

Differences in Patent Law   Governmental patent law policies also shaped the competitive advantage of the two national chemical industries. When Perkin discovered his aniline purple dye in 1856, he quickly filed for a British patent. Because many new aniline dyes were created by reacting an existing dye such as aniline red with other chemicals, the numerous dye firms that had sprung up in the wake of Perkin's success in the early 1860s became embroiled in litigation over who had the right to make what colors. By 1865 the courts had finally developed workable rules for the interpretation of patent claims in chemicals (Travis 1993, p. 43), but by then, the British industry had lost time in learning to compete against well-organized industrial firms.

Meanwhile, Perkin and other dye innovators had no compelling reason to try obtaining patents in Germany, because Germany did not have uniform patent protection. Unencumbered by patent litigation, German firms could focus on making the processes for these early aniline dyes more economically efficient. When German firms started to make dye innovations in the late 1860s, it was not uncommon for them to claim rights in Britain that British firms did not enjoy in Germany. They simply filed for dye patents in Britain, since it was the most lucrative market at the time (Zimmermann 1965). In 1868, for example, Caro together with Graebe and Lieberman claimed British patent protection of their alizarin dye one day before Perkin filed for

---

[11]As an alternate explanation of the meteoric rise of the German university system, many commentators have cited a German emphasis on cultivating a person's self through *Wissenschaft* (scholarship) for its own sake, and the inability of Germany to express her national identity other than through the cultivation of a shared culture (Paulsen 1906). While there may be an element of truth in the social psychological explanation, we are more persuaded by the arguments along the utilitarian lines of regional and national strength described in the text.

[12]Incidentally, it is telling of British management philosophy that the country did not find it useful to create schools of business until the 1960s, lagging behind Germany and the United States by roughly three generations. See Locke (1984) for the history of business education in Germany and Britain.

his own patent. Rather than engage in legal fights, Perkin and the three Germans signed cross-licensing agreements and later created an alizarin convention to maintain prices.

While the absence of a uniform German patent clearly helped the German dye firms get started, as organic chemistry gave German firms the possibility to profit from R&D laboratories, it became a burden not to have patent protection. Without patent protection, no individual German firm had a strong incentive to invest in R&D, since its competitors would simply copy the inventions (Wetzel 1991). Thus, no corporate R&D laboratory was created before Germany passed a uniform patent law in 1877. The first patent (BASF's) came in the same year, and by 1886 all major dyestuffs firms had R&D laboratories (Homburg 1992).

Once a concrete proposal for a German patent law was on the table in 1876, the German chemical industry became very active in shaping the law to their own competitive advantages. The ubiquitous August Wilhelm Hofmann was already heavily involved in helping a patent protection to come about — perhaps because his British patent royalties had shown him how an academic could nicely increase his income with patents. An industry consensus developed that only chemical processes, not chemical products, should be patentable (Johnson 1992), and the unified industry successfully petitioned the German parliament for this change. The chemical industry association also led a drive in 1891 for a revised patent law that fine-tuned the rules for interpreting the nature and scope of chemical patents, affording greater legal protection against imitation (Wetzel 1991).

The British chemical industry was less successful in influencing patent laws in its home country. The leading British dye manufacturer, Levinstein, was keen on adopting the same patent policy that had forced German dye firms to open production sites in France and Russia. These two states required that the holder produce within the boundaries of the nation for patents to be valid. Although Levinstein approached the lobbying effort more as the owner of an important British company rather than as an industry representative, he was successful at getting the Patent Act of 1907 passed by parliament. Hoechst and BASF immediately announced they would manufacture dyes from imported intermediates in Britain (Haber 1971). A 1909 court decision, however, drew the teeth from the new patent law. In the end, German manufacturers continued manufacturing dyes at home, and the British dye industry had not gained the hoped-for competitive advantage.

Tariff Policies and Macroeconomic Environment   Since trade and macroeconomic issues are examined in detail in Barry Eichengreen's contribution in this volume, we will only briefly touch on them here. Both factors tended to place British chemical firms at a disadvantage in the later decades of the 19th century.

Battered by a long depression that had set in around 1873, German soda manufacturers lobbied to impose tariffs on soda imports. They sought relief particularly from the British soda manufacturers who had captured a large share of the German market when the Imperial government granted Britain a most-favored nation status in 1865. The dye industry opposed tariffs, fearing retaliation against dye exports, and managed to keep dyes and intermediates free from tariffs. However, tariffs were enacted on soda in 1879 and 1882 as part of a larger wave of protectionism that was sweeping Germany. The higher tariffs certainly helped German soda producers to win their home market from British competitors. Since the spirit of free trade still dominated Britain, the British government did not retaliate — say, by imposing tariffs on German dyes.

The British chemical industry faced another disadvantage in the decades before World War I: It was embedded in an economy with a slow growth rate. Transportation costs can be significant in the chemical industry, especially for industrial chemicals, which are characterized by low unit values in comparison to dye stuffs and pharmaceuticals. Even without Germany's rising tariff walls, British producers could not benefit to the same extent as German producers from the rise of the German economy, because they were farther away.

Institutions, Individual Incentives, and Competitive Advantages   Before leaving this time period prior to the onset of World War I, it is useful to review the essential features of the era. Britain and Germany differed considerably in the organization of their societies (Veblen 1915). Britain was the role model of a liberal state, where virtually all political authority rested with parliament. Germany, especially after the formation of the empire in 1871, had much greater power vested in the crown and its bureaucracy. Once British industry was organized as a large number of relatively small, but extremely successful, personally controlled firms, it became difficult to change to a new model that could exploit economies of scale and scope in manufacturing and R&D. In Germany, unencumbered by an alternate model of industrial success, the state played a large role in encouraging the large industrial firms.

Young talent also faced very different opportunities in the two countries. British talent had relatively more opportunities to make a lucrative career in London, the leading financial center of the world. By about 1850, once Britain had built her capital-intensive railway network and mining enterprises, she used her banking capabilities to finance railways and mining enterprises all over the world. Each year from 1850 to 1913, Britain exported on the average about 38% of her net capital formation (Landes 1969, p. 331). For the most part, the financial institutions in London (the so-called "City") appraised the risk of foreign investments very successfully and provided British investors higher returns on underwriting foreign bonds than the yields available on domestic bonds (Cairncross 1953). The corresponding figure for German capital export was only 11%, indicating that in Germany a much higher proportion of new capital formation was invested domestically. In addition, the British empire afforded prestigious career opportunities in government and diplomacy.

Meanwhile, German universities were producing a much larger number of scientifically and technologically trained individuals, who developed the technology that made it possible for German entrepreneurs to capture new and existing chemical markets. In contrast, the firms in the British heavy-chemical industry had too small an output, first of all, to pay for the R&D and secondly, to foresee sufficient profit to justify R&D investments (Elbaum and Lazonick 1987). In the context of a British state that was less willing to invest large sums in science and technical education, British chemical firms typically tried to maximize their profits by spending as little on R&D and new plants as possible.

The reform of the existing British business, education, and governmental institutions would have required a severe national crisis. But Britain's economy as a whole was still performing relatively well at this time, even as the German chemical industry caught up with Britain in the inorganic sector and completely surpassed her in the organic sector. In the beginning of 1914, it looked as if German chemical firms would continue to strengthen their international position, since British firms could not match Germany's ability to create and absorb new technology. By November 1918, however, Germany had lost the Great War and with it a large portion of the competitive advantages that she had built up in chemicals since the 1870s.

THE GREAT WAR AND ITS DIRECT AFTERMATH

When war broke out in August 1914, nobody expected the horrendous consequences of what became known as the Great War: in five years, 8 million people dead and 19 million injured. To the surprise of planners and the leadership of the chemical industry, chemicals came to play a key role in the war (Haber 1971). The chemical industry was a key to producing modern explosives and pharmaceuticals. Furthermore, the industry was called on to manufacture a new class of weapons — poison gas — that was first developed by German chemical scientists under the leadership of Fritz Haber[13] and later used both by the German and British armies. The onset of the Great War also completely altered the environment in which the two national industries operated, as in both countries shortages of crucial chemicals forced rapid and extensive government intervention.

The outbreak of hostilities between Germany and Britain quickly stopped the supply of German dyes and pharmaceuticals to the British Isles. Besides making it difficult to dye the uniforms of British soldiers, this move cut off the British textile industry from its most important suppliers. More important for the military effort was the inability of British chemical firms to transform basic organic chemicals such as toluene and phenol into organic intermediates in the high quantities needed for production of TNT (trinitrotoluene) and TNP (trinitrophenol) explosives. Nobel Industries, the leading manufacturer of traditional explosives, was ill-equipped to make TNT and TNP because those explosives were, as coal-tar derivatives, more closely related to dyestuffs than to Nobel's nitroglycerine products (Reader 1970). As a result, the British state abandoned its passive role in regard to the chemical industry, took control of large parts of the economy, and reorganized its chemical industry to supply the needed chemical compounds for waging a modern war.

In Germany, a British naval blockade ensured that Germany would not receive natural nitrates from Chile, her traditional supplier. The nitrate problem was critical for the war effort, since nitrates were used as raw materials for a wide variety of chemical reactions, most importantly for making explosives and fertilizers. Given that Britain also blocked food imports from overseas, the production of an adequate amount of fertilizers took on extreme urgency as the Great War entered its second year. However, the strength of German technological capabilities overcame this crisis by developing chemical processes that could produce nitrates artificially.

During the war, both the British and German governments created large bureaucracies to coordinate the flow of raw materials, as well as intermediate and final products, among firms, industries, and the different branches of the military. In Britain, the intervention of the government was more sweeping than in Germany, partly because there was no well-developed trade association in the chemical industry that could have provided an administrative link between the government and the individual chemical firms (Haber 1971, p. 232). The autocratic rule of government officials was rather profitable for both British and German chemical firms, especially the manufacturers of

---

[13]For an excellent essay on Haber's life and his role in organizing the cooperation of science, industry, and government for the German war effort, see Stern (1987) and Teltschik (1992, pp. 43–45 and 56–58).

explosives (Plumpe 1990, pp. 94–95; Reader 1970, pp. 304–305). This high degree of government intervention in the affairs of the two national industries, particularly in the case of Britain, also paved the way for a much more active involvement by the British government in the affairs of the industry after the war.

The scarcity of dyes resulting from the cessation of German shipments convinced the British government of the need to create and protect an indigenous dyestuffs industry. British Dye Ltd. was formed in 1915 and financed with capital from the government and the textile industry. The British government also confiscated property held by German nationals in Britain and sold it to British firms (Plumpe 1990, p. 108). German chemical firms lost all their patents and plants in Britain from 1916 to 1917, just as they did in France and the United States during the war. While British Dye Ltd. did take over the plants of Read Holliday & Sons, other significant British firms like Levinstein and British Alizarin resisted giving up their autonomy, particularly because the disappearance of German competitors made these organizations very profitable. Only the threat of a German reappearance in the British market, the willingness of the British government to prohibit the import of dyes except with licenses, and a £2 million government grant to support industry R&D efforts convinced Levinstein in 1919 to become part of one large British dye firm. The new British Dyestuffs Corporation was almost as large as the collection of *Dreibund* and *Dreiverband* firms (200 million and 210 million Marks in terms of capital respectively) in 1914 (Plumpe 1990, p. 104). Its capacity of 16,000 tons per year, however, was much smaller than the collective capacity of the German dye firms.

Because the war effort required extensive interaction between the government and the firms in the chemical industry, the leaders of the British chemical industry came to know one another very well. The need to coordinate actions of individual firms brought into existence an industry trade association (Haber 1971, p. 227), the Association of British Chemical Manufacturers (ABCM). The war also inculcated habits of consultation between managers that carried over to peacetime and led to a remarkable transformation of the structure of the chemical industry in the interwar period.

British attitudes toward government support of research and development underwent dramatic changes because of the military conflict. In 1915, the Privy Council created the Committee for Scientific and Technical Research to coordinate efforts to solve pressing technical problems, which illustrates Schmookler's (1965) arguments that emphasize how social crises can provide a strong impetus for increasing the funding of academic research. In 1916, this Committee for Scientific and Technical Research was given more prestige and authority by its conversion into the British Department of Scientific and Industrial Research. The new department received an initial capital grant of £1 million, plus an annual budget (Grant, Paterson, and Whitston 1988, pp. 19–20). Under its control, a chemical research laboratory was created that worked on a variety of technical problems related to chemicals. Another group, under the leadership of H. C. Greenwood (who had studied under Haber in Karlsruhe), was formed at University College in London to develop a synthetic nitrogen process along the lines of the Haber-Bosch method. When the government stopped funding the research group after the war, the remaining scientists joined Brunner-Mond, which had taken over the government synthetic ammonia plant in Billingham in 1920. These research efforts marked the first significant joint problem-solving efforts between academic and industrial chemists in Britain. Thus, the Great War helped teach the benefits of research coordination and made very clear that academic scientists could be useful in solving industrial problems.

For the German dye industry, the Great War brought a dramatic challenge to its dominant position. When the fighting broke out, the German government immediately prohibited the export of chemicals. Since the German dye industry exported 80% of its production (Plumpe 1990, p. 50), it was faced overnight with a potential loss of the majority of its customers. Although the chemical industry convinced the government relatively quickly to lift the export ban on dyes and pharmaceuticals for friendly countries, German firms experienced great difficulties in reaching their overseas export markets, since Britain imposed a naval blockade on Germany. Some German firms went as far as sending dyes on submarines to reach their large U.S. customers (Steen 1995)!

To the surprise of managers in the dye industry, however, it became apparent that the dye-production facilities rendered idle by the British sea blockade could be converted readily to produce explosives. This technological possibility allowed the precursor firms of IG Farben to become the largest explosives manufacturers in Germany during the war, with a market share of 77.1% at the peak production in 1917 (Plumpe 1990, p. 84). As the war progressed, it became obvious that German dye firms would face new competitors, high tariff barriers, and possibly even closed markets when they returned to the world markets after the war. The prospect of not being able to run dye plants at full capacity, and particularly the perceived need to build up more production facilities overseas after the war, led the leaders of the large German dye firms to reconsider Carl Duisberg's original 1904 plan of combining all important German dye firms (Plumpe 1990, p. 97). However, the economic prospects during wartime were not bad enough for the individual firms to surrender their autonomy. As a result, the eight largest dye firms — the six firms composing the *Dreibund* and *Dreiverband*, as well as two independents, Chemische Fabrik Griesheim-Elektron and Weiler-ter-Meer — formed the so-called community of interests (*Interessengemeinschaft*) in 1916 (Teltschik 1992, p. 46). This combining of almost the entire German dye industry went beyond an effort to maintain high prices. A community board (*Gemeinschaftsrat*) was created to facilitate joint decision making. It had the authority to determine corporate strategy, investments, acquisitions, and other financial matters. Without the threat of increased competitive advantages that foreign firms came to enjoy as a result of the war, it is doubtful that the proud German dye firms would have agreed to give up their autonomy in this way.

Since Germany already had a well-developed infrastructure for interaction among science, industry, and government, the war did not constitute a radical departure from existing interactions, as it did in Britain. Fritz Haber, the director of the recently established Kaiser Wilhelm Institute for Physical- and Electro-chemistry, became a military officer and used his institute very effectively in coordinating the joint war research efforts between government and industry (Haber 1971, p. 223). The first and most controversial result of the war research effort was the development of poison gas, which altered the way armies would fight battles.

By the end of the long war, the future prospects of German and British chemical firms were intimately tied with the fate of the nations in which they resided. The peace treaty of Versailles specified that the German chemical industry would lose, as part of German war reparations, all production facilities in France; would be required to reveal all trade secrets to the allied powers; and would face destruction of all production facilities that were used to make military chemicals. Thus, the British firms received the right to manufacture dyes and pharmaceuticals formerly under patent protection. In addition, the British chemical firms were not shy about using their government's military victory to get access to BASF's trade secrets concerning the Haber-Bosch process, whose industrial significance in providing a cheap source of fertilizers for a growing world population had been recognized all over the world. Moreover, the governments of Britain, France, and the United States were not about to hand back patents and other important property owned by Germans before the war.[14] The Versailles treaty also required that one-half of all dye stuffs and pharmaceutical stocks were to be delivered immediately at no cost and one-quarter of new production for the next five years at the lowest market prices. Initially, the Allies also demanded that Haber-Bosch plants be destroyed. Carl Bosch, however, as a member of the German delegation to Versailles, argued that the plants were necessary for fertilizer production. He struck a deal with the French that BASF would be allowed to continue to operate its plants in Ludwigshafen and Leuna if it built a similar plant in France (Teltschik 1992, p. 55).

The wartime changes had taken Britain's chemical firms a long way toward catching up with the German firms (Reader 1970, p. 318). The era of complete German dominance in many segments of the chemical industry seemed to have ended.

THE INTERWAR PERIOD: COLOSSAL CORPORATIONS, CRISIS, CARTELS, AND CATCH-UP

The era between the two world wars stands out for a variety of reasons. Britain was now commited to becoming self-sufficient in chemicals. The industrialized world moved farther and farther away from free trade to protectionist economic policies (Plumpe 1990). It was the period when a large number of cartel agreements came into existence in the chemical industry, fixing prices and carving up the world into exclusive territories for individual cartel members (Smith 1992). As cartels were largely made illegal in Britain and Germany after the Second World War, the chemical industry also would never see such prominent cartel arrangements again. Furthermore, in this period the large corporation, which had arisen a generation earlier in the United States, finally arrived in Europe and reshaped the industrial landscape of Britain and Germany (Chandler 1990). With the formation of IG Farbenindustrie Aktiengesellschaft (1925) in Germany and Imperial Chemical Industries (1926) in Britain, two gigantic firms came into existence. Given their central role in each national industry, the competitive dynamics of the German and British chemical industries in this interwar period can be illuminated by examining the development of these two corporations.

While the industy became rather static in its structure—made up of a few large industrial players—it proved extraordinarily dynamic in terms of spawning new technologies that vastly expanded the boundaries of the chemical industry. Among the dramatic breakthroughs that promised enormous growth opportunities for the chemical industry were nitrogen fertilizers, synthetic alcohol, synthetic fuel, synthetic rubber, plastics, artificial fibers, petrochemical organic feedstocks, paints (for the dramatically

---

[14]The U.S. government did not sign the Treaty of Versailles, which specified that Germany had to forgo all claims to property that had been owned by Germans in the Allied countries before the war and confiscated during the war. As a result, the United States was the only country in which German firms and individuals could file property claims after the war. Because German chemical patents and production facilities had already been sold off by the U.S. Alien Property Custodian, I.G. Farbenindustrie Aktiengesellschaft, formed in 1925 (see later discussion), settled in 1929 for a financial compensation of 29 million Reichsmark for the lost property (Plumpe 1990, p. 109).

expanding automotive industry), plant protection chemicals, new pharmaceuticals, and photographic materials (Haber 1971; Plumpe 1990).[15]

The economic data for this period show that total chemical production grew faster in Germany than in Britain, despite Germany's wartime losses and Britain's newfound commitment to the chemical industry. In 1913, German chemical sales exceeded Britain's by a ratio of 1.5:1 on a per capita basis; by 1938, the ratio had risen to about 1.8:1. In absolute terms, German chemical industry sales were 2.4 versus 1.1 billion Reichsmark in 1913; in 1938, sales stood at 5.9 billion Reichsmark for Germany versus 2.3 billion Reichsmark for Britain.

These industry-wide data, however, do not reveal that British firms, particularly ICI, caught up technologically with the German chemical industry during the 1930s (Reader 1975; Plumpe 1990). A look at the organic sector, for example, reveals that German production of dyes in 1938 was down by over 60%, to roughly 57,000 tons, compared with 1913 levels, while British production had increased fourfold, to about 20,000 tons. As we shall see, the reasons why the British chemical industry, despite its substantial technological catch-up, did not grow as fast as the German during the interwar years are intimately connected with the macroeconomic developments in the two countries throughout this period.

Government Policies

Britain's technological catch-up was first made possible by the forced transfer of German know-how to Britain during and after the war. Later, however, the British government took concrete steps to assist the domestic chemical industry by affording it ever-higher levels of protection. Throughout the second half of the nineteenth century, Britain had been the torchbearer of free trade. During the interwar period, however, Britain joined the growing club of nations that sought to remedy their political and industrial problems by instituting import restrictions and/or higher tariffs (Morgan and Pratt 1938). Protectionism in Great Britain was clearly driven more by political and military, than by economic, concerns. While it is difficult to make a final judgment about the overall costs and benefits of this protectionism for the British nation, from the limited perspective of the chemical industry the protective measures clearly helped British chemical firms to win customers.

From 1921 until 1934, a series of measures were passed that affected the competitiveness of the British chemical industry. Britain implemented the so-called Safeguarding of Industries Act in 1921, imposing a 33.3% ad valorem duty on the products alleged to be vital to the national defense. In January 1921, the Dyestuff Import Regulation Act was passed, prohibiting the import of most dyestuffs for ten years. After three extensions, the British parliament made this import prohibition of dyestuffs and intermediates permanent in 1934 (Morgan and Pratt 1938, p. 366). By the end of 1921, dye imports were reduced by 71% and eventually they fell to 13% of the 1913 level (Plumpe 1990, p. 112).

The British government also targeted specific products with tax-relief measures. Following long-standing complaints that high taxes on industrial alcohol disadvantaged the British chemical industry (Miall 1931), the British government in 1921 applied a more favorable tax regime to manufacturers of industrial alcohol such as Distillers Co. Ltd. (Grant, Paterson, and Whitston 1988, p. 34; see also Haber 1971, p. 221). Since the

---

[15]For a detailed description of the innovations achieved in the British chemical industry during the interwar period, see Morgan and Pratt (1938), as well as Hardie and Pratt (1966). Plumpe (1990) provides an overview of the major technical advances in the German chemical industry during the same period.

British government wished to achieve independence from oil-producing countries, it supported the creation of a British synthetic fuel industry by passing the British Hydrocarbon Oils Production Act in 1934. The act was designed to give a tax preference for 10 years on light oils produced from indigenous raw materials or by-products, and to make imported oil more expensive. The most sweeping protective measure came with the Import Duties Act of 1932, which imposed tariffs on all imported goods except those given special exemption (Morgan and Pratt 1938).

As the German dye industry increasingly found that Britain and other large traditional dye customers were locked behind high tariff walls, it had to focus on selling to the countries with large textile industries but low tariff barriers. China, Japan, and British India became the three countries that constituted two-thirds of world dye imports. However, these markets in the Far East could not entirely make up the loss of traditional markets. Capacity utilization for the German dye industry in the 1920s dropped, and the German dye industry was caught in a downward spiral.

During the interwar period, as other industrialized countries raised their tariff walls substantially, Germany developed into a country with moderate import tariffs.[16] The mildness of Germany's protectionism was partly an outgrowth of her military defeat. The victorious nations had forced Germany to grant them lower tariffs by giving them a most-favored-nation status and to turn the Rhineland into a free trade zone. Furthermore, as a large exporter, Germany was reluctant to give other nations even more reasons to increase their tariffs by putting up higher tariff walls around her own economy.

However, the German government intervened in other ways during the interwar period. The end of the war brought great social discontent to Germany, including massive strikes that called for the nationalization of important industries. To soothe the social and economic turmoil, the Office of Economic Demobilization was charged with making decisions about the allocation of resources (Grant, Paterson, and Whitston 1988, p. 24). Furthermore, a Chemical Exports Office was established in 1919 with the statutory power to issue trade licenses and to regulate export prices. Because these coordinating committees provided the occasions for frequent meetings between the leaders of the various chemical firms, they paved the way for the multitude of mergers in the 1920s.

The experience of the chemical industry in the interwar period (as well as in earlier eras) strongly suggests that government can be a crucial ally when firms in a particular nation want to gain competitive advantages over their foreign rivals. The British chemical industry improved its competitive position to a large extent because of active government intervention. In the case of Germany, government intervention helped prevent an already bad economic situation from turning into a complete catastrophe for the chemical industry, as the industry returned to the production of non-military goods but, having lost its important overseas markets, faced enormous overcapacity problems.

Concentration

After the war, overcapacity plagued both German and British chemical firms. By 1920, the post-war boom had collapsed and led to a severe depression in the chemical industry that lasted until the end of 1921. In 1924, for example, German dye firms collectively produced only 70,000 tons of dyes, although they had the production

---

[16]Eichengreen's chapter in this volume gives more details on tariffs for selected chemicals in various countries.

capacity for 156,000 tons, and British firms made 15,000 tons of dyes in factories that could have produced 24,000 tons (Haber 1971, p. 253).

Manufacturers sought relief from the woes of shrinking markets and increased foreign competitors by merging with rivals or creating cartels (Haynes 1933). For instance, the fourth-largest German chemical firm, Rütgerswerke AG, took over Silesia and Fluoritwerke AG in 1920–1921. The producers of nitrogen had already formed a joint sales organization in 1919, marking the beginning of an era of extensive cartelization. The German dyestuffs cartel (the *Interessengemeinschaft*) reacted to this first postwar recession by trying to rationalize operations and cut costs. But these steps would not suffice for long.

The rising levels of protectionism in foreign markets during the early 1920s convinced the leaders of the German dyestuffs firms that a reorganization of German industry was required. It was pretty clear that German dyes and pharmaceuticals would probably become increasingly more unwelcome in many foreign markets as nations tried to build their own capabilities in organic chemicals. Against this backdrop, the head of BASF and already world-famous chemist Carl Bosch returned from a trip to the United States with the same conclusion that Carl Duisberg, head of Bayer, had reached after his journey to the States 20 years earlier. Bosch proposed a full merger of all German dye firms participating in the cartel to equip the German dye industry with the strength to face increased foreign competition (Holdermann 1953). On October 6, 1925, the decision was made to merge the eight members of the *Interessengemeinschaft* into a single legal and administrative entity to be called *I. G. Farbenindustrie Aktiengesellschaft*, or as we will often refer to it, IG.

Carl Bosch, the head of BASF, became chief executive officer of the new corporation and went on to dominate the affairs of the firm until his death in 1940.[17] Carl Duisberg, head of Bayer and tireless organizer of the industry-academic interface, was made chairman of the board, a committee that was given considerable powers in the organizational design of IG (Teltschik 1992, p. 77).[18] The two men trained and put in place a large hierarchy of excellent managers, who were called on to run the German chemical industry after World War II.

During the interwar years, IG became synonymous abroad with German industrial might. At its formation, IG had 70 subsidiaries and a capital of 1.01 billion Reichsmark. Sales grew from 1 billion Reichsmark (1926) to 3.1 billion Reichsmark (1942) and employment from 94,000 (1926) to 199,000 (1943).[19] In the mid-1920s, the German chemical industry was populated by 475 joint stock corporations; 1272 limited liability firms; 1261 proprietorships; and nine "associations," special corporate forms based on cooperative principles (Plumpe 1990, pp. 176–77). Although IG was "only" responsible for about 30% of German chemical sales, it dominated the industry. Over one-half the capital of the chemical industry was invested in IG, and its capital was over three times larger than the sum of the capital represented by the next 15 largest chemical firms in Germany (Stratmann 1985, p. 56). In 1937, IG was responsible for about 73% of all chemical investments. IG was also the most export-dependent large firm in Germany, even ahead of such important exporters as Siemens in electronics. During the years from 1926 to 1933, IG exported 57% of its output, almost one-half of which was still in the

---

[17]The life of Carl Bosch, documented in the biography by Holdermann (1953), offers a unique reflection of the turbulent history of the German chemical industry from the outbreak of World War I to the Second World War.

[18]For the initial corporate structure of IG and its development until 1945, see Plumpe (1990 pp. 145–65) and Chandler (1990). For a detailed account of the structure of IG's research organizations, see Marsch (1994).

[19]Our data on IG comes principally from Plumpe (1990), Stratmann (1985), Hayes (1987), and Teltschik (1992).

dyestuffs business. By the end of the interwar period, IG accounted for 8.7% of *total* German exports. The dominant position of IG in the German chemical economy was cemented by the fact that it produced many basic chemicals used as inputs by other firms, and because IG bought significant stakes in other chemical firms (Stratmann 1985, p. 58).[20]

In the eyes of the German chemical firms, IG was primarily a defensive act to cope with smaller foreign markets and more serious competition. But IG's British rivals could hardly see it that way. The large British chemical firms had great respect for German management and technological capabilities (Reader 1975, pp. 32–33); after all, BASF had created the Haber-Bosch process, something most experts had regarded as impossible in 1910. To meet the challenge presented by a unified, rationalized, R&D-intensive, and diversifying German chemical giant, the leaders of Nobel Industries and Brunner-Mond, Sir Harry McGowan (later Lord McGowan) and Sir Alfred Mond (later Lord Melchett), decided that the British chemical industry would have to create an organization of similar size. They worked out plans to bring about the largest merger so far seen in Britain, one that combined Brunner, Mond & Co. Ltd., Nobel Industries Ltd., the United Alkali Company, and the British Dyestuffs Corporation Ltd. to form Imperial Chemical Industries (ICI).[21] As the chosen name indicated, the new firm was to be the chemical firm for the British empire. The British government expressed considerable interest in forming a large national chemical firm and was more involved in the merger of ICI than the German state had been in the merger of IG (Grant, Paterson, and Whitston 1988, p. 26). The merger between the four firms was hardly one of equals. Brunner-Mond and Nobel more or less took over United Alkali and British Dyes (Reader 1975, p. 3), and their outlook as heavy-chemical firms dominated the corporate culture of ICI for the next decades. In the words of Haynes (1933), with the formation of ICI an "all embracing nation-wide chemical trust was created in Britain" (p. 241) that "controlled all the more important chemical branches in Britain" (p. 245).

On January 1, 1927, ICI was legally constituted with £72.8 million in capital. Plumpe's (1990, p. 184) calculations for 1929 show that ICI had about 20% more capital than IG ($502.2 million versus $403.9 million) but that the two firms were about equally profitable ($28.2 million versus $25.0 million). ICI, however, was considerably smaller when viewed in terms of sales. Again, Plumpe's calculations show that ICI in 1927 had sales of 550 million Reichsmark, rising to 715 million Reichsmark in 1929. The figures for IG in the same years are 1528 million and 1793 million Reichsmark. While ICI became the largest employer in Britain and generally set the tone for management-labor relationships in the interwar years (Reader 1975, pp. 57–58), it employed fewer people than did IG. Where ICI started with 47,000 employees in 1927, IG had 100,000. By 1933, as a result of the Great Depression, both firms had to reduce their payrolls; IG to 71,000, ICI to 42,000. Just before World War II, ICI employed 75,000 people and IG 136,000, a growth of 60% for ICI and 36% for IG since 1927.[22] It is important to note

---

[20]For an overview of other important concentration processes in the German chemical industry during the 1920s, see Plumpe (1990, pp. 176–82).

[21]Our account of ICI relies largely on the excellent corporate history of Reader (1975). The figures for the comparison of ICI and IG are taken directly or constructed from Reader (1975) and Plumpe (1990), unless otherwise specified.

[22]Roughly speaking, the output shares of ICI and IG in their respective countries were very similar in the interwar period. However, it is difficult to come by figures for ICI's share of the total British chemical output. Reddaway (1958, p. 224) reports that in 1948 ICI had about a 40% share of the output in a sample that looked at the most important branches of the British chemical industry. This figure is a good estimate of ICI's interwar market share.

that as a general statement, ICI was growing faster, when measured in sales, than IG from 1927 until 1945, particularly during World War II. ICI's faster growth was partly a result of management decisions but also partly possible because ICI started from a lower base and because the Great Depression did not affect Britain as severely as it did Germany.

IG and ICI had very similar product strategies during the interwar era (for more details, consult the Chandler, Hikino, and Mowery chapter). Both firms diversified heavily until the beginning of World War II. Since ICI in the late 1920s was still trailing technologically behind IG and also DuPont in the United States, ICI was very keen on developing cooperative relationships with these other leading chemical corporations. Besides the Patents and Process Agreement formed with DuPont in 1927,[23] ICI was party to some 800 agreements covering all aspects of chemical production (Grant, Paterson, and Whitston 1988, p. 27). IG was, in this regard, no different from ICI. Spitz (1988 p. 53) reports that IG operated with various degrees of influence in 2,000 cartels. The frequency of such agreements illustrates how both companies pursued during the interwar period the strategy of avoiding competition with existing firms whenever possible and closing off markets to new entrants.

IG went to greater length in trying to create an integrated whole, especially when the Great Depression forced the elimination of costly duplications across divisions. In 1927 to 1928, IG sold 33,000 different dyes under 50,000 different brand names; by 1932, IG offered only 6,000 different dyes (Marsch 1994, p. 47). At ICI, business units enjoyed great autonomy, and rationalization did not constitute a high priority. Although IG was clearly more successful in creating one unified firm, even within IG the corporate identities of its predecessor firms were maintained. This proved to become an important advantage in restructuring the German chemical industry after World War II.

Alfred Mond realized that a large technical gap remained between IG and ICI in 1926. As a true believer in the ability of science to solve industrial problems and to create new profitable products, Mond set up the Research Council in the autumn of 1927. This council brought together ICI's best scientists and those in academia to discuss promising new research. Mond also dramatically increased R&D spending and grants to university laboratories. While the British educational system was still trailing behind that of Germany in the training of engineers, technicians, and skilled craftsman in the interwar years, the highest levels of British science had caught up with Germany, making important contributions to the advancement of physics and chemistry (Webb 1980, pp. 556–61). British students no longer had to go to Germany for advanced chemistry training (Haber 1971, p. 367) but instead could rely on Cambridge, Oxford, and the University of London.[24] Mond clearly believed that it was in the interest of ICI to create strong links to universities by hiring professors as consultants, recruiting students, and supporting academic research and training. Other British firms such as Shell and Lever's were also very active in supporting universities (Sanderson 1972, p. 251).

A sizable number of people at the top of ICI, however, were not convinced that R&D investments would pay off for the firm, so ICI did not reach the level of R&D

---

[23]Taylor and Sudnik (1984) provide an excellent account of the DuPont-ICI relationship, as does Reader (1975).

[24]For a survey of the British University system and its relationship to industry, see Sanderson (1972, particularly pp. 243–64 and 276–94). Sanderson notes that, while the British chemical industry had a sufficient supply of chemists, it suffered from a chronic shortage of chemical engineers in the 1920s. New programs in chemical engineering at University College and Imperial College were created in 1923 to help relieve this bottleneck. Nevertheless, throughout the period from 1926 to 1939, whereas 4,167 students graduated with an honors degree in chemistry only 300 earned an honors degree in applied chemistry (p. 289).

investments acheved at IG. During the interwar years, ICI spent on average 2.4% of sales on research expenditures, only slightly less than DuPont's 2.8%. However, IG invested 9.9% of sales on R&D expenditures from 1926 to 1929, although this figure declined during the Great Depression to 5.2% from 1933 to 1938 and 4.0% from 1939 to 1945. IG's larger R&D investments gave it an impressive portfolio of 40,000 domestic and foreign patents (Stratmann 1985, p. 58). ICI also tried to create a strong patent portfolio, partly because it would give the firm a stronger position in negotiating licensing agreements with other chemical firms. Between 1932 and 1936, ICI hired 361 research chemists from universities bringing the research staff to 615 by 1938. The IG research organization, however, was still considerably larger during the 1930s, as it employed on average around 1000 chemists.

Initially, ICI's largest share of research funds went into high-pressure technology, where early successes in synthetic nitrogen were followed by largely unprofitable investments in coal hydrogenation. IG had followed a similar path — Hughes (1969) called it "technological momentum" — of pursuing technological success into unprofitable uses. Only in the mid-1930s did ICI begin to shift its R&D focus to the more successful organics division and to produce some stellar innovations, such as polyethylene, that could rival what came out of IG's research laboratories. Reader (1975, p. 77) concludes that after a decade or so of existence, ICI had the greatest concentration of scientific and technical manpower in British industry. It also had more capabilities by far in chemistry and chemical engineering than did any individual university department, making ICI the center of chemical innovations in Britain.

With Britain's building of ICI during the interwar period and the rise of petrochemical technology, the dominant leadership position of the German chemical industry before the Great War had been relegated to history (Plumpe 1990, p. 194). By forming a large corporation that could take advantage of scale and scope economies, relying on a background of government support, and running it similarly to the German chemical tradition of investing in R&D and other important firm capabilities, the British chemical industry created an organization that could compete with its German rivals.

Cartelization

Various branches of the chemical industry in Britain and Germany had tried to develop cartels in the 19th century (Haber 1958, pp. 225–29). Most broke down as soon as demand picked up or individual firms thought they could do better by violating the cartel agreement. In Britain, cartels such as the Sulfuric Acid Association, the Salt Trade Pool, the White Salt Association, the Salt Union Ltd., and the Bleaching Powder Association represent successive failed attempts to maintain the profitability of existing producers in branches of the heavy-chemical industry. Part of the impetus behind the merger of the 48 Leblanc soda firms that formed United Alkali was the failure of cartels to function effectively because they did not have any significant control over legally independent member firms.

In Germany, successful chemical cartels were also rare before the Great War. Except for the Alizarin Convention (see below), attempts to organize the German dye trade as stable cartels failed before the 20th century, according to Haber (1958, p. 229), because of the rapid pace at which new dyestuffs were created, repeatedly destroying the markets for existing dyes. The German soda manufacturers had tried many times before the 1890s to form cartels that would maintain a high price for soda, but with little success. An attempt to form a German soda cartel in 1891 was more successful, partly because the cartel formed a coalition with salt manufacturers and thereby controlled who could enter the soda industry (Haber 1958, p. 228).

In general, it was somewhat easier to form stable cartels in Germany because German cartels were endowed with greater legal authority. In Britain, restraint of trade was, in principle, prohibited by common law. But the German high court ruled in 1897 that price and quota agreements were enforceable in civil courts just like regular business contracts (Chandler 1990, p. 423). Not surprisingly, these legal differences led to less pressure in Germany than in Britain to use industry-wide mergers as a mechanism for reducing competition, since members of German cartels could go to court over the violation of the cartel agreement.

A few cartels, however, were very successful and became models for later cartelization efforts. From 1890s onward the British alkali industry faced no real competition from within, since the two large groups in the British alkali industry, United Alkali and Brunner Mond, had developed an understanding about how to avoid direct competition that lasted until the formation of ICI (Haber 1958, p. 226; Reader 1975). Probably the most successful venture in controlling markets in Germany was the Cyanides Convention, which was formed in 1886 and lasted until the 1930s, although it was not based on a contractual agreement (Haber 1958, p. 228). When the rate of introduction of radical new dyes dropped, the German dye industry was able to form the two lasting cartels, the *Dreibund* and *Dreiverband*, which merged in 1916 to become the *Interessengemeinschaft*, as already discussed.

A few attempts were made at international cartels to link German and British chemical firms before the Great War, but their success was very mixed. The Alizarin Convention of 1881, a union of nine German firms and one British, broke down after only four years (Haber 1958, p. 229). However, a second Alizarin Convention was established along different lines in 1900 and lasted until the Great War. To be sure, this was only one dye out of a great many sold in the world and the agreement involved only one British manufacturer, the British Alizarin Company. The makers of explosives in Germany and Britain set up the cross-national Nobel Trust in 1886. The profits of the Nobel Trust, however, averaged about 10% from 1887 to 1914 (Reader 1970, p. 501), making it less profitable than Brunner-Mond (15% average) and much less profitable than the German dye firms (20% average) during the same period. The Nobel Trust successfully cartelized the industry until the outbreak of the First World War made it politically impossible to maintain an Anglo-German union (Reader 1970, pp. 125–62). After initial failures at the turn at the century, agreements also were made between British and German soda industries on market share and sales prices in different countries of the world starting in 1906.

While cartels were not a new invention of the interwar period, their presence expanded rapidly during this time period. In 1905, Germany had 13 cartels in the heavy-chemical industry covering acids and alkalis and other products (Haber 1958, p. 229). Many branches of the inorganic trade were at this point highly competitive, although two cartels existed in dyes and seven in the pharmaceutical industry. By 1923 already, 93 cartels had already been developed in the German chemical industry, and the figure kept rising.

What were the reasons for the dramatic growth in the number of cartels? The observation made by 19th century economists that, as a general rule, cartels are children of bad times is clearly true for the interwar period. The world economy was not growing rapidly enough to absorb the additional production capacities that had come on-line during the Great War or were built after the Great War as nations embarked on chemical self-sufficiency programs. As the Great War had created close contacts between government and industry and between individual firms, it also had become much easier to set up cartels. In Germany, cartels were formed to deal with the overcapacities that came with the closure of foreign markets for German goods (Stratmann 1985, p. 67). Later, cartels became public policy. On July 15, 1933, the Nazi government passed a law that made it possible for the government to force industrial firms to form cartels, thereby helping to make cartels more stable. In Britain, overcapacity also constituted a large force behind the cartel movement. Eventually, the creation of ICI and IG reduced the number of players in the industry and thus made cartel agreements easier to administer and enforce.

The first important international cartel during the interwar period was formed among dye firms, who wanted to prevent other countries from entering dye production by agreeing to share technology only among the cartel members. From 1926 until 1932, all important European dye manufacturers joined the dyestuffs cartel. By 1929, German dye firms (by then IG), Swiss dye firms (the Swiss IG), and French dye firms formed a three-party cartel that produced 80% of all dyestuffs in the world. In 1932, this three-party cartel, acting as an administrative unit, formed a four-party cartel with ICI, incorporating all the important dye manufacturers of Europe into one cartel that set specific sales quotas for each firm. The German IG group was allocated 65%, the Swiss IG group 17%, the French group and ICI shares of 8.5% each (Schröter, 1992, p. 39). As late as 1938, despite growing competition from the United States, Japan, and Italy, 62.2% of world dye sales were still made by firms belonging to the dyestuffs cartel (Schröter 1992, p. 47).[25]

No less important in terms of size was the Nitrogen Cartel (Convention Internationale de l'Azote, 1930–31 and 1932–39). Overcapacity problems served also as the motor behind this international cartel. The British desire not to buy synthetic nitrogen from Germany after the war led to substantial domestic investments in Haber-Bosch process plants. Brunner-Mond overestimated how popular nitrogen-based fertilizers would become among the farmers in the British Empire and thus built up an overcapacity. In 1931, IG had a capacity of 804,000 tons annually. ICI was the second-largest producer in the world with 200,000 tons. Overproduction of synthetic nitrogen had already reached 221,000 tons from 1928 to 1929; a year later, the figure rose to 274,000 tons (Plumpe 1990, pp. 231–33). ICI and IG at this point concluded that the only solution to avoiding a costly price war was to set up an international cartel. Compared to the dye cartel, the international nitrogen cartel was less successful, as it was not able to incorporate all important members (Schröter 1992, p. 48); but it was effective in stopping overproduction, as the figures by Plumpe show (1990 p. 233), and in bringing stability to the industry.

The spirit of cartelization diffused from dyes and nitrogen to a large number of other product markets such as synthetic fuel, rayon fibers, potassium, and aluminum (Plumpe 1990; Reader 1975). By the late-1920s, all segments of British and German chemical industries had come to be organized by cartels or other forms of agreements, replacing free entry and market success as the organizing principle of the industry (Plumpe 1990, p. 200).

The interwar period of cartelization did not last long enough to warrant sound judgments as to whether cartelization reduced the innovativeness of the chemical

---

[25]IG also reached a separate cartel agreement with Japanese producers, carving up the Japanese markets into quotas for each firm (Schröter 1992, pp. 43–46).

industry. Too many new products were already in the pipeline when competition in the world chemical industry was systematically eliminated. Even after agreeing to limit competition, the German and British chemical industries in the interwar period experienced lower profitability than dye, explosives, and ammonia soda firms had experienced prior to World War I. Lower returns despite limited competition suggests that the overall economic climate during the interwar period was not as favorable for the chemical industry.

Macroeconomic Turbulence

The 21 years between the two world wars were characterized by extreme macro-economic turbulence (as Eichengreen explores in more detail in his chapter). Germany's war reparation payments and her high budget deficits in the years after the Great War triggered runaway inflation. In Britain, the return to the gold standard at prewar rates in 1925 overvalued the pound and thus made British chemical exports more expensive. As a result, British chemical exporters were put at a competitive disadvantage relative to their German rivals.

In the 1920s, Germany received large amounts of capital, in the form of short-term funds, from the United States and thus was more closely tied to the U.S. economically than was Britain. When the United States plunged into the Great Depression after the stock market crash, these short-term funds were called back to the U.S., which made capital scarce in Germany and helped lead to the collapse of German banks. As a result, the Great Depression turned out to be much more severely felt in Germany than in Britain. Since the German chemical industry still exported a large share of its production, large exporters such as IG suffered from a dramatic decline in demand for their products (Plumpe 1990). While Germany experienced the Great Depression as a great economic crisis, Britain was much less affected. As Britain abandoned the gold standard in 1931, and thereby devalued the pound sterling, her economy recovered rather quickly.

Clearly, macroeconomic factors have had a non-trivial influence on the development of competitive advantages in the two national industries. The generally faster growth of the German economy in the decades around the turn of the 20th century and the interwar period pulled chemical output along with it. This should be no surprise: After all, the chemical industry supplies its products to most industrial sectors in the economy and so it will be strongly affected by macroeconomic conditions.

PREPARATIONS FOR WORLD WAR II AND WARTIME, 1935–1945

Both Britain and Germany systematically prepared their chemical industries to be able to wage war starting as early as 1935 (Plumpe 1990; Reader 1975). World War II affected the comparative position of the two national chemical industries in almost exactly the same way as had World War I. The outbreak of fighting once more resulted in Germany losing her presence in the important overseas markets, although some of the lost exports could be made up by finding new markets in countries such as Italy, Hungary, and Sweden. Recapitulating another feature of the Great War, the property of German chemical firms was confiscated in many nations of the world, even in areas such as South America, which were not part of the conflict.

After the experience of the First World War, both countries were much better prepared to take control of the chemical industries. For example, this time the German government took full control of prices, making it difficult for chemical firms to earn high returns.[26] In neither country was the chemical industry to emerge from this war with large earnings that could have funded postwar expansion. After its experience with nitrate shortage in World War I, Germany pushed for self-sufficiency in World War II. The German government financed a great part of the costs of researching synthetic rubber and synthetic fuel, and much of IG's "private" investments were actually going into the autarky sectors. However, Plumpe (1990) has shown that traditional organic areas such as dyestuffs, pharmaceuticals, and photochemicals, not the investments in autarky projects, provided by far the largest share of profits for IG. While synthetic rubber and synthetic fuel received the largest part of IG's investments, both products yielded very low returns, largely because the government put a 5% cap on returns of investments that had any government participation (Plumpe 1990, p. 568).

While the First World War officially ended with a truce between all warring parties, World War II ended in May 1945 with the unconditional military surrender of Germany. Partly because the technologies of killing had become even more powerful, the Second World War was, by any measure, much more devastating than the already ghastly Great War. According to conservative estimates, 30 million were dead as the result of the hostilities: 20 million in the Soviet Union, 5.25 million in Germany, and 386,000 in Britain. The German state literally withered away as the country was divided into four zones that were put under the military rule of the U.S., British, French, and Soviet armies. All chemicals firms were subjected to the direct control of the military, which immediately ordered that all production be stopped. As the chemical industry, and particularly IG, had played a large role in making Germany's six-year aggression possible, the German chemical industry and IG had become intimately associated in the eyes of the world with the Nazi regime.[27] Many high-ranking corporate officers were to be tried at Nuremberg for crimes against humanity (Teltschik 1992, pp. 190–99). The Allies had already made clear during the war that IG, which possessed 344 plants and mines across Germany and occupied Europe (Hayes 1987, p. xi), would be broken apart. The only question that remained was how small the pieces would be.

The war programs — with the exception of artificial fibers and synthetic rubber — provided neither the British nor the German chemical industry with a lasting competitive advantage. In fact, the war forced both chemical industries, but particularly the German one, to focus on many chemical fields that would be of little future significance. Reader (1975) has called the period from 1935 to 1945 a time of big distraction for ICI.

With the end of the war, the future of the German chemical industry was uncertain. A full 85% of the chemical plant facilities were physically intact. But one proposal (the Morgenthau plan) called for turning Germany back into a purely agricultural state, which would mean the disappearance of the chemical industry as a serious competitor

---

[26]Because the Nazis also took control of the German universities and government research institutes, many leading scientists and engineers who were either Jewish or deemed ideological enemies were removed from their posts. Because most of these émigrés were unwilling or unable to return after World War II, Keck (1993, p. 132) concludes that the national-socialist period "left a damaging effect on the quality of German science for one or two decades."

[27]Plumpe (1990) and Hayes (1987) provide the best scholarly descriptions of the relationship between IG and the Nazi regime. In his published dissertation, Stratmann (1985, ch. 3) gives an account of the relationship of the German state and the entire chemical industry for the period from 1933 to 1945.

from the world market. The relative position of the two chemical industries in the postwar years would be shaped, to a large extent, by the policies of the victors. The German chemical industry, embedded in a fully defeated Germany, had lost the ability to be the master of its own fate.

HALF A CENTURY OF PEACE: THE GOLDEN AGE OF PETROCHEMICALS, 1945–1997

Up to this point, we have tried to identify the different factors in the matrix framework, more or less one after the other, as they have individually shaped the shift in competitive advantages between the British and the German chemical industries over a hundred years. The analysis up to the end of World War II has shown that investments in firm capabilities, strong interactions between firms and a well-developed university system, government policies toward the industry, a favorable macroeconomic environment, finance, and the like rather than domestic natural resources have played a key role in explaining the changing fortunes of the two national industries.

Instead of running through the various levels in the matrix, as we have done before, we will explore in this section how the various factors in the matrix dynamically interacted to produce the competitive outcomes in the post-World War II era. We will try to illuminate the development of the relative positions of the two national chemical industries by focusing on how the same factors that were important in previous eras combined, though in different degrees and different ways, to shape the competitive position of British and German chemical firms in the period after the Second World War.[28] Since the various levels in the matrix framework are dynamically related in very complex ways, we will attempt to do some justice to this complexity by examining the major developments after 1945 as they were simultaneously produced by the different levels in the matrix.

Over the last 50 years, the chemical industry has produced innumerable large and small innovations that created new petrochemical processes, improved plastics and fibers, important new pharmaceuticals, and even an entirely new class of products based on biotechnology (Landau 1989; Landau and Rosenberg 1992; Teltschik 1992, pp. 213–30 and 289–324; Aftalion 1991). Because of this pattern of innovation, firms wanting to secure their long-term future were required to continue making large investments in R&D capabilities.

---

[28]While the statistical materials available on all aspects of the chemical industry have improved dramatically in recent decades, no excellent secondary sources cover the history of the entire chemical industry after 1945. No one of the caliber of a Haber (1958, 1971) has written a comprehensive history of the European and American chemical industries for this period. Similarly, there are no firm histories such as Plumpe's (1990) for IG or Reader's (1970, 1975) for ICI covering the postwar period. Teltschik (1992) and Aftalion (1991) have written valuable works, but they are nowhere near to the depth and comprehensiveness of Haber's, Plumpe's, or Reader's works. Pettigrew (1985) has published a study on ICI for the period from 1963 to 1983, but it is concerned mainly with organizational change. Hofmann (1975) provides detailed financial data on German chemical firms and some financial data on other important chemical firms in the world, but only for the period from 1960 until 1974. Spitz (1988) has written a good history on one segment of the industry, petrochemicals. To commemorate its one-hundredth anniversary, the Society for Chemical Industry organized a conference at which academics, corporate scientists, chief executive officers of firms, and other important members of the chemical community gave presentations on the history, present, and future of the chemical industry. The papers of that conference (Sharp and West, 1982) offer much information, often on very specialized subjects, but do not constitute a systematic account of the post-World War II chemical industry.

During World War II, much of the German infrastructure and physical plant was destroyed. (All references to Germany in the period after World War II should be understood as applying to West Germany, unless we explicitly say otherwise.) In particular, the lack of transportation equipment and severe shortages of raw materials led the German economy to spiral downward (Stratmann 1985, p. 154) until the currency reform and other postwar economic reforms reversed the trend and once again set the economy on a growth path.[29] In 1950, British dye production surpassed German production for the first time since about 1875. Her output of 92,000 tons, compared to 30,000 tons in Germany, gave Britain a 3:1 lead on a per capita basis. Total chemical production in 1951 was 14.65 billion Deutsche Mark for Britain and 9.7 billion Deutsche Mark for Germany, which translates into a 1.5:1 lead for Britain on a per capita basis.

The Allies did allow Germany to become fully integrated after the war into the capitalist economies; they thereby gave the German chemical industry the opportunity to use its largely intact human capabilities to rebuild the successor companies of IG into world-leading chemical exporters[30] (Stratmann 1985; Stokes 1988). In 1994, of the top four chemical producers in the world, three were German successor firms of IG: Hoechst (#1), BASF (#2), and Bayer (#4) were individually larger than ICI had been before its demerger in 1993, indicating how much more rapidly the German chemical industry grew once the German economy was stabilized in 1948.

Britain could not sustain her early lead in the postwar era. Britain's chemical industry led Germany in the shift to petrochemical feedstocks, which gave Britain a technological first-mover advantage over Germany's technological strength in coal-based chemicals. But production figures for the chemical ethylene succinctly tell the story about the postwar performance of the two industries (just as production figures for sulfuric acid told the story in the 19th century). In 1955, Britain produced 106,000 tons of ethylene, West Germany only 35,000 tons. This 3:1 lead on a per capita basis was reversed in favor of West Germany by 1965, when West Germany produced 686,000 tons to Britain's 529,000 tons, or 1.2 times more than Britain when measured in terms of per capita production. In another 10 years, Germany produced twice as much ethylene on a per capita basis (3.36 million tons versus 1.56 million tons in absolute numbers). In the early 1970s, for the first time since the industrial revolution, Germany had moved on a par with Britain in terms of per capita gross domestic product; in both countries, per capita GDP was about $10,000 (in 1985 U.S. dollars).

Before the reunification of Germany in 1989, total chemical production for West Germany stood at 77 billion ECU and at 39 billion ECU for Britain (1 U.S. dollar is about 1.2 ECU), which made the German industry 1.8 times as large as the British on

[29]The currency reform of June 1948 clearly favored the holders of real values (stock, physical property, intellectual property) over the holders of monetary instruments (currency, loans, bonds, etc.). While the former basically retained their wealth, the latter found up to 93.5% of it destroyed. Of the new currency, every German individual received DM 40 (later an additional DM 20), and firms were given DM 60 for each employee so that they could continue paying their bills. This event marked the second time in a generation that many Germans had lost their entire wealth due to inflation, which helps explain why Germans have been some of the most conservative investors in the world during the post-World War II period. Later in this volume, Da Rin's chapter offers details on how German savings and investment behaviors have shaped the nation's system of industrial finance; Eichengreen's chapter will touch on the relationship between savings behavior and the macroeconomic environment.

[30]For scholarly studies of the policies taken by the Allies toward the chemical industry from 1945 until 1952, see Stratmann's (1985) *Chemische Industrie unter Zwang?* and Stokes's (1988) *Divide and Prosper.*

a per capita basis. Since the 1970s, the German chemical firms Hoechst, BASF, and Bayer—the three main successors of IG—have been among the largest chemical firms in the world, typically ranked 1, 2, 3 (Hofmann 1975, pp. 60–61).[31] The comeback of the German chemical industry since 1950 is a success story but no miracle. Instead, it was a matter of hard work, heavy investments into firm capabilities, good educational institutions, well-designed public policies, and favorable macroeconomic conditions.

By 1992, the unified German chemical industry accounted for 19.7% of world chemical trade, the British 10%. Germany also exported $1.1 billion more in chemical goods to Britain than she imported from Britain. However, in the last 15 years or so, Germany has shown less dynamism. By the mid-1990s, after reunification expanded Germany's population while adding little to the output of its chemical industry, the per capita output of both chemical industries has returned to its 1913 proportions of 1.45:1 in favor of Britain. Along with the absorption of East Germany, this gap also reflects a higher growth rate of Britain's chemical industry during the last 15 years. From 1980 to 1994, the British chemical industry grew at an average rate of 3.4% each year, while Germany's grew at 2% per year (CEFIC 1995). In absolute numbers, of course, both industries have grown tremendously since 1913, as their production volumes now stand at 91 billion ECU for Germany and 45 billion ECU for Britain.

In essence, these figures show that the German chemical industry enjoyed substantial competitive advantages until 1980. Since then, Germany has suffered from labor inflexibility and high labor costs, resulting in low productivity gains intermingled with an unfavorable macroeconomic climate that slowed domestic demand and growth. Meanwhile, the Thatcher government fundamentally altered Britain's economic policies, and the British chemical industry has rebounded somewhat.

The German Chemical Industry under Allied Control

Considerable uncertainty existed over what policies the Allies would take toward the chemical industry after World War II. In 1952, it was finally resolved how IG's vast holdings would be broken up, and the successor firms were legally incorporated.[32] The military authorities originally placed strict prohibitions on chemical production and research, but they soon realized that the reversal of German economic decline required allowing chemical firms to produce again. It had become obvious that chemicals are the lifeblood of a modern economy, absolutely essential for the functioning of every part of the system.

After World War II, the immediate prospects of the German chemical industry looked dismal. In July 1945, President Truman issued an executive order to make public all information already obtained and still to be obtained from the enemy. In practice, this meant that American authorities copied and disseminated the proprietary records of 400 important technology firms and research institutions.[33] In what the U.S.

---

[31]The journal *Chemical & Engineering News* (for example, 1995) publishes a list of the top chemical producers every year in May. For a list of the 250 largest chemical firms in 1989, see the appendix in Aftalion (1991).

[32]For a description of the different policies toward the chemical industry adopted by the four allied powers in their respective zones between 1945 and 1949, see Stratmann (1985).

[33]In the Paris agreement of 1955, Germany had to accept the confiscation of all these assets without compensation (Teltschik, 1992, pp. 188–90). It needs to be stressed that the record of how Germany dealt with occupied territories during the war suggests that she probably would not have behaved any differently if she had won the war. See also Spitz (1988, ch. 1) for an evaluation of how the American, British, French, and Soviet chemical industries benefited from this involuntary technology transfer.

Chamber of Commerce called the biggest treasure hunt the world had ever seen, the Allies confiscated know-how, trademarks and patents, and the total assets of IG in foreign countries without compensation (Teltschik 1992). But the technological shift toward petrochemicals made this loss much less significant than expected. After World War I, Germany had surrendered leading-edge technology such as the Haber-Bosch process. But after World War II, Germany surrendered mainly coal-based technology that was soon largely obsolete, as it became possible to derive feedstocks from oil.

The United States deemed it essential to abolish cartels in Europe and bring down tariff walls to ensure peace and economic prosperity. As a result, the military authorities required Germany to put stringent antitrust laws on the books. Whereas German chemical firms after World War I had found it increasingly more difficult to send their products abroad, they had the opposite experience after World War II. Germany also benefited from the cold-war conflict between the United States and the Soviet Union, as American policy makers decided it was in their interest to rebuild Germany into a powerful industrial economy and turn it into a strong ally against the Soviet Union. Thus, Germany was made part of the Marshall plan, whereby American capital was sent to Europe to ease the dollar shortage and enable purchases of key imports after the war.

By the standard of nations that start and lose wars, West Germany was lucky. She was allowed to rebuild her economy. Her technological losses ended up being minimal, since she would have had to convert to petrochemicals in any event. Germany was allowed to fully participate in the trade among the countries of the free world. By 1952, basically all restrictions were removed from the German chemical industry. The breakup of IG was finally completed, and new independent successor firms, created very much along the lines of the original founding firms in 1926, had taken over the plants and were ready to reenter the export market for chemicals.

Interactions between Firm Capabilities and the Larger Social and Economic Environment

British and German society emerged from World War II with completely different psychologies (see Horstmeyer chapter for details). World War II had destroyed the Nazi movement, many cities in Germany, and the lives of millions of people, but it had not destroyed the capabilities of the German chemical firms. Each successor firm of IG could start where IG had stopped: as one of the most sophisticated chemical firms in the world, with a long tradition of investing heavily into research. The destruction of Germany's plant and territorial ambitions also may have done more than any politicians could have achieved in creating a strong consensus in German society that what the country now needed was long hours of hard work. In contrast, working-class British soldiers returned from the war with the strong impression that the working classes on the continent had a much higher living standard and social security benefits than they enjoyed in Britain. The British working classes, after the long deprivation of the war, demanded what they perceived was their fair share of the wealth created in the British economy.

Britain and Germany thus set off on quite divergent paths in terms of how the postwar economy was to be organized. While in Germany the politics of the moment called for moderation in public spending and wage increases, to put the economy back on track and to create full employment, in Britain the politics of the moment called for higher wages and social-welfare spending, which made enormous new demands on the economy and the government budget. While the idea of nationalizing some of

Germany's important industries around the Ruhr was shelved in favor of creating a free economy, the British labor government put the nationalization of industries high on the agenda. Germany developed a system of cooperative labor relations, in which workers held 50% of the seats on the boards of all large stock companies, and there was little labor strife. In Britain, on the other hand, labor relations became even more adversarial, making Britain one of the most strike-prone industrialized countries from the 1950s to the 1970s.

Two other national developments had important effects on the competitiveness of British industry. Britain was too weak after the war to hold its empire together. As a result, British firms, which had not experienced much competition in British colonies and dominions before the war, now faced competition from other European and American firms in the markets of these newly independent countries. Furthermore, since Britain refused to join the European free-trade community (formed in 1957) until 1973, it was more difficult for British than for German firms to export to Europe.

Given these divergent national experiences, the chemical industries in both countries faced very different environments after the war. Demand for chemicals was growing at a much faster rate than GDP in the western world from the 1950s until the first oil price shock in 1973; for example, the German chemical industry grew by an annual average of over 10% from 1950 to 1973 (Hofmann 1975, p. 175). In this environment, chemical firms had great opportunities for making profits and expanding production capacities. The German firms, with their long experience with exporting dyes and other chemical products all over the world, and with a domestic environment of labor peace and macroeconomic stability, were in a much better position than their British rivals to take advantage of the petrochemical boom. German chemical companies became large exporters again in the 1950s, earning important foreign currency that in turn helped stabilize the macroeconomic balance of the German economy.

While ICI had a head start on the German successor firms in petrochemical technology, the three largest successors of IG were able to catch up in this new technology relatively quickly by creating joint ventures with British oil companies (BASF with Shell and Bayer with BP) and by importing American petrochemical technology (Stokes 1994).[34] Because the United States gave a military guarantee in the postwar era that all western European countries would have access to cheap middle eastern oil, British firms did not enjoy any advantages over German ones in terms of access to cheap resources. Thus, the capabilities of firms in the context of their nations' institutions would be decisive in determining success during the golden age of pet-rochemicals. In Britain, the large oil companies BP and Shell, much more than ICI, were able to take advantage of the petrochemical revolution and build very successful business lines in chemicals, just as U.S. oil companies did. The successors of IG, despite losing patents and trade secrets, had the advantage of excellent managements and technical staffs. By obtaining the petrochemical technologies, they could enter into fibers and plastics and acquire cheap feedstocks for their traditional lines in organic inter-mediates.

ICI, like other British chemical firms in existence before the petrochemical revol-ution, was hampered by the fact that it now faced new competition in the former colonies and dominions of the British Empire and by the fact that it had to negotiate with 38 unions rather than one, as did its German competitors. Like many other firms in Britain, ICI was overstaffed and suffered from lower productivity gains than its large

[34]One of us—Ralph Landau—was intimately involved in the transfer of American technology to West Germany as early as the 1950s.

German rivals in the 1960s and 1970s (Pettigrew 1985, p. 82). In addition, ICI had other historically rooted disadvantages over its German rivals that prevented the company from experiencing a growth miracle. For example, the 1929 agreements with DuPont kept ICI out of the U.S. market until the U.S. government forced cancellation of the agreement in 1952. German firms had already returned to the U.S. market, using their prewar contacts, at a point when ICI had to build up its U.S. organization from scratch. Similarly, the interwar cartel agreement had kept ICI out of the continental market, and it took the firm until the 1970s to build up a sizable business on the continent. By that time Bayer, BASF, and Hoechst were each individually larger than ICI.

The postwar breakup of IG unleashed energies and provided the flexibility needed among the successor firms to redeploy firm capabilities to take advantage of the petrochemical revolution and the booming world economy. However, postwar ICI suffered from weaker management capabilities and lack of access to the cheap capital that would have made it possible to exploit the technological opportunities of the petrochemical revolution and allowed the firm to move quickly into American and European markets. Thus, contrary to widespread expectations, it was not the British but the German chemical industry that profited more from the international regime born after World War II and the economic prosperity that ensued.

British Revival and German Competitive Troubles in the 1980s and 1990s

With the election of the conservative politician Margaret Thatcher as prime minister in 1979, Britain moved closer to the free market system that had reigned under conservative government in Germany until 1967, while Germany was moving closer to the policies of the British government after World War II. Thatcher broke the power of unions, reduced government spending, and deregulated the economy. Conversely, Germany enacted a large social welfare system that made labor very expensive and inflexible. By the 1980s, German workers enjoyed a shorter work week, longer vacations, and higher levels of job security than almost any workers in the industrialized nations. Higher wages are not a problem as long as productivity levels keep pace. But this was no longer the case in Germany during the 1980s. Since the first oil price shock in 1973 to 1974, Germany has not reached the level of growth that had turned the country from a field of rubble into one of the richest countries in the world. It became increasingly difficult for Germany to pay for the higher social benefits that unions achieved, even after a conservative government was elected in 1981. Thus the British chemical industry, which had finally learned how to trim its work force and achieve the productivity levels of her German rivals, gained significantly in competitive position vis-à-vis Germany's.

In recent decades, Germany has no longer enjoyed an advantage in the form of a superior system of higher education. While Germany's secondary education institutions, and particularly her systematic training of technicians and skilled craftsman, remains superior to that of Britain (Walker 1993), Germany's system of higher education has dramatically declined in relation to the British one. Keck (1993) writes that Germany's higher education sector is one of the "weak components in the country's innovation system," having been neglected since the mid-1970s (see also Kahn, 1996). While student enrollments have increased by more than 65%, neither expenditures nor the number of staff has grown since 1975 (Keck 1993, p. 141). By 1986 to 1987, Germany had fallen far behind Britain in terms of financial support per student, spending less than one-half of what Britain spent on a per student basis (p. 140). Britain's share of papers in the world's scientific literature—8.2% in 1991 versus 7.7% for Japan and 5.8% for West Germany (Walker 1993, p. 180)—is one indicator that at the highest levels of science, Britain has become at least a full equal to Germany.

In recent years, the German economy faces a number of other strains that have affected the chemical industry. The absorption of East Germany into the West German economy after reunification put enormous burdens on Germany's federal budget. Since billions of Marks needed to be transferred to the former East Germany to maintain the economy there, German taxes—already a strain on the German economy—climbed even higher. German regulations have also become increasingly costly to chemical firms, as Germany combines a high population density with one of the strongest environmental movements in Europe. (Esteghamat's chapter later in this volume analyzes in greater detail the impact of environmental regulations on the chemical industry.)

The recent American and British movement to increase shareholder value, coupled with the desire of Germany's large chemical firms to tap foreign capital markets, has pressured German chemical firms to pay higher returns to their shareholders. German chemical firms have traditionally paid lower dividends than their British and American rivals and have put much of the profits back into R&D and plant investments. Unless German firms can become as cost-efficient as their rivals throughout the world, this pattern will be difficult to sustain in a world of global capital movement, as Richards will discuss in his chapter in this volume.

There are signs that the German chemical firms are trying to make their businesses more profitable by organizing mergers, creating alliances, or seeking divesting of commodity chemical product lines. Unlike in the high-growth regions of Asia, the European commodity chemicals business is plagued by chronic overcapacity problems, requiring substantial restructuring of the industry. While British chemicals firms have responded to these economic pressures earlier (in part because of more-demanding shareholders), German chemical firms are beginning to focus their continuing activities on technologically more sophisticated areas, such as pharmaceuticals in the case of Bayer and Hoechst, that are likely to bring higher returns. The demerger of ICI into a commodity chemicals (now changed by the acquisition of Unilever's specialty chemicals and an aggressive program of divesting more commodity-type chemicals) and life science firm is widely seen as a successful model for narrowing product portfolios. Both British and German firms are likely to maintain their position in traditional European markets, but much of their future capital investment will probably go to the regions of the world that experience higher growth rates.

Germany of the late 1990s faces a situation similar to that of Britain in the late 1970s. There is a broad acknowledgment that the industrially cozy times of the cold war era are over and that German industry faces an enormous competitiveness problem. To what extent Germany will be able to scale back worker security and general benefits remains to be seen. If the last 150 years in the relationship between the two chemical industries and the two larger societies in which they were embedded tell an enduring lesson, it would be the notion that profound institutional changes are very difficult to make in the absence of a great crisis—whether the crisis be real or just imaginary.

CONCLUDING REMARKS

This study of the British and German chemical industries over the last 150 years has shown that social actions by managers and public policy makers were mainly responsible for creating the competitive advantages of firms. We want to close by extracting from this industrial history some lessons for corporate leaders and public policy makers.

Long-term competitive dynamics clearly cannot be explained with references to a single factor. The rise and fall of the two national industries were caused by the interaction of numerous variables (Porter 1990). The idea that it is useful to think about competitive advantages in terms of a matrix framework has received strong support in our study. Factors both external and internal to firms have shaped the successes and failures of the two national industries. We have seen that legal, social, and macro-economic environments are very important in supporting or harming the competitiveness of local firms. Governments have a wide variety of levers by which they can influence the development of industries. The British government only learned in the wake of World War I that inaction is also a form of action; wide-ranging government policies allowed British firms to regain ground against their German rivals.

One of the most striking features of the British and German chemical industries is that the leading firms in both countries have remained leaders for very long periods of time. The big three German firms and the forerunners of ICI have their origins in the 1860s and 1870s. In fact, a look at the list of the 25 largest chemical firms in the world in 1994 shows that 17 were already engaged in chemical production before World War I; the major exceptions are largely oil companies that entered the petrochemical market around the 1940s. It is striking how successfully incumbent firms have branched off into new chemical technologies and continued to grow into gigantic enterprises, rather than being displaced by innovative start-up firms.

What has been the basis for this remarkable industrial longevity? The record of the two national chemical industries shows that committed long-term investments in firm capabilities can create and maintain competitive advantages of firms even in the face of dramatic technological changes. The 19th-century German chemical industry surpassed the British one in the new organic sector by building up large organizations that not only created powerful marketing, management, and manufacturing capabilities but also were able to institutionalize innovation in the form of R&D laboratories. The petrochemical revolution coinciding with the military and economic destruction of Germany at the end of World War II seemed to provide the British industry with a new opportunity to leap ahead of Germany in the postwar period. But with equal access to cheap oil and the development of a market for basic petrochemical technology (as explored in the chapter by Arora and Gambardella), the capabilities that German firms had built up over 70 years proved to be a decisive competitive advantage in the postwar era. With a continued commitment to investing heavily in R&D, both British and particularly German firms have successfully branched off into new, high-growth product areas.

We have seen that the chemical industry displays enormous advantages in size due to economies of scale and scope. It is difficult, however, to judge where the limits to scale and scope economies may lie. IG did not exist long enough to provide a natural experiment of whether combining very successful independent corporate identities in one large corporation would have hurt the innovativeness and performance of the German chemical industry in the long run. However, the post-World War II success of IG's initially much-smaller successor companies at least suggests that independent decision making by managers who have time to know their business in detail may make firms more successful. There is plenty of evidence in the United States that focused companies outperform conglomerates (Teece, Rumelt, Dosi, and Winter 1994). The 1993 demerger of Britain's ICI into smaller corporate entities pointedly raises the question of whether the very big German chemical firms are not becoming too large to take full advantage of the firm capabilities that have developed within the individual units of these large chemical enterprises. As German chemical firms increasingly try to access foreign capital markets, the financial community will clearly pay close attention to the question of optimal scope. (See the chapter by Richards.)

As markets become increasingly global in many branches of the chemical industry, the large firms are under pressure to develop top management teams that are expert in much more than chemical technology. Forced to operate as truly multinational firms, top management teams increasingly need to be competent in such diverse fields as international finance, marketing, foreign investment, and government and public relations (see also the "Process of Innovation" chapter). Responding to these changed demands, the big three German chemical firms in recent years have found it necessary to break with the 100-year-old tradition of having a chemist as CEO, although the majority of members remain scientists and engineers by training.

While firms compete in the marketplace, nations play a substantial role in providing firms with an environment for building up firm capabilities. A top-notch system for education in science and technology clearly is vital for training the human resources that firms can draw on to create strong firm capabilities. It took Britain many decades to catch up in creating a system of higher education in science and technology that was comparable to the one in Germany. Once a system of higher education is in decline because of persistent underfunding, and perhaps because of a lack of competition between individual universities for the best researchers, teachers, and students, it may be impossible to improve the system dramatically over a short period of time. This may have a significant impact on a country's ability to enter "new tech" product markets at a stage when it is still undetermined which firms will become the successful players.

The tendency is to focus only on the highest level of the educational system, but it is worth remembering that the German chemical industry's rise to leadership in the dye sector benefited substantially from technical trade schools in the beginning of its existence. It appears crucial to ensure that a high-tech industry can draw not only on relatively mobile university-trained scientists but also on a pool of skilled workers. While top scientists are often internationally mobile, skilled workers typically are much more tied to the particular nation in which they grew up and were trained. Hence, a nation can benefit enormously from high-quality worker training programs.

The record of the chemical industry shows that whether or not a government will enact policies favorable for the growth of industry depends to a large degree on how well the industry and the appropriate governmental agencies are connected and interact on a routine basis. The interface between the German chemical industry and the government was well-developed in the last quarter of the 19th century, allowing such important collective policy initiatives as the passage of a patent law to be especially tailored to the needs of the chemical industry. The crises of World War I brought British government and industry together. There is no reason why governments and industries cannot create standing forums for consultation, where the needs of each party are seriously discussed on a regular basis. The probability of enacting the right set of policies is much greater if the relevant government agencies and industry participants educate one another about their needs and develop joint efforts to remove obstacles for firms to become competitive in international markets.

Government policy appears most effective when it creates and maintains a broad collection of institutions which set economic, competition, and regulatory policies that protect the overall interests of society, but let firms and markets make the actual business decisions.

The 150-year history of the two chemical industries points to one particularly enduring lesson: Significant industrial competitive advantages can be created or lost — but not overnight. While a crisis may create the sudden motivation to change, industrial capabilities are in essence the product of long-term investments.

REFERENCES

Aftalion, F. (1991), *A History of the International Chemical Industry*, University of Pennsylvania Press, Philadelphia, PA.

BASF (1990), *Chemistry for the Future: 125 Years BASF*, BASF Aktiengesellschaft, Ludwigshafen, Germany.

Beer, J. H. (1959), *The Emergence of the German Dye Industry*, University of Illinois Press, Urbana, II.

Ben-David, J. (1977), *Centers of Learning: Britain, France, Germany, United States*, McGraw-Hill, New York, NY.

Ben-David, J. (1984), *The Scientist's Role in Society*, University of Chicago Press, Chicago, IL.

Berghoff, H. and R. Möller (1994), "Tired Pioneers and Dynamic Newcomers? A Comparative Essay on English and German Entrepreneurial History, 1870–1914," *Economic History Review*, XLVII(2): 262–87.

Cairncross, A. (1953), *Home and Foreign Investment, 1870–1913*, A. M. Kelley, Clifton, NJ.

CEFIC (1995), "Facts & Figures, The European Chemical Industry in a Worldwide Perspective," European Chemical Industry Council, Brussels, November.

Chandler, A. D. (1990), *Scale and Scope*, Harvard University Press, Cambridge, MA.

Chemical & Engineering News (1995), "The Largest Chemical Firms in the World in 1994," May 8 issue, pp. 13–18.

Duisberg, C. (1933), *Meine Lebenserinnerungen* (My Life), P. Reclam jun, Leipzig, Germany.

Elbaum, B. and W. Lazonick, eds. (1987), *The Decline of the British Economy*, Clarendon Press, Oxford, U.K.

Flechtner, H.-J. (1959), *Carl Duisberg: Vom Chemiker zum Wirschafütsfhrer* (From a Chemist to an Industrial Leader), ECON Verlag GMBH, Düsseldorf, F.R.G.

Fruton, J. S. (1988), "The Liebig Research Group — A Reappraisal," *Proceedings of the American Philosophical Society*, 132(1): 1–66.

Grant, W., W. Paterson, and C. Whitston (1988), *Government and the Chemical Industry: A Comparative Study of Britain and West Germany*, Clarendon Press, Oxford, U.K.

Haber, L. F. (1958), *The Chemical Industry during the Nineteenth Century*, Oxford University Press, Oxford, U.K.

Haber, L. F. (1971), *The Chemical Industry, 1900–1930*, Clarendon Press, Oxford, U.K.

Hardie, D. W. F. and J. Pratt (1966), *A History of the Modern British Chemical Industry*, Pergamon Press, Oxford, U.K.

Hayes, P. (1987), *Industry and Ideology: IG Farben in the Nazi Era*, Cambridge University Press, Cambridge, MA.

Haynes, W. (1933), *Chemical Economics*, Van Nostrand Company, New York, NY.

Hoechst (1990). *Der Weg: Vom Farbstoff Fuchsin zu Hoechst High Chem* (The Road: From the Fuchsine Dyestuff to Hoechst High Chem), Hoechst Aktiengesellschaft, Frankfurt am Main, Germany.

Hofmann, A. W. (1863), *Colouring Derivatives of Organic Matter, Recent and Fossilized*, International Exhibition 1862. Report by the Juries on the Subjects in the Thirty-six Classes into which the Exhibition was Divided, Class II — Section A (Chemical and Pharmaceutical Products and Processes), London, U.K.

Hofmann, R. (1975), *Welt-Chemiewirtschaft* (World Chemical Economics), Westdeutscher Verlag GmbH, Opladen, F.R.G.

Hohenberg, P. (1967), *Chemicals in Western Europe 1850–1914: An Economic Study of Technical Change*, Rand McNally, Chicago, IL.

Holdermann, K. (1953), *Im Banne der Chemie. Carl Bosch: Leben und Werk* (Under the Spell of Chemistry. Carl Bosch: His Life and Work), Econ-Verlag, Düsseldorf, F.R.G.

Holmes, L. (1989), "The Complementarity of Research and Teaching at Liebig's Laboratory," *Osiris*, 5 (2nd series): 121–64.

Homburg, E. (1992), "The Emergence of Research Laboratories in the Dyestuffs Industry, 1870–1900," *British Journal for the History of Science*, 25: 91–111.

Hornix, W. J. (1992), "From Process to Plant: Innovation in the Early Artificial Dye Industry," *British Journal for the History of Science*, 25: 65–90.

Hughes, T. P. (1969), "Technological Momentum in History: Hydrogenation in Germany, 1893–1933," *Past and Present*, August(441): 106–32.

Johnson, J. A. (1990), *The Kaiser's Chemists*, University of North Carolina Press, Chapel Hill, NC.

Johnson, J. A. (1992), "Hofmann's Role in Reshaping the Academic-Industrial Alliance in German Chemistry," in C. Meinel and H. Scholz (eds.), *Die Allianz von Wissenschaft und Industrie: August Wilhelm Hofmann,1818–1892* (The Alliance of Science and Industry: August Wilhelm Hofmann, 1818–1892), VCH Publishers, Weinheim & New York.

Kahn, P. (1996), "The Decline of German Universities," *Science*, 273(12 July): 172–74.

Keck, O. (1993), "The National System for Technical Innovation in Germany," in R. R. Nelson (ed.), *National Innovation Systems*, Oxford University Press, New York, NY, pp. 115–57.

Landau, R. (1989), "The Chemical Engineer and the CPI: Reading the Future from the Past," *Chemical Engineering Progress* (September): 25–39.

Landau, R. and N. Rosenberg (1992), "Successful Commercialization in the Chemical Process Industries," in N. Rosenberg, R. Landau, and D. C. Mowery (eds.), *Technology and the Wealth of Nations*, Stanford University Press, Stanford, CA, pp. 74–119.

Landes, D. S. (1969), *The Unbound Prometheus*, Cambridge University Press, New York, NY.

Locke, R. R. (1984), *The End of Practical Man: Higher Education and the Institutionalization of Entrepreneurial Performance in Germany, France and Great Britain, 1880–1940*, JAI Press, Greenwich, CT.

Maddison, A. (1991), *Dynamic Forces in Capitalist Development*, Oxford University Press, New York, NY.

Marsch, U. (1994), "Strategies for Success: Research Organization in German Chemical Companies and IG Farben Until 1936," *History and Technology*, 12: 23–77.

Meinel, C. and H. Scholz, eds. (1992), *Die Allianz von Wissenschaft und Industrie: August Wilhelm Hofmann, 1818–1892* (The Alliance of Science and Industry: August Wilhelm Hofmann, 1818–1892), VCH Publishers, Weinheim & New York.

Miall, S. (1931), *A History of the British Chemical Industry*, Chemical Publishing Company, New York, NY.

Morgan, G. T. and D. D. Pratt (1938), *British Chemical Industry: Its Rise and Development.* E. Arnold, London, U.K.

Morrell, J. B. (1972), "The Chemist Breeders: The Research Schools of Liebig and Thomas Thomson," *Ambix*, 19: 1–46.

Paulsen, F. (1906), *The German Universities and University Study*, Charles Scribner, New York, NY.

Pettigrew, A. M. (1985), *The Awakening Giant: Continuity and Change in Imperial Chemical Industries*, Blackwell, New York, NY.

Plumpe, G. (1990), *Die I. G. Farbenindustrie AG: Wirtschaft, Technik und Politik 1904–1945* (I.G. Farbenindustrie AG: Economics, Technology, and Politics 1904–1945), Duncker & Humblot, Berlin, Germany.

Porter, M. E. (1990), *Competitive Advantage of Nations*, Free Press, New York, NY.

Reader, W. J. (1970), *Imperial Chemical Industries: A History. Vol. I*, Oxford University Press, London, U.K.

Reader, W. J. (1975), *Imperial Chemical Industries: A History. Vol. II*, Oxford University Press, London, U.K.

Reddaway, W. B. (1958), "The Chemical Industry," in D. Burn (ed.), *The Structure of British Industry*, Vol. 1, Cambridge University Press, New York, NY, pp. 218–59.

Sanderson, M. (1972), *The Universities and British Industry 1850–1970*, Routledge & Kegan Paul, London, U.K.

Schmookler, J. (1965), "Catastrophe and Utilitarianism in the Development of Science," in R. A. Tybout (ed.), *Economics of Research and Development*, Ohio State University, Columbus, OH.

Schröter, H. (1991), *Friedrich Engelhorn*, Pfälzische Verlagsanstalt, Landau/Pfalz, Germany.

Schröter, H. G. (1992), "The International Dyestuffs Cartel, 1927–39, with Special Reference to the Developing Areas of Europe and Japan," in A. Kudo and T. Hara (eds.), *International Cartels in Business History*, University of Tokyo Press, Tokyo, Japan, pp. 33–52.

Seefelder, M. (1994), *Indigo in Culture, Science and Technology*, Ecomed, Landsberg, Germany.

Sharp, D. H. and T. F. West, eds. (1982), *The Chemical Industry*, Ellis Horwood, Chichester, U.K.

Smith, J. K. (1992), "National Goals, Industry Structure, and Corporate Strategies: Chemical Cartels between the Wars," in A. Kudo and T. Hara (eds.), *International Cartels in Business History*, University of Tokyo Press, Tokyo, Japan, pp. 139–58.

Spitz, P. H. (1988), *Petrochemicals: The Rise of an Industry*, Wiley & Sons, New York, NY.

Steen, K. (1995), "Confiscated Commerce: American Importers of German Synthetic Organic Chemicals, 1914–1929," *History and Technology*, 12: 261–84.

Stern, F. (1987), "Fritz Haber: The Scientist in Power and in Exile," in *Dreams and Delusions: National Socialism in the Drama of the German Past*, Vintage Books, New York, NY, pp. 51–76.

Stokes, R. G. (1988), *Divide and Prosper: The Heirs of I.G. Farben under Allied Authority, 1945–1951*, University of California Press, Berkeley, CA.

Stokes, R. G. (1994), *Opting for Oil: The Political Economy of Technological Change in the West German Chemical Industry, 1945–1961*, Cambridge University Press, New York, NY.

Stratmann, F. (1985), *Chemische Industrie unter Zwang?* (Chemical Industry under Coercion?), Steiner, Wiesbaden, F.R.G.

Suchanek, P. (1992), "Farbstoffe zu Hofmanns Zeiten und Heute " (Dyestuffs during the Time of Hofmann and Today), in C. Meinel and H. Scholz (eds.), *Die Allianz von Wissenschaft und Industrie: August Wilhelm Hofmann (1818–1892)*, VCH Publishers, Weinheim & New York.

Taylor, G. D. and P. E. Sudnick (1984), *DuPont and the International Chemical Industry*, Twayne, Boston, MA.

Teece, D. J., R. Rumelt, G. Dosi, and S. Winter (1994), "Understanding Corporate Coherence: Theory and Evidence," *Journal of Economic Behavior and Organization*, 23: 1–30.

Teltschik, W. (1992), *Geschichte der Deutschen Grosschemie: Entwicklung und Einfluss in Staat und Gesellschaft* (History of the Large-Scale German Chemical Industry: Development and Impact on State and Society), VCH Publishers, Weinheim, Germany.

Titze, H. (1987), *Das Hochschulstudium in Preussen und Deutschland 1820–1944* (Higher Education in Prussia and Germany 1820–1944), Vandenhoek & Ruprecht, Goettingen, F.R.G.

Travis, A. S. (1992), "Science's Powerful Companion: A. W. Hofmann's Investigations of Aniline Red and Its Derivatives," *British Journal for the History of Science*, 25: 27–44.

Travis, A. S. (1993), *The Rainbow Makers: The Origins of the Synthetic Dyestuffs Industry in Western Europe*, Lehigh University Press, Bethlehem, PA.

Turner, R. S. (1989), Commentary: Science in Germany, in K. M. Olesko (ed.), "Science in Germany: The Intersection of Institutional and Intellectual Issues," edited, *Osiris*, 5 (2nd series: 296–304.

Veblen, T. (1915), *Imperial Germany and the Industrial Revolution*, Transaction Publishers, New Brunswick, NJ.

Verg, E., G. Plumpe, and H. Schultheis (1988), *Meilensteine* (Milestones), Bayer AG, Leverkusen, F.R.G.

Walker, W. (1993), "The National Innovation Systems: Britain," in R. R. Nelson (ed.), *National Innovation Systems*, Oxford University Press, New York, NY, pp. 158–91.

Webb, R. K. (1980), *Modern England: From the Eighteenth Century to the Present*, Harper Collins, New York, NY.

Wetzel, W. (1991), *Naturwissenschaften und Chemische Industrie in Deutschland: Voraussetzungen and Mechanism ihres Aufstiegs im 19. Jahrhundert* (Sciene and Chemical Industrie in Germany: Conditions and Mechanisms of its Rise in the 19th Century), F. Steiner, Stuttgart, Germany.

Wiener, M. J. (1981), *English Culture and the Decline of the Industrial Spirit, 1850–1980*, Cambridge University Press, New York, NY.

Wrigley, J. (1987), "Technical Education and Industry in the Nineteenth Century," in B. Elbaum and W. Lazonick (eds.), *The Decline of the British Economy*, Clarendon Press, Oxford, U.K., pp. 162–88.

Zimmermann, P. A. (1965), *Patentwesen in der Chemie: Ursprünge, Anfänge, Entwicklung* (Patents in Chemistry: Origins, Beginnings, Development), BASF AG, Ludwigshafen, F.R.G.

3 Chemicals: A U.S. Success Story

ASHISH ARORA and NATHAN ROSENBERG

The U.S. chemical industry is the largest of any nation today and has been since at least World War II. In 1996, sales of the U.S. chemical industry amounted to $372 billion, while those of all western Europe taken together amounted to $495 billion and of the Japanese chemical industry, $216 billion. On a value-added basis, the chemical industry accounts for about 11.3 percent of U.S. manufacturing and produces about 1.9 percent of U.S. gross domestic product. The preeminence of the U.S. chemical industry is not reflected in size alone. The chemical industry is one of the largest in terms of R&D, spending an estimated $18.3 billion in 1996 (CMA Handbook, 1997, p. 90). Since almost all (98%) of the R&D in chemicals is privately financed, chemicals are the leading industry in terms of privately financed R&D. In the last 40 years, many of the major innovations that have shaped the growth and direction of the industry were first commercialized by U.S. firms. Even more remarkably, contrary to a widespread belief that U.S. industry has tended to neglect process innovations, U.S. firms have pioneered most of the major process innovations as well. One list of major postwar, commercial chemical-process developments shows that U.S. firms commercialized 20 of the 41 major innovations listed, or just under 50%, and were among the pioneers in virtually every other innovation (Landau 1989).

The commercial success of the chemical industry in the United States is a story of a constellation of factors and the complementarities among them. Competitive success at the level of the firm has depended on successful application of new knowledge for commercial purposes, requiring substantial investments on many fronts: for example, in technological capability, production facilities, and marketing. These investments were made in a U.S. market that offered a number of favorable factors: superior technology, a large domestic market, and abundant natural resources. The relative importance of these factors has varied over time and across the different sectors of the industry, and no single factor has been sufficient by itself to confer permanent — or even long-lived — advantage.

The growth and development of the chemical industry in the United States bears a close relationship to the overall development of the U.S. economy. Hence, for analytical convenience, one may divide the history of the U.S. chemical industry into three phases: pre-World War I, the interwar period, and the post-World War II period, with the two World Wars marking periods of discontinuity.

The first phase, which lasted until about the beginning of World War I, witnessed rapid growth in the size of the industry. As discussed in the next section, the growth of

We are grateful to Ralph Landau and Alfonso Gambardella for helpful comments and suggestions.

the chemical industry in the United States before World War I was partially based on an abundance of natural resources. However, the abundant resources were not just good luck. In their discovery, transportation, and use, they were themselves a consequence of purposeful human endeavor (Rosenberg 1972; Nelson and Wright 1992; David and Wright 1991; Wright 1990). While the United States at this time lagged behind other leading countries such as Germany and Britain in generating new technologies, the gap in the exploitation of new technologies was much smaller.

During the interwar period, the chemical industry began to systematically invest in research and development and in building up broad-based technological capabilities. Initially, these capabilities were used to good effect in adapting and improving technologies imported from Europe. These investments in research truly paid off, however, during World War II and in the postwar decades. Petrochemicals, which were at the heart of the growth of the chemical industry after World War II, powerfully emphasize the complementarity between technology and resource endowments. The United States had abundant reserves of natural gas and oil, as well as firms with the ability to develop technology to exploit these raw materials for making the basic chemical building blocks. But petrochemicals, which started out as a quintessentially American phenomenon, soon became global. The development of a market for crude oil and the widespread diffusion of petrochemical technologies enabled other countries to develop a petrochemical sector as well and considerably narrowed the gap between the United States and other nations. Thus, the post-World War II period contains the peak of U.S. preeminence, as well as a time of catch-up by other countries.

PHASE I: THE DECADES BEFORE WORLD WAR I, 1860–1914

Growing with the U.S. Economy

The output of the U.S. chemical industry grew rapidly from 1860 to 1914. Although changing industry definitions mean that the numbers in Table 3-1 must be interpreted with some caution, the overall picture they paint of growth is unambiguous. Between 1849 and 1914, the total number of establishments grew from 170 to 633. The number of employees per establishment increased from about 8 to about 70, while value added per establishment increased, albeit in nominal terms, from about $10,000 to $150,000, indicating that value added per employee nearly doubled during this period. On the eve of World War I, the U.S. chemical industry was already the second largest in the world, after Germany, and if one were to include oil refining in the definition of the industry, it was the largest in the world. (Unless otherwise stated, all dollar figures are in nominal terms.)

The development of the chemical industry in the United States was surely related to overall industrial growth, which had risen rapidly during this period. The United States' share of world manufacturing output passed Britain's somewhere between 1880 and 1900, and U.S. manufacturing per capita surpassed that of Britain by 1900 (Bairoch 1982). Moreover, both the chemical industry and manufacturing depended heavily on the use of nonreproducible natural resources. Wright (1990) shows that after the Civil War, the relative material costs in the United States decreased over time. The rate of discovery of oil fields increased, and by the turn of the century the United States was the leading producer of coal, copper, iron ore, zinc, phosphate, molybdenum, lead, tungsten, and many other minerals.

TABLE 3-1. Growth of the Chemical Industry, 1849–1939: United States Chemical Industry

Year	Number of Establishments	Wage Earners: Thousands	Value of Products $ Millions	Value Added by Manufacture: Millions	Employees per Establishment	Value Added per Establishment, in Thousands of Dollars
1849	170	1.4	5	17	8.2	10
1859	299	2	5.4	2.3	6.7	7.7
1869	444	6	24	9.9	13.7	22.3
1879	649	11	44	17	17.1	25.4
1889	632	17	71	32	27.5	49.8
1899	530	24	80	35	44.3	65.9
1904	572	30	122	53	52.8	92.0
1909	597	35	177	79	58 6	132.8
1914	633	44	222	96	70.1	152.0
1919	1053	90	731	368	85.8	349.2
1921	981	59	503	255	59.8	260.8
1923	1039	90	788	387	86.7	372.8
1927	1028	68	746	374	65.9	363.5
1929	1676	79	959	508	47.2	303.3
1931	1556	61	687	394	39.5	253.0
1933	1367	65 .	582	324	47.4	236.7
1935	1580	82	821	430	51.6	272.2
1937	1667	98	1145	606	58.7	363.3
1939	1931	90	1198	668	46.6	345.7

Source: From "Census of Manufacturing," in *Chemical Facts and Figures* (1940), Manufacturing Chemists Association, Washington, DC, p. 1.

Notes: The figures include compressed and liquefied gases, explosives, insecticides, and fungicides, and wood-distillation products. They do not include figures on drugs and medicines. The value of output in this category for 1937 was $335.8 million.

Not surprisingly, the U.S. chemical industry during this time was largely inorganic and strongly based on mineral and other natural resources. Britain also had a largely inorganic chemical industry, but this was geared toward supplying bleach, soda ash, and caustic soda to textiles, soap, glass, and other consumer-goods industries. In the United States, fertilizers and explosives were important products, as were mineral acids such as sulfuric acid and nitric acid. Soda ash and caustic soda, along with bleach, were imported in substantial quantities from Britain. After the 1880s, however, the adoption of the Solvay soda process and the electrolytic cell, and a period of tariff protection helped domestic production of alkalis and electrochemicals grow rapidly to the point of virtual self-sufficiency. In addition, a number of affiliated sectors processed available natural resources such as wood, animal tallow, or animal bones to produce waxes, gums, casein, and black pigment (carbon black).

Although there was an organic-chemical sector that included dyestuffs and coal-tar compounds, it was quite small. Synthetic dyestuffs production had been started in the United States in 1864 by Read Holliday & Sons, a British firm, and in the 1870s a number of firms entered the industry due to a 50¢ plus 35% ad valorem tariff (Haynes, vol I, p. 311). But after the tariff was lowered, in 1883, to 10% below the average American tariff level in the 19th century, only the established firms such as Read Holliday, Heller and Merz, and Schoellkopf managed to survive, in part because they were able to produce coal-tar intermediates such as phenol.

By 1914, the total share of organic chemicals (including items such as plastics, coal-tar products, and nitrogen compounds) was about one-quarter of the total value of chemical-industry production. By contrast, acids and electrochemicals alone accounted for over one-quarter of the total value (Haynes, vol. II, p. 277). As late as 1921, despite the war-induced boom in organic intermediates and dyestuffs, inorganic chemicals still accounted for a little over one-half the total value of output of about $444 million. Of these, the largest share was of sodium compounds, which accounted for a little less than one-fifth of industry output, followed by acids which accounted for one-eighth. The major organic categories consisted of coal-tar products (less than one-fifth), and plastics (less than one-tenth). The natural-resource dependence of the U.S. chemical industry is further underscored by the fact that electrochemicals accounted for just under one-eighth of the total output of the chemical industry.

An important factor limiting the growth of the organic-chemical industry (especially coal-tar chemicals) was the widespread use in the United States of the beehive oven for producing town gas and coke until World War I (Tarr 1994). In 1855 there were 106 beehive ovens in the country; by 1909, at their peak, there were almost 104,000 ovens. The beehive oven was cheaper and easier to build than the by-product ovens used in Germany and Britain. However, by-product ovens had two major advantages. First, they produced superior grades of coke and did so more efficiently. More pertinently, by-product ovens captured the volatile by-products of the coking process: about 70% tar and 30% ammonia, as well as aromatics such as benzene, xylene, toluene, and naphtha. In beehive ovens, these volatile and valuable by-products were vented into the air, causing extensive environmental damage. Since land and high-quality bituminous coal were abundant, however, beehive ovens were attractive.[1] The more densely

---

[1] A number of interrelated factors led to the substitution of recovery ovens for beehive ovens after 1914. First, the cutting off of German coal-tar chemical supplies induced the growth of an indigenous coal-tar chemical industry. Second, as Tarr (1994) points out, a number of cities switched to the use of municipal gas produced by by-product oven plants. By 1929, the coke ovens were supplying 75% of the coke manufactured in the United States. Between 1909 and 1940, the number of beehive ovens in the country shrank from 104,000 to about 15,000, by which time petroleum and natural gas had become the basis of the organic chemical industry.

populated European countries did not enjoy this luxury, but this is another instance of the resource intensity of the U.S. chemical industry.[2]

The U.S. chemical industry relied on simple techniques or imported technology during this period. Important segments of the industry such as dyestuffs, pharmaceuticals, and fine chemicals were dominated by subsidiaries and affiliates of foreign firms, especially German firms (Steen 1995). In fact, after World War I many of these subsidiaries — such as Rohm and Haas, Merck, and Schering — transformed themselves into American firms and contributed to the development of the organic-chemical industry. Even in inorganic chemicals, the major processes for the production of sulfuric acid, ammonia soda, and electrochemicals were imported from Europe. In some cases, technology came with capital, as it did initially in electrochemicals. In other cases, it came with people: The U.S. chemical industry benefited greatly from the services of trained and skilled chemists, engineers, and entrepreneurs who migrated from Germany, Switzerland, the Netherlands, and Britain. Viscose rayon fiber was commercialized by American Viscose, a subsidiary of the English textile firm Courtaulds, in 1911; the Celanese corporation was organized by the Swiss Dreyfus brothers, using British capital to produce cellulose acetate rayon in 1918.

Large as the chemical industry was, the consumer-goods sectors of the economy were much larger. Even after the First World War, when employment in the chemical industry was estimated at between 60,000 and 100,000 (depending on the definition of the industry), employment in the automobile sector was about 200,000.[3] Clearly, growth of demand played a crucial role in the growth of the chemical industry. Population growth, geographical expansion, and the development of a transport infrastructure (primarily railways) were important ingredients of this growth. Fertilizers, oil-refining, explosives, rubber, paper and pulp, and glass, soap, and textiles were important sources of demand for basic industrial chemicals of various sorts. By 1914, the United States produced about 4 million (short) tons of sulfuric acid, more than twice the combined output of Britain, Germany, and France, and about 80% of total European production.

For most of the 19th century, with the exception of large producers of fertilizer and explosives, and somewhat later, sulfuric acid, the typical U.S. chemical firm was a small, family-owned and -managed single plant, and geographically localized. Geographical localization was the consequence of high transport costs and the need for customer service. But as transport costs fell, growth was often accompanied by an increase in the size of the establishment, and such consolidations often occurred during economic downturns.[4] The average number of workers per establishment grew from 6 in 1850 to about 16 in 1870, while the average capital per establishment increased from $13,000 to $43,400. The number of establishments initially increased with output, but around the turn of the century, growth in production was actually accompanied by a decline in the number of establishments, as shown in Table 3-1. The number of establishments grew once again around the start of World War I.

With rapid economic growth and geographical integration, the U.S. chemical firms had a very large market. The large size of the domestic market encouraged production on a large scale, providing U.S. producers with valuable experience in large-scale

---

[2]Of course, environmental concerns have become far more important in the industry today, and these are discussed at greater length in the chapter by Esteghamat in this volume.

[3]See table 2.3 in Mowery (1981, p. 60) for comparisons of shares at this time in industrial output and industrial research.

[4]One of the earliest notable mergers was the one that led to formation of General Chemical Company, in 1899. General Chemical was formed by combining 12 companies and 19 plants for the production of sulfuric acid based on the "contact" process.

production and the opportunity to exploit economies of scale, initially in explosives, fertilizers, and sulfuric acid. The large size of the market also provided an inducement for some firms to invest in in-house technological capabilities. However, systematic investments in technological capability, in the form of corporate research laboratories, would be more fully evident only after World War I.

Patterns of Competitiveness

To provide some perspective, it is instructive to briefly review corresponding developments in two other leading chemical-producing countries of that time, Britain and Germany, whose experience is discussed in depth in the chapter by Murmann and Landau. As they note, the growth of the German chemical industry from the 1860s onward was impelled specifically by the development of the dyestuffs industry, and the organic-chemical industry more generally. By the end of the century, the Germans dominated the organic-chemical-based sectors of the world chemical industry. The British chemical industry, which had been largest in the world prior to the German ascendancy, had become largely inorganic in character. In inorganic sectors such as soda ash, caustic, and bleach, Britain remained the leading producer in the world, and a major exporter, until nearly the eve of World War I. Hardie and Pratt (1966, p. 69) note that in 1913, Britain imported £1.893 million worth of dyestuffs and exported £14.34 million worth of heavy inorganic chemicals.

The United States imported large quantities of soda and bleach from Britain until the last decade of the 19th century. Around that time, a combination of protective tariffs and the rapid growth of domestic-production capacity caused the imports to taper off sharply.[5] Imports did not recover, even after the tariff protection was reduced, because the U.S. alkali and bleach industry was now based on superior technologies (ammonia soda, and electrolytic chlorine and caustic soda). By the 1920s, in barely two decades, the United States had become largely self-sufficient in these important industrial chemicals. Dyestuffs continued to be imported in substantial amounts from Germany. A domestic dyestuffs industry did develop in the United States behind tariff barriers, but the lowering of tariff protection in the 1880s, and the consequently renewed supply of German dyestuffs, caused many domestic producers to exit.

These patterns point to the differing areas of technical and commercial leadership of the various countries. The German advantage lay in the organic-chemical industry, grounded in the rapidly developing science of organic chemistry, and the ability of the leading German companies to commercialize this new knowledge. The British continued to enjoy a competitive advantage in the production of high-volume inorganic compounds, as well as in the coal-tar-based intermediates sector. However, it seems fair to say that by the end of this period, the ability of British chemical firms to compete rested heavily on their superior access to markets in colonies and dominions, rather than on cost, quality, or technology.

U.S. leadership clearly was based on the exploitation of its abundant endowment of mineral and natural wealth for a large and prosperous market. But it would be a mistake to see the U.S. industrial performance as arising from resource abundance *as opposed to* the effective use of technology. Rather, prospecting, mining, and refining and smelting technologies were used to increase the effective supply of mineral

---

[5]Eichengreen's chapter in this volume discusses in greater depth the pattern of tariff barriers during the 19th century.

resources.[6] Even when initially borrowed from Europe, the technologies had to be adapted and modified to suit local conditions. Similarly, while a large market helped reduce unit costs, there was nothing automatic about it; "scaling up" production to take advantage of economies of scale required considerable technological sophistication and ingenuity, and eventually led to the development of an entirely new discipline of chemical engineering.

Electrochemicals, whose role in the U.S. chemical industry Haber (1958) has been likened to the one dyestuffs played for the German industry, exemplifies the interaction among technology, resource endowments, and market size. The production of electrochemicals, the most prominent of which were chlorine and chlorine-based compounds such as bleach, by its nature required cheap electrical power. The United States was abundantly endowed with sources of cheap hydroelectric power. In turn, the commercial exploitation of these natural resources presumed a certain level of technological sophistication and large-scale production to offset the large fixed cost of hydroelectric power. Process technology was sometimes European in origin but adapted to U.S. conditions, most importantly to allow for production on a much larger scale than had originally been envisaged.[7] The initial exploitation of hydroelectric power for electrochemistry involved both foreign capital and foreign and domestic technology. In 1895, the Dow cell for producing chlorine was commercialized. Soon, in 1897, Mathieson bought the American rights to the Castner-Kellner chlorine process for a plant in Niagara. In 1906, Hooker developed an alternative process, while the Gibbs cell was introduced in 1908.[8] These investments in new technology helped the United States rapidly achieve near self-sufficiency in this sector.

Sulfuric acid provides another compelling illustration of the factors that form the basis of competitive advantage, and that foreshadowed the development of the petrochemical industry. The "contact" process was a very significant innovation in production of sulfuric acid commercialized by BASF around the turn of the century. In 1899, General Chemical (later merged into Allied Chemicals) tried to secure the American rights from BASF. In the face of very stringent licensing fees ($1 million lump sum and 25% royalty on value of output) General Chemicals decided to commercialize a domestically developed version of the process. After legal proceedings initiated by BASF, both firms settled. Haynes vol. I, p. 265) claims that the American plant outperformed the German one. Indeed, he notes that General Chemical provided the plant design and engineering services for four sulfuric acid units for BASF in Ludwigshafen. Each unit was for 20,000 tons, four times the size of the largest existing BASF facility. This accomplishment is even more remarkable because BASF was renowned for its engineering and development prowess, as evidenced later in the development of the Haber-Bosch process for synthetic ammonia.

The general picture for this time period is of a U.S. chemical industry that lagged behind the leading European ones in technical sophistication and experience—

---

[6]The development of the Frasch process, which allowed the exloitation of brimstone deposits, dramatically increased the supply of sulfur. After the commercial introduction of the Frasch process in 1901, exports of elemental sulfur grew from 3,000 long tons in 1904 to 30,000 tons in 1907 (Haynes, vol. 1, p. 267).

[7]Commenting on Haynes's characterization, Trescott (1981, p. xxviii) says: "Thus the statement that 'electrochemical advances in theory and practice were mainly American in origin' must be challenged.... That these advances were 'almost wholly American in their large-scale commercial exploitation' is more nearly true, and here the stress, as will be shown later, should be on *large scale....*" (emphases in original).

[8]These examples also illustrate that globalization of technology is not entirely a modern phenomenon. The Castner-Kellner process was developed by an American but first commercialized in Britain. Gibbs, on the other hand, was an Englishman who migrated to America.

Germany in organic chemicals, Britain in inorganic chemicals, and both in engineering skills. But the U.S. industry was rapidly closing the technological gap in the simpler inorganic chemicals and growing rapidly in size. This pattern of development is reflected in the historical experience of other sectors, as well. For instance, Wright (1990) argues that as early as 1890, U.S. industries such as machine tools, engineering, and steel had either become, or were close to becoming, internationally competitive in terms of cost and quality. By the end of the 19th century, chemical firms in the United States began to display considerable evidence of technological dynamism, as well. A number of important innovations made in areas such as electrochemistry and sulfur extractions were American in origin.

PHASE II: THE SYSTEMATIC APPLICATION OF SCIENCE TO INDUSTRY, 1914–1939

Growth and Development

The First World War marked an important watershed for chemical industries in all the major industrial countries. In the United States, the war had provided a sharp increase in demand for chemical products. German and British demand for munitions and other war materials was high, and traditional sources of imports were adversely affected by dislocations in supply routes. The experiences during the war also drove home the strategic importance of chemicals because of their role in producing explosives, dyes, drugs, and fertilizers and prodded the government into policies to promote development of the chemical industry. Tariff protection, which was provided for much of this period, was used strategically. But unlike the modern history of protection in many developing countries, protection in this case did not lead to cozy domestic oligopolies. The large economic size of the U.S. market and its vast geographical expanse, together with the absence of any overwhelmingly dominant incumbent firms, implied that competition remained keen even with reduced pressure from foreign competition.

The U.S. chemical industry expanded dramatically during the war. The total value of output increased from $221 million in 1914 to $730 million in 1919, while value added increased from $96.2 million to $367 million, as shown in Table 3-1. Even accounting for price inflation, there was a very significant increase in output. The number of establishments rose from 633 to 1053, as did employment. Sulfur output more than doubled from 417,000 long tons in 1914 to 1,190,000 long tons, while that of potash increased nearly threefold, from 40,000 short tons in 1915–16 to over 116,000 short tons in 1919 (Haynes 1947, vol. II, pp. 375 and 372).

The end of the war proved difficult for the chemical industry and the economy as a whole. The United States had been fully on a gold standard since 1879, but that was suspended during the First World War. It was revived during the period 1919–1921 but because of the war and its aftermath, effective stabilization of the system did not return until 1923. The typical result was sharp but short recessions that often resulted in consolidations in the industry through mergers and acquisitions of smaller firms. The sharp reduction in German supply of dyestuffs and other organic chemicals during the war had provided a boost to the domestic organic-chemicals industry. With the end of the war and the resumption of German exports, the U.S. organic-chemicals industry was hard hit. Acetone output fell from over 10 million pounds in 1914 to a little over 4 million pounds; carbon tetrachloride output fell by a similar margin, from nearly 10 million pounds to a little over 4 million pounds (Haynes, vol. II, p. 521).

The war caused qualitative changes as well. As one observer put it, "The war left American manufacturers with overgrown plants, an excess of raw materials, and an arsenal of new ideas. The last have probably influenced the following era more than either of the former" (Victor Clark, cited in Haynes, vol. III, p. 422). The war had heightened the awareness of the potential offered by chemistry, especially organic chemistry. Supernormal profits earned during the war provided companies such as DuPont with the financial resources to invest in chemical research. The breathing space offered by the war was important because U.S. chemical firms needed time to learn. Hounshell (1995) points out that, while DuPont's initial investments in dyestuffs or ammonia can hardly be called successes, they proved invaluable in the successful development of later innovations. Research in dyestuffs enabled DuPont researchers to develop a process to manufacture tetraethyl lead and neoprene, as well as leading directly to the discovery of Teflon (polytetrafluoroethylene). Similarly, DuPont's long struggle to produce ammonia had provided it with capabilities in high-pressure catalytic reactions that proved invaluable in the production of nylon intermediates. One of these intermediates was not available on the market except in minuscule quantities, and the other was produced only in Germany.

The U.S. chemical industry grew rapidly during the 1920s, benefiting from the buoyant economy. Between 1900 and 1930, the compound annual growth rate was 6.3%, higher than even the 4.8% annual growth rate of the German chemical industry and far higher than the 2.1% annual growth rate of the British industry. Although industry growth slowed drastically during the 1930s, due to the Great Depression, the successful introduction of new products ranging from quick-drying lacquers, rayon, and cellophane to pesticides and fertilizers earned the industry the distinction of being "depression proof." This is corroborated by Table 3-2, which shows the index of

TABLE 3-2. Index of Employment, Selected Sectors: 1923–40 (1923–25 Avg. = 100)

Year	Fertilizers	Paints and Varnish	Rayon and Allied
1923	100.5	95.6	87.3
1924	93.1	97.6	93.1
1925	106.4	106.8	119.6
1926	12.8	n.a.	n.a.
1927	100.8	117.5	164.8
1928	107.6	n.a.	n.a.
1929	113.4	122.3	244.4
1930	111	n.a.	242.2
1931	78.8	94.4	241.9
1932	56.6	87.3	214.3
1933	70.7	95.9	276.7
1934	93.8	110.4	292.3
1935	94.6	116	315.7
1936	95.1	122.7	320
1937	113.1	132.7	344.1
1938	101.6	117.4	284.7
1939	106.1	122	298.5
1940	105.1	1213.5	313.5

Source: From *Chemical Facts and Figures* (1940), Manufacturing Chemists Association, Washington, DC, pp. 23–24.

employment in three prototypical chemical sectors. The fertilizer sector displays the classical cyclical pattern, falling to almost one-half at the nadir, and does not grow appreciably between 1923 and 1940. Paints and varnishes are less affected, but they too increase by only 23% over this period. In contrast, rayon and allied products grow rapidly untill the end of the 1920s and recover quickly after a brief decline during the early 1930s. Overall, employment in this sector more than doubles from 1923 to 1940.

Accumulation of Technological Capability

The post-World War I period also marks the start of systematic attention to utilization of scientific knowledge. For the most part, the underlying fundamental contributions remained largely European in origin, and the purpose of American investments in R&D was to be able to utilize existing knowledge. In the course of the century, American scientific contributions in this area would increase in prominence. Perhaps of greater significance at this time, however, is the development of commercialization capabilities. When, beginning in the late 1930s, the development of polymer chemistry opened a window for product innovations in synthetic fibers and plastics, these commercialization capabilities would pay rich dividends.

The expropriation of German intellectual property in the aftermath of World War I may have played an important role in catalyzing U.S. investments in research and development. German firms had aggressively patented abroad and used patenting as a tool of business strategy, to deter entry by local producers in organic chemicals. But in 1917, after the entry of United States into the war, German property — including intellectual property — was confiscated, and valuable German chemical patents were licensed to U.S. firms on nominal terms. In addition to dyestuffs patents, these patents also included a number relating to the Haber-Bosch process for nitrogen fixation. Nitrogen fixation was the key to a number of strategic products such as nitrogen fertilizers and explosives. The United States government set up a number of chemical factories, the most important of these being an ammonia plant at Muscle Shoals, Alabama. But although about 250 German patents related to nitrogen fixation were filed in the United States, BASF had kept the catalyst and the process for its preparation a secret, and constructing the plant was a difficult task as well (Haynes, vol. II, p. 87). Thus, even though the patented component of the Haber-Bosch process was available, making the plant work proved a difficult and time-consuming affair.

In general, the loss to the German firms from the expropriation of patents tended to exceed the value gained to the licensee, because the latter lacked the experience and the know-how to operationalize the knowledge disclosed by the patent.[9] Product and process patents had been prominent in the industry since the 1800s (Travis 1993), and German firms are said to have been especially skillful in using patents and trade secrets to protect their dyestuffs, pharmaceuticals, and fine-chemicals markets. In some instances, the patents were framed in a manner calculated to deceive and mislead would-be imitators (Arora 1996). In general, the "thicket of patents" with which German firms protected their innovations had tended to intimidate their U.S. counterparts. Thus, even though the availability of patents may not have helped the U.S. firms a great deal in imitating the technology, their expropriation removed a major barrier to their willingness to invest in those areas (Mowery 1981). Moreover, German firms could not export products covered by their own patents, which had been licensed to U.S. producers.

[9]Steen (1995, p. 272) reports that Sterling Products Company paid $5.3 million for Bayer's U.S. assets, including the famous Bayer Aspirin trademark.

The U.S. government occasionally went further and made aggressive use of "strategic trade" policies, by threatening to deny access to the large United States domestic market as an inducement for German and Swiss firms to part with technology. During the 1920s, using access to the U.S. market as a bargaining chip, Treasury Secretary Andrew Mellon induced companies such as Durand & Huegenin of Basel to part with their technology and trade secrets and dissuaded any attempts to seek legal redress under existing trade-secrecy laws (Haynes, vol. IV, p. 14). A somewhat similar story is told about DuPont's "poaching" of IG Farben chemists in the 1920s (Hounshell and Smith 1988).

The increased willingness to invest in scientific and technological capability is shown by the trends in the formation of corporate research laboratories. Table 3-3 shows that the number of industrial R&D laboratories accelerated during the period from 1919 to 1928. Similar trends emerge if one examines the ratio of scientific personnel employed to total employment in the manufacturing sector. Throughout the interwar period, the chemical industry remained the most research-intensive manufacturing industry, and its research intensity grew over time, as indeed it did in a number of other sectors. During the early part of the 20th century, a number of American chemists went to German or Swiss universities for their graduate education and then returned to work in U.S. industry and academia. A number of prominent scientists — including Little, Teeple, Whitaker, Milner, and Sadtler — became industrial consultants (Haynes, vol. III, p. 423). University researchers, and universities themselves, found their services in demand. The Mellon Institute was founded in Pittsburgh before the war as an experiment in providing industry with access to trained university researchers. DuPont established 18 fellowships and 33 scholarships at colleges and universities across the country in 1918. This was also the period when chemical engineering was being established at MIT, Columbia, and the University of Wisconsin (Haynes, vol. III, p. 395). The role of universities in the development of the chemical industry is further discussed in Rosenberg's chapter later in this volume.

TABLE 3-3. Formation (Number) of Corporate R&D Laboratories in the United States, Selected Sectors: 1899–1946

Sector	−1899	1899–1908	1909–1918	1919–1928	1929–1936	1937–1946	Total (%)
Chemicals	40	56	88	178	146	107	615 (26.7)
Food and beverages	11	20	32	50	48	40	201 (8.7)
Textiles	3	4	11	16	28	17	79 (3.4)
Paper	4	6	15	38	26	13	102 (4.4)
Petroleum	5	3	15	25	31	10	89 (3.9)
Fab. Metals	0	17	24	53	37	28	159 (6.9)
Non elec. mach	6	14	49	65	63	30	227 (9.9)
Elec. mach	12	18	28	53	64	44	219 (9.5)
Instruments	6	4	17	23	32	36	118 (5.1)
Total (All manufacturing	112	182	371	660	590	388	2303

Source: Extracted from David Mowery (1981), table 2.2, Ph.D., Stanford University.

The initial payoff from the increased technological capabilities came in the form of the utilization of externally developed technologies. During the interwar period, a number of new technologies were imported and utilized by U.S. firms. A variety of technologies for nitrogen fixation were introduced. Even before the government-sponsored Muscle Shoals project, based on the Haber-Bosch process, American Cyanamid had been formed to exploit the cyanamid process for producing ammonia and had begun production in 1909. Thermoset plastics based on urea formaldehyde — for instance, Bakelite — were brought to the United States by the Belgian inventor Leo Baekeland. Catalytic techniques in refining, although developed by a Frenchman, Eugene Houdry, were first commercialized in the United States in the late 1930s and soon spawned a number of greatly improved refining processes. Other prominent innovations that were imported into the United States include rayon (by DuPont, as well as Courtaulds and Celanese) and cellophane. By the 1930s, the larger U.S. firms had become innovators in their own rights, with innovations such as neoprene, tetraethyl lead, and antifreeze.

Recently available patenting data for the interwar period, presented by Cantwell (1995), illustrate the point. While one must interpret statistics based on patent counts with caution, the trends displayed by the patents statistics are interesting. They show that Europe continued to be an important source of chemical technology through the interwar years, but that U.S. technological capabilities in chemicals were growing rapidly, as well. Table 3-4 suggests that patenting activity in the United States was related to overall economic activity. Note that chemical sectors, especially synthetic resins and fibers, pharmaceuticals, coal and petroleum products, chemical processes (such as liquid air purification and separation, and refrigeration), and organic chemicals typically show much higher rates of patent growth than the 2% annual overall growth rate in patenting between 1890–96 and 1920–24. From 1920–24 to 1933–39, chemical-sector patenting grew much more rapidly than overall patenting, and more rapidly yet than the other "high-tech" sectors such as motor vehicles, aircraft, metallurgical processes, and office equipment. Moreover, as the "depression proof" label suggests, many of the chemical fields registered strong growth even when overall patenting was actually declining. Since over 80% of U.S. patents were granted to U.S. residents, these trends persist even if one confines oneself to patents granted to U.S. residents.

Table 3-4 also shows the trends in patents granted to European residents immediately underneath the italic figures for total U.S. patents during each time period. Patents to European residents grew at 2% during 1890–96 to 1920–24 and over 4% annually during 1920–24 to 1933–39. In part, this growth reflects the increasing internationalization of industry and perhaps also the growing recognition of the importance of the U.S. market. However, the trends also reflect the dominant position that European countries, and Germany in particular, enjoyed in the chemicals sector, as Cantwell's (1995) more-detailed analysis confirms. On the other hand, the figures suggest that over this period the United States improved its relative position substantially in two sectors: food and tobacco, and coal and petroleum products (Cantwell 1995, p. 310).

Complementarities in Demand and Supply

The revealed competitive advantage in petroleum products points to the important and growing linkages among chemicals, automobiles, and petroleum refining. The links between oil refining and chemicals, tenuous to begin with and confined largely to the use of certain chemicals such as sulfuric acid, became progressively stronger over the

TABLE 3-4. Annual Average Percentage Growth in Total U.S. Patents and Patents Granted to European Residents over Selected Periods (Classified by Selected Field of Technological Activity)

Sector	1890–96 to 1920–24	1920–24 to 1927–29	1927–29 to 1933–39	1920–24 to 1933–39
Inorganic chemicals	*5.20*	*−1.39*	*6.21*	*2.88*
	2.27	7.10	3.12	4.80
Agricultural chemicals	*3.04*	*−4.55*	*9.74*	*3.37*
	4.56	−0.65	12.06	6.42
Photographic chemistry	*4.69*	*4.32*	*13.11*	*9.26*
	3.21	22.22	11.91	16.22
Synthetic resins and fibers	*6.74*	*17.40*	*11.02*	*13.71*
	5.07	21.63	18.68	19.95
Bleaching and dyeing	*2.64*	*13.08*	*5.70*	*8.80*
	3.54	11.38	5.45	7.95
Pharmaceuticals	*2.88*	*7.93*	*6.78*	*7.27*
	2.07	19.15	2.76	9.49
Rubber and plastic products	*2.52*	*−10.30*	*−0.67*	*−4.91*
	0.65		6.60	4.13
Coal and petroleum products	*5.82*	*13.38*	*10.94*	*11.98*
	5.46	8.21	11.10	9.85
Telecommunications	*3.73*	*7.47*	*0.61*	*3.49*
	3.44	24.34	3.08	11.73
Office equipment	*9.28*	*−6.77*	*5.16*	*−0.13*
	10.00	3.10	10.92	7.50
Motor vehicles	*5.75*	*−4.24*	*−3.78*	*−3.97*
	10.5	−3.19	−0.91	−1.89
Metallurgical processes	*4.27*	*−2.91*	*2.52*	*2.69*
	3.73	3.38	4.95	4.27
Aircraft	*12.61*	*1.15*	*−2.54*	*−0.98*
	14.33	−1.15	0.45	−0.24
Total (All manufacturing)	*1.95*	*1.74*	*−0.39*	*0.51*
	2.04	5.02	3.24	4.00

Source: Extracted from John Cantwell (1995), "The Evolution of European Industrial Technology in the Interwar Period," tables 3 and 4, pp. 298–99, 302–303.

Notes: The italicized figures represent total patenting; the figures below them represent growth rates of patents granted to European residents during the time period.

period, until the important segments of chemicals and oil refining converged into what is now known as the petrochemical sector. The linkages with the automobile sector are perhaps less evident now. But, as discussed more fully in Rosenberg's chapter in this volume, in the 1920s the automobile sector had important spillovers for the chemical sector. The automobile sector provided an important source of demand for synthetic and semi-synthetic polymer-based products such as thermoplastics and, more importantly, for lacquers and paints. It also stimulated the search for improved refining techniques to produce motor fuels with higher octane ratings. The search for antiknock agents resulted in tetraethyl lead, which was soon produced in very large quantities. The search for higher-octane gasoline had a huge impact on refining technology and, indirectly, on the chemical industry.[10]

The case of the automobile sector is only one example, albeit a very important one, of a general phenomenon: the important role of the rapidly growing domestic market in the United States. During the 1920s, in addition to automobiles, consumer goods such as radios, records, photographic film, and tires relied on new chemical materials and oil refining. To meet demand for such products, entirely new types of products and materials had to be forthcoming. The relationship between overall economic growth and the growth of the chemical industry in the United States appears to have strengthened over this period, although firm quantitative evidence is lacking. Also, the chemical industry itself is a major consumer of chemical products. Unlike during the pre-World War I era, when many major inputs and intermediates for the chemical industry were imported, now growth in demand from consumer-goods sectors had a larger multiplier effect on the industry. To paraphrase Haynes, industry was becoming "chemicalized." For instance, a number of general-purpose chemicals — such as the organic intermediates including phthalic anhydride, required by the rapidly growing chemical industry — were commercialized in the United States.

Changing Industry Structure: Relaxing Price Competition

The interwar period was characterized by a number of mergers and consolidations in the chemical industry. The forces leading to consolidation in the industry are well illustrated by the experience of the sulfuric acid market. During 1921 to 1929, the output of sulfuric acid nearly doubled, the price decreased by 20%, and the number of producers decreased by 15%. Similarly, in dyestuffs between 1923 and 1929, while the total volume of production increased by 20%, prices dropped by more than that, so that in terms of sales revenue, production actually decreased. Most small dyestuffs producers exited, and only those that had specialized niches survived (Haynes, vol. IV, pp. 17 and 227).

Antitrust concerns precluded the possibilities of American firms forming cartels and trusts of the sort that were common in Europe. Instead, consolidation took place through mergers, and some rather large companies were formed through such processes. Before World War I, the chemical industry had seen only two mergers of note, both in response to competitive threats from new technologies: General Chemical was formed in 1899 in response to the development of the "contact" process for sulfuric acid; and Mutual Chemical was formed in 1906 by the threat from the Schultz chrome tanning process. The mergers after 1920 would leave a lasting impression on the U.S. industrial structure. The more prominent results of the consolidation process at this time include Allied Chemical and Dye Corporation, and Union Carbide and Carbon Corporation; but firms such as American Cyanamid, DuPont, American Home Products, Mathieson Alkali, Monsanto, and Hercules also made a number of acquisitions during this period. Haynes (vol. IV; p. 46) estimates about 500 such mergers during the 1920s but observes that this number is lower than in other industries such as steel, automobiles, textiles, and even banks and hotels.

---

[10]Similarly, the rapid growth of the U.S. steel industry during this period (and earlier) provided the chemical industry with valuable coke-oven by-products. The growth of the steel industry stimulated the growth of specialized machinery producers, who later played an important role in the chemical industry as well. The rise of the metallurgical industries such as aluminum had similar beneficial effects (Trescott 1981, pp. 136 and 150).

Economies of scale and scope undoubtedly played a major role in the rationale for consolidation. Since the 1850s, the British inorganic producers had demonstrated the cost advantages of large-scale production. The Leblanc process had also pointed to the ever-present potential economies of scope in the form of potentially useful waste products (in this case, chlorine). The new technologies — Haber-Bosch, electrolytic soda and bleach, and rayon — were even more scale-intensive. To see this process as merely a response to technological imperatives alone, however, is to miss much of what is important about industry evolution. A number of mutually reinforcing forces played an important role.[11] As the different regional markets became more integrated economically and products became standardized, price competition intensified (Sutton 1991). Firms responded by trying to differentiate their products: sometimes through variations in the products themselves; or by differences in service, delivery, and so on; or by creating brand names. In turn, that differentiation required closer control over the marketing and distribution of their products. Hitherto, most firms had relied on retailers and jobbers to distribute and sell their products. Many of those retailers had started out as importers of chemicals; their knowledge of the local market, as well as their links with the local customers were valuable to the U.S. chemical producers as well, especially when the U.S. producers were relatively small and inexperienced. The reliance on retailers and jobbers was problematic for companies that were introducing new products, since successful commercialization of product innovations requires strong links with customers. Thus, a number of U.S. chemical firms invested in downstream marketing and distribution. This further raised the fixed costs required to enter these markets, and for a given size of the market, the number of firms that could survive fell.

Knowledge-based economies of scope provided an important impetus for corporate growth through diversification. The central R&D laboratories played an important role in diversification into new products. For individual corporations, these knowledge-based spillovers proved to have an enduring influence on their future development path, as Chandler, Hikino, and Mowery document in their chapter in this volume. For instance, DuPont became familiar with nitrocellulose through its explosives business. This knowledge helped in DuPont's move into cellulosic products such as rayon and cellophane. The experience gained in the textile market with rayon later proved invaluable in the commercialization of nylon and Dacron after World War II. Similar stories can be told about a number of companies such as Dow (chlorine, petrochemicals), Union Carbide (petrochemicals, especially ethylene, and air-separation gases), and Air Products (air-separation gases, catalysts).

Wall Street had little to do with financing these ventures. Indeed, as the chapter by Da Rin shows, lack of financing had constrained the growth of the industry in the United States, as it had in Britain. For the most part, companies such as Dow and Monsanto had to rely on retained profits to finance their expansion. DuPont was an exception, in that the wealth of the DuPont family and the very profitable explosives business allowed DuPont to raise money on the capital markets. American Cyanamid was another exception, in that Wall Street investment banks participated actively in its expansion through mergers and acquisition. Interestingly, American Cyanamid appears to have grown with little regard for technological synergies. It started as a fertilizer company and grew rapidly by acquiring a variety of companies in heavy chemicals

---

[11]Chandler (1990) masterfully shows the importance of the three-pronged investments in manufacturing, management, and marketing in the rise of the large corporate organization. The point of view taken here is complementary, focusing as it does on the market forces that impelled this process.

(Kalbfleisch), engineering (Chemical Engineering Corporation), solvents and industrial chemicals (Selden), dyes (Calco), explosives (Burton), and pharmaceuticals (Lederle). From 1910 to 1916, the share of fertilizers in its output dropped from 100% to 75%. By 1929, American Cyanamid was the fourth-largest, and the most diversified, U.S. chemical company (Haynes, vol. IV, p. 46).

Consolidation in industry structure had taken place somewhat earlier in Germany and Britain. In these countries, however, the process had resulted in the rise of two mammoth companies, IG Farben and ICI, which absolutely dominated their respective industries. Although DuPont, Allied Chemicals, Union Carbide, and American Cyanamid were the largest U.S. chemical firms in the pre-World War II period, their situation was different. The sheer size of America's home market, coupled with antitrust restrictions on horizontal mergers, prevented any company from becoming very large in relation to the industry as a whole.[12] This feature has important implications for the development of the U.S. industry, for it implied greater diversity and competition in the U.S. industry and is a recurring theme in our story.

PHASE III: THE SECOND WORLD WAR AND AFTER, 1941 TO THE PRESENT

Innovation, Growth, and Maturity

World War II profoundly shaped the evolution of the chemical industry. In the decades after the war, the chemical industry continued to grow strongly. Value added in the industry more than doubled, from about $5.3 billion in 1947 to $12 billion in 1957. The number of employees per establishment grew during this period, but much more slowly than value added per establishment, which doubled from $500,000 to more than $1 million. Table 3-5 traces the development of the industry relative to U.S. manufacturing as a whole from about 1960 to 1990. Growth in this period was now firmly driven by the commercial exploitation of new technologies. As Figure 3-1 shows, plastics, synthetic rubber, and synthetic fibers (SIC 282) and drugs (SIC 283) were among the sectors that grew the most rapidly after the war.

Innovation was not simply a matter of research. The successful introduction of a synthetic fiber, or a new plastic, involved much more than the laboratory discovery of a new polymer with the appropriate tensile strength or melt-flow properties.[13] Such a discovery was only the beginning of a long and costly process of development. Hounshell (1988) describes in rich detail the process for developing a synthetic fiber. A new synthetic fiber had to be modified so that it could be colored. If necessary, new coloring and fixing agents had to be developed. Often, the production of the yarn

[12]By contrast, in Germany individual markets dominated by single companies and cartels were pervasive. In potash and nitrogenous fertilizers, there were government-sponsored cartels. In the period immediately following the First World War, the German ministry of interior found 47 cartels in the chemical industry. Thirteen were concerned with inorganics and two (major ones) with dyestuffs (Haber 1971, p. 267). IG Farben is said to have been a part of some 2000 national and international cartels. In Britain, the efforts of the Leblanc-based Alkali cartel are well known. Haber also points out that Brunner-Mond and the United Alkali Company did not attempt to entice each others' customers and entered into long-term contracts.

[13]Like most product innovations, the newly discovered polymers initially had limited applicability. Product innovations in the United States were, however, followed by process innovations that substantially reduced costs and transformed "specialty" products into mass products. Once again, complementarities were key: Absent large potential markets, such process innovations would be less attractive, and absent the expectation of a large volume of output, firms such as DuPont would not have invested as enthusiastically in searching for new products.

TABLE 3-5. Index of Production, the U.S. Chemical Industry: 1959–93
(1957–1959 Avg. = 100)

Year	All Manufacturing	Chemicals and Allied Products
1959	105	114
1960	108	121
1961	110	123
1962	118	136
1963	124	149
1964	132	160
1965	139	167
1966	153	189
1967	156	204
1968	165	224
1969	173	245
1970	167	245
1971	170	261
1972	187	293
1973	204	321
1974	201	332
1975	181	303
1976	200	339
1977	217	365
1978	230	390
1979	240	407
1980	228	384
1983	230	383
1984	251	400
1985	256	400
1986	258	414
1987	271	438
1988	283	464
1989	288	478
1990	288	489
1991	282	488
1992	289	501
1993	301	515

Sources: Chemical Statistics, Manufacturing Chemists Association (1971), Washington, DC, pp. 434–35; *Chemical Statistics*, Manufacturing Chemists Association, Washington, DC, 6th edition (1966), pp. 363–65; *U.S. Chemical Industry Statistical Handbook* (formerly *Chemical Statistics* (1984), Chemical Manufacturers Association, Washington, DC, p. 27; *SRI Chemical Economics Handbook*, 1990.

Note: The cumulative index was formed by chaining four sets of production indices with different bases.

involved development of new techniques for generating, drawing out, twisting, and even spinning. These process developments required active cooperation and feedback from the users — in this case, textile firms — which in turn necessitated further changes to the product. These systematic iterations made large demands on the managerial ability to organize and manage the process between research laboratories and downstream users, an aspect at which the U.S. chemical firms appear to have excelled.

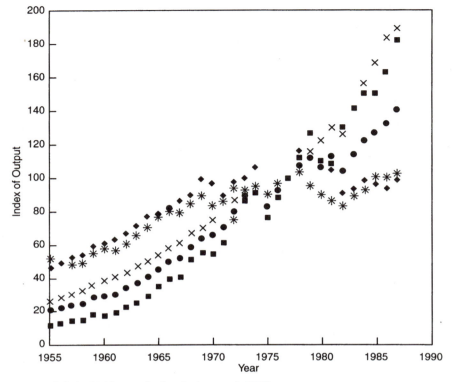

Figure 3-1. Index of Chemical Output in the U.S. Selected Sectors: 1955–87 (1977 = 100).
Source: SRI Chemical Economics Handbook, SRI, Menlo Park, CA.

This aspect of successful innovation was relatively new to many chemical companies, especially those used to producing chemicals in large volumes. ICI offers a prime example: When presented with the opportunity to commercialize nylon in Britain (thanks to its long-standing patent agreements with DuPont), ICI hesitated because of the perception that textile fibers were too close to "the fashion trade," a market segment with which ICI was unfamiliar.[14] As another example, DuPont had a superior process technology that allowed it to commercialize polyester even before ICI, which invented polyester, could do so. DuPont was successful because of the substantial investments it made in being able to understand and respond to user needs. Its researchers were

---

[14]Eventually ICI entered into a joint venture to commercialize nylon with a textiles company, Courtaulds, that had commercialized rayon in the U.S. ICI did develop the next major textile fiber, polyester, on its own, with assistance from DuPont, as discussed in greater detail in "The Process of Innovation" chapter by Landau in this volume.

knowledgeable not only about chemistry but also about the technologies used by its customers to transform chemicals into household products. A DuPont executive named D. H. Dawson (quoted in Backman 1964, p. 44) put it this way:

> Much of the work in these (DuPont) laboratories is directed toward the cultivation of markets once or twice removed from our own. In the case of synthetic yarns, we have found it productive, and frequently necessary, to study not only their weaving or knitting into fabrics, not only the dyeing and finishing of these fabrics, but their conversion into garments — cutting and sewing techniques, yes, and consumer care of the garments. In plastics, especially the newer types, it is often necessary to work out the design of a plastic component for use in an automobile or a washing machine and only then go to work with our immediate customer, the supplier of molded or extruded parts, on methods of producing the parts.

In their attention to downstream commercialization activities, the U.S. chemical innovators of the 1930s and 1940s appear similar to their German counterparts of the 1880s. Contrary to what some believe, whereby the German success in organic dyestuffs is attributed completely to their access to superior science, commercialization capabilities played an important role in German success as well. Haber (1958, p. 174) notes the attention paid by the German chemical companies to the needs of their users. Their investments in training personnel in the use of dyes (especially the fast vat dyes) foreshadows the more modern use of "technical services" as a competitive strategy. It is also interesting to note that Germany did not trail the United States in polymer science or its applications. Apart from the development of polymer chemistry by the contributions of prominent scientists, a number of commercially important polymers were first developed, and in some cases, even first commercialized in Germany.[15] For example, polyvinyl chloride, polyvinyl acetate, and polystyrene were first introduced in Germany, but it was left to American companies such as Union Carbide (polyvinyl chloride) and Dow (polystyrene) to undertake the development and improvements required to turn them into commercial successes.

During the interwar years, the energies of IG Farben — which dominated the German chemical industry in both quantitative and qualitative terms — were occupied by a state-led drive for self-sufficiency. Enormous resources were devoted to high-pressure technologies for coal hydrogenation and gasification. These projects were technically demanding and extremely costly, with low commercial payoff. The opportunity cost in terms of the missed opportunities in the industrial and consumer applications of polymer science may have been even higher. The dominance of IG Farben meant that few other firms in Germany could have pursued these opportunities. After World War II, the successors to IG Farben did aggressively invest in polymers, but by that time U.S. firms held the lead.

The close relationship between overall economic growth and the growth of the chemical industry continued after World War II. Clearly, the stable macroeconomic climate in the two decades after World War II played an important role in the growth

---

[15] Methyl rubber was commercialized during World War I, and even though it proved to be inferior to neoprene and buna rubber, developed later, it testifies to the German capabilities in this field. As another example, one of the most important developments in polymers — that of stereo-regular polymers of ethylene and propylene — derives from a German research institute. Despite that, German firms are not the leading producers of high-density polyethylene and polypropylene, which are commercially the most important polymers today. As the example suggests, it was not weakness in German science or technology that appears to have caused them to lag in the commercialization of polymers.

of the chemical industry. Economic growth was rapid, and inflation and interest rates tended to be low, favoring investment and growth. The OPEC-engineered oil price increase in 1973 came at a time when overcapacity had already begun to develop in many industry segments. The immediate effect of the oil shock was the sharp increase in input prices for the chemical industry, but the more serious effects were related to the slowdown in overall economic growth. Table 3-5 clearly shows the slowdown in chemical output in response to the oil shocks of the 1970s, as well as the close relationship between chemical output and aggregate manufacturing output. Our own estimates suggest that during the 1960s, an increase of 1 percent in the growth of GDP increased the output growth of the U.S. chemical industry by 2.2 percent. But in the 1970s, the multiplier dropped to 1.7, and the figure for the 1990s is estimated to be 1.2, reflecting the falling share of manufacturing and construction in the economic activity of the United States.[16]

The second oil shock in 1979 exacerbated the effects of the earlier price increase, and the industry had to continue finding ways of saving on energy costs. For instance, modifications in ammonia process design have reduced requirements for natural gas, and new heat-transfer fluids have improved the energy efficiency of many processes by increasing the recovery of waste heat. Similarly, research in new catalytic processes has also yielded a number of new processes that are energy efficient and also friendlier to the environment (MIT 1989, p. 25). Even though the real price of oil gradually declined in the 1980s and the industry adapted to higher raw-material and energy costs, the chemical industry in the United States and other developed countries had clearly entered a new phase of its development cycle.

As the industry matured, opportunities for major product innovations became scarcer, and firms reined in R&D spending. Baily and Chakrabarty (1985) found that the average number of new chemical products introduced per year dropped from 322 during the period 1967–73 to 39 during the period 1974–79, with a modest increase to 64 during 1980–82. Similar declines are reported in process innovation (MIT 1989). Figure 3-1 indicates that even the "new" chemical sectors, which had pulled the industry through the depression in the 1930s, were no longer "recession-proof." Sectors such as synthetic fibers and synthetic rubber—areas that had powered industry growth earlier—were now vulnerable to economic slowdown, and other sectors were even more profoundly affected. Paints and varnishes output hardly increased over the 10-year period from 1977 to 1987, even though it had increased by a little less than 40% in the preceding 10 years. During this period, employment was affected even more strongly than output as the industry underwent significant restructuring, described more fully in the chapter by Arora and Gambardella.

The Narrowing Gap

The German chemical industry emerged from World War II with a manufacturing capacity at least as great as in 1939, despite the substantial destruction of chemical facilities during the war. Nonetheless, the division of Germany after the war, and the removal of a substantial amount of capacity as reparation, had a substantial, if

---

[16]From log–log regressions of chemical output on total manufacturing output (not reported here), we find that the elasticity of chemical output with respect to total manufacturing output is 1.44 for the period 1959–93. The elasticity for 1959–73 is 1.58, while for 1970–93 the elasticity drops to 1.12. Thus, while a 1% increase in manufacturing output was associated with an increase of 1.6% in chemical output before the oil shock, after the oil shock the figure drops to a little more than a 1.1% increase.

short-lived, impact on German production (Eichengreen, this volume). From 1938 to 1951, the U.S. share of world chemical production rose from 30% to 43%, while that of West Germany fell from 22% to 8%. The British industry managed to hold on to its share of world output, while that of the Japanese industry dipped slightly. Between 1950 and 1960, while Germany and Japan recovered, the British share dropped.

By the end of the 1960s, the European countries and Japan had successfully rebuilt their chemical industries and had closed much of the gap with the United States. Since the end of the 1960s, relative national shares in world output have largely remained constant, with only a small decline in the share of the United States and a rise in the share of Japan. The oil shock and the unsettled macroeconomic conditions of the 1970s slowed down industry growth in all of the developed nations. But by 1980, Japan had overtaken Germany as the world's second-largest chemical industry, accounting for 11% of the global output, to Germany's 8%. The pattern of a swift ascendancy of the United States immediately after the war, a narrowing of the gap thereafter with Western Europe and Japan, and the strong growth of the German, Japanese, and, to a somewhat lesser extent, British industries is also reflected in the trends in shares in world exports (Eichengreen, this volume, Table 2). U.S. net exports of chemicals declined from $14.5 billion in 1980 to $8.3 billion in 1985. Although a number of factors, from exchange rates to domestic macroeconomic conditions, also affect net exports, these figures clearly reflect the increase in chemical production and exports by non-OECD countries.

The chemical industry became global, not just in terms of the flow of goods and capital but also in technology, which diffused internationally through licensing and direct foreign investment. Specialized engineering firms from the United States were central to the international diffusion of process technologies, in which chemical firms from the U.S. and Europe also participated actively (and will be discussed at greater length later). A number of countries in Asia, Latin America, and the Middle East invested heavily to build national chemical industries. Between 1970 and 1985, the installed-ethylene capacity in the Arab world increased from nothing to 2.5 million tons per year, and installed-methanol capacity rose by 2 million tons per year (MIT 1989, p. 26). In the mid-1980s, a cheaper dollar and declining growth opportunities in their home markets prompted European chemical firms (and, to a lesser extent, Japanese firms as well) to expand heavily into the U.S. market. The expansion, accomplished through direct investments, acquisitions, and alliances, underlined the globalization of the industry, as well as the declining U.S. dominance.

The Role of the U.S. Government

U.S. government policies during the interwar years had had only a modest effect on the development of the chemical industry, with a few notable exceptions such as the occasional expropriation of the industrial property of firms from competing nations, direct investment in nitrogen fixation, and somewhat fitful tariff protection. Antitrust enforcement was lax during the interwar years. The most important effect was perhaps indirect, to the extent that government macroeconomic policies helped bring on the Great Depression.

World War II changed the picture considerably. The United States government played a particularly important role in two critical areas during the war: petroleum refining and synthetic rubber. Research into synthetic rubber in both Germany and the United States had begun well before World War II. During the military expansion of the 1930s, Hitler's government put a very high priority on a substitute for natural rubber as a part of its drive towards self-sufficiency. At considerable expense, IG Farben, which had produced the first synthetic rubber during World War I, managed to develop a much better product based on copolymers of butadiene with styrene, and acrylonitrile, referred to as Buna-S and Buna-N, respectively. In the United States, DuPont had commercialized a high-performance, but expensive, rubber substitute, neoprene. Standard Oil had made substantial progress in developing a process for producing a key input, butadiene, from oil rather than coal, as in the IG Farben process. Unlike German research, however, American research in synthetic rubber was driven more by commercial than by political motives.

During the war, the U.S. government took a more active part. It organized a cooperative program of research and development involving the four major rubber companies and Standard Oil, with a number of major chemical companies also participating (Morris 1994). After Buna-S was chosen as the target, Dow was put in charge of the styrene production (although Monsanto, Union Carbide, and Koppers also were involved). A number of oil and chemical companies shared the production of butadiene. Rubber companies carried out the copolymerization and fabrication. The federal government invested approximately $700 million in the construction of plants, which private companies operated as government contractors. After the war, most of the government-owned plants were sold to private firms, usually those operating them during the war, and usually on very favorable terms. The rubber program was a major success. Not only did rubber consumption rise substantially (from less than 800,000 long tons in 1941 to over 900,000 long tons in 1945), but no less than 85% of the total in 1945, was synthetic rubber, compared with less than 0.5% in 1941. But the rubber program provided another, equally important result: After the war, a number of companies in the United States had technological expertise, experience, and production capabilities in synthetic rubber and, therefore, in synthetic polymers more generally.

In petroleum refining, the government played an indirect role through its procurement policy. Between 1939 and 1945, operating capacity in refining increased by 29% (Chapman 1991, p. 74). But more important was the effect on the composition of output. Hitherto, gasoline for motorcars had been the major output of the refining industry in the United States, produced through fractionation and, in the 1930s, by thermal cracking as well. Now, in addition to butadiene, the oil industry was also asked to provide high-octane aviation fuel, as well as aromatic products such as toluene. To produce the required quantities of aviation fuel, as well as aromatics, new technologies such as catalytic cracking and alkylation (both vital to the production of high octane fuel) and catalytic reforming (to produce aromatic compounds) had to be adopted. These techniques had been developed over the 1920s and 1930s, but without the stimulus of the war would have diffused only slowly.

In addition to their effect on direct subsidies and procurement, government policies on the diffusion of technical information and expertise had a strong leveling effect. Cooperative research was actively encouraged. The synthetic rubber research project was a major project in which rival firms shared information and personnel. The heightened interfirm mobility of personnel led to the breakdown of many secrecy barriers. After the war, the government also compelled the compulsory licensing of some major product innovations; for example, DuPont was forced (albeit on attractive terms) to license nylon to Monsanto.[17] The net effect of the entry of oil firms into chemicals and the wider availability of chemical-process technology from specialized engineering

[17]Similarly, the ICI-DuPont patents and processes agreement was terminated in 1952 as part of a consent decree, and the ICI polyethylene patent was licensed to a number of companies.

firms was to increase, in virtually all sectors, the number of firms with the capability to innovate in the future. These included the hitherto lesser-known chemical companies, such as Dow and Monsanto, as well as oil and rubber firms. There was widespread entry into organic chemicals, synthetic rubber, and especially plastics and resins, and concentration in these sectors decreased substantially in the first two decades after the war (MCA 1971). By 1967, concentration (measured as the market share of the top 4 producers) was lower in organic chemicals than in inorganic chemicals such as alkalis or inorganic pigments.

Petrochemicals: The Rise of an Industry

The Second World War also resulted in the expropriation of German technology, but as in the case of the First World War, the loss of proprietary technology proved to have a limited impact. After World War I, however, the limited impact was caused by the inability of the recipient firms in the United States and Britain to make use of such technology. After World War II, the impact was limited because German technology was largely coal-based. IG Farben had developed an extensive and sophisticated capability for using acetylene — derived from coal — as a basic feedstock. As noted earlier, IG had also developed the technology for the high-pressure hydrogenation and liquefaction of coal to produce oil and other basic feedstock chemicals. But although they were a major technological accomplishment and played a significant role in the German war effort, coal hydrogenation technologies proved to be of limited commercial significance;[18] the discovery of abundant reserves of oil and natural gas in North America and in the Middle East in the 1940s and 1950s laid to rest any fears about a short-term world shortage of oil. Other firms, principally in the United States, had developed techniques for producing the same basic feedstock chemicals from oil and natural gas.

The end of the war confirmed the petrochemical basis of the industry (Stokes 1994) in Europe and Japan. But in the United States the process had begun earlier, and the early lead and experience of the U.S. industry was to prove a source of technological advantage for a limited time. The United States had abundant reserves of oil and natural gas and a large oil-refining industry by the early part of the 20th century. In contrast, Germany and Britain had abundant coal (which the United States had as well) but little oil. In 1940, 71% of world petroleum-refining capacity was located in North America, with only 7% in western Europe. German supremacy in organic chemicals was based on coal and coke-oven by-products, and initially oil was seen primarily as an energy source. But the relative scarcity of coke-oven by-products in the United States and the relative abundance of natural gas and refinery gases made the latter attractive as inputs. The move was led by Union Carbide and Dow among the chemical companies and by Standard Oil and Shell among the oil companies.[19] Over time, oil companies proceeded to integrate farther downstream — for instance, from ethylene into polyethylene. By the early 1960s, oil companies accounted for a significant portion of

---

[18]Technologies such as hydrogenation did have important spillovers for certain refining techniques such as reforming. But even here, the systematic development of chemical-engineering technologies in the United States proved to be economically far superior, as Hoechst found out when it tried to develop an in-house design for a cracker: While technologically feasible, it proved to be commercially inferior to the crackers available from U.S. engineering firms (Stokes 1994).

[19]The implications of this "convergence" for industrial structure are more fully discussed in the chapter by Arora and Gambardella.

the productive capacity in a number of basic petrochemical products. It is no coincidence that two oil companies—Amoco, and Phillips Petroleum—played such an important role in the development of the two major polymers, polypropylene and polyester, as discussed in Landau's chapter on innovation in this volume.

The spread of petrochemicals before the Second World War was slow and limited to the United States. Between 1921 and 1939, the volume of organic chemicals not derived from coal rose from about 21 million pounds to about 3 billion pounds, while increasing in value from $9.2 million to about $400 million (Haynes, vol. V, ch. 15). By 1939, coal-based organic compounds totaled 300 million pounds in weight and $260 million in value.[20] By 1950, one-half of the total U.S. production (by weight) was based on oil and natural gas, and by 1960 this figure had reached 88%. But the early U.S. dominance was not long-lived. The rise of petrochemicals was swift in western Europe after the Second World War. In Britain, where only 7% of the chemical production (by weight) was based on oil in 1949, the figure was 63% in 1962. The technological lead of the United States in petrochemicals was eroded as the technology diffused through oil companies and engineering-design firms.

The oil and natural-gas endowments of the United States did not prove an overwhelming source of comparative advantage after World War II because the world trade in crude oil, largely based on Middle East oil supplies, had grown greatly. Government regulation of oil imports in the United States also played an important part. Since the late 1930s, the oil industry had been regulated by the government. Among other things, the production of individual companies was regulated to prop up the domestic price of oil. After World War II, the regulations were extended to restrict imports of oil. The net effect, according to some scholars, was that the crude oil acquisition costs for U.S. refineries were 60% to 80% higher than the costs in western Europe through the late 1950s and 1960s (Chapman 1991, p. 188). It should be emphasized that U.S. firms did have access to another cheap source of light hydrocarbons such as ethane, propane, and butane—namely, natural gas. Thus, the differential in the price of oil was not as significant as it appears on the surface. The differential sharply narrowed in the 1970s as a result of the oil shock, which raised the world prices of oil. But the differential, while it lasted, served to reduce the advantage provided to the United States by its natural resource endowment. After the oil shock, the situation was reversed for a time, and oil prices in the United States were actually lower than those faced by firms in western Europe and Japan. On the whole, the mineral wealth of the United States was instrumental in enabling the United States to develop an early technological and commercial lead in the newly emerging petrochemical sector.

Although the intimate linkages between the oil and the petroleum industry may seem natural from a modern perspective, they are in reality a human creation, fashioned in the United States. The centrality of petroleum to the chemical industry was not just because the by-products of the oil-refining industry were used as feedstocks by the chemical industry, although the feedstock link has been crucial; there are many other subtle but important linkages as well. In effect, oil refining became almost as much a chemical process, and chemical production benefited from advances made in the design and engineering of refineries.

---

[20]This growth took place alongside a rapid growth in the more efficient type of coke ovens (Tarr 1994). In other words, in the initial period at least, coal-tar-based and petroleum- and natural-gas-based feedstocks were used for somewhat different types of products. Over time, the development of techniques such as catalytic reforming for producing aromatic organic compounds from petroleum fractions were developed. Together with the abundant reserves of oil and gas, such techniques caused the raw-material basis of the chemical industry to change from coal to oil.

As noted, chemical techniques of catalysis, hydrogenation, and the like were finding application in the oil industry, which had hitherto relied on straight distillation but now required the other techniques to produce higher-octane fuels, as well as to convert heavier oil fractions (such as gas oil) into more valuable gasoline-range hydrocarbons. In this respect, the pattern of demand in the U.S. market was substantially different from that in Europe, where the demand for the lighter fractions was substantially lower than that for medium fractions such as naphtha. In the United States, the growth of the automobile industry increased the demand for gasoline and made it a more valuable product than kerosene, demand for which dropped with the introduction of electric lighting. Oil refining has always been extremely capital-intensive, and therefore puts a premium on large-scale production. Moreover, since opportunities for product differentiation in gasoline and similar products were more limited than in chemical products such as dyestuffs, the payoffs for cost reduction and, hence, process innovation were greater than in traditional chemicals. It therefore is not surprising that the oil-refining sector played a major role in the development of chemical engineering — the science and practice of developing and operating large-scale, continuous-flow, chemical processes. Such knowledge proved to have much broader applications in the chemical industry, as well.

It is unlikely that petrochemicals would have become so ubiquitous in the absence of the major polymer-based product innovations, which were produced out of petrochemicals. Today plastics and synthetic fibers account for about 12% and 4%, respectively, of the total value added in the U.S. chemical industry; since they are farther downstream than petrochemicals, their share in the value of sales would be even higher. But on the eve of World War II, their share was fairly small, with only some thermoset plastics and a semi-synthetic fiber (rayon) of any note. Many polymers would have been unable to gain widespread acceptance but for the fact that they were made available at very low prices. This point, which is obvious but inadequately appreciated, is worth stressing. A great deal of the growth in synthetic polymer products came from substitution for existing natural substances: wool, cotton, and silk; and also of steel and aluminum; and wood. In some instances — synthetic fibers and engineered plastics being prime examples — the synthetic substances provided better functionality. But on the whole, the lower cost of synthetic polymers played a very important role in their success.

A constellation of factors, rather than any single one, proved decisive in this respect. Lower unit costs depended on cheaper feedstocks and on technological changes in areas such as catalysis and chemical engineering. Repeating a story that characterizes much of the history of U.S. industrialization, these three forces fed on one another, producing a virtuous cycle of innovation, growth in demand, and then more innovation. Innovations in refining technologies made cheap raw materials from previously unused products of oil distillation available as feedstocks for the production of polymers. Systematic research and development enabled firms to produce synthetic polymers with the desired mix of characteristics. Like most product innovations, the newly discovered polymers initially had limited uses. Product innovations, however, were followed by major improvements in the production processes that transformed specialty products into products produced in high volumes and with diverse uses. In the absence of a large potential market, incentives to invest in such innovations would have been limited. As Hounshell (1988) has carefully documented in the case of tire cord, the substitution of one material for another in the production process requires long and expensive experimentation. Without the reasonable assurance that process innovation and economies of scale would indeed drive down the price of the polymers considerably, many users, especially industrial users, would not have been prepared to switch to the new materials — in which case, a potentially large market might never have been realized.

Expanding Markets and Division of Labor

The large size and rapid growth of the market for chemicals afforded opportunities for exploiting the benefits to be derived from large-scale production. From an early stage, U.S. firms were introduced to the problems involved in the large volume production of basic products. Trescott (1981) notes that U.S. electrochemical plants tended to be much larger than European plants and that this factor was important in reducing unit costs. The ability to deal with large-scale plants and (eventually) continuous process technology was to become a critical feature of the chemical industry.

But there was more to expansion than a large and growing market inducing a disposition to build large plants. A chemical plant is a complex system that requires careful balancing of tradeoffs along several dimensions. "Scaling up" — going from bench-scale design to commercial scale, or from a smaller-scale plant to a larger-scale plant — is by no means trivial. The development of chemical engineering and its applications to the problem of large-scale production played a critical role in the reduction of unit costs and the consequent rapid growth of plastics and other polymer-based chemical products. The story of the development of chemical engineering is told in Rosenberg's chapter in this volume. Here, we want to point to the crucial role of the specialized process-design and engineering construction firms, commonly known as *SEFs*.

These firms played a crucial role in the development of new and improved processes, as well as in their diffusion. As one might expect, given its comparative emphasis on large-scale production, the United States enjoyed an early lead in developing the chemical engineering of plants. In practical terms, this meant strong capabilities in handling higher temperatures and pressures. In commercial terms, it meant selling such services, even though the German firms pioneered the commercialization of such processes (Landau and Rosenberg 1992). The first SEFs were formed in the early part of the century, and their clients were typically oil companies. However, SEFs also started operating in some bulk chemicals such as sulfuric acid and ammonia. The Chemical Construction Corporation built sulfuric acid and other plants, while the Chemical Engineering Corporation targeted synthetic ammonia and methanol processes. Both corporations attained some success in project exports to Europe, as well.[21] Later, most SEFs would operate in designing large-scale plants for refineries, and petrochemical building blocks. Prominent among the early SEFs are companies such as Kellogg, Badger, Stone and Webster, UOP, and Scientific Design.

Initially, the chief market for the SEFs were U.S. firms which could contract out design and engineering services. Apart from oil companies, however, established chemical firms were less eager to outsource such functions, in part because of a tradition of secrecy, and in part because the chemical processes tended to be smaller and more complex than was the case in refining. Established chemical companies were especially reluctant to "buy" technology from SEFs or to enter into alliances with SEFs to develop new technology. But the rise of synthetic polymers united what had been a number of disparate markets such as fibers, plastics, rubber, and films. Consequently, the number of potential entrants increased, thereby increasing the demand for the services of SEFs.

American SEFs also looked overseas for potential clients, especially after the end of World War II, when many European firms were suffering from the effects of the war.

---

[21]As Haynes (vol. IV, p. 46) wrote: "These engineering enterprises created a unique phase of Cyanamid's business. The building of chemical plants, chiefly sulfuric acid, synthetic ammonia, and nitric acid operations, on a 'turnkey' basis, that is, delivered in running order, soon grew into a profitable, world wide enterprise."

TABLE 3-6. Trends in Manufacturing Concentration in Selected Chemical Sectors: 1947–1967

	Number of Firms	Total Value of Shipments, in $ Million	% Accounted for by:	
			4 Largest	8 Largest
ALKALIS				
1967	19	719	63	88
1947	18	208	70	93
INDUSTRIAL GASES				
1967	113	588.7	67	84
1947	69	93.5	79	87
INORGANIC PIGMENTS				
1967	65	549.3	59	78
1954	73	37.4	67	83
INDUSTRIAL ORGANIC CHEMICALS				
1967	339	6377	45	58
1954	202	2198	55	70
PLASTICS AND RESINS				
1967	508	3473	43	64
1947	94	478	63	89
SYNTHETIC RUBBER				
1967	33	926	61	82
1954	13	361	53	81

Source: Chemical Statistics Handbook, 7th edition (1971), Manufacturing Chemists Association, Washington, DC, p. 432.

American SEFs took advantage of this opportunity to combine their superior know-how about chemical-process design with proprietary technologies, to offer technology packages to customers in Europe and the United States.[22] By supplying technology and know-how for a wide variety of chemical processes, the SEFs facilitated a substantial amount of entry. Entry of new producers took place for most petrochemical products, as reflected in Table 3-6. Often entrants were not "new" firms but rather firms that had operated in other sectors of the chemical industry and wished to exploit some real or perceived competitive advantage. This had a major impact on industry structure, both in the United States and abroad.[23]

SEFs provide a vivid illustration of the economies of scale—or economies of specialization—that operate at the level of the industry rather than of the plant or the individual firm. By specializing in process design and engineering, and by working for a number of clients, SEFs could accumulate skills and expertise that no single chemical

---

[22]The technology transfers were not all from the United States to Europe. Spitz (1988) points out that many European chemical firms, particularly technology-rich but cash-poor German firms, were willing to license their technologies for revenue in the 1950s. For much of this period, however, Japan remained a net importer of technology.

[23]The issue of evolving industry structure is taken up at greater length in the chapter by Arora and Gambardella in this volume.

company could match. As independent developers of technology, SEFs were similar in some respects to today's biotech companies, allying with a number of chemical firms in developing new technologies.[24] European and Japanese firms (and later, firms in the Middle East and East Asia) benefited greatly from technology transfer by the SEFs. Freeman (1968, p. 30) noted that for the period from 1960 to 1966, nearly three quarters of the major new plants were "engineered," procured, and constructed by specialist plant contractors. This estimate is confirmed by a more recent study, which finds that for the period from 1970 to 1990, over three-fourths of the petrochemical plants built all over the world were "engineered" by SEFs (Arora and Gambardella, 1996). By providing technology licenses to firms the world over, SEFs played a major role in the diffusion of chemical, especially petrochemical, technologies.[25] The important point is that, although initially the benefits of the division of labor between chemical producers and SEFs accrued to U.S. chemical firms, over time these benefits became available to chemical producers in other countries, as well. Thus, the very factors that underpin the U.S. success were also responsible for enabling other countries to catch up.

Dynamics of an Industry's Competitive Advantage

The experience of the U.S. chemical industry points first and foremost to the multi-faceted nature of competitive advantage. While initially the nation's raw-material endowment was crucial to the industry's growth, over time the size of the market and the technological advances, both in products and processes, came to be very important.

A first major theme of our discussion is that no single source of advantage appears to have lasted very long. The development of new global markets has seen to that. Alternative sources of raw materials, or synthetic substitutes, were developed when raw materials became a major bottleneck. The shortage of Chilean nitrates led first to the cyanamid process and then to the famous Haber-Bosch process for nitrogen fixation. The shortages of sulfur caused by Sicilian attempts to restrict supplies led to the alternative sources of sulfur, such as pyrites, as well as to such innovations as the Frasch process for hot-water mining of brimstone, which is today the leading process for sulfur. The U.S. endowments of oil and natural gas, while crucial in getting the petrochemical industry off the ground, did not prove to be a source of overwhelming advantage, as others sought and developed the vast oil and gas deposits in the Middle East. However, thanks partly to its first-mover status with oil- and natural-gas-based feedstocks, the United States does remain to this day a leader in petrochemical technologies (Arora and Gambardella, 1996).

Initial dominance in technology and know-how often seems short-lived when rivals are willing to make the necessary investments to acquire and adapt. Britain's 19th-century lead in coal-tar intermediates and dyestuffs ended in a matter of a decade and a half. The experience and know-how gained by the U.S. firms in petrochemicals and large scale production were, to a large extent, made available to non-American firms after World War II by the SEFs. Mobility of personnel, wars, and other historical

---

[24]For instance, Badger used its fluidized-bed catalytic process to develop processes for phthalic anhydride with Sherwin Williams, ethylene dichloride with B.F. Goodrich, and acrylonitrile with Standard Oil of Ohio. UOP similarly had a number of strategic partnerships: with Dow (Udex — aromatics extraction), Shell (Sulfonale — benzene extraction), Ashland oil (Hydeal — dealkylation of tolune), Toray (Tatoray — disproportionation of toluene and C4 to benzene and xylene), and BP (Cyclar — reforming of LPG into aromatics).

[25]Specifically, Freeman showed that for the period 1960–66, SEFs as a group accounted for about 30% of all licenses (for processes), a figure consistent with the findings reported by Arora and Gambardella (1996).

accidents have also disrupted long-lived leadership by individual countries. The more general point to be made here is that leadership at the level of the firm can last, provided the firm makes the required investments in technology, production, and marketing. History matters more for understanding the fortunes of firms than of national industries, because the development trajectories of industries (taken as a whole) appear to be more successful than those of individual firms in shaking themselves free from the grip of the past.

The differences between the fortunes of firms and of the industry as a whole are even less surprising when one appreciates the global nature of the chemical industry. Although globalization has become a catchword today, chemical firms were systematically investing abroad well over 100 years ago. The Solvays and the Nobel-Dynamite group of companies were the first truly multinational network companies. German chemical companies also located subsidiaries and branches abroad, and a significant fraction of the U.S. pharmaceutical companies today trace their heritage to German fine-chemical companies. Foreign capital and technology flowed to the United States early in the century, to take advantage of the large market and abundant raw materials. After World War II, many U.S. companies such as Dow and Monsanto ventured abroad.

Today, the interpenetration of national markets is truly impressive. In 1996, the book value of direct foreign investments by foreign chemical companies in the United States was $74.8 billion, compared to $69.4 billion of U.S. investment in foreign chemical companies. A great deal of international trade consists of the intra-firm transactions of such companies. In 1994, the exports of U.S. chemical company parents amounted to $61.8 billion, of which over two-thirds went to their foreign affiliates. Likewise, 56% of U.S. chemical imports were channeled through the U.S. affiliates of foreign companies (CMA 1995). This impressive interpenetration makes it increasingly harder to identify the fate of national industries with one, or even a handful, of firms. Unlike 19th century organic chemicals, or petrochemicals earlier in the 20th century, the fruits of U.S. biotechnology research are available today to any firm willing and able to make the appropriate investments. This is not to argue that U.S. firms do not have a technological advantage, but simply to stress that any such advantages are much smaller than in the past.

The second major theme suggested by the U.S. experience is the importance of complementarities. Technology, market size, resource endowments, and supply of entrepreneurial capital all combined to produce the U.S. success story. It is tempting, but ultimately futile, to pick one or the other factor as being primarily responsible for successful performance. Partly this is because the relative importance of a given factor changes over time. But more to the point, picking one or the other factor as the source of comparative advantage is misleading because the experience of the U.S. chemical industry emphasizes the importance of synergies.

This idea of synergies resonates with the current advances in growth theory. Researchers are now finding that complementarities between capital (of both the physical and human variety) and technology appear to be of great importance for understanding economic growth. To this list we would like to add the complementarities between resource endowments and technology and those between market size and technology. As the petrochemical sector clearly shows, while U.S. technology was natural-resource intensive, U.S. natural-resource endowments themselves were increased, in an economic sense, by technology. In the United States, the large market focused the attention of domestic producers on the importance of large-scale production at a time when the payoffs to such a strategy were less apparent than they are today. But without the development of chemical-engineering techniques large, continuous-process chemical plants would not have been possible and neither would have been the widespread substitution of lower-cost synthetic fibers and plastics for natural materials. In other words, the market was, in a dynamic sense, enlarged by technological progress.

The presence of a large domestic market is important enough to qualify as a third major theme. The value of a large market is not restricted to the large size it enabled individual firms to attain. The advantages of bigness at the level of the firm have been well-documented (Chandler 1990). No less important, but not well understood, are the advantages of bigness at the level of the industry (Rosenberg 1976). These economies of specialization, exemplified by SEFs, have played an important role in the growth and development of the U.S. chemical industry.

The large size of the market also meant that a great number of firms could survive and attain large size. The ensuing diversity in approaches and specialization has been an important source of vitality. Large as it was in the 1920s, DuPont did not have the kind of influence that its counterparts IG Farben and ICI had. Even in explosives, antitrust policy had limited DuPont's market power. Companies such as Allied Chemical, Dow, Union Carbide, American Cyanamid, Air Products, and Monsanto came to occupy leading positions in a number of important chemical sectors. In each of these sectors, the leading firms faced significant domestic competition, ensuring that new technological opportunities, when they arose, were tried out. When the experiments or strategies occasionally resulted in failures, the outcomes were not devastating for the industry as a whole.

Although it may seem difficult to believe today, Allied Chemical in the 1920s was as large as DuPont. Unlike DuPont, however, it adopted a strategy of staying in large-volume inorganic chemicals (and in ammonia and fertilizers). The Allied Chemical strategy was one of low-cost, high-volume production, with considerable divisional autonomy and minimal attention to investments in in-house technological capability. Later events proved this to be a somewhat short-sighted strategy for the prevailing conditions. What is important to note is that, unlike the effect on Britain's ICI (and its predecessor companies), which took a long time to come around to accepting the importance of in-house research capabilities, the effect of Allied's failure on U.S. industry as a whole was very limited. In contrast, the reluctance of the few leading British chemical companies in the 1870s and 1880s to invest in the new organic chemistry proved to have long-lasting effects on the British chemical industry.

The point is not to debate the virtues of this or that strategy. Rather, it is to emphasize that, in an industry where new technologies and change are of central importance, a competitive diversity of approaches can be quite valuable.

REFERENCES

Arora, A. and A. Gambardella (1996), "Domestic Markets and International Competitiveness: Generic and Product Specific Competencies in the Chemical Engineering Sector," mimeo, Stanford University, April; forthcoming, *Strategic Management Journal.*

Arora, A. (1996), "Patent, Licensing and Market Structure in the Chemical Industry," working paper, Heinz School, Carnegie Mellon University; forthcoming, Research Policy.

Backman, J. (1964), "Competition in the Chemical Industry," Manufacturing Chemists Association, Washington, DC.

Baily, M. N. and A. K. Chakrabarty (1985), "Innovation and Productivity in U.S. Industry," *Brookings Papers on Economic Activity*, 2: 609–39.

Bairoch, P. (1982), "International Industrialization Levels from 1750–1980," *Journal of European Economic History* (Spring), 11: 269–310.

Cantwell, John (1995), "The Evolution of European Industrial Technology in the Interwar Period," in Francois Caron, Paul Erker, and Wolfram Fischer (eds.), *Innovations in the European Economy between the Wars* Walter de Gruyter, Berlin, Germany.

Chandler, A. (1990), *Scale and Scope*, Harvard University Press, Cambridge, MA.

Chapman, K. (1991), *The International Petrochemical Industry*, Basil Blackwell, Oxford, U.K.

CMA (1995), Chemical Manufacturers Association, Washington, DC.

David, P. A., and G. Wright (1991), "Resource Abundance and American Economic Leadership," *CEPR Publication No. 267*, Stanford University, Stanford, CA.

Freeman, C. (1968), "Chemical Process Plant: Innovation and the World Market," *National Institute Economic Review* (August) 45: 29–51.

Haber, L. F. (1958), *The Chemical Industry during the Nineteenth Century*, Oxford University Press, Oxford, U.K.

Haber, L. F. (1971), *The Chemical Industry, 1900–1930*, Clarendon Press, Oxford, U.K.

Hardie, D. W. F. and J. Pratt (1966), *A History of the Modern British Chemical Industry*, Pergamon Press, Oxford, U.K.

Haynes, W. (1945–1954), *American Chemical Industry*, Vols. 1–6, Van Nostrand, New York, NY.

Hounshell, D. (1995), "Strategies of Growth and Innovation in the Decentralized DuPont Company," in Francois Caron, Paul Erker, and Wolfram Fischer (eds.), *Innovations in the European Economy between the Wars*, Walter de Gruyter, Berlin, Germany.

Hounshell, D. (1988), *Synthetic Textile Fibre Industry*, unpublished manuscript, Carnegie Mellon University.

Hounshell, D. and J. K. Smith (1988), *Science and Corporate Strategy: DuPont R&D, 1902–80*, Cambridge University Press, Cambridge, U.K.

Landau, R. (1989), "The Chemical Engineer and the CPI," *Chemical Engineering Progress* (September): 28–29.

Landau, R. and N. Rosenberg (1992), "Successful Commercialization in the Chemical Process Industries," in N. Rosenberg, R. Landau, and D. Mowery (eds.), *Technology and the Wealth of Nations,* Stanford University Press, Stanford, CA.

Manufacturing Chemists Association (1971), *Chemical Statistics Handbook, 7th Ed.* MCA, Washington, DC, p. 432.

MCA (1971), Manufacturing Chemists Association, Washington, DC.

MIT (1989), *Working Papers of the MIT Commission on Productivity*, Vol. 1, MIT Press, Cambridge, MA.

Morris, P. J. T. (1994), "Synthetic Rubber: Autarky and War," in F. T. I. Mossman and P. J. T. Morris (eds.), *The Development of Plastics*, Royal Society of Chemistry, Cambridge, U.K.

Mowery, D. (1981), "The Emergence and Growth of Industrial Research in American Manufacturing, 1899–1946," Ph.D. dissertation, Stanford University.

Nelson, R. R. and G. Wright (1992), "The Rise and Fall of American Technological Leadership: The Postwar Era in Historical Perspective," *Journal of Economic Literature* (December): 1931–64.

Rosenberg, N. (1972), *Technology And American Economic Growth*, Harper-Row, New York, NY.

Rosenberg, N. (1976), *Perspectives on Technology*, Cambridge University Press, Cambridge, U.K.

Steen, K. (1995), "Confiscated Commerce: American Importers of German Synthetic Organic Chemicals, 1914–1929," *History and Technology*, 12: 261–84.

Spitz, P. (1988), *Petrochemicals, The Rise of an Industry*, John Wiley & Sons, New York, NY.

Stokes, R. G. (1994), *Opting for Oil*, Cambridge University Press, Cambridge, U.K.

Sutton, J. (1991), *Sunk Costs and Market Structure*, MIT Press, Cambridge, MA.

Tarr, J. A. (1994), "Searching for a 'Sink' for an Industrial Waste: Iron-Making Fuels and the Environment," *Environmental History Review* (Spring) 18(1): 9–34.

Travis, A. S. (1992), *The Rainbow Makers: The Origins of the Synthetic Dyestuffs Industry in Western Europe*, Lehigh University Press, Bethlehem, PA.

Trescott, M. M. (1981), *The Rise of the American Electrochemical Industry, 1880–1910*, Greenwood Press, Westport, CT.

Wright, G. (1990), "The Origins of American Industrial Success, 1879–1940," *American Economic Review* (September) 80(4): 665–68.

4 The Japanese Puzzle: Rapid Catch-Up and Long Struggle

TAKASHI HIKINO, TSUTOMU HARADA,
YOSHIO TOKUHISA, and JAMES A. YOSHIDA

Japan's chemical industry poses an intriguing puzzle: In spite of its past impressive growth and present sheer size, the industry basically remains invisible on the international economic scene. Standard books on the global chemical industry usually devote only a few paragraphs to Japan (Haber 1971; Spitz 1988; Aftalion 1991; Chapman 1991; Quintella 1993). Even many Japanese cannot name the biggest companies of the Japanese chemical industry. This makes a sharp contrast to such familiar names in the world chemical industry as DuPont, ICI, and Bayer. The anonymity of Japan's chemical giants is also different from the nation's big businesses in other industries such as automobiles and electronics; for instance, Toyota, Nissan, Sony, and Panasonic. When the major industries of Japan are compared to those of other nations, particularly the United States, authors usually consider such examples as steel, shipbuilding, automobiles, consumer electronics, and semiconductors. In spite of its vast scale, the chemical industry is commonly ignored, even in most influential books on Japanese industries (Patrick and Rosovsky 1976; Okimoto, Sugano, and Weinstein 1984; McCraw 1986; Hart 1992).

This chapter aims to present an economic and historical assessment of the evolution of the chemical industry in Japan and, in so doing, to shed light on the reasons behind its rapid growth in the past and nagging problems and facelessness at present. Particular emphasis will be placed on the post-World War II development of petrochemicals, when Japanese industry established itself in global markets. The chapter is especially concerned with the economic, political, and organizational strengths and weaknesses of the Japanese chemical industry and how they relate to the international competitiveness of the industry.[1]

Useful comments and criticism were given by other participants of this project, particularly Ashish Arora, Alfred Chandler, Alfonso Gambardella, Ralph Landau, Peter Murmann, and Tim Taylor. Other people who contributed to this chapter in various capacities are Alice Amsden, Kunioki Kato, Hirokazu Sago, and Shuji Tamura.

[1] For general industrial and business developments of modern Japan, see Hirschmeier and Yui (1981), Nakamura (1981, 1983), Ito (1993), and Miwa (1996). A concise outline of the nation's technological development can be available in Morris-Suzuki (1994) and Odagiri and Goto (1996). For the Japanese chemical industry in particular, Itami (1990) is the only academic monograph on the industry in Japanese. Tokuhisa (1995) summarizes various viewpoints of the business executives actually working in the industry. In English, Ward (1992) conveniently summarizes the state of the Japanese chemical industry of the early 1990s. Molony (1990) covers the Japanese industry before World War II.

Compared to the common development patterns of other advanced chemical economies in the United States and Europe, three factors particularly contributed to the basic structural characteristics of the Japanese chemical industry: government industrial policy; the nature of corporate control and group organization; and the evolution of national systems of technology learning and capability building. While focusing on these three issues, the chapter analyzes the Japanese chemical industry in three phases: the period of electrochemicals and coal chemicals before the 1950s; the prime years of petrochemicals, up to the early 1970s; and the overall struggle to restructure starting in the 1970s, particularly after the first oil-price shock in 1973–74. (The chapter by Arora and Gambardella offers many structural details of the Japanese chemical industry. For the various specific aspects of the Japanese industry, Da Rin's chapter on finance is also useful, as are Eichengreen's on macroeconomics and trade and Esteghamat's on regulation.)

ECONOMIC PROFILE OF THE JAPANESE CHEMICAL INDUSTRY AND ENTERPRISES

Japan's chemical industry has played a significant role in the nation's manufacturing economy. Table 4-1 shows that in terms of the contributions to the national economy measured by value added, the chemical industry occupies second place, behind only the electrical and electronic-machinery industries. Given the capital-intensive nature of the chemical industry, employment generation by the industry is understandably small compared to that in such labor-using sectors as electrical machinery, general machinery, and transportation equipment. On the other hand, the chemical industry leads all other major industries in terms of value added per worker.

The Japanese chemical industry has been a leading player internationally. Japan has been an important exporter of chemical products, while the country has been a significant customer for chemical products from the United States and Europe. Table 4-2 illustrates that Japan is now the seventh-largest exporter, as well as importer, of chemical products. The trade balance of Japan's chemical products has been in surplus in recent years.

TABLE 4-1. The Largest Industries in the Japanese Manufacturing Economy: 1995

Industry	Value Added (billion yen)	Persons Engaged (1,000)	Value Added per Person (million yen)
Manufacturing total	117,204	10,321	11.4
Electrical machinery	19,643	1,750	11.2
Chemicals	16,194	841	19.3
Transportation equipment	12,494	914	13.7
Foods and beverages	12,373	1,259	9.8
Nonelectrical machinery	12,131	1,087	11.2
Fabricated metals	7,970	817	9.7
Primary metals	6,936	458	15.1

Source: Compiled and calculated from unpublished data in *Kogyo Tokei Chosa*, 1995, courtesy of Japan Ministry of International Trade and Industry.

TABLE 4-2. Ten Largest Chemical-Trading Nations in the World: 1995

Country	Exports ($ million)	Imports ($ million)	Balance ($ million)
Germany	70,477	43,263	27,214
United States	61,701	40,378	21,323
The Netherlands	32,327	20,988	11,339
Belgium-Luxembourg	35,889	25,339	10,550
Switzerland	20,332	10,978	9,354
France	40,821	32,673	8,148
Great Britain	33,585	27,343	6,242
Japan	30,077	24,548	5,529
Canada	8,882	12,405	− 3,523
Italy	19,575	27,553	− 7,978
World total	436,810	436,810	—

Source: Compiled and calculated from "Facts and Figures for the Chemical Industry," *Chemical and Engineering News*, June 24, 1996, p. 69.

TABLE 4-3. Largest Chemical Companies Listed in *Fortune Global 500*: 1995

Rank	Company	Country	Sales ($ million)	Profit ($ million)	Profit Margin	Founding Year
58	Du Pont	United States	37,607	3,293	9%	1802
63	Hoechst	Germany	36,409	1,193	3	1863
71	Procter & Gamble	United States	33,434	2,645	8	1837
78	BASF	Germany	32,258	1,724	5	1861
87	Bayer	Germany	31,108	1,671	5	1863
156	Dow Chemical	United States	20,957	2,078	10	1897
198	Ciba-Geigy	Switzerland	17,509	1,824	10	1884
206	Mitsubishi Chemical	Japan	17,074	233	1	1934
207	Rhône-Poulenc	France	16,996	665	4	1928
227	ICI	Great Britain	16,206	844	5	1926
247	Kodak	United States	15,269	1,252	9	1884
286	Akzo Nobel	Netherlands	13,383	818	6	1899
327	Norsk Hydro	Norway	12,578	1,125	9	1905
328	Asahi Chemical Industries	Japan	12,538	96	1	1931
388	Fuji Photo & Film	Japan	11,241	755	7	1934
413	L'Oreal	France	10,698	631	6	1909
445	Henkel	Germany	9,907	302	3	1876
448	Sumitomo Chemical	Japan	9,862	192	2	1925
453	Toray Industries	Japan	9,753	189	2	1926
475	Solvay	Belgium	9,268	417	4	1863
489	Dai Nippon Ink & Chemical	Japan	8,996	72	1	1908
492	Monsanto	United States	8,962	739	8	1901

Source: Compiled and reorganized from "The Fortune Global 500," *Fortune*, August 5, 1996. Date on founding years comes from various company publications.

In spite of the fact that many of Japan's chemical companies play a large role in global markets, they are not well-known internationally. Table 4-3 lists the 22 chemical companies (and their founding dates) that appeared on the *Fortune Global 500* list for 1995. Japan has greater representation on this list of the largest chemical manufacturers in the world, with six companies, than does either the United States or Germany, which have five and four companies, respectively. Many readers will recognize the names of all (or nearly all) of the U.S. and European chemical firms. Yet, as noted earlier, we suspect that the names of the Japanese firms will be largely unfamiliar or unknown, although all of those Japanese enterprises are well-established and international in scope.

Table 4-3, however, also offers a clue as to one source of the invisibility of Japanese chemical enterprises. The profit margin of large Japanese chemical companies was almost uniformly lower than that of competitors in other nations, although it is also true that the low figures in the table are partially a reflection of the weak Japanese economy in the 1990s. Table 4-4 attempts a more systematic comparison of profitability of the largest Japanese chemical companies against those of U.S. firms. The table shows first that for both diversified firms and specialized companies, profitability is visibly higher for U.S. enterprises. This difference is particularly notable because 1995 was a prosperous year for the Japanese chemical industry in general. Table 4-4 also illustrates that in the United States diversified chemical giants enjoy profitability as high as that of specialized firms. Although the table is somewhat ambiguous on the Japanese side in this regard, Japan's diversified firms have historically suffered from poor performance, as is discussed in the strategy chapter by Chandler, Hikino, and Mowery of this volume.

TABLE 4-4. Profitability of the Largest Chemical Companies of Japan and the United States: 1995

Japan	Return on Equity	Profit Margin	United States	Return on Equity	Profit Margin
Diversified Firms					
Mitsubishi Chemical	1.9%	0.8%	Du Pont	29.3%	9.1%
Suitomo Chemical	5.9	1.5	Dow Chemical	23.0	8.9
Mitsui Toatsu	2.9	1.0	Monsanto	22.0	8.0
Showa Denko	0.1	0.0	Union Carbide	59.3	15.4
Ube Industries	5.3	1.4	FMC	49.4	4.9
Specialized Firms					
Asahi Chemical Industries	3.6	1.4	Eastman Chemical	41.9	10.9
Sekisui Chemical	4.8	2.2	W.R. Grace	16.7	6.0
Kao	7.6	4.0	Air Products & Chemical	16.9	9.5
Toray Industries	4.1	3.2	Morton International	19.3	8.8
Dainippon Ink & Chemical	2.7	1.3	Sherwin-Williams	18.6	6.1

Note: The five largest companies, in terms of sales, are selected for both diversified and specialized firms of Japan and the United States. Return on equity is defined as net income divided by equity. Profit margin is defined as net income divided by sales.

Source: For Japanese companies, compiled and calculated from *Kaisha Shikiho, 1997 Shunki*; for United States Companies, compiled from "Annual Survey of American Industries," *Forbes*, January 1, 1996, pp. 94–95.

TABLE 4-5. Return on Equity of Japanese Chemical Companies, by Product Specialty: 1986–1995

	1986	1987	1988	1989	1990	1991	1992	1993	1994	1995
Diversified chemicals	4.6%	8.7%	10.8%	11.0%	8.0%	5.2%	2.1%	−0.8%	−0.2%	3.0%
Chemical fertilizers	6.4	4.3	2.7	3.0	2.9	−3.0	1.0	−1.9	3.9	2.6
Soda-based chemicals	4.9	8.0	9.0	7.1	0.3	4.3	−2.8	−2.9	2.0	7.0
Petrochemicals	7.8	9.8	13.5	13.2	9.2	5.1	1.2	−2.4	1.4	4.6
High-pressure gas	5.3	6.7	6.2	6.5	7.1	8.8	3.4	2.5	4.6	3.0
Synthetic resin	7.7	8.8	9.3	9.0	7.4	5.6	2.6	−0.6	2.7	4.0
Plastic processing	7.4	8.6	8.7	8.1	6.8	5.4	3.2	3.1	3.1	4.0
Fine chemicals	7.6	7.4	7.0	5.9	4.8	4.5	3.2	3.2	1.6	3.7
Paints	6.7	7.5	10.7	6.9	6.1	5.5	4.4	2.8	2.3	2.1
Ink and pigments	7.1	7.4	7.0	5.9	4.9	2.8	0.8	−0.2	−1.4	1.0
Oils, fats, and detergents	7.4	7.9	7.8	8.0	7.5	5.1	4.3	4.1	3.9	3.9
Cosmetics and toiletries	7.7	5.6	4.1	4.5	5.4	6.3	6.4	6.4	6.3	6.2
Photographic film	11.3	11.0	10.9	10.8	10.3	10.2	8.3	6.4	6.9	5.4
Insecticides and pesticides	5.1	4.0	3.6	3.6	4.3	2.5	2.2	1.8	2.2	2.7
Pharmaceuticals	7.1	8.6	9.1	8.5	8.0	7.4	6.9	7.3	7.4	7.2

Note: Return on equity is defined as net income divided by equity.

Source: Compiled from *Anarisuto Gaido*, various years (Tokyo: Daiwa Shoken Chosabu, various years).

Table 4-5 further demonstrates Japan's characteristics of poor and unstable performance of diversified enterprises integrated in upstream, basic petrochemical production. The table shows that companies in basic commodity chemicals such as general chemicals (basic petrochemicals and derivatives), chemical fertilizers, and petrochemicals have experienced huge fluctuations of profitability. Firms in downstream fine and specialty chemicals, on the other hand, have sustained a reasonable level of profitability, even in the economic downturn of the early 1990s.

BASIC CHARACTERISTICS OF THE JAPANESE CHEMICAL INDUSTRY

For the illustrative purpose of international comparison, the basic strategic and structural traits of Japan's chemical industry can be summarized in five points, which outline its major achievements and struggles.

Competencies and Industry Structure

The strength of Japanese chemical companies lies in quick learning and incremental process-innovation capabilities. In other words, firms possess high commercialization and customization competencies. However, Japanese chemical firms have not generally shown great technological competencies in radical product or process innovation. The most comprehensive study of the radical innovations in the chemical industry, that of Achilladelis, Schwarzkopf, and Cines (1990), investigated 203 international cases of radical innovation and found Japan's representation to be marginal. When questioned,

Japanese chemical engineers generally support this conclusion (Tokuhisa 1995a, ch. 5). Until recently, most Japanese chemical companies concentrated their efforts in process and product development, at the expense of basic research.

The characteristic Japanese technological competencies have resulted in a skewed distribution of Japanese chemical companies toward capital- and resource-intensive ends and away from knowledge-intensive parts of the chemical industry. A telling case of Japan's relative weakness in knowledge-driven industrial growth is the pharmaceutical industry. There are 10 pharmaceutical companies in the *Fortune Global 500* for 1995: Johnson and Johnson, Merck, Bristol-Myers Squibb, American Home Products, Sandoz, Roche Holding, Glaxco Wellcome, SmithKline Beecham, Pfizer, and Abbott Laboratories (6 U.S. firms, 2 British, and 2 Swiss). But in sharp contrast to the case of the industrial chemical sector, none of the largest firms in this part of the chemical industry is Japanese.

Some prominent Japanese chemical companies are located on the downstream end, particularly in consumer chemicals. However, upstream commodities such as basic chemical producers have performed badly, and highly research-oriented areas such as pharmaceuticals have not yet developed into world prominence. Although it is common in the chemical industry in most industrial economies to find more profitable companies in downstream parts of the industry, Japan's case is an extreme one in that *all* upstream producers have been suffering from a structural problem of excess capacity and cyclical downturn for a quarter of a century.

Productive and Operational Efficiency

A variety of productivity calculations have found that the Japanese chemical industry as a whole has enjoyed relatively high productivity in the narrow technical and economic sense. Table 4-6 comes from perhaps the most detailed comparison of industry-level productivity across nations, carried out by Dollar and Wolff (1993). The table exhibits the high labor productivity of the Japanese chemical industry relative to its counterparts in the United States, Germany, Great Britain, and France. It is probably misleading to take these numbers literally and to naively assume that Japan's efficiency in chemicals is similar to or higher than that of other nations, because productivity calculations across countries are always subject to complicated issues related to industry classifications, measurement problems, and exchange rates. Nonetheless, the relatively high figures for Japan's chemical industry do have a story to tell.

According to the estimates of Jorgenson and Kuroda (1995), the multifactor productivity of the Japanese chemical industry has grown much more rapidly in recent decades than that of its American (or German) counterparts. According to their calculations, shown in Table 4-7, productivity growth of the Japanese chemical industry was consistently higher than that of the American chemical industry from 1960 to 1985. In fact, these researchers find that Japanese productivity in chemicals surpassed the German level in 1967 and the American level in 1974. Jorgenson-Kuroda's conclusion seems to suggest that Japan's chemical industry has been one of the best performers of productivity catch-up. Additional estimates also find that the relative level of productivity of Japan's chemical industry is not, by international standard, low at all; nor is it low when compared to other Japanese industries (Fecher and Perelman 1992). This implies that a narrowly-defined productivity or efficiency problem is not a cause of the ongoing troubles in Japan's chemical industry. In other words, in broadly manufacturing-operational terms, Japanese chemical plants have actually excelled in maximizing

TABLE 4-6. Relative Labor-Productivity Levels for Japan, Germany, and the United States, by Sector and Industry: 1985 and 1988

	Japan/U.S.		Germany/U.S.	
Sector and Industry	1985	1988	1985	1988
Total	0.64	0.66	0.67	0.65
Manufacturing	0.63	0.59	0.61	0.54
Chemicals and allied products	1.03	0.94	0.60	0.53
Primary and fabricated metals	0.74	0.72	0.56	0.52
Electrical and electronic products	0.73	0.63	0.71	0.57
Transportation equipment	0.60	0.54	0.56	0.47
Textiles	0.42	0.41	0.75	0.73
Food and kindred products	0.68	0.79	0.49	0.51
Agriculture, forestry, and fisheries	0.19	0.20	0.30	0.35
Mining	0.21	0.22	0.18	0.11
Construction	0.65	0.84	0.65	0.72
Transportation	0.50	0.50	0.51	0.47
Electric, gas, and sanitary services	0.96	0.85	0.55	0.46
Wholesale and retail trade	0.55	0.56	0.58	0.58
Finance and insurance	1.11	1.22	1.46	1.31
Other services	0.72	0.79	0.94	1.01

Note: Purchasing-power-parity exchange rates were used in the conversion of Japanese and German figures to U.S. dollars.

Source: Compiled from David Dollar and Edward N. Wolff, *Competitiveness, Convergence, and International Specialization* (Cambridge: MIT Press, 1993), Table 9.2, p. 181.

productive efficiency, thanks to their learning and incremental process-innovation competencies discussed above.[2] Any problems must then be sought in firm- and industry-level sources such as the overall capabilities and strategies of the firms, and the organization and structure of the industry.

Specialization and Diversification

Historically, the largest enterprises in Japan's chemical industry have stayed largely within the chemical industry proper and have diversified only narrowly. Except for a few notable cases such as Showa Denko and Ube Industries, firms have rarely operated in two major industry categories. The limited extent of vertical integration into upstream and downstream areas is also notable. While large U.S. and European chemical companies commonly receive 50 to 70 percent of their revenue from chemical

[2] Aggregate-productivity measurements must be disaggregated in order to pin down what has really been the relationship between the impressive productivity catch-up and the structural troubles of the Japanese chemical industry. Certainly, the choice of petrochemicals as the key strategic area for massive expansion contributed to the quick rise of productivity, since this particular industry is labor-saving and capital-intensive, which is suitable to Japan's competencies for efficiency growth. As is usual with these international comparisons of productivity, a note of caution is necessary, since these figures are critically dependent on many assumptions, particularly those related to purchasing-power-parity exchange rates.

TABLE 4-7. Productivity Growth of Japanese and U.S. Industries: 1960–1985

	Average Annual Growth Rate of Productivity					
	1960–1970		1970–1980		1980–1985	
Industry	Japan	U.S.	Japan	U.S.	Japan	U.S.
Chemicals	3.34%	1.501%	0.731%	−1.517%	2.671%	1.630%
Electrical machinery	3.304	0.093	3.663	0.693	3.222	0.500
Machinery	2.212	0.809	0.377	0.693	−1.073	0.785
Primary metals	0.915	0.088	0.781	0.534	0.624	−2.294
Textiles	0.526	1.437	−1.220	0.187	0.188	0.309
Foods	−0.155	0.556	0.370	0.208	−0.917	0.800

Source: Compiled from Dale W. Jorgenson and Masahiro Kuroda, "Productivity and International Competitiveness in Japan and the United States, 1960–1985," in Dale W. Jorgenson, *Productivity: Volume 2: International Comparisons of Economic Growth* (Cambridge: MIT Press, 1995), Table 9.8, pp. 404–405.

sales (with some exceptions), for the large Japanese companies the share more often ranges from 70 to 100 percent. One historical reason for this difference is that larger firms are more likely to diversify beyond their original industry, and until recently, Japan's chemical firms were relatively small compared to the global industry's giants.[3]

The *industry* focus of Japan's chemical enterprises, however, does not necessary mean *product* specialization. On the contrary, Japanese large enterprises are usually active in many product areas within their original industry. (See the chapter by Arora and Gambardella for a more systematic analysis of this issue.) The Japanese firms are also renowned for their product proliferation, which creates multitudinous varieties of a single product. The pressures for product proliferation come mainly from the demand side: Japanese customers, industrial and final consumers, have historically preferred a wide variety of choices. Because Japan's upstream producers of basic chemicals have not possessed the market power to establish their own product standards, they have had to customize their products to these market needs, although creating wide varieties and differing specifications usually raised their cost by around 15 percent. The absence of market power results mainly from two sources. First, the basic chemical industry, particularly petrochemicals, developed *after* the major user industries such as electrical machinery, textiles and automobiles had adopted their own standards. Second, Japan's chemical producers simply do not possess the necessary size and technological sophistication in the product market they serve to enforce their own specifications.

Patterns of Internationalization

The internationalization of Japan's chemical industry takes different forms depending on the specific foreign market. Although Japanese companies have exported to advanced economies, their manufacturing investment in such nations still remains limited. However, in emerging markets, Japanese companies both have strong export activities, mostly through trading companies, and often engage in direct manufacturing investments by establishing subsidiaries or joint ventures.

---

[3] Fruin (1994) provides a historical and comprehensive set of lists of the largest Japanese enterprises. For a comparable set for the United States, Great Britain, and German, see an appendix in Chandler (1990).

TABLE 4-8. Japanese Exports and Imports of Chemical Products: 1995

	Exports (million yen)	Imports (million yen)	Balance (million yen)
Taiwan	366,099	53,415	312,684
South Korea	391,074	90,838	300,236
Hong Kong	216,311	3,742	212,569
China	192,079	124,270	67,809
Great Britain	52,310	121,535	−69,225
Switzerland	30,935	109,473	−78,538
France	51,266	151,004	−99,738
Germany	103,180	286,342	−183,162
United States	452,928	663,247	−210,319
World total	2,829,276	2,309,160	520,116

Note: For exports and imports, five largest trading partners are included. Nine nations are listed because the United States is both the largest exporter and importer.

Source: Compiled and calculated from Somucho Tokeikyoku, *Nippon Tokei Nenkan, 1997* (Tokyo: Nippon Tokei Kyokai, 1996), Tables 12-8 and 12-9.

Table 4-8 lists the largest trading partners of Japan's chemical industry, in terms of exports and imports. The table reveals three patterns. First, the destination of Japan's chemical exports is concentrated in the United States (particularly with high-polymers such as plastics composites and optical and carbon fibers) and a group of emerging economies in East Asia. Second, imports come mostly from the prominent chemical producers of the United States and Europe. Third, the United States is by far the most significant trading partner of chemical products for Japan. Table 4-8 tells an interesting story when it is examined against the trade patterns of the United States. The United States (and Europe) are strong exporters of chemicals, while emerging economies, particularly those of East Asia, import chemicals from major industrial economies, including Japan. Japan, then, stands somewhere in the middle: an active importer of chemicals from advanced chemical industrial economies of the United States and Europe; and a large exporter of chemical products to emerging economies.

The ambiguous status of Japan's chemical industry, as the follower of advanced chemical economies and the leader of emerging chemical economies, explains the relative absence of direct manufacturing investments by Japan's chemical producers in the United States. Of the top 50 chemical producers in the United States, 10 are the subsidiaries of foreign firms, all of them based in western Europe; none is owned by a Japanese firm (Peaff 1996). In spite of many recent attempts by Japanese firms to enter into the U.S. market, their presence remains limited.

The delay and underrepresentation of chemical manufacturing operations by Japanese companies in the United States and Europe is also related to two other issues. First, many Japanese firms were restricted from selling in foreign markets as a stipulation of patent and technical-licensing agreements. When Japan's chemical companies became active in importing foreign (particularly American) technology in the 1950s, the Japanese enterprises had to agree to long-term export restrictions. Foreign companies seldom inserted the restrictive clause in the main contract, possibly for antitrust reasons. But usually they attached a side letter, which articulated geographical areas that Japanese parties were prohibited from entering.

Second, even after Japanese companies started their export activities, their long-term reliance on the networks of general trading companies hindered their direct contact with foreign customers, particularly those of advanced industrial economies. The extensive use of general trading companies historically had its advantages, of course. The trading companies had a cost advantage over chemical manufacturing firms in locating potential customers, because they already had established comprehensive sales networks that could be utilized at a minimum incremental cost. As long as the Japanese companies sold basic commodity chemicals, this lower cost was a distinct advantage. But for all of these cost savings, in the short run Japanese chemical enterprises put themselves in an isolated position. The lack of their own sales and distribution branches left them without direct contact with their customers, whose feedback on products and services became critical for product development. After the 1970s, as they tried to upscale into high value-added products such as fine and specialty chemicals, Japanese companies often could not meet the specific demands and needs of foreign clients.[4]

Group Affiliation and Strategic Complexities

Most major Japanese chemical enterprises are members of diversified business groups customarily called *kigyo shudan*. And all such Japanese business groups have at least a few prominent chemical enterprises as their members. This is true not only for the three biggest groups, Mitsubishi, Mitsui, and Sumitomo but also for other groups, such as Fuyo, DKB, and Sanwa, which are organized around large commercial banks. Figure 4-1 illustrates the case of the Mitsubishi Group, the most powerful *kigyo shudan* in Japan since World War II. *Kigyo shudan*, or group enterprises, originated after World War II, when pre-war groups called *zaibatsu* were dissolved as a reform carried out under Allied occupation, and newly independent operational companies established various structural ties to each other without a clear center of strategic command. The groups of Mitsubishi, Mitsui, and Sumitomo became the prototypes of *kigyo shudan*. Later in the 1950s, some large commercial "city" banks such as Fuji and Sanwa formed similar groups around them, and this type of cluster also became regarded as *kigyo shudan*.

The corporate groups are denoted as *kigyo shudan*, while *keiretsu* literally means "relational lineage." Even in Japan, the terms are often used interchangeably, but for structural and historical analyses it is better to treat them separately. For analytical purposes, *keiretsu* is an array of enterprises connected through long-term transactional relationships and usually controlled by the largest entity of the group. Toyota, a member of the Mitsui *kigyo shudan*, for instance, is the center of a huge hierarchical *keiretsu* network of suppliers and dealerships, which engages in relational, rather than transient, transactions. The term *keiretsu* was originally employed during World War II as a systematic arrangement of machinery parts supply for military purposes.[5]

In Japan's chemical industry, however, these relationships in terms of *kigyo shudan* and *keiretsu* are actually more complicated than in the case of other industries for reasons related to actual plant operation as well as ownership and transaction. First,

---

[4] After the 1970s, environmental regulations in Japan accelerated direct manufacturing investments overseas in various "pollution-prone" industries. The chemical industry was one of the representative cases. See Darby (1997).

[5] For general and detailed discussions, see Bisson (1954), Hadley (1970), Nakamura (1981), Gerlach (1992), Shimotani (1993), and Landau (1995, esp. pp. 69, 79, 82, 130).

The Mitsubishi Group Kinyo-kai President Club

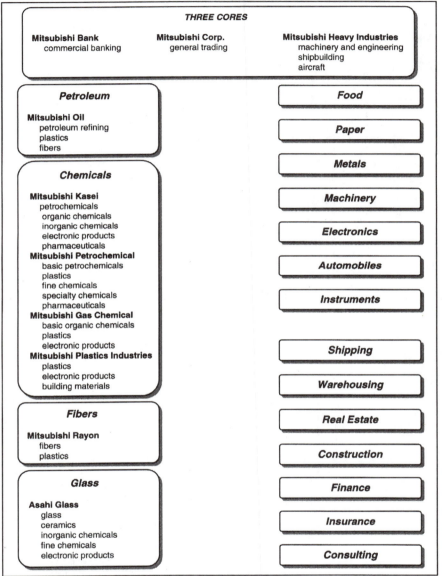

Figure 4-1. The basic structure of the Mitsubishi Group in the early 1990s.

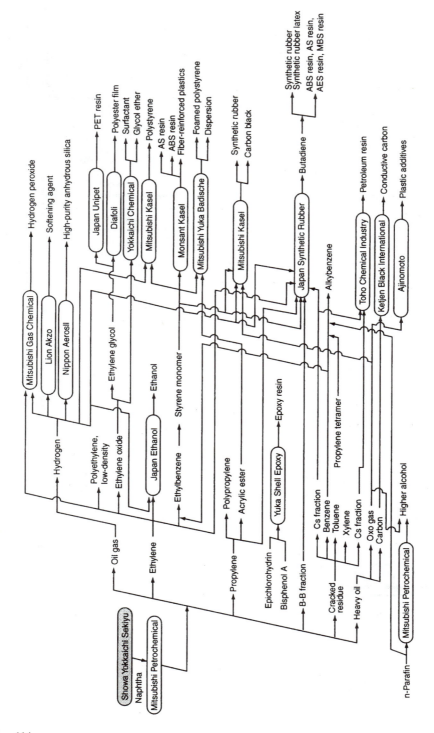

Figure 4-2. Mitsubishi Petrochemical Complex in Yokkaichi. Courtesy of *Japan Chemical Week* (Chemical Daily Co., Tokyo).

the establishment of large-scale petrochemical complexes since the 1950s necessitated the financial participation by such stakeholders as petroleum companies, chemical firms (sometimes including foreign giants), downstream chemical customers, commercial banks, and trading companies, which individually did not have the financial capacities to found such giant complexes. At least the Big Three *kigyo shudan* of Mitsubishi, Mitsui, and Sumitomo could manage to establish operating subsidiaries and plants within their own groups. For others, however, ownership participation had to include non-group enterprises, which made strategic decisions of operating subsidiaries and joint ventures more difficult. Second, in the resulting petrochemical center illustrated in Figure 4-2, the critical issue often became logistical coordination for the efficient and steady operation of constituent plants involving many specialized subsidiaries and joint ventures (companies are circled in the figure): Once the entire system went into operation, vertically linked chemical manufacturers had to attune their procedures to keep their processing flow constant (Tokuhisa 1995b).

The complexity of the ownership, transactional, and operational ties among firms forming a petrochemical complex became a significant structural rigidity opposing flexible strategic moves, particularly when downsizing and exit became necessary in the 1970s. The efficient operation of such complex systems was very difficult, even within a single firm; when many enterprises were involved, the process and strategic coordination became further demanding. Because many of operating companies were established to manufacture a single product, downsizing often meant the liquidation of those enterprises, a strategy that their management (and some of their parent companies) vigorously opposed. Particularly when the companies of different enterprise groups were mixed up in the same complex, any strategic decisions became problematic.

For yet another reason, since the early 1970s, intergroup strategic alliances in the chemical industry have become more frequent than in other industries. Surplus capacity and structural depression in the industry eventually forced chemical companies to cooperate to rationalize manufacturing operations, a move often initiated and endorsed by MITI (Ministry of International Trade and Industry). The ministry also actively encouraged intergroup cooperation in terms of the distribution networks of a common product.

DETERMINANTS OF INDUSTRY STRUCTURE AND PERFORMANCE

The basic characteristics of the Japanese chemical industry, in terms of its economic performance, technological weakness, growth strategies, and group affiliation, as outlined earlier, are critically influenced by four factors: the activist industrial policy of Japanese government; the managerial control of individual firms within a group structure; the late development of higher education in chemical engineering; and the limitation of firm competencies in technological generation.

Government Industrial Policy

Among the many factors that have influenced the development of the Japanese chemical industry, perhaps the role of government policy has been the most distinct and controversial from a comparative perspective.[6] In part because of Japan's lateness in

[6] For comparative perspectives on the various aspects of government policies in the chemical industry, see Haber (1971, ch. 8), Bower (1986), and Eichengreen's chapter in this volume. For Japan's industrial policy in general, see Johnson (1982), Mason (1992), Calder (1993), Tilton (1996), and Kikkawa and Hikino (1998).

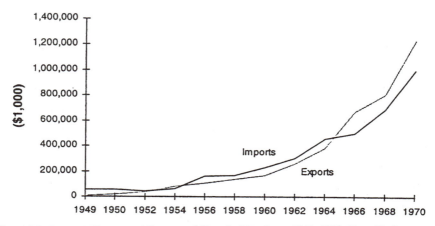

Figure 4-3. Japanese Export and Imports of Chemical Products: 1949–1970. *Note:* Exchange rate throughout this period was $1 = Y360. *Source;* Compiled and calculated from Japan Ministry of Finance, *Quarterly Bulletin of Financial Statistics*, various years.

developing a chemical industry, the Japanese government played a significant role, negative as well as positive, in determining the speed, direction, and pattern of the industry's evolution.[7] The intervention of government has not affected all branches of the chemical industry uniformly. Before World War II, the role of the government was small and indirect, mostly concentrated on the establishment of educational and research facilities and other infrastructure. There was no direct or systematic industrial policy to speak of. The most important contribution of Japan's government to the chemical industry occurred when MITI guided the growth of petrochemicals, beginning around the mid-1950s (Kikkawa 1995).

MITI worked closely with the Ministry of Finance in developing industrial policy instruments, mostly because of their mutual concern about the balance of payments and foreign exchange (Nippon Kagaku Kogyo Kyokai 1978; Sekiyu Kagaku Kogyo Kyokai 1989; Landau 1995, pp. 128–29). When Japan started to reindustrialize in the 1950s, after the destruction of World War II, chemical imports, particularly those of feed-stocks, soared substantially, as Figure 4-3 illustrates. This is because such user industries as textiles, machinery, electronics, and automobiles started growing rapidly. Increasing imports became an acute concern for the Japanese economy, because foreign exchange was scarce and could not continue to finance such huge imports. This balance-of-payment concern was the most immediate impetus for the Ministry of Finance and MITI to cooperate in devising an industrial policy to encourage the import substitution of chemical products, particularly feedstocks (Kikkawa 1995). When these macro-economic issues became less pressing, by the end of the 1960s, the policies advocated by MITI and the MOF (Ministry of Finance) started to show some conflict.[8]

---

[7] The issue of government policy should better be understood in the historical context of the evolving international economy, rather than as unique characteristics of Japanese industrialization. The effective and instrumental use of microeconomic intervention, particularly industrial policy, is common among a successful group of late-industrializing nations, of which Japan can be considered a pioneer. See Hikino and Amsden (1994).

In the 1950s, MITI developed a useful general rule in deciding whether to grant permission for the construction of individual petrochemical facilities. MITI realized that when a bureaucratic organization tries to monitor the performance of private businesses, it often encounters an information barrier. A political entity does not typically possess technical capabilities equal to those of private businesses, which makes it hard to evaluate the performance of the firm. However, MITI came up with a clear and neutrally determined rule for judging the efficiency of individual operations: whether it could make money selling at the international price then prevailing. MITI permitted the construction of particular facilities only when the submitted plan promised a high probability of being able to sell their products at prevailing international prices. This imposition of performance standards was significant, because only with a clear target of international prices could government guide the industry to grow to achieve efficiency as well as size (Kikkawa and Hikino 1997).

In the chemical industry's catch-up process of the 1950s and 1960s, MITI (and the Ministry of Finance) thus skillfully utilized strong competition among business groups whose rivalry worked positively during the expansionary, import-substitution phase. Ironically, though, the same competitive forces led to excess investment, chronic overcapacity, and depressed prices in basic petrochemical products beginning in the 1970s. MITI tried to rescue *all* the constituent enterprises by encouraging depression cartels, a move that could have worked in the long run only if some of those firms had strategically downsized or exited by finding a new source of growth in other industries and products. Unfortunately, those enterprises usually became locked in their original industries for two reasons: the limited strategic choices within group structure and the confined nature of technological competencies, as will be reviewed next.

Group Structure and Managerial Control

The Japanese group enterprise organization—as it developed after World War II— affected the competitive behavior of large chemical producers in several ways. On the positive side for firms, group membership to a certain degree ensured stable growth for the individual constituent companies. Intragroup product sales and the mutual shareholdings among member firms created a comfortable setting in which managers could formulate long-term growth strategies. Formal and informal networks within a group lowered transaction costs particularly related to market and technology information (as in the case of main bank system).[9] But to enjoy these and other possible benefits, the firms had to accept some limitations. In particular, chemical firms soon found that whenever they attempted to enter another product area, another member of the group likely was already in that area, and they were eventually blocked from trying to enter into the product territories of other companies within their own group. The petrochemical companies of the Mitsubishi and Mitsui groups, for instance, could not integrate vertically into synthetic fibers, simply because Mitsubishi Rayon and Toyo

---

[8] But with the exception of petrochemicals, MITI's influence was not critical in forming the basic structure of the chemical industry, nor was its impact uniform across sectors. Synthetic fibers, for instance, administered by the MITI textile division, had little in common in its development with petrochemicals (Suzuki 1997). The pharmaceutical industry was regulated tightly by the Ministry of Welfare, whose price and safety regulations are much different from MITI's systems. When Japanese chemical companies tried to enter the pharmaceutical industry, they soon found that working with the Welfare Ministry in addition to MITI was very difficult and cumbersome.

[9] For a convenient summary of group structure, see Miwa (1996).

Rayon (Toray), respectively, had already been operating in the industry. This restrictive situation certainly limited the possibilities for corporate growth by diversification and integration.

Often, the group structure has also acted as an obstacle for external corporate growth through merger and acquisition, mostly because intergroup rivalry makes the merger of companies across groups almost impossible (except in a few severe cases of depressed industries such as ocean shipping). Even within a group, the merger of constituent enterprises is often difficult because of noneconomic reasons such as executive and company rivalries. The case of Mitsubishi Kasei and Mitsubishi Petrochemical merging in 1994 and the planned amalgamation of Mitsui Toatsu and Mitsui Petrochemical in 1997 are two good examples of the lengthy time required to realize the widely held view that a merger was necessary for reorganization, rationalization, and survival.

The resistance of group structure and rivalry to restructuring is further compounded by Japan's corporate governance structure, in which management has a very high degree of autonomy, with little responsibility to equity-holders. Many economists have argued that, in general, management-controlled firms exhibit a strong drive to grow, regardless of the uncertainty and profitability of new investment (Marris 1969; Baumol 1959; Williamson 1967). When a company's senior management has only a small equity stake, it can even take high risks in committing new investments, as Joseph Schumpeter (1934) suggested. On the other hand, a too-autonomous management tends to ignore and resist downsizing requirements and continues to invest a firm's resources in projects whose expected return is relatively low: Managers try to create more opportunities for their own promotion and prestige, while profitability remains a secondary concern for them. This orientation toward sales growth and market share may work in a positive way as long as growth opportunities exist within the boundaries of a firm's capabilities. But when core capabilities no longer offer growth opportunities, the momentum for expansion does not work in a positive manner (Hikino 1997).

Japanese firms offer a vivid example of this expansion-oriented investment behavior. As long as opportunities for profitable investments existed in petrochemicals—as was the case during the 1950s and 1960s—this managerial drive worked in the positive manner for chemical companies as well as all of industry and the Japanese government. Beginning in the 1970s, however, when restructuring and exit became necessary, this growth orientation became dysfunctional and harmful. Without the need to face active shareholders and other disciplinary mechanisms of capital markets, the management control of large chemical producers did not make industry reorganization any easier. (For details see the chapter by Da Rin in this volume.)

Higher Education in Chemical Engineering

The United States has a long-standing commitment to the educational discipline of chemical engineering, and as Rosenberg describes in this volume, relationships between engineering programs at American universities and chemical companies have been established in a systematic and cooperative manner. The pattern in Japan is very different. Higher education in chemical engineering actually started relatively early, around 1905, as a form of applied chemistry. Some of the most dynamic of the chemical entrepreneurs, such as Jun Noguchi, held an engineering degree. However, when compared to the case of the United States, Japan's commitment to Ph.D. education in chemical engineering was quite limited until well after World War II.[10]

---

[10] Ralph Landau (1997); see also his chapter in this volume.

One of the reasons for the late development in Japan of doctoral studies in chemical engineering is that historically the Japanese chemical industry did not need a large supply of such engineers. As long as Japanese firms relied on the introduction and operation of foreign technology, what they acutely craved was an abundant supply of plant-level engineers. Particularly in demand were *mechanical* engineers, who could construct and operate complex chemical plants, not chemistry specialists.

The relationships between university engineering programs and private industry in Japan have traditionally been precarious, for several reasons. First, Japanese academia has held to an ivory-tower image of university institutions, in which teachers and students alike are skeptical of any academic research that might be put to the service of the profit motive. From this perspective, the very idea of "industry-academic cooperation" sounds negative. Second, Japanese laws regarding public employees have long prohibited professors and researchers of national and public universities from being employed on a regular basis by private companies. Because many prominent universities with strong chemical and chemical engineering programs (such as University of Tokyo, Kyoto University, and Tokyo Institute of Technology) are national universities, this legal constraint has hindered the smooth flow of information and knowledge between universities and firms. Although the Japanese government has somewhat loosened these restrictive practices in recent years, public employees still are prohibited from participating in founding companies or making strategic decisions for existing firms.

Within these societal and legal constraints, Japan's chemical industry suffered a backlash from the surge of concern over pollution and environmental issues in the late 1960s (discussed in the chapter by Esteghamet). The popularity of chemical engineering in Japanese universities became especially low in the late 1960s and early 1970s, precisely when chemical firms started feeling an acute need for a huge supply of well-educated and capable chemical engineers.[11]

Limited Competencies and Constrained Strategies

Japanese chemical companies have strong competencies in certain areas, but they often lack more comprehensive organizational capabilities, which they rely on other parties to provide. For example, financing for the huge projects in Japan's chemical industry has taken place largely through group banks and, directly or indirectly, the government. New technology has usually been generated by foreign firms—either chemical companies, petroleum enterprises, or specialized engineering firms—and bought by Japanese chemical companies. Trading companies contributed by marketing and exporting the products of chemical companies outside a group, while other enterprises within a group bought their raw materials from the chemical companies within the same group (Landau 1995, p. 105).

Japanese chemical companies have concentrated on two basic functions: project-execution capabilities and operational capabilities. They excelled in finding an appropriate source of technical know-how and making the knowledge workable for their production facilities. The companies also invested in improving various operational

---

[11] Japan's weaknesses in higher-level chemical engineering education certainly contributed to the struggles of its chemical industry to nurture technological capabilities. However, it would be misleading to single out this educational issue as *the* key variable behind Japan's troubles in chemicals, because a shortage of Ph.D. engineers and scientists is common in most Japanese industries. As Japan's major industries have approached the world technological frontier, a short supply of highly qualified engineers has become a universal concern for all knowledge-intensive industries.

aspects for incremental gains in productivity, performance, and costs. Given the level of technical knowledge of most Japanese chemical firms earlier in this century, the specialization and concentration on a narrow function made perfect sense.

A basic trait of the Japanese system of innovation throughout the twentieth century has been the importation and incremental improvement of the latest foreign technology (Hikino and Amsden 1994). This common pattern across modern industries has certainly been present in the chemical industry (Saffer and Yoshida 1980). In this regard, Japan's history is in sharp contrast to that of the United States, Germany, and Great Britain. Particularly since World War II, those three nations have acted as technology generators and innovators, while Japan (and many other late-industrializing nations) developed as a technology learner and commercializing specialist. This difference in the nature of technology acquisition had a substantial impact on the long-term growth strategies of Japan's chemical enterprises.

The continuous success of the U.S. and West European chemical industries at pushing out the technological frontier has made it difficult for a latecomer such as Japan to catch up. As Chandler (1994, 1997) emphasizes, in many other industries such as automobiles and consumer electronics, in which Japan caught up with the world leader after World War II, one finds a failure to innovate and commercialize new processes on the part of established oligopolistic firms in the rest of the world. By contrast, large chemical enterprises from the United States, Germany, and Great Britain have continued to exploit opportunities in radical and incremental product and process innovations, although in general the overall level of technological breakthroughs substantially slowed down since the 1970s (Achilladelis, Schwarzkopf, and Cines 1990; Smith 1994). (See the chapters by Arora-Rosenberg and Murmann-Landau.)

The Japanese overall level of technological competency at the time of the development of the nation's chemical industry resulted in a dual character for the evolution of technology and enterprise. On one hand, the level was high enough to absorb the advanced technical achievements of the United States and Western Europe. On the other hand, Japan's chemical industry was not mature enough to compete directly head-to-head with the leading global enterprises.

Because of the immaturity of Japan's technological competence, it was probably efficient and economical for each firm to concentrate in relatively narrow product areas and to commit to incremental process innovations as a way of lowering costs and improving product and service quality (Iijima 1981; Hamasato 1994, ch.3). Although Japan's enterprises have been successful along these dimensions, it has meant that the technology-learning process became localized and that the technical competencies of the industry stayed uneven (Harada 1995, 1997).

As Japanese companies have struggled to keep up with the continuous developments of chemical technology and the foreign enterprises embodying them, they have also been challenged by firms from developing nations, such as South Korea, which have been quickly catching up in basic commodity chemicals. Today, the Japanese chemical industry consists of many unsystematically diversified and integrated firms struggling to reorganize themselves by developing new technologies in very similar areas of fine and specialty chemicals, pharmaceuticals, and new materials.

The weakness of Japan's technological capabilities in chemicals has become an especially significant factor since the early 1970s. At that time, in addition to project execution and operational capabilities, considerable restructuring and innovation were needed. The Japanese chemical companies have been slowly catching up with their U.S. and European counterparts in terms of research and development, including basic

TABLE 4-9. Research and Development Activities of Representative Japanese Industries: 1995

Industry	Number of Companies	Number of Researchers	Intra-Firm Expenditures on R&D (billion yen)	R&D Expenditure as Percent of Sales	R&D Expenditure per Researcher (million yen)
Manufacturing, total	12,019	362,360	8,365	3.4%	23
Chemicals	1,559	61,257	1,549	5.3	25
Industrial chemicals and fibers	534	21,177	551	4.2	25
Oil and paints	270	9,004	156	4.4	17
Drugs and medicine	386	20,091	633	7.8	32
Electrical machinery	2,134	145,367	3,065	5.9	21
Transportation equipment	427	35,668	1,220	3.2	34
Machinery	2,070	34,127	697	3.2	21
Instruments	555	18,267	334	5.5	18

Note: Five industry groups listed are the five largest in terms of their R&D expenditures.

Source: Compiled and calculated from Somucho Tokeikyoku hen, *Nippon Tokei Nenkan, 1997* (Tokyo: Nippon Tokei Kyokai, 1996), p. 726.

TABLE 4-10. Ten Largest Chemical-Patenting Countries: 1985–1995

Country	1985	1990	1995	Annual Change, 1985–1995
United States	11,557	13,122	15,259	2.8%
Japan	3,733	5,473	6,138	5.1
Germany	2,179	2,706	2,486	1.3
France	716	917	1,020	3.6
Great Britain	854	953	903	0.6
Canada	358	432	564	4.6
Switzerland	458	475	435	−0.5
Italy	296	409	414	3.4
Taiwan	18	52	305	33.0
South Korea	9	52	303	42.0
World total	21,257	26,017	29,433	3.3

Note: Countries are ranked by the number of U.S. patents issued in 1995. Origin of patents is based on address of inventor whose name is first on patent application.

Source: Compiled from "Facts and Figures for Chemical R&D," *Chemical and Engineering News*, August 26, 1997, p. 71.

research. As Table 4-9 tells, R&D expenditures in the entire chemical industry account for 18.5 percent of the total sum of the manufacturing sector, while the employment of researchers in the chemical industry has risen to 16.8 percent. In both measures, the chemical industry comes second after the electrical-machinery industry in terms of research-and-development investments in Japanese industries.

TABLE 4-11. International Technology Transfer of Japanese Industries: 1975–1994

	1975	1980	1985	1990	1993	1994
			(billion yen)			
Manufacturing industries, total						
Exports	58.9	133.3	205.6	320.7	394.1	452.6
Imports	164.9	233.2	288.6	368.3	359.6	367.8
Balance	−106.0	−99.9	−83.0	−47.6	34.5	84.8
Chemical industry						
(including pharmaceuticals)						
Exports	21.5	31.9	38.2	58.2	59.3	64.1
Imports	26.9	39.3	37.4	54.0	61.4	59.0
Balance	−5.4	−7.4	0.8	4.2	−2.1	5.1
Electrical machinery industry						
(including computers)						
Exports	7.3	23.0	59.5	97.0	127.4	140.5
Imports	38.2	61.7	84.2	159.9	159.2	177.4
Balance	−30.9	−38.7	−24.7	−62.9	−31.8	−36.9
Transportation equipment industry						
Exports	6.3	21.8	32.4	92.0	127.7	164.2
Imports	35.7	40.3	59.7	52.3	40.4	35.6
Balance	−29.4	−18.5	−27.3	39.7	87.3	128.6

Source: Compiled and calculated from Somucho Tokeikyokuhen, *Nippon Tokei Nenkan, 1996* (Tokyo: Nippon Tokei Kyokai, 1996), p. 728.

These R&D investments have actually resulted in rising technological capabilities of the Japanese chemical industry. Table 4-10 illustrates that Japan has been very active in chemical patenting, now surpassing established nations in chemicals such as Germany, France, Great Britain, and Switzerland, although the nature and significance of various patents differ substantially. One outcome of Japan's rising technological capability in chemicals is its technology trade balance, shown in Table 4-11. In contrast to the case in the 1950s and 1960s, when Japan mostly imported chemical technology, Japan's chemical industry now exports a substantial amount of technology, which almost equals imports. In spite of these achievements, however, nothing the chemical firms have done so far has resulted in the sort of technological leadership capabilities that might have led these companies out of their long struggle.

THREE EVOLUTIONARY PHASES OF THE JAPANESE CHEMICAL INDUSTRY

The distinctive characteristics of Japan's chemical industry can best be understood by examining the origins and developments of the petrochemical industry since the mid-1950s: Petrochemicals pushed the Japanese chemical industry into international prominence in the 1960s and petrochemicals also symbolize the ongoing struggles of the industry since the 1970s. Furthermore, the petrochemical industry is the most significant target of various industrial-policy instruments exercised by the Japanese government.

Before Petrochemicals: To the 1950s

The first phase of Japan's chemical industry, from early in the 20th century until the mid-1950s, was the period of electrochemicals and coal chemicals. New entrepreneurial groups, as well as old *zaibatsu* groups, played a significant role, while government's policy remained limited (Miyajima 1997). Agriculture was the primary customer for the chemical industry, with the textile industry playing some role with its demand for dyes.

Many of today's largest chemical companies of Japan are actually well-established enterprises whose histories go back to the pre-World War II years. Those large chemical enterprises were already of considerable importance to Japan's economy at the time of World War I although, given the relatively small size of Japan's overall economy at the time, the Japanese chemical companies were not large by international standards (Landau 1995, p. 136). Many of Japan's leading chemical companies of today were firmly established by the interwar years. Table 4-12 also illustrates that the pattern of large industrial enterprises across Japan's economy in 1918 was not dissimilar to the one found in such advanced economies as the United States, Great Britain, and Germany, although the labor-intensive textile industry was relatively more important in Japan's pattern (partially because of the later industrial and economic development of the nation).

TABLE 4-12. Industrial Distribution of the 200 Largest Industrial Enterprises in the United States, Great Britain, Germany, and Japan in the Period of World War I

SIC	Industry	United States (1917)	Great Britain (1919)	Germany (1913)	Japan (1918)
20	Food	30	63	23	31
21	Tobacco	6	3	1	1
22	Textiles	5	26	13	54
23	Apparel	3	1	0	3
24	Lumber	3	0	1	3
25	Furniture	0	0	0	0
26	Paper	5	4	1	12
27	Printing and publishing	2	5	0	1
28	Chemicals	20	11	26	23
29	Petroleum	22	3	5	6
30	Rubber	5	3	1	0
31	Leather	4	0	2	4
32	Stone, clay, and glass	5	2	10	16
33	Primary metals	29	35	49	21
34	Fabricated metals	8	2	8	4
35	Nonelectrical machinery	20	8	21	4
36	Electrical machinery	5	11	18	7
37	Transportation equipment	26	20	19	9
38	Instruments	1	0	0	9
39	Miscellaneous	1	3	1	1
	Total	200	200	200	200

Note: Ranked by total assets, except for Great Britain, where ranking is based on the market value of quoted capital.

In the 1920s and 1930s, the old *zaibatsu* groups such as Mitsui, Mitsubishi, and Sumitomo were reluctant to commit themselves to such emerging new fields as chemicals and heavy industries, mainly because of the entrepreneurial conservatism of the family owners and senior managers (Morikawa 1990). It was left to a small group of aggressive entrepreneurs to take this opportunity to advance their interest in chemicals. Industrial groups such as Nissan, Nichitsu, Nisso, and Mori became organized around large electrochemical firms (Udagawa 1984). By aggressively and successively importing the latest technology available, those enterprises became leaders of Japan's chemical industry (Daito 1980; Mikami 1980; Uchida 1980; Oshio 1989; Molony 1990; Shimotani 1992).

World War II had a destructive impact on the newly emerged groups in two ways. First, starting in the late 1920s many of the groups, seeking a cheaper source of energy, aggressively transferred their operations to Korea and China. In the aftermath of World War II, these groups not only lost a substantial part of their production facilities but also became a negative political and social target, because of their colonial aggression. Second, the organizational design was not coherent enough to weather the turmoil resulting from the war. Therefore, these groups simply collapsed as groups, although many constituent companies such as Asahi Chemical, Hitachi, and Nissan Automobile survived because of special technical and organizational competencies. Since the war all but stopped the inflow of new technical information, the old *zaibatsu* chemical firms also had a chance to catch up with the newer leaders by nurturing their coal-based chemical technology.

Catch-Up through Petrochemicals: Mid-1950s to the Early 1970s

The opportunities for petrochemical development in Japan were ripe in the early 1950s, as summarized in Figure 4-4. The most important of the major factors influencing the emergence and subsequent development of petrochemicals was that, while Japan was basically isolated from various commercial and technical developments during World War II, chemical—and particularly petrochemical—technology was extensively developed, tested, and accumulated in Germany and the United States. Another factor on the supply side was the reconstruction of Japan's petroleum-refining facilities by the end of the 1940s. On the demand side, many chemical-user industries were rapidly growing in the early 1950s, starting with plastics and synthetic-fiber industries, with the automobile-manufacturing and electronics industries coming slightly later. The coal-based chemical industry could not meet the demand for chemicals at a reasonable price (Shimono 1987). Many young Japanese managers were eager to take advantage of this disequilibrium by investing in the development of petrochemicals.

The rapid rate of replacement of coal by petroleum as the key raw material has characterized the development of the Japanese chemical industry since the 1950s. The timing of the reorganization of the petroleum-refining industry after World War II was critical in its relationships to the petrochemical industry. The petroleum-refining industry reconstructed itself along a structural plan that was originally made by the corporate members on the advisory panel for the General Headquarters of Occupied Forces. Five major international oil companies were represented on the panel: Standard-Vacuum Oil, Shell, Caltex, Tidewater Oil, and Union Oil. Fearing the loss of the vast markets in mainland China due to political and economic uncertainties, the company representatives sought to expand their own interest in the Japanese market. In place of tariff protection and industrial targeting, which were usual for other strategic

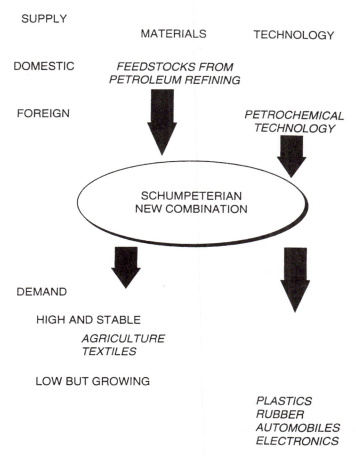

Figure 4-4. Opportunities in Japanese Chemical Markets around 1950.

industries including petrochemicals, the petroleum-refining industry relied on free trade and foreign capital for its postwar development (Saito 1990). When the petrochemical industry in postwar Japan started, around 1955, the petroleum-refining industry already had been reorganized by 1951 under the overwhelming influence of occupation policies (Kudo 1990).

Petrochemical technology was completely novel in Japan in the late 1940s, but some enterprises were eager to initiate production. But the initial process through which industrial enterprises, MITI, and the Ministry of Finance tried to establish the petrochemical industry took several years. MITI established a consulting committee, whose recommendations became formalized in the Petroleum Industry Development Plan (The First Plan) in July 1950, in order to evaluate and coordinate various schemes and possibilities (Sekiyu Kagaku Kogyo Kyokai 1989).

The First Plan clarified the overall purpose of petrochemicals: The new industry was to supply basic petrochemical products to domestic users at internationally competitive prices. These criteria were significant for two reasons. First, the basic target was the domestic market, which resulted in import substitution, not a drive for exports, at least at this stage. Second, the government's reliance on international prices as a feedback mechanism contained a key element of efficiency and price competitiveness, which is intriguing given the subsequent history of MITI's policies and the chemical industry's troubles. The MITI plan specified three directions for the introduction of petrochemical production: (1) to secure basic domestic supplies of benzol, organic acids, and acetone for immediate needs in the plastic and synthetic fiber industries; (2) to introduce the domestic manufacturing of ethylene and derivatives; and (3) to achieve low-cost production of basic petrochemicals to raise the international competitiveness of chemical and related industries (Sekiyu Kagaku Kogyo Kyokai 1989).

The First Plan contained substantial financial and other incentives for investment. However, MITI adopted strong administrative guidance in following the guidelines of the First Plan to screen, alter, and choose specific proposals submitted by various enterprises. Proposals were judged by MITI on instrumental and practical, rather then political, grounds, and the ministry accepted *all* schemes that met the plan's objective criteria. This rule—sometimes called "Equal Opportunities for Worthy Competitors"— encouraged oligopolistic competition for facility expansion among large firms. In spite of all the instruments of administrative guidance, MITI would later find that it was not able to regulate excess competition and capacity expansion.

With MITI's approval, the first petrochemical complex was thus launched in Kawasaki, near Yokohama, in 1957, when Nippon Petrochemicals, a subsidiary of Nippon Oil, constructed an ethylene-producing facility with a capacity of 25,000 tons a year. The complex had other specialized, downstream chemical producers such as Showa Petrochemical, Furukawa Petrochemical, Asahi Electrochemical, Nippon Soda, Asahi-Dow, Nippon Olefin, and Nippon Zeon. The intergroup combination of constituent enterprises became a prototype of subsequent complexes (Kudo 1990).

MITI approved three more ethylene centers: Mitsui Petrochemical in Iwakuni, which started operation in 1958; Sumitomo Chemical in Niihama, in the same year; and Mitsubishi Petrochemical in Yokkaichi, in 1959. (The fully developed structure of the Yokkaichi complex is demonstrated in Figure 4-2 on page 114.) By this time the three old *zaibatsu* groups, Mitsui, Mitsubishi, and Sumitomo, had recovered financially from their postwar reorganization, although Mitsui and Mitsubishi still had to establish those two new companies as a joint venture within each group. These joint ventures would later became problematic when the restructuring of petrochemicals became necessary in the 1970s. Sumitomo Chemical avoided this trouble by making its own investment, although the company had to restrict its product portfolio to polyethylene and ammonia.

The most profitable and dominant product in this initial period of Japanese petrochemicals was polyethylene. Manufacturing technology for producing polyethylene was imported from various sources, and even within one complex two different methods sometimes competed against each other. Polyethylene and other polymer materials became new industrial materials that found wide markets in housing construction, machinery, transportation equipment, and other uses (Kawade and Bono 1970; Tokuhisa 1995b).

By the end of 1960, all facilities approved by the First Plan had started operation, but the petrochemical industry was still a field into which many companies desired to enter, including not only chemical companies but also petroleum, synthetic fiber, and even steel manufacturers. MITI thus adopted the Second Plan to meet the demand for investment in petrochemicals. The first phase of the Second Plan consisted of expanding four ethylene centers and complexes. MITI also approved five new ethylene centers, four of which were organized by petroleum companies: Tonen Petrochemical in Kawasaki, Daikyowa Petrochemical in Yokaichi, Maruzen Petrochemical in Chiba, and Idemitsu Petrochemical in Idemitsu. The only nonpetroleum company whose ethylene production was approved was Mitsubishi Kasei. Because Mitsubishi Petrochemical, a joint venture of several Mitsubishi enterprises including Mitsubishi Kasei itself, had already entered ethylene production in the First Plan, those two companies within the same group became sorts of competitors.

Many of the companies participating in these complexes were organic-chemical enterprises that still possessed manufacturing technologies, such as fermentation and carbide manufacturing, which had become obsolete with the emergence of petrochemicals. Some of these companies smoothly moved to new technological bases, while others (such as Chisso) experienced difficulties. In general, compared to the almost uniform success of the enterprises of the First Plan, the outcome of the Second Plan was mixed. This mixed response was mostly because the participants in the Second Plan were much more diverse, smaller, and more independent; therefore the coordination of their businesses was more difficult.

In addition, MITI was seriously concerned at this time with the international competitiveness of Japan's petrochemical industry, particularly because the liberalization of capital markets in 1967 had potentially opened the Japanese market to foreign producers. MITI's response was to raise the minimum production capacity of ethylene-producing facilities to 300,000 tons a year. By setting the entry barrier relatively high, the ministry aimed to discourage new investment proposals and facilitate the concentration of production into a few efficient producers. But by this time Japan already had a number of ethylene crackers that were less than world-scale in size and efficiency.

Given the patterns of oligopolistic rivalry and autonomous expansion-seeking managers, however, MITI's new guideline had the ironic effect of encouraging new investment. For existing producers, participation in the 300,000-ton projects became a matter of survival, because the economies of scale of new facilities could wipe out conventional producers with high production costs. Existing producers thus were forced to add ethylene-producing capacities, which they did mainly through joint ventures among themselves, regardless of group affiliation. For instance, Mizushima Ethylene was established as a joint venture of Mitsubishi Kasei; Asahi Chemical, a member of the DKB group, and Nippon Mining. All in all, nine ethylene centers actually began their operations by 1972, while MITI originally expected that, at most, four of such investments were possible. Although the scale of 300,000 tons certainly lowered the production costs of ethylene as MITI had expected, the sudden, drastic rise of supply created excess capacity as early as 1971. Regardless of their different origins and entry strategies, however, profitability of the "full-line" petrochemical companies had stayed high and stable up to the late 1960s. As Table 4-13 illustrates, petrochemical processes had become dominant for the production of basic chemicals, and the import substitution of upstream chemical products had been achieved as MITI and MOF had originally intended. This period of growth and prosperity is crucial in understanding the subsequent development of the Japanese chemical industry. As downstream chemical users such as synthetic fibers, plastics, and synthetic rubber developed rapidly in the domestic market, demand for upstream basic chemicals remained high. Moreover, supply stayed relatively tight at this time, thanks to the government's control over entry,

TABLE 4-13. Growth of Petrochemical Processes by Product: 1958–1971

Product	Production by Petrochemical Processes	
	1958	1971
Xylene	75%	99%
Toluene	41	94
Ammonia	16	96
Benzene	8	81
Acetic acid	0	100
Octanol	0	100
Phthalic anhydride	0	66

Source: Japan Ministry of International Trade and Industry.

so that companies operating upstream could enjoy high profitability throughout the 1960s. Japan's chemical and engineering firms were rewarded for developing their technical competencies.

Technological Impediments and Lengthy Struggles: Early 1970s to the Present

The time since the early 1970s has been a difficult period of reorganization and financial and strategic struggles for Japan's chemical industry, particularly its basic petrochemical segments. MITI has faced substantial difficulties in guiding the industry through the turbulent time, a recent result of which has been a radical shift of basic policies, toward a reliance on free-market forces for restructuring.

One basic and nagging trouble with petrochemicals has been excess capacity resulting from aggressive investments in large-scale facilities. Domestic capacity for the production of ethylene rose from around 2.3 million tons in 1968 to more than 5.0 million tons in 1972. The imbalance of supply and demand was made worse when the growth of such industries as consumer electronics, automobiles, and synthetic fibers slowed in the early 1970s, reducing their demand for chemicals. Another factor in the slower growth of ethylene demand was that petrochemicals had by then replaced conventional organic chemicals as a raw material, so that replacement demand gradually declined (Watanabe and Saeki, 1984).

The first oil-price shock of 1973 drastically worsened Japan's excess capacity of petrochemicals by driving up the price of the key input. After the second oil-price shock in 1979, the Japanese petrochemical industry, particularly its upstream part, became structurally depressed. But by the mid-1990s, Japan's petrochemical industry had not recovered to reasonable profit stability. This long turmoil in the industry, unusual compared to the experience of other industries, had several economic reasons in addition to the huge excess capacity that had been built up in the early 1970s. First, while American producers could utilize natural gas as a substitute raw material for petrochemicals, Japan was completely dependent on petroleum, and the competitive position of the Japanese industry deteriorated as a result (Tokuhisa 1995b).

Furthermore, there was the so-called naphtha problem. Naphtha, a petroleum fraction similar to kerosene, is used as the feedstock for ethylene crackers in many countries where natural gas liquids are not available. As a part of the policy to establish Japan's own petroleum-refining industry, the importation of naphtha was tightly controlled by the government, and a heavy tax was levied on imports. Japan's domestic-petroleum prices were set higher than international prices, and Japanese chemical producers could not import petroleum freely. In the end, therefore, Japanese chemical companies had to bear the high costs of feedstock.

On the demand side, there were two additional issues. First, as the Japanese economy started experiencing business slumps in the 1970s and in early 1990s, domestic demand did not grow rapidly and smoothly. Foreign markets also were uncooperative, because some developing nations built their own petrochemical facilities and took over markets that had been Japan's traditional export outlets.

The "naphtha problem" had to be solved. In the mid-1980s, finally, MITI sided with the petrochemical interests. In 1982, MITI allowed the freer import of petroleum, and the price declined (Tokuhisa 1995b). But even as petroleum imports rose substantially in the early 1980s, the depression of the petrochemical industry worsened. Because the industry itself seemed incapable of finding a solution to its excess-capacity problems, MITI responded by organizing a group for the chemical industry within the Industry Structure Council whose mission was to come up with concrete plans to reduce the domestic capacity of a few major, basic petrochemicals (Bower 1986). Based on the recommendation of the Council, a depression cartel was formed, through which the capacity of ethylene-producing facilities was supposed to be cut by 36 percent, from 6.35 million tons a year to 4.06 million tons, and the cartel achieved 88 percent of the targeted reduction. Other products targeted for capacity reduction were polyolefins, vinyl chloride, ethylene oxide, and styrene monomer.

The common strategy of Japan's chemical companies since the early 1970s has been a combination of restructuring and rationalization of petrochemical facilities, expansion into overseas markets, and diversification into fine and specialty chemicals and pharmaceuticals (Watanabe and Saeki 1984). The process of industry restructuring and reorganization, however, has been difficult in all these areas. Ironically, all of the reasons underlying difficulties with reorganization—growth-oriented aggressive management, long-term and committed employees, industrial groups, involved commercial banks, silent capital markets, and so on—had worked positively in the initial developmental phase of the Japanese petrochemical industry since the mid-1950s.

The exit from petrochemical facilities whose products faced depressed prices and excess capacity did not proceed smoothly and rapidly. In principle, financially troubled companies could go out of business, but this did not happen on a large scale, for several reasons. The autonomous management of Japanese companies rarely even considered this option, because senior managers wanted to keep their own employment opportunities more than anything else. Many of the petrochemical companies are joint ventures within, and even across, groups, a fact that made the decision-making related to fundamental reorganizations extremely complicated and tentative. Discipline from financial markets was very weak. Japan has almost no active shareholders who force management to adopt certain policies. Commercial banks, given their huge financial stake in large-scale facilities, do not want chemical companies to default on their debt obligations. If anything, the commercial banks were often instrumental in advancing more money rather than in forcing reductions, in order to maintain their businesses. In addition, policies of life-time employment naturally worked as a huge exit barrier.

Ever since the economic necessity for the reorganization and consolidation of the basic petrochemical industries became evident in the early 1970s, therefore, the major

Japanese chemical enterprises have labored in their operational rationalization and strategic redirection. In spite of the basic and structural problem of inefficiency and overcapacity, the macroeconomic environment, particularly in the 1980s, was too growth-oriented to cause any industry to downsize, because demand for petrochemical products was reasonably strong and very low interest rates encouraged expansion rather than contraction and exit. It thus took more than 20 years to consolidate two of Mitsubishi's petrochemical-related companies, Mitsubishi Kasei and Mitsubishi Petrochemical. Refer to Figure 4-1 on page 113 for the overlapping product portfolios among various Mitsubishi chemical-related enterprises. Even worse, the Mitsubishi situation is actually better than that of the Mitsui group, which has been taking more time to solve a similar situation. When the Mitsui reorganization finally happens, it will take place as a defensive reaction to the Mitsubishi move. The substantial delay of these reorganizations symbolizes the difficulty of restructuring Japan's petrochemical industry within the country's business and institutional context (Tokuhisa 1995a). This difficult situation is in sharp contrast to the restructuring process that has occurred in the United States and Europe, where government influence is far less and private enterprises, induced by the capital market, have much greater flexibility.

Japanese chemical companies recently have been forced to diversify into many new product areas, mainly because they often do not possess core competencies in terms of a particular product. Remember, Japanese chemical enterprises have historically relied on the capabilities of finding innovative and commercializable technologies, mostly in the United States and Europe, and improving their processes and products. As Japanese chemical firms face mature markets for basic-commodity chemicals, many have sought to use their project-execution and improvement capabilities to find growth opportunities in new product areas. The product portfolio of the companies resulting from such entry strategies has turned out to be extensive, but few enterprises have found a new growth core to distinguish them from other Japanese and foreign firms. Moreover, Japanese firms using this strategy typically have been limited to new products in their existing part of the chemical industry. Moving to other growing industry areas was next-to-impossible, because other areas were almost always already covered by another firm in the group. Except for Sumitomo Chemical, even moving within chemical fields proved difficult because many groups have some large chemical enterprises specializing in narrow niches.

Considering these institutional forces, the most trouble-free strategy for reorganization and growth is product development based on technological capabilities. But historically this area is not the forte of Japanese chemical companies: Japanese companies have been technology learners, not generators. Japanese chemical enterprises thus now face a formidable task of upscaling themselves to become real product innovators, when overall technological progress in the chemical industry has slowed down worldwide, and established chemical firms of the United States and Europe are ready and capable to commercialize any technological opportunities available.

CONCLUSION

The basic characteristics of Japanese petrochemical companies were formed during the boom years from the mid-1950s to the late 1960s. While continuously trying to identify and import the latest product and process innovations in the United States and Europe, Japanese companies invested heavily in their own incremental process-improvement capabilities. With their sharpened process-learning and improving competencies, these companies competed against each other by building ever-larger facilities for the production of ethylene and derivatives to exploit cost savings resulting from scale economies. This strategy was made possible by substantial operating profits and also by the availability of low-cost loans from the government and the so-called main banks of industrial groups.

The impressive growth of Japan's chemical industry was the result (and also the cause, for that matter) of the nation's overall economic development during the 1950s and 1960s, which in turn was critically influenced by public-policy choices. The relative invisibility of Japan's chemical industry today can be explained by looking at the post-World War II period in its entirety. The huge industry created by the early policy decisions was instrumental in rapid catch-up, but after the 1970s the industry structure became less suitable, as Japan approached the world technological frontier. Thus, Japan's chemical firms have a fairly secure domestic position, but they are less than impressive in terms of their technological leadership and profitability.

Just as in the case of Germany, the past had a strong hand in the present and future of the Japanese chemical industry after World War II. While Germany could rely on large established firms that had built up innovation and exporting capabilities for over 80 years (for details, see the Murmann and Landau chapter), Japan did not possess firms with strong technological capabilities and marketing skills. In Japan, old trading companies, not chemical firms themselves, had created the commercial capabilities that would link Japanese producers with the world market. Since direct foreign investment typically follows trade, Japanese companies, like British firms, were not able to quickly establish a foothold in the rapidly growing markets of North America and Western Europe. Thus, while German enterprises could readily provide competitive products to the dramatically growing world market in the 1950s, Japanese chemical firms were forced to confine themselves to importing American and European chemical technology and serving the domestic market.

Unlike the government in Germany, which focused largely on creating a favorable macroeconomic environment, the Japanese government after World War II assumed a strong microeconomic role in the Japanese chemical industry. In trying to facilitate and coordinate the chemical industry's catch-up process, the government utilized and fostered fierce competition among the various industrial groups that succeeded the *zaibatsu* system. This rivalry worked positively during the import-substitution phase, when domestic capacity increased rapidly in the 1950s and 1960s. The same competitive forces, however, led to overcapacity and weak prices, depressing the profitability of producers for decades after the 1970s. In contrast, the large German chemical firms for the most part were able to coordinate their investment decisions in accordance with their accumulated human capital and technological capabilities. As managers (most of whom had worked together at IG Farben before and during World War II) kept each other informed on product areas in which each firm would expand, they paved the way for a growth pattern that allowed firms to be profitable and to provide the resources for large, new research-and-development expenditures that made German firms innovative leaders in the postwar years.

Japan's chemical industry is an enormous success when judged solely in terms of its providing the chemical inputs that allowed such downstream users as automobile and electronics industries to flourish. The biggest challenge for the Japanese chemical industry then seems to lie ahead. To become a world leader, it must become a leader in innovation and develop firms with distinct capabilities. This step may require a drastic restructuring of the industry—a formidable task, given the entrenched position of the present group system. Hence, the history of the Japanese chemical industry shows, once again, that history has long-term consequences, and that the most difficult problems facing industrial growth are institutions and organizations, not just technology. Economic growth is very much a path-dependent process in which yesterday and today determine to a large extent the economic options of tomorrow.

REFERENCES

Achilladelis, Basil, Albert Schwarzkopf, and Martin Cines (1990), "The Dynamics of Technological Innovation: The Case of the Chemical Industry," *Research Policy* (February) 19: 1, pp. 1–34.

Aftalion, Fred (1991), *A History of the International Chemical Industry*, University of Pennsylvania Press, Philadelphia, PA.

Baumol, William J. (1959), *Business Behavior, Value, and Growth*, Harcourt, New York, NY.

Bisson, Thomas (1954), *Zaibatsu Dissolution in Japan*, Princeton University Press, Princeton, NJ.

Bower, Joseph L. (1986), *When Markets Quake: The Management Challenge of Restructuring Industry*, Harvard Business School Press, Boston, MA.

Calder, Kent E. (1993), *Strategic Capitalism: Private Business and Public Purpose in Japanese Industrial Finance*, Princeton University Press, Princeton, NJ.

Chandler, Alfred D., Jr. (1990), *Scale and Scope: The Dynamics of Industrial Capitalism*, Harvard University Press, Cambridge, MA.

Chandler, Alfred D., Jr. (1994), "The Competitive Performance of U.S. Industrial Enterprise since the Second World War," *Business History Review* (Spring) 68(1): 1–72.

Chandler, Alfred D., Jr. (1997), "The United States: Engines of Economic Growth in the Capital-Intensive and Knowledge-Intensive Industries," in Alfred Chandler, Jr., Franco Amatori, and Takashi Hikino (eds.), *Big Business and the Wealth of Nations*, Cambridge University Press, New York, NY.

Chapman, Keith (1991), *The International Petrochemical Industry: Evolution and Location*, Blackwell, Oxford, U.K.

Daito, Eisuke (1980), "The Development of the Ammonia-Soda Process in Japan, 1917–1932," in Akio, Okochi and Hoshimi Uchida (eds.), *Development and Diffusion of Technology: Electrical and Chemical Industries*, University of Tokyo Press, Tokyo, Japan.

Darby, James (1997), "The Environmental Crisis in Japan and the Origins of Japanese Manufacturing in Europe," *Business History* (April) 39(2): 94–114.

Dollar, David and Edward N. Wolff (1993), *Competitiveness, Convergence, and International Specialization*, MIT Press, Cambridge, MA.

Fecher, Fabienne and Sergio Perelman (1992), "Productivity Growth and Technical Efficiency in OECD Industrial Activities," in Richard E. Caves (ed.), *Industrial Efficiency in Six Nations*, MIT Press, Cambridge, MA.

Fruin, W. Mark (1994), *The Japanese Enterprise System: Competitive Strategies and Cooperative Structures*, Clarendon Press, Oxford, U.K.

Gerlach, Michael L. (1992), *Alliance Capitalism: The Social Organization of Japanese Business*, University of California Press, Berkeley, CA.

Haber, L.F. (1971), *The Chemical Industry, 1900–1930: International Growth and Technological Change*, Clarendon Press, Oxford, U.K.

Hadley, Eleanor (1970), *Anti-trust in Japan*, Princeton University Press, Princeton, NJ

Hamasato, Hisao (1994), *Ronshu Nippon no Kagaku Kogyo*, Nippon Hyoronsha, Tokyo, Japan.

Harada, Tsutomu (1995), "Institutionalization of Technological Change and Paradox of Technical Transfer: The Japanese Chemical Industry, 1910–45," working paper, Standford University, October.

Harada, Tsutomu (1997), "Technological Change, Learning, and Institutional Adjustment," Ph.D. dissertation, Economics Department, Stanford University, March.

Hart, Jeffrey A. (1992), *Rival Capitalists: International Competitiveness in the United States, Japan, and Western Europe*, Cornell University Press, Ithaca, NY.

Hikino, Takashi and Alice H. Amsden (1994), "Staying Behind, Stumbling Back, Sneaking Up, Soaring Ahead: Late Industrialization in Historical Perspective," in William J. Baumol, Richard R. Nelson, and Edward N. Wolff (eds.), *Convergence of Productivity: Cross-National Studies and Historical Evidence*, Oxford University Press, New York, NY.

Hikino, Takashi (1997), "Managerial Control, Capital Markets, and the Wealth of Nations," in Alfred D. Chandler, Jr., Franco Amatori, and Takashi Hikino (eds.), *Big Business and the Wealth of Nations*, Cambridge University Press, New York, NY.

Hirschmeier, Johannes and Tsunehiko Yui (1981), *The Development of Japanese Business*, 2nd edition, George Allen & Unwin, London, U.K.

Iijima, Takashi (1981), *Nippon no Kagaku Gijyutsu: Kigyoshi ni miru sono Kozo*, Kogyo Chosakai, Tokyo, Japan.

Itami, Hiroyuki (1990), *Kagaku Kogyo ni Mirai wa Aruka*, Nihon Keizai Shimbunsha, Tokyo, Japan.

Ito, Takatoshi (1993), *The Japanese Economy*, MIT Press, Cambridge, MA.

Johnson, Chalmers (1982), *MITI and the Japanese Miracle: The Growth of Industrial Policy, 1925–1975*, Stanford University Press, Stanford, CA.

Jorgenson, Dale W. and Masahiro Kuroda (1995), "Productivity and International Competitiveness in Japan and the United States, 1960–1985," in Dale W. Jorgenson, *Productivity: Vol. 2: International Comparisons of Economic Growth*, MIT Press, Cambridge, MA.

Kawade, Tsunetada and Mitsuhisa Bono (1970), *Sekiyu Kagaku Kogyo*, new edition, Toyo Keizai Shimposha, Tokyo, Japan.

Kikkawa, Takeo (1995), "Enterprise Groups, Industry Associations, and Government: The Case of the Petrochemical Industry in Japan," *Business History* (July) 37(3): 89–110.

Kikkawa, Takeo and Takashi Hikino (1998), "Industrial Policy and the Competitiveness of Japanese Industries," in Hideaki Miyajima, Takeo Kikkawa, and Takashi Hikino (eds.), *Competing Policies for Competitiveness*, Oxford University Press, Oxford, U.K..

Kudo, Akira (1992), "Sekiyu Kagaku," in Shin'ichi Yonekawa, Koichi Shimokawa, and Hiroaki Yamazaki (eds.), *Sengo Nippon Keieishi, Dai Ni Kan*, Toyo Keizai, Tokyo, Japan.

Landau, Ralph (1995), "Strategy for Economic Growth: Lessons from the Chemical Industry," mimeo, Center for Economic Policy Research, Stanford University, July.

Landau, Ralph (1996), "Strategy for Economic Growth: Lessons from the Chemical Industry," in Ralph Landau, Timothy Taylor, and Gavin Wright (eds.), *The Mosaic of Economic Growth*, Stanford University Press, Stanford, CA, pp. 398–420.

Landau, Ralph (1997), "Education: Moving from Chemistry to Chemical Engineering and Beyond," *Chemical Engineering Progress* (January): 52–65.

Marris, Robin (1969), *The Economic Theory of "Managerial Capitalism,"* Basic Books, New York, NY.

Mason, Mark (1992), *American Multinationals and Japan: The Political Economy of Japanese Capital Controls, 1899–1980*, Harvard University Press, Cambridge, MA.

McCraw, Thomas, ed. (1986), *America vs. Japan*, Harvard Business School Press, Boston, MA.

Mikami, Atsufumi (1980), "Old and New Zaibatsu in the History of Japan's Chemical Industry: With Special Reference to the Sumitomo Chemical Co. and the Showa Denko Co.," in Akio Okochi and Heshimi Uchida (eds.), *Development and Diffusion of Technology: Electrical and Chemical Industries*, University of Tokyo Press, Tokyo, Japan.

Miwa, Yoshiro (1996), *Industry and Business in Japan*, New York University Press, New York, NY.

Miyajima, Hideaki (1997), "MITI's Reluctant Involvement in the Chemical Industry during the Inter-War Years," mimeo, Waseda University, Tokyo, Japan.

Molony, Barbara (1990), *Technology and Investment: The Prewar Japanese Chemical Industry*, Harvard University Press, Cambridge, MA.

Morikawa, Hidemasa (1990), *Zaibatsu: The Rise and Fall of Family Enterprises*, Tokyo Daigaku Shuppankai, Tokyo, Japan.

Morris-Suzuki, Tessa (1994), *The Technological Transformation of Japan: From the Seventeenth to the Twenty-First Century*, Cambridge University Press, Cambridge, U.K.

Nakamura, Takafusa (1981), *The Postwar Japanese Economy*, University of Tokyo Press, Tokyo, Japan.

Nakamura, Takafusa (1983), *Economic Growth in Pre-war Japan*, Yale University Press, New Haven, CT.

Nippon Kagaku Kogyo Kyokai (1978), *Nippon no Kagaku Kogyo: Sengo 30-nen no Ayumi*, Kyokai, Tokyo, Japan.

Odagiri, Hiroyuki and Akira Goto (1996), *Technology and Industrial Development in Japan: Building Capabilities by Learning, Innovation, and Public Policy*, Oxford University Press, New York, NY.

Okimoto, Daniel, Takuo Sugano, and Franklin B. Weinstein, (1984), *Competitive Edge: The Semiconductor Industry in the United States and Japan*, Stanford University Press, Stanford, CA.

Oshio, Takeshi (1989), *Nichitsu Konzern no Kenkyu*, Nippon Keizai Hyoronsha, Tokyo, Japan.

Patrick, Hugh and Henry Rosovsky (1976), *Asia's New Giant*, Brookings Institution, Washington, DC.

Peaff, George (1996), "Dow Replaces DuPont to Lead Top 100 U.S. Chemical Producers," *Chemical and Engineering News*, (May 6): 15–17.

Quintella, Rogerio H. (1993), *The Strategic Management of Technology in the Chemical and Petrochemical Industries*, Pinter Publishers, London and New York.

Saffer, Alfred and James A. Yoshida (1980), "Sources of Technology," *Chemtech* (November): 670–673.

Saito, Tomoaki (1990), "Sekiyu, in Shin'ichi Yonekawa, Koichi Shimokawa, and Hiroaki Yamazaki (eds.), *Sengo Nippon Keieishi, Dai Ni Kan*, Toyo Keizai, Tokyo, Japan.

Schumpeter, Joseph (1934), *The Theory of Economic Development*, Harvard University Press, Cambridge, MA.

Sekiyu Kagaku Kogyo Kyokai (1989), *Sekiyu Kagaku Kogyo 30-nen no Ayumi*, Kyokai, Tokyo, Japan.

Shimono, Katsumi (1987), *Sengo Nippon Sekitan Kagaku Kogyoshi*, Ochanomizu Shobo, Tokyo, Japan.

Shimotani, Masahiro (1992), *Nippon Kagaku Kogyoshi Ron*, Ochanomizu Shobo, Tokyo, Japan.

Shimotani, Masahiro (1993), *Nippon no Keiretsu to Kigyo Grupu*, Yuhikaku, Tokyo, Japan.

Smith, John Kenly (1994), "The End of the Chemical Century: Organizational Capabilities and Industry Evolution," *Business and Economic History* (Fall): 23(1): 152–161.

Spitz, Peter H. (1988), *Petrochemicals: The Rise of an Industry*, John Wiley & Sons, New York, NY.

Suzuki, Tsuneo (1997), "Industrial Policy and the Development of the Synthetic Fiber Industry in Postwar Japan," Paper presented to the Fuji Conference, Japan Business History Society.

Tilton, Mark (1996), Restrained Trade: Cartels in Japan's Basic Materials Industries, Cornell University Press, Ithaca, NY.

Tokuhisa, Yoshio (1995a), *Kagaku Sangyo ni Mirai wa Aruka*, Nippon Keizai Shinbunsha, Tokyo, Japan.

Tokuhisa, Yoshio (1995b), *Sekiyu Kagaku Kogyo shi Kankei Ronbunshu*, unpublished manuscript, Fujisawa, Japan.

Uchida, Hoshimi (1980), "Western Big Business and the Adoption of New Technology in Japan: The Electrical Equipment and Chemical Industries, 1890–1920," in Akio Okochi and Hashimi Uchida (eds.), *Development and Diffusion of Technology: Electrical and Chemical Industries*, University of Tokyo Press, Tokyo, Japan.

Udagawa, Masaru (1984), *Shinko Zaibatsu*, Nippon Keizai, Tokyo, Japan.

Ward, Mike (1992), *Japanese Chemicals: Past, Present and Future*, Economic Intelligence Unit, London, U.K.

Watanabe, Tokuji and Yasuharu Saeki (1984), *Tenki ni tatsu Sekiyu Kagaku Kogyo*, Iwanami Shoten, Tokyo, Japan.

Williamson, Oliver E. (1967), *The Economics of Discretionary Behavior: Managerial Objectives in a Theory of the Firm*, Prentice-Hall, Englewood Cliffs, N.J.

PART III
Innovation at the Heart of the Industry

5 The Process of Innovation in the Chemical Industry

RALPH LANDAU

Science and technology lie at the heart of the evolution of the chemical industry, but except in broad strokes such as the rise of the synthetic dyestuffs industry, the development of petrochemicals, or the evolution of pharmaceuticals, there has been relatively little said in the other chapters of this volume about *how* science and technology actually evolve, and from *where* and by *whom*. In an industry with over 70,000 different products, virtually none of which were commercially utilized in the early 1850s, there have to be many detailed stories of how these products were conceived, developed, and introduced. Many such stories, probably most, have been lost in the shroud of oblivion, and the living witnesses have long disappeared. In more recent years, serious efforts, such as those by the Chemical Heritage Foundation, have been made to record history by oral interviews with prominent innovators (primarily those in the United States) and by certain publications such as *Chemical Enterprise* (1994). Some of the volumes dealing more broadly with the industry also touch on aspects of the process of innovation (Spitz 1988; Aftalion 1991; Plumpe 1990; Teltschik 1992; Haber 1971). One particularly well-documented innovation is the discovery by Fritz Haber, in 1908, of the fixation of atmospheric nitrogen over a catalyst, by reaction with hydrogen, to produce ammonia, the base for a very large worldwide fertilizer industry. Its commercialization under the leadership of Carl Bosch at BASF in Ludwigshafen, Germany, is well described in both Plumpe and Teltschik in particular, as well as in other publications.

My purpose in this chapter is to describe how two major postwar chemical innovations came about: polyester fibers and plastics, and polypropylene fibers and plastics. Because one was intimately related to my own work and the other was brought about by people and companies I knew, this chapter serves the purpose of both recording the facts and also illustrating the process by which so much innovation in this industry takes place. Inevitably, there were many more facts about their case histories than I, or any one observer, could possibly have known, and even written records do not always reveal the true facts. I am reasonably confident, however, that the general lines of the story are sound.

Thanks are due to SRI, Tecnon (U.K.), Hoechst, and Chem Systems for data and information used in this chapter.

The two case histories addressed here will emphasize a point I have made elsewhere: that the word "innovation" is widely misunderstood (Landau 1979, 1994; Rosenberg, Landau, and Mowery 1992). It is not synonymous with invention or discovery. Innovation is the *process* whereby any scientific discovery or invention or idea is converted into a *commercial* product, process, or service. It is in the process of commercialization that economics enters into the picture, hence justifying inclusion in this volume, which deals with the economic aspects of the chemical industry.

The process of innovation in the chemical industry is multifaceted, and it is frequently difficult to identify where breakthroughs actually occurred. But in general, the creation of fundamental or radical innovations (the literature uses several terms to describe essentially the same phenomenon) is easier to recognize (if not to attribute) than the many gradual—or evolutionary—or incremental—or "ladder"—innovations (Gomory 1992) that often constitute the major long-term contribution to successful commercialization. The two cases chosen for this chapter illustrate these distinctions and show how both types of innovation have contributed to these histories. They also illustrate that the first to invent may not be the first to gain strong commercially important competitiveness. It still takes multipronged investments over an extended period of time, as described by Chandler (1990), for first movers to remain strong competitors. Indeed, successful commercialization can also be partly luck, partly the result of systematic search, and above all due to persistence in the necessary investments. It often takes a very long time to develop a successful product or process after a reaction works in the laboratory.

In most of the chemical industry, investment combined with a mastery of chemical engineering are the critical elements in the conversion of the laboratory reaction to a successful commercial product (see the chapter by Nathan Rosenberg). The case histories also show how dependent many new products and companies are on complementary innovations made by a number of firms and by the users of the products. Finally, the chapter also demonstrates how vital to the chemical industry the American petroleum industry has been, not only in providing economic raw materials for the manufacture of petrochemicals, but for influencing the direction of chemical research and commercialization, and even in the evolution of the unique scientific discipline of chemical engineering. Other institutional features of the industry also are pointed out: for instance, the importance of intellectual property and finance, as well as the research institution-industry interface.

POLYESTER FIBERS AND PLASTICS

The case of polyester fiber is one of the great breakthroughs in the chemical industry of the postwar period. The true initial breakthrough in synthetic fibers was nylon, initiated by Wallace Carothers and his team at DuPont (Hounshell and Smith 1988), followed by polyester and, shortly thereafter, acrylic. These are the big fiber innovations, all occurring either just before or just after the second world war. Polyester fiber accounts for 38 percent of total United States synthetic fiber production; it is in a class with cotton as the giants of the textile fiber world. In fact, in recent years, polyester has often been preferred to cotton when the price of cotton rises excessively due to shortages (*Chemical Week*, 3 May 1995, p. 20). Blends of cotton and polyester are widely used, especially in Asia. The discovery of this fiber has been one of the most important products of today's consumer society, and a tremendous success story. In fact, excess

TABLE 5-1. World Production of Textile Fibers: 1993–1995 (Measured in 1000 Metric Tons)

	1993	1994	1995
Cotton	16,845	17,450	18,303
Synthetic fibers	19,800	21,248	22,464
of which:			
Polyester	10,350	11,229	11,975
Polyamides (nylons)	3,710	3,858	4,034
Polyacrylics	2,323	2,515	2,557
Polypropylene	2,962	3,106	3,288
Other synthetics	455	540	610
Wool and silk	1,701	1,720	1,800
Cellulosics	2,727	2,840	2,930
Miscellaneous*	8,096	8,200	8,530
Total fibers	49,169	51,458	54,027

*Flax, hemp, jute, Ramie, Sisal, and hair.

Source: Tecnon (U.K.) Ltd.

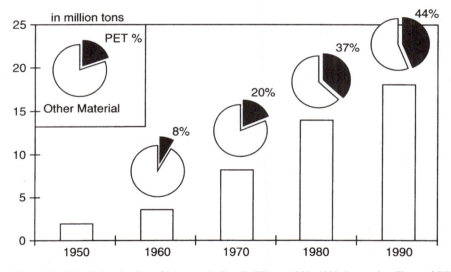

Figure 5-1. World Production of Man-made Textile Fibers: 1950–1990, Increasing Share of PET Fibers. Source: HOECHST.

production capacity is being built, particularly in Asia. As a result, although capacity usage in 1995 was over 90 percent, it may well drop to 75 percent later in this decade as many new plants come on stream.

Table 5-1 shows world production of textile fibers, natural and synthetic, for the years 1993 to 1995. Figure 5-1 gives this information in a graphic form, since 1950. Table 5-2 gives world production figures for textile fibers since 1900. Figure 5-2 gives

TABLE 5-2. World Production of Textile Fibers: 1900–1990 (Man-made Fibers, Cotton, and Wool)

Year	Population (billion)	Man-made Fibers	Cotton	Wool	Sum Textile Fibers
1900	1.6	—	3.2	0.7	3.9
1950	2.5	1.7	6.6	1.1	9.4
1960	3.0	3.3	10.1	1.5	14.9
1970	3.7	8.1	11.8	1.6	21.5
1980	4.4	13.7	14.3	1.6	29.6
1990	5.3	17.7	18.6	2.0	38.2

Source: Hoechst.

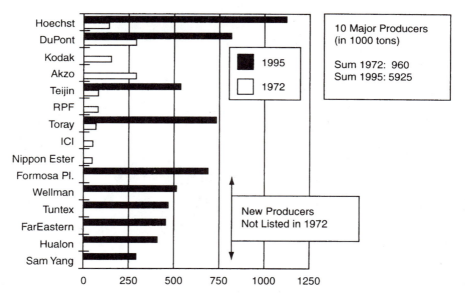

Figure 5-2. 10 Major PET Fiber Producers (Predominantly Textile): H1972/1995. Source: HOECHST.

the major polyester fiber producers and their capacities since 1972. The U.S. production of synthetic fibers from 1993 to 1995 is given in Table 5-3.

Furthermore, the polymer that constitutes polyester fiber is polyethylene terephthalate, a product that has found increasing utility as a plastic product, particularly as recyclable bottles for soft drinks. Thus, in 1994, about 1.5 million tons of polyester plastic were produced in the United States, almost as much as the fiber. Thus, this versatile and highly valuable polymer is both dominating the fiber market and growing rapidly as a plastic. The story of its rapid evolution to its present status is not only interesting in its own right but also one with which I have been intimately involved, and I can therefore illuminate a history that has been touched on only briefly in the existing literature.

TABLE 5-3. U.S. Production of Synthetic Fibers 1993–1995 (Measured in 1000 Metric Tons)

	1993	1994	1995
Polyester	1754	1750	1775
Nylon	1249	1267	1286
Olefins (primarily polypropylene)	1093	1095	1131
Acrylics	201	201	200
Cellulose	226	230	240
Total	4523	4543	4632

Source: Tecnon (U.K.) Ltd.

The Development of Polyester Fiber

The development of polyester fiber also illustrates a general feature of innovation: Real breakthroughs often come from outsiders, perhaps lone individuals or small organizations that are not as hierarchical or bureaucratic as the typical larger company (Landau 1994; Gomory 1992). Such was the case with Rex Whinfield and J. T. Dickson who, working in 1940 with the small Calico Printers Association in England, realized that DuPont's broad patent coverage of nylon left open a possible loophole. Nylon is a polyamide polymer, and in its best-known incarnation, as nylon 66, is a polymer made from a simple dibasic acid, adipic, and a simple diamine, hexamethylene diamine. The chemical structure of adipic is as follows:

Adipic Acid

where C is a carbon atom, H a hydrogen atom, and O an oxygen atom. The two ends are acidic groupings , thus forming a dibasic or two-functional acid. Hexamethylene diamine is

Hexamethylene Diamine

where N is a nitrogen atom. The two end groups $-C-NH_2$ are amines; hence the name *hexamethylene* (for six CH_2 groups) diamine. When these two compounds are reacted together, the acid groups at either end react with the amine groups to form long chains of essentially linear, straight-chain aliphatic polymers, which when properly drawn and treated make nylon fiber.

What Whinfield and Dickson explored during the period from 1939 to 1941 was whether an *aromatic* dibasic acid (one that has a benzene ring in it instead of straight links of carbon and hydrogen) not only could circumvent the Carothers patents but also might possess different properties. They worked not with amines but with dihydric alcohols, such as ethylene glycol:

$$OH-\underset{\underset{H_2}{|}}{C}-\underset{\underset{H_2}{|}}{C}-OH$$

The hydroxyl (OH) groups at the end could react with the acid groups at either end of the aromatic dibasic acid to propagate long, linear polymer chains. Of course, Whinfield and Dickson could only surmise what utility such polymers might have, but they could patent them, and did. Later, DuPont worked with aromatic dibasic acids and amines to produce aramid products more like nylon, but these do not have the widespread applicability that polyesters eventually achieved.

Whinfield and Dickson discovered that their polymers had high melting points, unlike the usual lower-melting fibers, and could be melt-extruded and stretched to produce an interesting fiber. "Crimping" the fibers, which is easier to do than with nylon, causes them to become resilient like wool, again not easy to do with nylon. "Heat setting" these fibers (at higher temperatures, but below any thermal degradation limit) gives them a permanent set, which makes the final garment crease- and wrinkle-proof— a boon to travelers and busy people the world over. The simplest of these polymers was polyethylene terephthalate. The aromatic dibasic acid utilized was terephthalic acid:

The six-member ring shown above is the benzene ring, which has three unsaturated (double) bonds, typical of benzene. If this benzene nucleus were to be hydrogenated, it would remove the double bonds and produce cyclohexane, a molecule of little reactivity (my company worked with this compound in developing its process for manufacturing cyclohexanol as a key precursor of the adipic acid for nylon; see Landau, 1994, p. 27). The three double bonds in benzene make it a more reactive molecule, but in this technology the benzene serves simply as a stable aromatic nucleus, which imparts the high-melting-point characteristic.

Of course, these discoveries took place during Britain's fight for existence against Germany, and Calico Printers was much too small an organization to exploit such a potentially valuable patent. The British government declared the patents to be secret and decided that no work should be done on these inventions during the war, as it could not affect the outcome of the war, and other projects were of higher priority. However, two of the national laboratories showed that the product was a crystalline superpolyester similar to those made by Carothers. The fiber was named Terylene by its inventors.

Because of the foregoing difficulties, Calico Printers sold the patent rights to Imperial Chemical Industries (ICI). Under the DuPont-ICI agreement of long standing, DuPont acquired the exclusive rights in the United States and ICI in the rest of the world. In 1953, ICI formed its Terylene Council as a start toward a new division. It was at this time that I first met Dr. A. Caress, its chairman, and George F. Whitby, its managing director. Before this, ICI had made some fiber, and the first sale of Terylene filament yarn was made on October 4, 1948 (ICI 1954). Whinfield joined ICI, and I met him there frequently. I remember him as a quiet, reserved scientist.

Further work by ICI and others, of course, developed many other variants of this basic structure, using different dibasic acids and dihydro compounds in various proportions. The biggest initial problem confronting ICI and DuPont, however, was how to manufacture economically the two molecules that composed the fiber, namely ethylene glycol and terephthalic acid.

Ethylene Glycol. Ethylene glycol had been made since World War I by an inelegant process that involved reacting ethylene (the basic building block of the future petrochemical industry) with chlorine in an alkaline solution. By-product wastes were substantial and in those days were dumped into rivers or large bodies of water. Yields were relatively low. This was the industrial situation until E. Lefort (1933) in France patented a process to convert ethylene to ethylene oxide with oxygen over a silver catalyst. For a time no one seemed to notice, but the Union Carbide and Carbon Corporation, which was one of the pioneers in petrochemicals (Spitz 1988), bought the rights for the United States and, after further experimentation and some additional inventions, secretly built a plant at Charleston, West Virginia. Ethylene oxide has many uses in its own right, but it can be relatively readily hydrated with water to ethylene glycol. Thus, when ICI was contemplating a significant scale of manufacture for Terylene, it needed a glycol process.

By this time my firm, Scientific Design Co., Inc., had already developed a competing process for the direct oxidation of ethylene over a silver catalyst and had commercialized it in 1953 in an 8000-metric-tons-per-year plant for Naphtachimie in Lavera, France. This was the first time outside Union Carbide that such a process was commercially proven (Landau 1994, p. 5). Shortly thereafter, the Royal Dutch Shell Group in the Hague announced their own version, using oxygen instead of the air my company employed. Under certain circumstances, oxygen could be the more economical feedstock, since the air compressors we required were not necessary if someone also had a large air liquefaction plant nearby. Soon Scientific Design had an oxygen version, and Shell and Scientific Design have competed to this day in licensing this technology around the world. Shell and its licensees now constitute about 40 percent of world capacity, Scientific Design's licensees over 30 percent, Carbide's production about 15 percent, and a group of miscellaneous technologies the residue. Scientific Design, unlike Shell, did not manufacture the product itself and thus did not compete with its customers, a fact which helped Scientific Design in the competitive bidding that took place over the years between Shell and Scientific Design. Scientific Design also specializes in supplying a steadily improving catalyst and plant design service.

Since the initial plant in 1953, the scale of installation has gone up by at least twentyfold; the costs have come down, thanks to skillful chemical engineering, and the yields have improved from perhaps two-thirds of those theoretically possible to approaching 90 percent. This improvement resulted from steady incremental innovations by Scientific Design and its clients (through a "club" that exchanges corporate experiences) and the strong competitive pressure of the powerful Shell organization, which was also improving its process performance and also had such a "club." No doubt Union Carbide achieved similar results. This illustrates the important point that chemical engineers contribute substantially to incremental innovation, even after a plant is built, by studying the manufacturing operations, while chemists strive to improve the catalyst.

Thus, the breakthrough invention by Lefort was steadily and incrementally improved by various organizations, chemists, and engineers. Catalysis lies at the heart of the modern chemical industry, and a commercially important catalyst is both a scientific and an engineering accomplishment. What may work well in a laboratory tube at atmospheric pressure may not be at all satisfactory when commercial conditions of high production per unit volume of catalyst are required, often under corrosive conditions, with short catalyst life and possible poisons, and the need for economic recovery and purification of the reactor effluent. Furthermore, reactor design also offers many variants (such as fixed or fluidized beds and heterogeneous or homogeneous systems), with many problems relating to good uniform diffusion of the feedstock across a multitube reactor or the necessity of avoiding back-mixing of feed and product in a fluidized bed, where catalyst attrition and poisoning often also appear.[1] Catalytic-reaction systems, as Haber and Bosch discovered, are particularly the province of the chemist and the chemical engineer (Landau and Brown 1965).

ICI therefore licensed Scientific Design's ethylene oxide and glycol process and built it at its Severnside location in the early days of Britain's petrochemical industry development. But in truth, when I cofounded Scientific Design in 1946, I had no idea about the future role of ethylene glycol in fiber manufacture. At that time, its use as an automobile antifreeze was the principal one, and I expected that the postwar period would see a great increase in automobile demand, so that this seemed like a worthwhile process to research. Furthermore, as a result of the mobility of personnel brought about by the war, I knew that Union Carbide was in fact making ethylene oxide by a direct oxidation technology of their own development. My wartime employer, M.W. Kellogg Co., had become the leading engineering company in fluidized-bed catalytic processes, especially fluidized cracking of petroleum. Kellogg sought to apply the technique to chemicals manufacture and indeed built a small commercial size plant to make phthalic anhydride from napthalene by such a technique. Since I was largely employed for chemicals applications by Kellogg, I participated in this project. Although Kellogg thought about adapting it to ethylene oxidation, the idea did not go far. So, when I left to go to Scientific Design, this was one idea I carried with me.

Our experts, however, soon decided that a valuable catalyst like silver was not really suited for a fluidized bed, and furthermore, we knew such a development would have many theoretical and practical problems. So we decided very early to focus on advanced fixed-bed-reactor research. Indeed, my first patent in 1956 dealt with a novel inert support style for the silver catalyst which, as far as I know, is still in use. The further history of our efforts to develop this process is described in Landau (1994). We never really deviated from the fixed bed concept and by 1948 had a pilot plant running in England, with the help of Petrochemicals Ltd. As a company with limited resources, we were unwilling to lose track of what looked like a feasible commercial position by

---

[1] A fluidized bed is a suspension of finely divided catalyst particles in reaction gases, which acts much like a fluid. A heterogeneous catalyst is a catalyst generally fixed on a solid support base through which reaction gases are passed and from which products issue. A homogeneous catalyst is a chemical compound dissolved in a liquid into which gases are passed, resulting in products issuing from the liquid mass and that must be separated out.

TABLE 5-4. Ethylene Glycol-Ethylene Oxide: Producers and Year-End Capacities (Thousand Metric Tons)

Country	No of Companies	1995	
		EG	EO
Americas			
United States	13 companies	3337	4230
All others		1399	1191
Europe			
Germany	5 companies	276	785
United Kingdom		sold	240
All others		675	950
Asia/Pacific			
Japan	7 companies	685	888
Korea	3 companies	405	360
Saudi Arabia		1140	855
All others		363	300
World total		8280	9799

Source: SRI International.

exploring many other pathways, and we proved correct. We encountered competition from Shell soon thereafter, and this fierce competition drove both companies for many years to strive for greater efficiency and economy, as they still do.

The ethylene glycol part of the Terylene polymer hence was developed without any stimulation from the invention of this fiber and by a small, outsider group working solely within commercial parameters, building on an invention by the Frenchman Lefort who also had stimulated Union Carbide. Table 5-4 gives recent capacity data for both ethylene oxide and ethylene glycol, which is made from the oxide.

Terephthalic Acid. The story is quite different for the terephthalic acid part of the molecule. This was made on a small scale by ICI and by DuPont, the exclusive licensee in the United States under their long-term agreement, by the brute force method of oxidizing para-xylene with concentrated nitric acid. This reaction was messy, dangerous, and corrosive, with many byproducts, effluents, and pollutants. The yields were poor. Furthermore, *para*-xylene was not yet an article of commerce. *Para*-xylene is:

$$
\begin{array}{c}
CH_3 \\
| \\
C \\
HC \diagup \ \diagdown CH \\
\| \qquad | \\
HC \diagdown \ \diagup CH \\
HC \\
| \\
CH_3
\end{array}
$$

This is a benzene ring with two methyl groups situated on opposite carbon atoms in the benzene ring. In this respect it is, of course, evident that if the groups at the ends can be converted to acid groups we would have terephthalic acid. But first, where could *para*-xylene be found?

The answer was found in petroleum fractions made for the aviation gasoline demands of the war in about 1938 by reforming processes that converted the normal paraffin hydorcarbons of typical petroleum fractions to aromatics with much higher octane numbers. This gave Americam airplane designers greater scope for combat performance and, indeed, gave the allies a great advantage in air power, which proved decisive, as Teltschik (1992) has described. The first breakthrough in this sequence was the hydroforming process of Standard Oil Co. (Indiana), based on the I.G. Farben patent rights it had acquired with Exxon and others before the war, through the efforts of Frank A. Howard, first head of Esso Research and Engineering Company. Hydroforming involved treating a heavy gasoline fraction with hydrogen in the presence of a base-metal (alkaline) catalyst. Instead of cracking the gasoline, as would happen in conventional thermal reforming, the hydrogen treatment converted the mostly parafinic (straight-carbon-chain) hydrocarbons into the mostly branched chain compounds of higher octane number. This process became an important contribution to the Allied war effort. Parenthetically, my first assignment at the M. W. Kellogg Co. in 1941 was in the process design of hydroforming units.

Because of the government encouragement of wartime interchanges of patent rights among oil companies, one of the participants became the Universal Oil Products Co. (UOP). This was an engineering and research organization for the oil industry founded in 1914 with the original Dubbs thermal-cracking process for gasoline production. UOP became, in the subsequent years, a primary source of technology and innovation for the independent oil refiners of the United States aand later for those abroad. Kellogg, by contrast, was largely associated in this period with the major oil companies, who frequently put pressure on the independents.[2]

As soon as the war ended, UOP began independent research on improvement of the reforming process for the expected postwar boom in automobile production. The major refineries combined catalytic cracking units with thermal reforming, which permitted them to achieve satisfactory octane levels. Smaller independent refiners, however, could not afford large catalytic crackers and, therefore, could not produce a satisfactory octane level with thermal cracking and thermal reforming. This led UOP into postulating that some improved version of hydroforming might alleviate the waning competitiveness of the independent refiners.

The young MIT chemical engineer Vladimir Haensel was put in charge of this research problem. The search was complex and often involved hair-raising risks by UOP and Haensel to demonstrate a laboratory process on a commercial scale, in a race against time. The process chosen involved small quantities of the very expensive noble metal platinum on an inert support. The first patents were filed late in 1947. When low-octane gasoline was passed over this catalyst, more that 90 percent of the feedstock was reformed to an octane rating in the 80s, whereas thermal reforming typically produced only a 70 percent yield, with octane numbers in the 70s.

This process was dubbed "Platforming™" and became an outstanding commercial success. It proved to be a key for the large-scale production of aromatic chemicals, including benzene, toluene, xylenes, and the like, which in turn were basic feedstocks for

[2]The whole UOP history is itself almost a story of the evolution of the oil-refining industry from the first days of Henry Ford until the post-War II era and is told in Remsberg and Higden (1994)

the burgeoning petrochemical industry. Haensel, now a recognized catalyst expert, ultimately received the Perkin Medal and the National Medal of Science as well as the Draper award. UOP then added the Udex™ process (developed with the Dow Chemical Co.) to extract the desired aromatic chemicals from the platformer effluent. UOP said that the Platforming and Udex™ processes "enabled one refinery, for example, to produce more benzene, toluenes, and xylenes than the entire coal tar industry had produced during World War II" (Remsberg and Higden 1994). These aromatic feedstocks for fibers, plastics, and rubbers probably created the basis for further products for *one-half* of the postwar petrochemical industry production. The petrochemical industry currently produces about 90 billion pounds per year of these basic aromatic chemical feedstocks. Shell Oil Co. led the major refiners in adopting this Platforming™ process, and an early licensee was also the Standard Oil Co. (California)—now known as Chevron. This is significant, because Chevron became the first producer of the key raw material for terephthalic acid—that is, *para*-xylene.

The xylenes produced by the Platforming™ process were mixed xylenes; that is, the methyl groups were primarily on two adjacent carbon atoms (*ortho*-xylene), or at positions once removed (*meta*-xylene), with a smaller amount, perhaps 20%, constituting *para*-xylene. But separating these was not easy, as the three xylenes had similar properties, and only *ortho*-xylene (useful for phthalic anhydride manufacture in recent years) could be distilled out. Fractional crystallization was used, and Chevron in the early 1950s was the sole producer in the United States. Needless to say, the process was

TABLE 5-5. World Capacities: 1996 (000 MT)

Country	Number of Companies	p-Xylene	Terephthalic Acid (TPA)	Polypropylene	Polyester Bottle Resins
Total—Africa		36		337	12
Japan	39	2255	1710	2604	173
Korea, South	22	1705	2240	1917	224
Taiwan	12	420	2440	450	102
All others		1996	3406	3391	196
Total—Asia		6376	9796	8362	695
Total—Eastern Europe		852	45	1295	30
Total—Middle East		232	70	600	120
United States	38	3579	2609	5581	1441
All others		608	570	574	468
Total—North America		4187	3179	6155	1909
Total—Australia				320	
Total—South America		218	130	1054	203
Germany	12	340	82	1099	100
United Kingdom	5		550	325	235
All others		105	770	4965	665
Total—Western Europe		445	1402	6389	1000
Total—World		12346	14622	24512	3969

Source: SRI International.

expensive. DuPont obtained a lock on the output for the purpose of manufacturing its version of polyethylene terephthalate, which it called Dacron. This situation soon changed, as many more companies entered the business (see Table 5-5). Here, another UOP process called "Parex™" proved an important contribution. It involved selective adsorption, using solid Zeolite adsorbents invented by Union Carbide, in a continuous liquid-phase operation to separate high-purity *para*-xylene from the other xylene isomers (*ortho*-xylene could then be separated by distillation from the *meta*-xylene). This process is now the key to production of the raw material for polyester fibers, and UOP has licensed 54 plants in 20 countries since the introduction of the Parex™ process in 1971.[3]

I perhaps have devoted excessive space to this phase of the polyester story because it also illustrates the extraordinary role American specialized engineering firms have played in the rapid build up of the world's petrochemical industry.[4] A comparable role was played by other such firms as Kellogg, the Lummus Company, Foster-Wheeler, and Stone & Webster in the development of the olefin side of the petrochemical industry— the crackers and separators that produce ethylene, propylene, butanes and butylenes, and others. These chemical-engineering triumphs also include close collaboration between chemists and chemical engineers. Haensel once said (in Remsberg and Higden 1994): "UOP was successful because we reduced the amount of time it took to go from the laboratory to the marketplace. You could do that as long as you had cooperation between the chemists and the chemical engineers." I have said much the same in my own papers (Landau 1958; Landau and Brown 1965).

Another point, of course, is that the oil, chemical, and automobile companies were closely related and supported each other in the postwar boom that followed. With so much new technology creating so many new products, the question economists must puzzle over is which came first, the creation of new innovation on the supply side, or the demand created by a growing population and world recovery? This question echoes through the new information age—after all, no one really demanded a computer for the home and business before it was invented. Barry Eichengreen (this volume) explores these issues from the macroeconomic point of view and finds that one of the most important effects is on the demand side of the economy. As national GDP and demand rises, the chemical industry will also grow. This correlation is especially notable in the post-World War II era (when the story of this chapter applies).[5]

This demand, in itself, is probably not surprising, but it does not explore further the consequences of the spurt in technological innovation that occurred after the war, when the industry grew at a rate at least twice as great as that of the GDP in the first decades

---

[3] Further developments by others have included processes to isomerize the unwanted large quantities of other isomers back to an equilibrium mixture, from which more p-xylene could be extracted. Later, processes were even developed for isomerization and disproportionating reactions to convert toluene to xylene. So great has been the demand for p-xylene that there are not always enough xylene fractions occurring in refineries. These details, however, are of peripheral interest here.

[4] The role of specialized engineering firms is also described in the chapters by Arora and Gambardella and by Arora and Rosenberg in this book.

[5] Eichengreen's appendix applies a Granger causality test to this era and finds that the evidence points to GDP growth leading the chemical industry, rather than the chemical industry leading GDP growth. (It should be noted here that, for chemical industries that depend on substantial exports, as is the case for Germany, the demand needs to be measured to include the foreign demand. Data for this additional complication are not readily available.) The appendix to this chapter analyzes Granger causality from a microeconomic-feedback point of view.

after the war, and at a rate faster than it has since then, not only in the United States but elsewhere, which suggests an important supply-side component. This relative rise in the growth of the chemical industry must have reinforced the growth in overall GDP by a feedback process. Since most of the products of the industry go to other industries, it is probable that the feedbacks are quite complex loops, as in Kline and Rosenberg's (1986) chain-link model of innovation. (See also the chapter by Arora and Rosenberg on this multiplier effect.)

Thus, the higher relative growth of this industry contributes to the GDP rate of growth itself, to improvement in export sales of the country, and to the creation of new consumer demand that would not exist had not the new products appeared. It creates demand in the products of other industries that use chemicals and stimulates innovation in other industries, while it is, in turn, stimulated by the developments in other industries (such as in oil or in equipment manufacture) to greater innovation and growth rates itself. Since, as shown earlier in this volume, this industry is only 1.9 percent of value added in the total economy, such feedback effects on the national GDP are necessarily small in the long run, in relation to all other factors affecting demand, but they may be significant in the shorter run.

Partially because of our contacts with ICI, and partially because of another episode altogether, we became interested in terephthalic acid manufacture and in finding a better and cheaper process for making it. The story is told in Landau and Saffer (1968), but a brief summary is appropriate here.

My company actually encountered the problems of making this then-exotic compound in 1952, before meeting ICI, and before we had any idea of the future of polyester fiber. At that time we networked widely, looking for market needs that we thought our team of chemists and chemical engineers could address. We also looked abroad, as we were regarded with suspicion as "nouveau" by the established, large chemical companies. However, Hercules, an American company of medium size, asked whether we could help them in recovering aluminum chloride from an alkylation process they proposed to use for an initially undisclosed process. We soon learned they had been working on a way to synthesize a potential precursor for terephthalic acid. They were looking for non *para*-xylene starting materials, as they had discovered that DuPont had tied up Chevron's supply. Furthermore, we knew they had acquired an exclusive license from the German company Witten, which had invented a four-step oxidation process to eliminate the need for nitric acid (it subsequently turned out that a patent by Chevron actually preceded it). However, lacking *para*-xylene, this process was not commercially feasible for Hercules. This stimulated me to think of other possible *para*-substituted precursors for this acid, and indeed, possibly, for their process of manufacture.

We soon focused on a compound we thought we could synthesize, not the dimethyl benzene (xylene) but the isopropyl equivalent, *para*-diisopropylbenzene, synthesized from benzene and propylene. We had commercial experience with a similar process, namely the production of cumene from benzene and propylene to make phenol, in which only one isopropyl group was substituted instead of two. In fact, we had contracts for cumene plants with Allied Chemical and British Hydrocarbon Chemicals (a joint venture of British Petroleum and the Distillers Co.). The process, an alkylation in the presence of aluminum chloride, was likely, in our view, to permit a more intense alkylation, which would make the *para*- and *meta*-substituted diisopropylbenzenes (*ortho* was sterically hindered from forming). The problem would be to separate the two and recycle the unwanted *meta* products back to the reactor, where it would be reconverted to the mixed products and so on to extinction. We were sure this would work, but the literature was very misleading regarding the physical characteristics of these hitherto obscure compounds.

We persevered and proved, within a few months, that we could separate the compounds by relatively simple but very intensive distillation. With our limited resources, we had to build the largest distillation column in Long Island at the time, which while not expensive by today's standards, still was expensive for us. We had a process! With enthusiasm, we reported this to our prospective client, carefully avoiding mention of which compound we were going to propose until we could see a contract possibility. Alas, this opportunity vanished immediately: they had found a source of *para*-xylene, and wanted to proceed with their Witten process development. So having a possible cheap *para*-feedstock (the *para*-xylene was very expensive in those days), we decided to see if air oxidation of the *para*-diisopropylbenzene would be any easier than the *para*-xylene oxidation. This proved to be the case, and we had a terephthalic acid process not using nitric acid, with air as the oxidant, and only a single step oxidation with decent yields. The raw materials were cheap. We were excited and looked immediately for a possible client, whom we found abroad.

About this time, the U.S. Patent Office declared a patent interference (a conflict on the date of first invention) between us and Shell (again!) on our oxidation process. We eventually won, but in the meantime we studied the process much more carefully, including many variations and combinations of catalysts. In so doing, we discovered something truly sensational—a process for the oxidation of *para*-xylene to terephthalic acid in one step with very high yields! We knew, as I said once, that we had a tiger by the tail. The *para*-diisopropylbenzene process was abandoned as too expensive by comparison, and we focused on how best to exploit rapidly this truly major break-through. This episode also illustrates the unpredictability of technological change, as discussed by Rosenberg (1996).

Our first effort was, of course, at DuPont, to whom we must have seemed an extraordinarily unlikely bunch of revolutionaries! What they wanted was a secrecy agreement, and this we were reluctant to do, as we were still filing patents on many other potential applications of our bromine-assisted catalyst and knew DuPont could invent around us by sheer size alone. There were probably other internal reasons at DuPont for this rebuff. The terephthalic acid was made by their explosives department, because the nitric acid was a traditional product of that department. It was a very profitable product for them. If our process had been acquired, a large internal market for such a profitable product would have disappeared, and some division heads would have looked bad. This issue of transfer pricing within the firm has vexed large chemical companies for a long time, and only a vigilant top management can override divisional managers where that is necessary for looking after the interests of the company as a whole. DuPont was perhaps the first chemical company to develop the modern system of divisional structure, which it later transferred to General Motors. But as it and other companies have wrestled with these internal conflicts of interest, there is a growing recognition of the advantages of having independent technological suppliers. On the whole, the perils are now much better understood, and a frequent test is whether a more favorable outside supplier exists for at least part of the requirements, a form of limited competition that tends to keep the internal supplier honest.

Another factor in DuPont's rebuff to us may have been the feeling that the acetic acid solvent we used, which contained bromide ions, could not be handled at reaction temperatures in any economically viable materials. Ironically, that situation was changed by the explosives department's own innovation—the explosive cladding of thin plates of titanium onto steel, thus offering a cheaper corrosion-resistance possibility. It turned out that the subsequent order for titanium-clad steel for the first terephthalic acid plant using our process was one of the largest that DuPont received for this new material of construction. In due course, this innovation helped DuPont itself abandon the nitric acid technology.

This episode led us to the Standard Oil Company of Indiana (Amoco), which had decided to go into aromatic chemicals, recognizing that the aliphatic chemical path utilizing ethylene and propylene was already well addressed by other companies. When they heard of what we could do with the xylenes, they were very interested, but we wanted a joint venture. During the early negotiations, their vice president expressed the view that this would be like the "mating of an elephant and a mouse." Unfortunately, they were right, so we settled by selling them the domestic rights exclusively but retaining licensing rights abroad. At that time, Amoco was a largely domestic company and, like so many others, did not have a feel for future potentialities abroad.

It did not take us long, experienced as we were in foreign markets, to license the process to Mitsui Petrochemical Co., who actually built the first commercial plant. They also bought a license on our ethylene oxide process. This was only one of many occasions in which the Japanese chemical industry bought immensely valuable western know-how to build up its industry (as the separate chapter by Hikino et al. in this volume narrates).

Amoco, of course, built a plant, although by a somewhat circuitous route, since they did not have pure *para*-xylene as yet. But during this period from 1954 to 1957, we were engaged in continuing discussions with ICI, the owner of the fiber rights outside the United States. In 1955, however, ICI's management broke off negotiations and convinced themselves they had found a process of their own. After all, when we told them it was possible to oxidize *para*-xylene with air in one step, half of the problem was solved. They knew it could be done!

It was only in early 1957 that ICI saw our first published patent in Australia and realized that we had beaten them to the same invention, but only by a matter of weeks! They now eagerly sought a license, and with Amoco's approval, we did sign a contract. We also found other licensees abroad, in France, Japan, and England. Indeed, as the years have gone by, ICI became the second-largest producer in the world of terephthalic acid, after Amoco, and Mitsui is right up there. Both played a significant role in the improvement and evolution of this process. As a result, they became fierce competitors of Amoco, who is gradually losing its dominance in this field. ICI sold its terephthalic and polyester businesses to DuPont in 1997.

The story does not end here, however. Our terephthalic acid was over 98 percent pure, with a light tan appearance. But fiber manufacture requires a product 99.7 percent pure or better. This purity could only be attained by converting the acid to its dimethylester by reaction with methanol, which could be purified. Terephthalic acid is an extraordinarily high-melting solid, very difficult to purify further by any known means. This esterification, of course, added to the cost and brought the price of the final product into the range of that made by Hercules using the more-complicated Witten process. Nevertheless, production by our process increased steadily in a number of countries.

Here, the contribution of my company ceases and Amoco's begins. Its commercialization efforts were expensive and time consuming, but its first plant started up in late 1958! Such progress is almost impossible today, with the numerous regulatory and capital-cost obstacles all companies encounter. There were severe corrosion problems and difficulties in handling the solid terephthalic acid product that precipitated from the reaction mixture. Furthermore, the reactor design was initially a batch one, and conversion to continuous operation of this novel reaction system took time and money.

Fortified by this commercial success, Amoco then tackled the most pressing commercial potentiality—obtaining fiber-grade, pure terephthalic acid directly, so as to obviate the necessity for the esterification step. This was indeed a technical challenge, which was solved by Dr. Delbert H. Meyer and his colleagues, who produced a product known as PTA (fiber-grade, pure terephthalic acid). To gain market share and induce the usually cautious fiber producers to substitute PTA for the dimethylester, Amoco priced its product at the same level as the methyl ester, which gave a 17 percent price advantage to the product. Also, by 1961, the basic polyester fiber patents of Whinfield and Dickson were about to expire, and more potential producers here and abroad desired to compete with DuPont and to differentiate their product using this price advantage. Furthermore, ICI had felt it necessary, to license its Terylene process to others, as explained below.

TABLE 5-6. World Production and Consumption of PTA and DMT (000 MT).

	Production					
	PTA			DMT		
	1975	1985	1995	1975	1985	1995
North America	400	1213	2475	1260	1525	1060
Latin America	—	332	604	35	255	488
W. Europe	235	740	1175	690	984	873
E. Europe	40	47	2	275	535	586
M. East/Africa	—	—	68	—	64	170
S./S.E. Asia	—	—	770	15	56	207
East Asia	—	783	4690	20	144	315
Japan	75	854	1600	590	328	360
World total	750	3969	11384	2825	3891	4059

	Consumption					
	PTA			DMT		
	1975	1985	1995	1975	1985	1995
North America	410	720	2093	1200	1496	1129
Latin America	45	249	500	135	168	247
W. Europe	160	380	1036	640	821	801
E. Europe	20	62	14	300	597	587
M. East/Africa	10	94	209	20	133	249
S./S.E. Asia	neg.	276	1635	55	106	300
East Asia	55	1400	4792	105	212	367
Japan	65	650	927	470	340	349
World total	765	3831	11206	2925	3873	4029

Source: Tecnon (U.K.) Ltd.

Hoechst used the Witten process at that time and still does to a large extent, although they purchase PTA as well.

DuPont finally recognized the situation and negotiated a license for the United States with Amoco, from which my company received nothing, unfortunately. Eventually, DuPont moved to using PTA as well.

At the award ceremony at which Delbert Meyer accepted the Perkin Medal in 1995, James E. Fligg, Amoco's then chief chemical executive, gave the following statistics: In 1970, Amoco's capacity for PTA was only 500 million pounds, while the global capacity for dimethyl terephthalate (DMT) was about 7 billion pounds. By 1990, the annual PTA capacity had reached 17 billion pounds, while that of the DMT was about 10 billion pounds. By 1995, however, almost all of global growth for polyester was based on PTA, and the DMT production had shrunk. It is expected that PTA will reach a capacity of about 40 billion pounds by 2000. Amoco now accounts for over 40 percent of world capacity, and it is a bedrock business of Amoco Chemicals today. In fact,

TABLE 5-7. World Production and Consumption of *Para*-Xylene and Polyester Fiber (000 MT)

	Para-Xylene			
	Production		Consumption	
	1975	1985	1975	1985
North America	1065	1962	1110	1616
Latin America	25	241	25	362
W. Europe	610	988	650	1089
E. Europe	220	494	215	366
M. East/Africa	—	18	—	38
S./S.E. Asia	10	22	10	38
East Asia	20	435	15	602
Japan	440	705	460	759
World total	2390	4865	2485	4870

	Polyester Fiber	
	Production and Consumption	
	1975	1985
North America	1382	1586
Latin America	183	384
W. Europe	660	843
E. Europe	290	610
M. East/Africa	45	248
S./S.E. Asia	50	438
East Asia	170	438
Japan	440	645
World total	3220	6429

Source: Tecnon (U.K.) Ltd.

Amoco is the only major *p*-xylene producer forward-integrated into PTA, giving it the lowest cost position in the major polyester-film and bottle-polymer intermediates. Amoco has taken more than 20 years to position itself in the major product lines of polyester fiber and polypropylene (the two case studies in this chapter), but it is just the right commercial position to continue for many years as a significant contributor to solid corporate earnings. Meyer, in his Perkin Medal address, gives a further account of this history. Table 5-6 gives current and past production and consumption data for PTA and DMT; Table 5-7 gives data for *para*-xylene and polyester fiber production and consumption.

Polyester Fibers And Other Products

As mentioned before, when ICI acquired the rights to Terylene outside the United States, it set up a smaller (compared to that of DuPont) developmental group at Harrogate, in Yorkshire, which it called the Terylene Council (not yet a division of the company). There was also additional research work going on in the plastics and dyes divisions. The petrochemical era was beginning in the United Kingdom, and there were large capital demands for construction of the first such complexes, coupled with demands for many other existing products such as ammonia and alkalis, as well as new research-intensive products such as pharmaceuticals, crop products, plastics, and films among others, all of which product fields ICI was eager to enter. In fact, ICI entered earlier than DuPont and Hoechst in some cases, which made it difficult for ICI to commit all of the large resources needed to further the development of this unique and valuable property. Furthermore, ICI made the courageous decision to enter this fiber field alone, while the know-how from its previous nylon experience with British Nylon Spinners, a 50:50 joint venture with Courtauld's, remained locked up inside the joint venture. As a result, ICI undertook the development of the sophisticated machinery to make both filament and staple fiber without any previous experience and succeeded in a remarkably short period of time. Then, ICI had to decide whether to seek an exclusive position in its territory (Europe and the Commonwealth) or to earn royalties and gain technical feedback from a few selected licensers. Furthermore, customers in other nations were also clamoring for this exciting new product and could not be expected to wait. Additionally, there were supply problems for the raw materials and a product which needed a lot of work before it could become accepted by the essentially conservative British consumer.

Meanwhile, DuPont, under its exclusive license from ICI, also proceeded slowly at first, but the much larger availability of petrochemical raw materials and intermediates in the United States permitted the company to introduce Dacron (its trade name for polyester fiber) at a more rapid rate than was possible for ICI. DuPont also had more experience in fiber development, from its work on nylon and earlier cellulose fibers. Furthermore, the successful antitrust suit by the U.S. government to dissolve the traditional ICI-DuPont technical exchange agreement made it more difficult for the two pioneers to help one another.

It is, therefore, not surprising that ICI decided to issue some licenses, not only to gain the advantages cited above but also to gain some revenue before the basic patents would expire. Because of the long gestation time needed to develop such a radically new product in postwar Britain, capital constraints alone made it clear that a dominant position in Europe could not be built up by the time the patents expired, and so ICI, as mentioned above, licensed Hoechst and Glanzstoff in Germany, Rhône-Poulenc in

France, and Toyo Rayon in Japan. However, ICI entered into a successful joint venture in the United States with the Celanese Corporation and competed successfully with DuPont. (Celanese later acquited ICI's half and then merged with Hoechst, now the largest producer in the world.) In the end, these events proved fatal to ICI's own position in polyester fiber, as will be shown later on. Spitz (1988) quotes ICI's former chairman, Sir John Harvey-Jones, as saying "we certainly gave more licenses than were wise, resulting in overcapacity, from which it took years for ICI to recover. Yet there was no way ICI could have kept an invention like that to itself.... We couldn't have commanded the capital. The risk would have been enormous." Perhaps it could have proceeded by more joint ventures abroad, but again, capital requirements would have been large.

This statement, however, conceals a major change in the composition of ICI's main board. The rise of petrochemicals brought into the board people from that area, and the more research-oriented managers of the immediate postwar era were phasing out. The petrochemical management, like their predecessors in alkalis and ammonia, thought in terms of commodities and failed to see clearly that a new fiber required large resources of research and development, as well as heavy marketing and customer service. In their defense, one might say that Britain was undergoing severe economic problems, so that their customers in the United Kingdom were suffering, and manufacturing moves abroad were necessarily lagging behind this reality.

Thus, polyester fiber came to be treated as a commodity. ICI's competitors thought otherwise and came to dominate the field. However, vast technical problems had to be overcome before anyone could make a consistent profit. The early product was not very attractive to the consumer. It pilled, and dyeing was very limited. The product seemed too hot in summer and too cold in winter, primarily because, unlike cotton and wool, it did not absorb body moisture. It was soon established that medium- and high-tenacity filament yarn should be produced along with staple fiber of various lengths and fiber color to suit the kinds of spinning machinery existing in the textile industry (Nunn 1979). Staple fibers are drawn to give medium tenacity, but the pilling can be improved by using a lower average molecular weight, although abrasion resistance is somewhat reduced. The range of products made depended on the facilities and practices of the individual textile manufacturer.

The fibers (both filament and staple) are usually smooth in appearance and circular in cross section, although some other shapes can be made. The staple fibers are usually crimped, and a variety of filament yarns have been texturized in one manner or another. Thus, the long experience of Britain's textile fiber industry could be used to aid in adapting this new product to the existing technologies for spinning, weaving, and finishing textiles. The process of adaptation was very costly, and it was difficult to establish a strong, profitable position in any short period of time. But over the years, this process of adaptation has been perfected by a stream of continuous incremental improvements, made by both the textile industry and the fiber manufacturers. It is probably impossible to arrive at any rational estimate of the total cost of reaching today's superior product and dominant market position. Several examples based on DuPont's progress may illustrate this evolutionary process:

1. Unlike nylon, which when substituted for silk in women's hose signaled a clear market opportunity, Dacron was a much more complicated and expensive product to develop. Where was the market? And might it not impinge unfavorably on the markets for nylon and orlon (polyacrylic fiber)? Initially, then, Dacron was seen as a wool

competitor because of its resilience (the ability of the fiber to recover from deformation). While Dacron did penetrate the wool market to some extent, this did not prove to be its ultimate large-scale use. To avoid excessive competition with its other fibers, DuPont decided to focus on tire cords made of polyester fiber as their first large-scale application. The attempt was a failure, and indeed, it took 15 more years to find out how to make successful tire cords. But once the problem was solved, polyester fiber became the tire cord of choice.

2. Hounshell and Smith (1988) describe some of the subsequent extensive and expensive research efforts that DuPont made to move Dacron into the leading position among synthetic fibers. The major use that developed was in cotton-dacron blends. Aside from the desirable properties that this blend possessed (moisture transparency, washability and shape retention, dyeability, and so on) the blend also conferred a tremendous increase on the productivity of textile machinery, because addition of the strong polyester thread strengthened the cotton fibers, which otherwise tore much more easily. As a result of this dramatic increase in productivity, yarn that previously could be made at perhaps 80 meters a minute, could now be made at 1000 meters a minute. Instead of making sheets that are two feet wide, producers are running 15-feet-wide sheets at five times the speed. This speed is the factor that has made these materials so inexpensive. The synergy between these two products is truly astronomical. The textile companies reengineered the fiber to have exactly the same stress-staying properties as cotton, so that when the cotton and fiber are mechanically mixed together, they don't separate. Therefore, the elongation plus the strength of the fiber is matched to that of the cotton so that they can blend exactly and perform as a homogeneous mixture. Some of DuPont's textile customers contributed significantly to these developments.

3. For a long time, polyester fiber was considered unsuitable for the demanding requirements of sports clothing. Because athletes perspire, their clothes become heavier, which increases fatigue and diminishes performance. Natural fibers absorb water and thus put the burden of excess weight on the athlete. But engineering and technology have allowed the development of quick-drying, breathable synthetic-fiber clothing that permits rapid evaporation of perspiration and maintains a cool, dry comfort level (Aneja 1994). Polyester so-engineered is now widely used in such sports as baseball and tennis. As Aneja points out, comfort means many things: clothing that does not constrict movement, fabric that is soft to the touch, clothing that doesn't become clammy from perspiration or that can protect the wearer against cold and rain. Fibers have to be constructed to achieve these different goals, a process that requires careful study of human physiology, of heat and moisture comfort, of the diversity of the apparel market, and of the manufacturing process for the fiber. The result of all these efforts has been segmentation of sports apparel into categories, such as clothing for cycling, mountain biking, climbing, and hiking, among others, as well as for the better-known markets mentioned above. Aneja gives a detailed account of how this engineering has progressed at DuPont. He also cites potentially new applications such as in fibers with antimicrobial material incorporated into the fiber itself during polymerization, making it useful for hospital and medical facilities and for applications in which mildew or infection can occur.

4. The creation of microfibers from polyester permits the production of a finer product than the silk standard. These fibers constitute the most important fiber breakthrough in a long time. The process is adaptable to both nylon and polyester and can produce garments, such as sports shirts, with the feel and drape of silk but the

durability, suppleness, breathability, comfort, shape-retention, washability, and crease-resistance of the polyester fiber. Of course, these garments are still expensive and so go into the higher value-added end of the apparel market, but their application will spread.

5. For many years, especially as the environmental movement strengthened, there was strong consumer resistance to the use of polyester clothing, especially the double-knits, resulting in snobbish disdain for people too lacking in taste or income to afford the all-natural clothing favored by the fashion designers. As the quality and variety of synthetic fibers has steadily increased, there has come about a gradual reversal of this attitude on the part of the fashion industry. As one fashion writer put it (Menkes 1995): "O brave new year that has such fabrics in it! For the hippie generation, despising artificial fabrics came naturally, but technology is now hip. So, after 20 years of revering natural materials, designers have finally taken the sin out of synthetics, embracing the new materials that are sleek, shiny, techno-chic and supercool." Thus, innovative fabrics, rather than new silhouettes, have attracted designers and fashion houses. This transformation did not come cheaply, but it is here to stay. Of course, in many parts of the world, especially in Asia, polyester is esteemed as a blend with cotton for its excellent crease-resistance and washability, and its relatively low cost (cotton is often more expensive in many parts of the world). Fashion is not the criterion here, but functionality.

6. Another major series of improvements involves the dyeing of polyester fibers and garments made from them. Because of the inertness and chemical stability of polyester, it was initially very difficult to prepare colored fabrics that could even remotely match the brilliance of the natural fibers or of rayon or acetate. Gradually, a class of dyes known as *disperse dyes* appeared on the market, developed by dye manufacturers such as Bayer, Hoechst, DuPont, and ICI. Furthermore, when the polymerization process is modified by adding small amounts of copolymers, the fiber gains enhanced affinity for some dyes. The dyes must be light-fast, stable, and washable; capable of rapid dyeing; and come in a wide variety of colors. The disperse dyeing process is described by Broadhurst (1979) and Burkinshaw (1995), but research and new product developments have continued since those publications. To see how complex this whole process is, it is only necessary to visit a textile plant and observe the many carefully controlled steps that lead to the finished product.

7. Polyester fibers have found a substantial market in carpeting, tire cords, and transparent films such as Mylar, as well as in ICI's Melinex (developed at the same time and competitive with DuPont's Mylar) and other industrial applications.

Many lessons could be drawn from this singularly important commercialization of the postwar period. Breakthroughs often came from outsiders, not from main players. The market is the best guide for innovation. No government direction could have produced this outstanding result. The market for chemicals and chemical technology has been international for a very long time, and this industry has pioneered in internationalization. Diffusion of technology is the key to wealth creation at the aggregate world-economy level, but from the standpoint of the innovator, monopoly or oligopoly is essential to pay for the high cost of development by the private sector.

A final lesson is that when it comes to large-scale commercialization, only big companies can play. I wanted a joint venture but could not get it (we succeeded later in our propylene oxide development, but that is another story). Big companies alone have the money to pay for extensive testing and the risk of commercialization, and they therefore have every incentive to create continuing incremental improvements so as to gain a competitive position. Unlike the basic ideas, which can be protected for a limited time by patents, these incremental improvements depend on confidential know-how. It is by such manufacturing and marketing skills that large companies retain their competitive edge.

The Development of the Polyester Polymer as a Plastic

While polyester-fiber applications are the principal outlets for the polymer of PTA and ethylene glycol, in recent years another large application has arisen, that of polyethylene terephthalate (PET) as a bottle-grade resin, familiar to all as the container of choice for large bottles of carbonated soft drinks, where it has replaced glass almost completely. Its impermeability and resistance to the constituents of the beverage make PET safe and convenient to use. Amoco was among the pioneers in this application, as might be expected, as was ICI in Europe, but others have joined in, and production of this polymer is now at the level of over 3 billion pounds in 1994 in the United States, as stated earlier in this chapter. Table 5-5 (previously shown) gives the figures for the world as a whole. Demand for this product has grown rapidly as it continues to find new packaging applications, not only in bottles but in various types of films. However, global capacity has been increasing, which will probably depress prices.

Eastman Chemical and Hoechst Celanese now are the two largest North American PET producers. By 1998 Eastman will have a capacity for about 900,000 metric tons per year and Hoechst Celanese for about 725,000 metric tons per year. Total North American capacity is projected to grow from 1.1 million metric tons per year now to 2.5 million in 1998. Europe has been slower to convert to these applications, as has been Asia, and so there still may be ample room for growth in the long run.

The developments and improvements that have taken place over the years in these applications have involved many companies and individuals in many countries. It would be virtually impossible and unnecessary to trace the various streams of gradual innovation that have led to the present standing of this polymer as a fiber and a plastic and that have so affected both our economies and our degree of consumer satisfaction. For me, it is enormously gratifying to know that out of our small laboratory in the early 1950s came one of the key steps in a key innovation of the postwar era, known in its end products to virtually every consumer, and continuing to undergird the steady improvements that have been built on it. This development is one of the striking innovative sequences in the chemical industry. It is now so seemingly ordinary that it captures virtually no publicity.

We can also extract from this story some of the themes that occupy other sections of this volume. Chandler's thesis of the first mover's advantage in making the investments necessary to gain a dominant, or at least leading, position is clearly illustrated by Amoco with PTA and by DuPont with Dacron: In neither case were they the original inventors of the technology, but they had the willingness to invest patiently over many years. Amoco had the earnings of its oil and gas business to support its chemical arm during many years of losses. Now, the chemical business contributes a significant proportion of Amoco's overall profits and is a division of the parent company, where its performance is more visible to the shareholders and analysts. DuPont, after the war, had a profitability twice the chemical-industry average. It gained substantial sums from the divestiture of its General Motors holdings; and it had the success and experience of nylon to bolster its development of a new fiber, as in the case of polyester, which again took many years. Furthermore, for many years DuPont was a family-controlled company, which could take risks. Since then, the DuPonts have further strengthened their fiber business by the success of the Lycra Spandex fiber business and certain others such as Kevlar. Nevertheless, DuPont's CEO, John Krol, stated recently that the profitability of Dacron is not really satisfactory because of competition from companies with lower research and other overhead costs, and for other reasons. However, in July 1997, as mentioned above, DuPont purchased ICI's terephthalic acid and polyester position, among other activities, for $3 billion, thus assisting ICI as it purchased Unilever's specialty chemical business and completed its transformation away from a relatively loosely related commodity business. By this move, DuPont became a very strong producer of terephthalic acid (PTA), to bolster its fiber business. Ironically, ICI's process is still the one they purchased from Scientific Design in 1957 and have improved over the years. Having bought a license from Amoco years earlier, DuPont now can capitalize on Scientific Design's invention, while Scientific Design received no income from these events. (ICI's royalty payments to Scientific Design ceased many years ago.)

Hoechst in Germany is also a case of making the necessary investments. By acquiring the Celanese Co., with its strong market position, and using its polyester fiber license from ICI, Hoechst moved up in the race to become the largest polyester-fiber producer in the world, owing in large measure to the profitability of its other chemical businesses, its strong position in the European markets, and the financial-corporate governance situation in Germany described in this volume by Richards.

But what of ICI, the holder of the keys to this kingdom? It could not continue to make the necessary investments to obtain a leading position. ICI was late in moving into continental Europe and the United States. It also had been adversely affected by the loss of its Empire and by the creation of the postwar welfare state in Britain, which while not directly relevant to ICI's operations, nevertheless restricted the capital and customer markets. ICI simply did not have either the management or the financial resources to take the requisite long-term, patient investment strategy centered on a strong technological basis. Its R&D spending was always less than that of the other companies, as Marco da Rin shows in his chapter. The result is that this pioneer has gone out of the polyester-fiber business completely! This was a conscious decision by ICI. It focused its research and capital expenditures on pharmaceuticals, paints, plastics and film, and certain chemicals, which later were combined in a spin-off of Zeneca in 1993. ICI has sold its ethylene oxide and glycol facilities at Wilton in England to Union Carbide. Yet, as is so often the story of innovation in this complex industry, although it is true that technology can be bought, a strong commercial position must be fought for, with much capital expenditure.

The Principal Themes of this Story

Scientific Design (later Halcon) sold out long ago. The Witten process for DMT is obsolete. Newcomers have entered the business, and it is strongly competitive. The broader themes of this volume are clearly visible in this microeconomic story of how a technological- and innovation-based industry develops as a global powerhouse, and how growth by adding value is the strategy for raising a nation's, and the world's, standard of living.

Another theme that emerges from this story is that many companies and individuals, including textile and apparel companies, producers of the raw materials, and the developers of the processes underlying them, contributed to the ultimate establishment of a new synthetic fiber that has challenged the traditional natural cotton for supremacy. This history also displays various kinds of spillovers ("externalities," as economists call them), which mean there are benefits to others not captured by the producers of a particular product, who are deemed to have priced their product to reflect the more direct benefits to their customers.

But other spillovers cannot be contained within the pricing system. Scientific Design would never have entered the terephthalic acid development if Whinfield and Dickson had not invented polyester fiber, but Whinfield and Dickson in turn built on the work of Carothers at DuPont, and certainly the price of nylon could never include the benefit to society that resulted from the innovation of polyester fiber. Furthermore, polyester fiber could not have reached the stature it occupies today if the companies that pioneered petrochemicals (such as Union Carbide, Dow, Exxon, and Shell) had not pursued their own researches and if the great American specialized engineering firms (e.g., Kellogg, Lummus, Stone & Webster, Lummus, Braun, Fluor, and others) had not developed the petroleum cracker and distillation technologies that brought the price of the fiber into the range for cotton. If one digs further, the oil industry itself, which lay at the heart of the petrochemical conversion, could not have grown so large had it not been stimulated by the rise of the automobile, as Rosenberg's chapter in this volume describes in the history of the development of the chemical-engineering discipline. Chemical engineering also was an essential component of the polyester success story, and indeed, DuPont early succeeded in applying chemical engineering to its high-pressure, toxic, corrosive, and explosive operations in a very safe manner, which gave it a particular competitive advantage.

POLYPROPYLENE PLASTICS AND FIBERS

The second major postwar innovation I have chosen to address in this chapter deals with an even simpler molecule than those involved in polyester fibers and plastics. This molecule is propylene, a 3-carbon hydrocarbon with a structure as follows:

$$
\begin{array}{c}
CH_3 \\
| \\
HC{=}CH_2
\end{array}
$$

Again, this invention was not made by any of the major firms in the industry but rather by two outsiders, a recurrent theme in this industry. However, the full commercial development had to be made by large companies. A look at current production levels, shown in Table 5-8, helps one to visualize the present magnitude of this innovation.

There are a number of other producers of polypropylene in the United States, such as Huntsman, Aristech, Formosa Plastics, Phillips-Sumika (a joint venture of Phillips with Sumitomo Chemical), Quantum, Epsilon, Rexene, and more. Total U.S. capacity is about 13 billion pounds per year. The total capacity of the world is now estimated at 24 million tons per year (with an approximate market value of $20 billion per year) and will probably reach 28.3 million tons by the year 2000. Many more plants are under construction around the world. In fact, the scale of production for polypropylene has

TABLE 5-8. Production Capacity of Largest Producers of Polypropylene: Estimated Year End 1997 (Capacity—k Tonnes)

1	Montell	3560
2	SINOPEC—multiple small unit	1367
3	Targor (BASF/HOECHST)	1300
4	Amoco Chemical	1205
5	Fina Chemical	1115
6	Borealis	983
7	Exxon Chemical	938
8	Mitsubishi Chemical	880
9	Grand Polymer	847
10	Solvay Polymers	730
11	Huntsman Chemical	640
12	Sumitomo	602
	Estimated total world capacity	27,000

Source: Montell Polyolefins bv, The Netherlands.

already considerably exceeded that of polyester fiber and plastic, but in dollar terms they are not far apart. Whereas polyester fiber is the most important synthetic fiber, polypropylene has not yet quite overshadowed polyethylene in the United States, but it is growing more rapidly. Polypropylene has averaged a growth of over 10 percent per year for the past two decades, but this may slow down to about 6 percent per year (on a much larger base).

Both ethylene and propylene are produced from refinery operations or from natural gas and also from naphtha cracking (primarily used in Europe and Japan to create these key building blocks of the petrochemical industry). Naphtha is a petroleum fraction somewhat like kerosene in its boiling point.

How did this remarkable commercial success originate? The story is not as intimately involved with my own experiences, but I knew a number of the key players and had many contacts with them over the years, so that I have observed this development with keen interest. Polypropylene development is also an area in which catalysis plays a critical role, and its story illustrates the larger theme that catalysts lie at the heart of the modern petrochemical and petroleum refining industries. As mentioned before, the industrial design of a chemical-catalyst system represents the closest possible relationship between chemistry and chemical engineering and is an area in which American technology has played a leading role in many reactions. Nevertheless, polypropylene was discovered in Europe independently of the discovery in the United States; but the exact dates of discovery proved to be very controversial, resulting in a tremendous, nearly 30-year battle for patent rights in the United States. Discoveries often do take place almost simultaneously and independently when the soil is ripe. The paths to them may be very different, however.

The Origins of Polymer Chemistry

Polymers as a field of chemistry began with Hermann Staudinger in 1922. The German chemists disputed the exact nature of these products, however, and thus Carothers and his group at DuPont became the inventors of the first truly important polymer. The German Paul Schlack, an I.G. Farben chemist, independently had hit on a loophole in the Carothers patents and was able to develop a similar compound from caprolactam, which yielded a nylon called nylon 6 (as distinguished from DuPont's nylon 66, reflecting the fact, described above, that two 6-carbon-chain molecules were hooked together, whereas nylon 6 had only one 6-carbon chain-molecule, with an amine group on one end and an acid group on the other).

Not long after Staudinger, in 1933, ICI accidentally found that ethylene could be polymerized at very high pressure (30,000 pounds per square inch) to produce a useful plastic. This established the key fact that ethylene could be polymerized—a critical fact—which led to products that played a role during the war as cable insulators and for radar applications. This second major polymer[6] thus enticed many researchers after the war to look into further areas of polymer research, and many plastics, fibers, and rubbers were the outcome. Those developments served as the major forces for the rise of petrochemicals and the chemical industry more generally. It was in research toward more methods of making polymers that first, a new form of polyethylene and soon thereafter, polypropylene appeared.

The first public revelation of these new approaches came from Dr. Karl Ziegler of the Max Planck Institute for Coal Research in Mühlheim, Germany. The new discovery occurred in a very few days, only in late 1953: namely, that ethylene gas will polymerize rapidly with certain organometallic catalysts to form a polyethylene, although one with different characteristics than that made by ICI's high-pressure process. The high-pressure product tends to have a low density and a low melting point. The Ziegler product, with a higher density and a higher melting point, is more rigid, making it suitable for new applications. I well remember visiting Ziegler at his laboratory and seeing his dramatic demonstration of bubbling ethylene through a flask and obtaining a precipitate of polyethylene. Others who saw this demonstration were dazzled as well. Compared with the high-pressure process, Ziegler's discovery seemed to have enormous economic advantages. The difference seemed to Ziegler, at once, to be that the molecules of ethylene ($CH_2{=}CH_2$) were able to join together linearly, without interruption, whereas the high-pressure process, which was more random, resulted in a strongly branched polymer.

Ziegler's team, of course, soon studied many other organometallic catalysts and filed other patents. Many companies sought to obtain licenses. Ziegler had a very unusual deal with the Institute that employed him: He could set the terms of the license agreements and keep a large part of the resulting royalties for himself. As a result, he began signing agreements without having a commercial process, thus requiring his licensees to do their own development work. Accordingly, the technology diffused widely and rapidly, which undoubtedly spurred the ready adoption of this process by the petrochemical industry. This sequence of events also made Ziegler a very rich chemist indeed. I used to hear stories of the fine mansion he had built for himself along the Rhine, and that he called "Polyethylenaeum."

Of course, this invention did not just spring into his mind one day in November 1953. He had experimented with organometallic mixed catalysts for the preceding ten years. This work, in turn, had been stimulated by his early interests in the theory of free radicals, and in 1923 he had discovered a new method for synthesizing organic

---

[6] Polyvinyl chloride and polystyrene were also developed in the interwar years in the United States and Germany but grew rapidly thereafter. Buna rubber was developed first in Germany; then the United States worked out its own version in the synthetic rubber program of the war. See Howard (1947).

compounds of the metals sodium and potassium, which led him to the study of metal alkyls as a broad synthesis possibility. The new catalysts developed in 1953 grew out of his lifelong interest in this field. His was truly fundamental research, propelled only by his curiosity and financed by the Institute, which was in turn supported by the German government and by coal-mining companies. Once more, this support illustrates the traditional willingness of the German government to support basic research, regardless of who was in charge. Thus Ziegler came to Mühlheim under Hitler in 1943, with the world war raging, but the postwar Adenauer government continued this dedication to research. Only the name of the Institute had been changed, from Kaiser Wilhelm Institute to Max Planck Institute!

The scene now shifts to Italy and Professor Giulio Natta at the Milan Polytecnico. Although he and Ziegler had both been professors and researchers, of the two of them, Ziegler was more extroverted, hearty, sometimes even aggressive, while Natta was more scholarly, more shy. Natta had spent his career in a variety of systematic researches, which are well summarized by Carra, Pariso, Pasquen, and Pino (1982). In fact, during his career he published over 700 papers. One of his specialties was the study of molecular (particularly crystalline) structures by X-ray diffraction and electron diffraction. These techniques stood him in good stead when he began to work in the field that led to his discovery of crystalline polypropylene. Although Natta is described as being a chemical engineer, in those days (he graduated in 1924) his field was more applied chemistry. He also became an excellent applied chemist, in the mold of Fritz Haber with his ammonia process. Like Haber, Natta needed a large organization to develop the ideas he conceived in his laboratory.

First Steps Toward Commercialization

That organization proved to be Montecatini, then the largest of the Italian chemical companies. In 1947, Piero Giustiniani returned from disgrace, owing to his association with the Mussolini regime during the war, to head this company, which was floundering. In 1947, Natta and Giustiniani traveled for the first time to the United States to investigate the developments in petrochemicals that were not yet understood in Europe. I knew Giustiniani as well, having visited him also in 1947, in connection with other business. He was a forceful and dominant leader, and he soon moved Montecatini into this area, as well as others. His research and technology head was Dr. Bartolomeo Orsoni, who became a friend of mine, and who turned out to be the key figure in the relationships of Natta with Montecatini and Ziegler, as described later.

Thus, by the end of the 1940s, a close association had formed between Natta at the Milan Polytecnico and Montecatini. This relationship resembled, in many respects, the situation in Germany during the rise of dyestuffs chemistry—a close link between academia and industry. Of course, Montecatini had complete rights to any invention, unlike the position of Karl Ziegler at Mühlheim.

Ziegler and Natta crossed paths for the first time in 1952, when Natta attended a lecture in Frankfurt by Ziegler, who described his astonishing results in dimerizing olefins in the presence of a metalloorganic complex (alkyl aluminum compounds). Natta at once recognized the importance of Ziegler's research and realized that Ziegler had found a new principle for synthesizing polymer chains.

Natta was enthusiastic and proposed to Montecatini that they enter into a collaboration agreement with Ziegler. Arrangements were then made to send a few researchers to work with Ziegler's team. Natta's laboratory soon verified Ziegler's results. In early

1954, under the agreement, Ziegler (who seems to have been quite naive about patent matters) gave a copy of his patent application to Natta. By now, Natta's group was fully up to the technical level of Ziegler's team and started to test whether propylene could also be polymerized by such systems. By March 1954, the polymerization of propylene became the most important project to both Natta and Montecatini. It was soon crowned with success. Natta obtained a crystalline polypropylene with a high melting point, which he realized would have extensive commercial applications, unlike the case with polyethylene. In this, he was truly prescient. By May 1954 he had obtained a fiber from polypropylene as well. Natta's experience with crystal structure stood him in good stead, because he now recognized how to characterize the product, as well as define it. It took only three months from the time crude polypropylene was first synthesized until the first patent application covering all data about the stereospecific (isotactic) crystalline polypropylene. Isotactic polymers are linear, with the CH_2 groups positioned in the same direction; thus:

$$-CH_2-\overset{\overset{\textstyle CH_3}{|}}{CH}-CH_2-\overset{\overset{\textstyle CH_3}{|}}{CH}$$

Meanwhile, Ziegler got around to reexamining propylene polymerization by his own methods. Unfortunately for him, he filed his own patent shortly after Natta's and was thunderstruck to learn that he had been, as he thought, betrayed by a colleague to whom he had entrusted his own results in good faith. He expected that Natta and Montecatini would acknowledge their indebtedness to him. But of course, his initial patent filing did not cover polypropylene, and his later discovery came after Natta's filing. The cordial relations between the two shriveled immediately and remained distant until 1963, when they shared the Nobel prize for chemistry. This event was very unusual; before then, only Haber and Bosch had been Nobel laureates with a major industrial chemical success to their credit (excluding those in pharmaceuticals). The two chemists were partially reconciled by this occasion, and ultimately Montecatini agreed to share 30 percent of the royalties on polypropylene with Ziegler. The source of Ziegler's wealth is obvious! But both developments followed the typical European tradition of careful fundamental chemical research along many paths.

The contrast between the two success stories is evident. Ziegler sold patent licenses; Montecatini selectively sold full licenses (that is, including commercially useful know-how) to other companies and proceeded to develop their technology and commercial position.

U.S. Firms—Independent Path to Polymer Technology

A completely different story was unfolding in the United States. Before any of these accomplishments became generally known, and clearly before crystalline polypropylene had been discovered, several companies were working in the same area by very different routes. Standard Oil of Indiana (Amoco) had started work in this area in 1950 and Phillips Petroleum Co. in 1951. DuPont began in 1953, before Natta's discovery. Thus, Natta's disclosure only accelerated the American efforts, resulting in a race to the patent office. Here, the character of the then-prevailing U.S. patent law is relevant. If two or more inventors file for patents, the U.S. patent office may declare an interference, a proceeding whereby it seeks to determine which American company was actually the first to reduce the invention to practice and recognize its utility. A foreign company is limited to its filing date abroad, on the presumption that foreign law and practice are so different that, normally, there would be no significant delay between conception and publication, unlike in the United States: There, the patent office would seek to determine whether the invention filed by an American company was distinguishable or obvious from the prior art, thus engaging in a validity search that could consume years.

Of course, an interference had to be declared between these three American companies and Montecatini. The declaration made history. The interference commenced on September 9, 1958, and a nearly 30-year battle in the courts ensued before final victory was awarded to the Phillips Petroleum Co. The key claim of USP 4376851 (March 15, 1983) was broad and simple: "normally solid polypropylene, consisting essentially of recurring propylene units, having a substantial crystalline polypropylene content." It was a composition-of-matter claim, not one to any particular process—the best kind of claim. This may well have been the most expensive patent battle in history, but the stakes were extraordinarily high, because so much capacity had been built by Montecatini and its licensees, as well as by the other companies. The victory was, therefore, especially profitable to Phillips, which could start collecting royalties from that point on until the year 2000, at a high level of production. To Phillips, the battle has been worth probably $50 million or more; but Phillips has not won the production race.

It is not within the scope of this chapter to describe the details of this historic patent fight, which went all the way to the Supreme Court, where a suit by DuPont against Phillips on the issue of validity was finally settled in 1989. Ziegler and Natta had both died (in 1973 and 1979, respectively) before the final adjudication. One of the key reasons that Montecatini (now Montedison) lost the patent was the finding of the courts that Montecatini had committed "fraud" against Phillips. This was not a scientific issue, but one of fraud allegedly committed by Montecatini's attorneys.

Phillips' Polypropylene Development

Let us now trace the evolution of the Phillips invention, which resulted from the close collaboration of two men, Robert Banks and J. Paul Hogan. Both men were graduates of smaller, less well-known universities: Banks from the University of Missouri-Rolla, with an M.S. degree from Oklahoma State University and both degrees in chemical engineering; Hogan from Murray State University in Murray, Kentucky, with a B.S. degree in chemistry and physics, 1942. Note that neither had a Ph.D. degree. The pedigrees are not important in further illustrating the role of "outsiders" in key innovations and the contributions of many little-known institutions of higher learning, where—particularly in those days, before the postwar G.I. bill for veterans existed—the cost of education was such that few could afford to attend the more famous universities. My own company had much the same experience: Many of our top scientists and engineers came from relatively obscure institutions, although we had our share of M.I.T. graduates, since I remained connected with that great university after my graduation from there. I make this point especially in light of the recent trend in the United States to reduce federal and state support for graduate study and research. As the chapter by Rosenberg elsewhere in this book discusses, the university-industry relationships have been a critical source of competitive advantage for a science-based high-tech industry such as chemicals.

In 1987 Banks and Hogan received the Perkin medal, the leading award for outstanding commercial chemical innovation offered by the industry and its professional societies and named after its first recipient, in 1906, William Henry Perkin, the discoverer of synthetic dyes in 1856. This was the first and only time the medal has been awarded to more than one individual. As noted in the case of Delbert Meyer, as well as in others, the Perkin medal selection committee, which consists of representatives from all six of the professional societies in chemicals, unlike the Nobel committee has managed to keep a close eye on the commercially important innovations.

In the mind of the founder of Phillips Petroleum, Frank Phillips, the importance of a strong research department was clear in the later 1920s, and he concluded that such a department should include research on chemical products that could be made from Phillips' abundant supplies of liquid and gaseous hydrocarbons. The Second World War provided a major impetus to these efforts, because the petroleum industry was perhaps the best situated to meet American military requirements. Phillips made key wartime contributions by the production of carbon black for tires; synthetic rubber; and aviation gasoline, which originated from research conducted in the 1930s. In this period, process research at Phillips was fundamentally catalysis research. Part of this work involved dimerizing or trimerizing (adding two or three together to form a longer carbon chain molecule) simple olefins to make aviation gasoline components, and this had entailed, among other processes, the use of nickel oxide catalysts on an inert support.

By 1951, work was ongoing to increase the life of the catalyst and improve the yields of the products boiling in the motor fuel range. As part of this work, catalysts were modified with various other metal oxides, mostly from the so-called transition metals. The one metal oxide that truly altered catalyst behavior was chromium oxide. Quite unexpectedly, researchers found that in trying to dimerize propylene they had obtained a significant quantity of solid polymer, which turned out to be a mixture of amorphous and crystalline polypropylene. They further established that it was the chromium oxide that was making this polymer. Phillips quickly filed a patent application on January 27, 1953, more than a year before either Natta or Ziegler made polypropylene, but the patent didn't issue until 1983 because of the patent-office dispute described earlier. Phillips also proved that ethylene could exhibit behavior similar to that of the polyproylene process. It took only a few months of experimentation in 1951 to corroborate the invention. This process was commercialized by Phillips in less than six years as the Marlex Polyolefin Process.

Why did Phillips proceed with the polyethylene version rather than polypropylene? It seems clear that when they did their early work on polyethylene and polypropylene their main interest was on the polyethylene side. Since their work preceded that of Ziegler and Natta by two years, they knew at that time that polyethylene could be made into a polymer by the high pressure process which ICI had invented in the 1930s. However, no one had ever seen polypropylene before. It was natural, therefore, to focus their research-and-development efforts on polyethylene, which they thought might make a product different from that developed by ICI, and with more favorable economics. They did not see much future for a large market with an unknown product like polypropylene, and in that era of limited capital availability it was clear that polyethylene would get the nod. In fact, it has been the view of Phillips that real growth in polypropylene has only occurred recently when the new-generation catalysts have become available to greatly amplify the palette of products that can be made at lower cost. In addition, they could not be certain that the polypropylene patent would be granted, another reason their choice for commercialization fell on high-density poly-ethylene.

In actuality, polypropylene invention came first, and then high-density polyethylene, the latter patent issuing in 1958. By this time, the first Phillips polyethylene plant had been operating for over a year, and Phillips had already boldly issued licenses for several other plants by other companies. Unlike the case with polypropylene, the polyethylene patent situation was clear, and by 1962 twelve manufacturers were operating plants in seven countries. Phillips conducted regular, closed licensee meetings, much as my company did on ethylene oxide. No doubt, other such arrangements have been in place for other widely licensed technologies, a fact that greatly aids the diffusion of technology while still maintaining at least a partially oligopolistic control over the rate of diffusion and the improvements developed and thus justifying the expense of continuing R&D.

Shortly after these events, Phillips saw the growing interest in polypropylene manufacture elsewhere, but, because of capital constraints, felt that a joint venture with another company would be advantageous. Since they were already in the high-density polyethylene business, they sought as a partner a company making low-density polyethylene by the ICI high-pressure process; namely, the U.S. Industrial Chemicals Division of National Distillers Corp. In this way, Phillips could be involved in all three of these basic new polymers. This deal was opposed by the Federal Trade Commission, however, and Phillips entered into a consent decree that obligated them to divest their direct interest in the AB Chemicals Co., the joint venture, in the early and mid-1960s, and build their own polypropylene capacity. At the time of the plant start-up, the markets for polypropylene were very difficult, and this business was not very encourag-ing. Furthermore, the plant could only make homopolymers and a limited quantity of random copolymers, but could not produce block copolymers. These difficulties, combined with the success of the polyethylene business, further steered Phillips to stay in the high-density polyethylene business and avoid the polypropylene business until more recently.

In his Perkin medal address, Hogan stressed that by participating in teams com-prised of engineers, manufacturers, marketers, and research people—all focused on improving and developing these processes—he had gained immense breadth of experi-ence and could help integrate all of the company efforts into early successful commer-cialization. This was also my company's practice, and it yielded enormous advantages over the practices of many large companies which, at least in the past, compartmen-talized their functions. The pressures of international competition are forcing newer and more flexible systems along such lines. Hogan gave some more general observations that deserve quotation:

> Phillips became a major factor in the production of thermoplastics by the 1970s. Total production of thermoplastics in the USA from all manufacturers passed 20 billion annual pounds by 1980. These synthetic resins are derived almost totally from petroleum. Although only about two per cent of the US consumption of petroleum goes into thermoplastics, the impact of these resins upon lifestyle in the USA is tremendous. Thermoplastics find applications in clothing, housing, packaging, transportation, communication, and recreation, to name a few.

> During an eight year inflation period in the 1970s, the price of US synthetic resins increased only 128 per cent, compared to 300 per cent for refined petroleum products. At the same time, the price of natural products which synthetic resins replaced was increasing 137–329 per cent. Thus, the value of synthetic resins to the US economy far outstrips its share of the total petroleum consumption, and has a stabilising influence on inflation as well.

To this quotation might be added the comment that fibers and plastics have fueled the great postwar petrochemical expansion.

The Proliferation of Polypropylene Producers

The industrial application of the technology involved in these polymers, especially polypropylene, has had a tangled history, inevitable when a 30-year litigation is involved. For a number of years, Montecatini issued licenses to others under their patents, which were not in dispute outside the United States. One such company was Hoechst, which was licensed by Ziegler (semiexclusively for Germany) in 1953 under Ziegler's patent for polyethylene. The British start-up firm of Petrochemicals Ltd. also had obtained exclusive rights in the United Kingdom on this chemistry. Then Hercules in the United States obtained a license. Therefore, within a very short time four companies already were committed to this technology (including Montecatini, of course).

Hoechst was able to make polypropylene as early as March 1954 but decided not to encroach on Ziegler's research area. Petrochemicals Ltd. likewise succeeded, as did Hercules. Thus, Ziegler's catalyst-system invention led quickly to a new and very important product, but Ziegler failed to follow up with propylene, as described earlier. My own company faced a comparable situation when we found that bromine assisted catalytic oxidation to terephthalic acid, but being much more aware of the patenting potential, we refused to disclose the process until we had filed everything we could and had obtained our basic commercial position. It is also interesting to note that the same Petrochemicals Ltd. had obtained an exclusive license for the United Kingdom for our nascent ethylene oxide process but had proved financially unable to pursue their plan to be the first major petrochemical company in the U.K. They eventually sold out their assets to Shell, and this of course caused us great complications, as Shell was competing with us (Landau 1994, p. 53). The deep pockets of the Shell group thus ultimately led to their world domination of polypropylene.

Hercules, as I mentioned before, was also the initial contact we had on our ultimate terephthalic acid process. Hercules clearly was looking at new product diversification away from explosives and defense work, and their managers were searching the world for new technology. It was in pursuit of this that they bought the Ziegler rights and also the Witten process rights for DMT (dimethyl terephthalate). I have already described how the PTA technology of Amoco ultimately prevailed, and Hercules is out of this business, having backed the wrong technology in DMT, and not really having had the strength or position to capitalize on its hopes. Hercules also made polypropylene in early 1955 and filed patents on it, but it was too late to have any hope of prevailing in the patent-interference litigation.

Undaunted by the litigation, and probably confident that Natta was the true inventor, Hercules went ahead with a high-density, Ziegler-type polyethylene plant by September 1957, and a polypropylene plant three months later. The polyethylene found use in housewares, wire coatings, bottles, fibers (for seat covers and other uses), toys, chemical ware, and so on. The big market was in housewares and bottles. Polypropylene, although somewhat more expensive, had a specific gravity of <1 so that it became a very light, durable plastic with a higher melting temperature (above 300F) and thus could withstand boiling water. It was also strong in flexing, so it could be used, for example, in hinge manufacture for household goods. Having seen these properties, Hercules decided to enter into manufacture of this polymer in an initial plant of

50-million-pounds-per-year capacity, soon followed by another. But the profitability was ebbing fast. By 1961, at least nine American producers had a total capacity of 470 million pounds per year. Hercules then integrated further into fibers and film.

The problem, of course, was that the technology was diffusing rapidly, and the end of the prewar cartel system encouraged overbuilding and an excessive number of entrants. Also, the raw materials were abundant and cheap in the United States, and petrochemicals were glamorous. This had also encouraged DuPont to research this area, but it prudently had refrained from premature investment and is not a manufacturer. Hercules became one of 17 companies, at various times, that made polypropylene commercially. It remained the leading producer in the United States until 1983, when it formed a joint venture with Montedison (formerly Montecatini) in order to exploit a new Montedison process. By the end of the decade, Montedison became the sole owner, and Hercules was out of the polypropylene market. Its only (temporary) success was that Hercules' chief executive Alexander Giacco became head of the joint venture (named Himont) and that when Montedison was taken over by the Feruzzi group in Italy (Raul Gardiner was the chairman but he knew nothing of chemicals), Giacco was recruited for a brief while to run Montedison as well as Himont. The Feruzzi group, and Montedison with it, fell into further financial problems in the 1990s, after Gardiner had committed suicide; it had to be rescued by a syndicate organized by the leading bank Mediobanca, which would give the Agnelli family and Fiat effective control of the $24-billion-per-year chemical giant (this move has lately been rendered questionable by financial problems). One consequence of the change was that the deep-pockets Shell group then effectively forced the integration of the Montedison-Hercules polypropylene heritage into the Shell group via the formation of Montell, although Montedison continues as a part owner. Montell, with chief executive Peter Vogtländer (who was head of chemicals for the worldwide Shell group), is now by far the largest producer in the world. Shell is currently buying out Montedison's 50% interest because Montedison needs cash—and so, its long pioneering history in polypropylene development ends with its exit from the business.

Amoco did not experience the same sad ending, and it also had the motivation and financial strength to acquire Avisun's polypropylene fiber position and make polypropylene the second major product in their chemicals line, after PTA, as I have described earlier in this chapter.

The final outcome of the polypropylene patent suit, of course, involved immense contractual and financial complications. Phillips offered a license to all of the Montedison licensees, some of whom (particularly Phillips and Hercules) in turn sued Montedison for back royalties paid. Very complex settlement negotiations, controlled heavily by lawyers, followed until these issues were resolved, but not until the true end of litigation in 1989, as mentioned before.

In Europe, where this patent drama was not at issue, Hoechst moved forward to become the largest European producer, until Montell, and others such as BASF (which acquired ICI's position) entered the business as likewise did Shell.

The story of Phillips, however, does not end happily. In 1984, T. Boone Pickens, the CEO of Mesa Petroleum Co., announced a takeover bid for the company. As Figure 5-3 shows, Phillips Petroleum's earnings were declining prior to this event, and the stock market price was languishing. In addition, Michael Milken and his Drexel Burnham Lambert company were creating "junk bonds"—high-yield, unsecured debt with high leverage, which helped many companies build up their business but also gave the tools for takeover bids when the target company was perceived in the financial

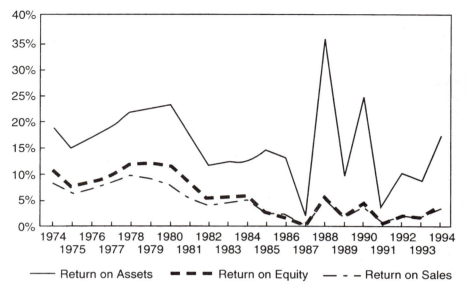

Figure 5-3. Phillips Petroleum Company.

markets as underperforming. This was the case at Phillips which, located in Bartlesville, Oklahoma, may have become too provincial in its management outlook in the face of such momentous developments on Wall Street, as well as in the oil industry more generally.

A bitter battle ensued, but by March 4, 1985, Phillips emerged as a still-independent company by exchanging roughly one-half of its stock for debt of $4.5 billion—the largest transaction in its history. Ultimately, the company would have to overcome nearly $9 billion in debt. Phillips became the first oil company to survive a takeover battle (Unocal followed a short while later). Six major oil companies, including Gulf, Cities Service, and Marathon, were less successful. If Pickens won, he would sell off parts of the business and give the shareholders a once-only enormous gain, but the long-range growth potential of an organization such as Phillips would be lost forever, worth more dead than alive, in the short run.

Phillips, however, was the most innovative of all the companies, and Phillips Chemical had a splendid record in such fields as synthetic rubber. The company took 10 years to reduce its debt burden and improve its market valuations. Phillips survived, but the chemical company and the corporate R&D efforts are much smaller than before, with fewer triumphs, and the company never really capitalized on its polypropylene victory, having no margin to gamble on building large capacities while the patent litigation dragged on. Today, Phillips is listed as having only 520 million pounds per year of capacity, involving the Japanese group Sumitomo, which supplied the latest gas phase process for the plant. In their annual report for 1994, however, Phillips asserts that with the new plant under construction, the partnership will have a capacity of 750 million pounds per year (more recently said to reach 880 million pounds per year), making it the sixth-largest polypropylene producer in the United States.[7] Phillips,

---

[7] It should be remembered that capacity figures are always approximations. Companies don't like to reveal their secrets, and de-bottlenecking and expansions, often underway, are not reported until later.

however, is still a small factor in the overall market place. True, royalties are substantial, but as I have repeatedly written, licensing never (or very rarely, as in the case of Carbide's UNIPOL polyethylene process) has the long-term profitability for new technology that self-manufacture can confer (Landau 1994). If sharing with others is necessary, it is far preferable to enter into alliances or joint ventures, and this trend is now very noticeable in the chemical industry. The "Not Invented Here" syndrome that had been so prevalent in United States firms during our earlier years as licensors has been replaced by the necessity, given financial and capital constraints, to use whatever technology and business positions can further the core competencies of the firm. The chemical industry has been a leader in this trend.

Of course, technology continues to advance. Much innovation has been taking place in the heart of the process, the catalyst systems, by a large number of the companies. The most recent innovations, pioneered by Exxon and Dow, are "Metallocene" catalysts of a special configuration different from the original organometallic catalysts of Ziegler and Natta (Ewen 1997). Thus, significant improvements are constantly underway and may even become qualitatively, as well as quantitatively, significant. The newer processes, which tailor products more exactly to the customer's needs, also are environmentally more benign, with fewer or no discharges into the atmosphere and water.

In recent discussions with major European and American chemical companies, I have been told repeatedly that these newer catalysts are perhaps the most important new developments in the polymer field. These so-called second generation technologies are expected to cause some truly revolutionary changes in polyethylene and polypropylene polymers. Recently, two Stanford chemists have disclosed a catalyst system of this class to permit the manufacture of a rubbery type of polypropylene with varying elasticity (Cortes and Waymouth, 1995). Amoco, the exclusive licensee worldwide, is seeking to exploit this discovery.

All of these products have enhanced performance in purity, clarity, and tailoring to varied end uses, especially in packaging materials. In these applications, they confer sharper, stronger heat-sealing characteristics at lower temperatures; they have superior strength; and they have excellent purity and low extrudables (for taste and odor). Other applications will follow in extrusion coating, adhesives, injection and blow molding, wire and cable coating, and so on. Mixed polymers of unusual properties are also made attractive, and in all likelihood, the advantages will be seen most particularly in the various types of polyethylene, which may be able to penetrate some currently unavailable markets in which more expensive polymers have been employed. Many existing facilities may be retrofitted with the new catalyst systems, so that new capital-investment requirements may be modest. Although metallocene catalysts are still more expensive than conventional Ziegler-type catalysts, which they resemble in being metal-organic compounds, this differential will probably disappear within ten years as their use spreads. In 1996, metallocene demand is expected to be about 150 million pounds, but in ten years, almost two-thirds of the demand for all second generation technologies will come from single-site, metallocene-based, linear low-density polyethylene alone, at a level of 2 billion pounds demand. Most of this will go into film production, as estimated by Robert Bauman at Chemsystems.

Second Generation Polypropylene Products

Second-generation polypropylene developments also are advancing, but because of higher catalyst costs, the polyethylene metallocenes will be more exploited than polypropylenes, so that initially, at least, specific applications of better performance applications will be developed. Some of these applications could include high-ethylene random copolymers and non-wovens. These products may well permit market expansion into fiber applications now currently not being served. All in all, these new second-generation technologies will result in a more intense level of interpolymer competition. It is abundantly clear that polymers, as a whole, represent a continuing blend of new technologies and intensive market and product developments. Because so many major companies are already in polymer manufacture, it is inevitable that extensive licensing and cross-licensing will spread, led by the examples of Exxon and Dow. The latter company has just announced an arrangement with Montell, the largest polypropylene producer, to enter the polypropylene business in a major way, expecting that the demand, growing up to 10 percent per year worldwide—the highest rate of the principal polymers—will sustain a planned capacity of 3 billion pounds per year. The world capacity in 1996 is estimated by Dow to be about 48.5 billion pounds per year. In the United States, current capacity is about 10.6 billion pounds, at about a 90 percent operating rate. Exxon has announced an alliance with Union Carbide in polyethylene. Dow and British Petroleum are also going to work in polyethylene. These moves signal large expectations for these new technologies, but the costs of development seem to argue for such unusual alliances. These arrangements will evidently be more likely to arise as the globalization of the industry, the high costs, and its cyclicality argue for spreading the risks. Licensing to third parties also will be featured.

This trend has most recently been underlined by further major moves in Europe. BASF has decided to merge the polyethylene operations with Shell and its polypropylene business with Hoechst. BASF and Shell will form a 50:50 joint polyethylene venture, making it the third-largest producer in Europe after Borealis (itself a joint venture between Neste of Finland and Statoil of Norway) and Polymeri Europa (itself a venture between Enichem of Italy and Union Carbide), and will have a total capacity of about 1.4 millon tons per year (out of a total European production of 12 million tons per year and a worldwide capacity of 23 million tons).[7] The new BASF-Hoechst venture in polypropylene will be Europe's second largest after that of Montell (*Chemistry & Industry*, 21 October 1996, p. 725). Other such arrangements will not be surprising.

These moves by Dow, following closely after the sale of its pharmaceutical business to Hoechst, demonstrate anew Dow's strategy of concentrating on areas in which it has long had a solid base. Polypropylene nicely augments Dow's existing polyethylene capabilities. However, there are many players in this field, and profits may be quite elusive until the next big cyclical boom in the world economy, especially as Asia develops. Polypropylene prices and operating capacity have been quite volatile in the last decade, but the level of commitment by so many companies demonstrates that investments are being made for the long term. Technology and market development are still the driving forces in this innovative industry.

As a concluding comment on these stories, my association with both of these major innovations is encapsulated in the story told in Landau (1994), as when the head of Shell chemicals (E.G.G. Werner) grouped me with Ziegler and Natta as the "three outstanding individuals in the postwar petrochemical industry." Although flattering, this analysis is not the reason I choose these examples: the true reason lies in their commercial importance. Moreover, the real innovations are team efforts by many organizations throughout the industrial world in this highly globalized industry, as

---

[8] Malone (1995) gives another graphic illustration, in the case of the company Silicon Graphics.

Werner also pointed out. The fact that my small, flexible, energetic company could compete with the greats in many areas is a major justification for the support of entrepreneurship in this country as an ongoing public policy. We see this every day in the computer and software industries and in biomedicine and biotechnology. But, as these stories have indicated, at a certain size technology is not enough. Then it is a matter of business wherewithal to grow into a long-term survivor in a world of giant competitors.[8] Few such entrepreneurial companies can endure in the face of adverse macroeconomic and institutional policies, however, as discussed elsewhere in this book. In the chemical industry, an older and more capital-intensive industry than those mentioned, entrepreneurs are now uncommon in the main lines. The few exceptions— for instance Gordon Cain (1997) and Jon Huntsman, who have specialized in leveraged buyouts of units no longer desired by large companies—only emphasize this situation.

However, as this book illustrates, the larger companies often endure for a century or more. Nevertheless, a brief look at management styles and personalities shows that large-company growth may be considerably affected by shifting trends. Thus, in the pre-World War I era, when organic chemistry was the source of growth in the chemical industry, the German companies were led by Ph.D. chemists, usually originating in the research organizations. Because of the strong links with university-based science, these leaders sought affiliation with universities, and the title of Professor Doctor was esteemed by many. This trend continued until about 1980, when world conditions changed and market and financial considerations became more critical (as explored by Richards in this volume). By the mid-1990s, all three big German chemical companies were led by non-technologists or scientists.

In the United States, with its large domestic markets and the need for large-scale continuous process plants, where the economics of mass production are essential and the basic chemistry is simpler, engineers in management have had a prominent, often dominant, role for most of this century.[9] Unlike, for example, in the steel industry, the financial community understood little of the underlying technologies in the chemical industry and made little effort to dominate it. In the case of Allied Chemical, where the financial community did dominate, long-lasting competitive advantage could not be sustained. Only in the 1980s, when financial constraints working through the equity markets encouraged takeover bids and/or acquisitions or mergers, did technology become less important as the industry matured. The case of Union Carbide illustrates this, as does the case of ICI spinning off Zeneca (cases discussed by Chandler, Hikino, and Mowery in this volume). With that trend, managements in the United States, while continuing to have a strong manufacturing-engineering tradition, have made their financial operations much more important than, say, R&D. In the United Kingdom, the dominance of research executives waned after the Second World War, as recounted earlier in this chapter.

In Japan, the unique *keiretsu* structure and the strong control by government led to leadership that was not primarily focused on R&D, or even on finance, but more on managing intergroup relations "harmoniously" and dealing successfully with the government agencies. I do not think I ever met a CEO in Japan who had much, if any, research experience or inclination. This does not imply that the Japanese companies

[9] The previously mentioned Frank Howard, who made the I.G. Farben-Esso agreements and then had a prominent role in America's synthetic-rubber expansion during World War II had this to say about the critical role of American chemical engineers and their superiority (Howard 1947, p. 241): "Twenty five years of intimate contact with the industrial science of Europe left this observer with the conviction that in chemical engineering America has no close second."

have not been making steady improvements in the technologies they have licensed, and they are important players, especially in Asia, in these fields. Companies seek the leaders their external environments require. The record shows that approximately one-third of the Japanese inside directors are technical (chemistry, engineering, mining) and two-thirds are "liberal" (economy, finance and accounting, law and commerce) The profile of Japanese chemical company presidents is about in this ratio also.

SOME GLEANINGS FROM THESE CASE HISTORIES

1. Growth and comparative advantage for industries and countries obviously take place at the firm and the individual level. The creation of so much additional wealth by the activities of the innovators described in this section, however, has made a significant contribution to the growth of the overall, aggregate economy. That is, growth proceeds from the microeconomic level to the macroeconomic level through a feedback mechanism. However, the macro level influences the micro performance, as will be further described below.

2. The experiences of this industry and of the case histories clearly show how little direct intervention of government has mattered and how important has been the functioning of markets. This industry has clearly, from early days, tended to be international in its investment and sales activities, and the case histories described reflect that situation markedly. Other newer industries are somewhat slower to follow suit, but they should.

3. The case histories described are a fair representation of the complexity of the innovation process. It cannot be viewed from the aggregate level alone, as further described in the introduction to this book.

4. Two types of innovation process usually go on in the chemical industry, one of which we have described as radical or breakthrough; the other is incremental. Radical innovation is very often a function of what I call "outsiders"—that is, individuals or firms not directly involved in the field that ultimately evolves out of their activities. This process underlines the need for encouraging innovation at all levels of society, which includes making provisions for the entrepreneurial world to function as efficiently as possible and to provide maximum opportunities for all kinds of organizations. It would appear from these stories, and from many others that could be substituted for them, that in an industry as capital-intensive and as large in scale as chemicals, incremental improvement is accomplished by large global companies that can benefit from constant searches for greater productivity, product quality, and customer satisfaction. On the other hand, it is often smaller groups that are most likely to engage in breakthrough work; this is especially feasible now, in the field of software and biotechnology.

5. The extraordinary influence of the American petroleum industry on the evolution of the world chemical industry has many ramifications, some of which are detailed elsewhere in this book.

Finally, following the general concept of the matrix of comparative advantage discussed in the introduction to this book, these case histories illustrate the following, from my personal perspective.

In terms of macroeconomic influences, while we have not dwelt on the macroeconomic environment during these innovations, except in Britain, it is clear that they would have succeeded only in the very favorable macroeconomic conditions that prevailed in the United States after the war and which also prevailed under the "Wirtschaftswunder" of postwar Germany. This topic is discussed further in the chapter by Eichengreen. Trade policies clearly have been particularly important, in that the innovators enjoyed the world market which encouraged innovation by raising the potentiality of profiting from an individual innovation. Tax policies have not been involved explicitly in these stories, but they have influenced the companies and individuals in different ways. In particular, for our work the availability of a favorable tax on capital gains was a very important incentive to commit capital to high risk research and development. That incentive is also very important to the newer and younger technological enterprises in computers, software. and biotechnology.

The institutional setting for innovation has also been important. The financial aspects of the industry are clearly involved in the rapid build up by so many companies of these new products. How they were financed is not a part of the story of this chapter, but it is involved in the other chapters in this book by Da Rin and Richards. However, in describing the differences between Ziegler's and Natta's financial conditions, I have underlined that large companies are needed not only for marketing and manufacturing skills but also to provide the finances for research and development and for plant construction. This was true in our own case in the organization of our Oxirane joint venture with Arco to make propylene oxide. The intellectual-property issue is a particularly important part of this industry, as illustrated earlier. As previously mentioned, the overall legal system, particularly that involving antitrust and torts, is explored in the next chapter, written by Tom Campbell and me, with an emphasis on the United States, which is the most litigious country in the world. The role of the intermediating institutions such as the Max Planck Institute is graphically illustrated in the story described above.

The structural and supportive policies of government can also make a substantial difference. For example, the educational system of the countries involved has been critical to the success of these innovations, in particular the close relationship between universities and industries. We have also stressed that what has been happening in the post-World War II era is the dominance of the chemical engineer over the chemist, which was a traditional German specialty for almost 100 years prior to the Second World War. The truth is that the molecules that have reached large-volume production, as described in this chapter, and in many other instances that make up the modern chemical industry are much simpler than the molecules that powered the dyestuffs and, eventually, the pharmaceutical industry. As a result, the design of plants, the optimization of operating conditions, the large scale of operation, and the like are the hallmarks of the good chemical-engineering approach to modern chemical manufacture. The history of this emerging dominance of the chemical engineer over the chemist is told in this book in the chapter by Rosenberg. This judgment is not entirely true for the pharmaceutical sector of the industry, but capital, marketing, and government regulation are leading to similar overall developments.

The chemical industry clearly is characterized by many individual variations among companies. The ease of entry is not as great as it once was, when the oil industry moved in en masse and helped create the petrochemical industry of the world. Yet, there is ample opportunity for would-be companies to buy technology and enter the industry, especially in other countries that are emerging economies. There are many spillovers, interactions, and stimuli for further innovation in this complex industry with 70,000 products and almost infinite possibilities for still further molecular innovations, as the pharmaceutical and biotechnological industries are revealing every day. The speed of commercialization has become an increasingly important factor. The integration of process and product has likewise become important, as the case histories reveal. Catalysis and chemical engineering can be identified at the heart of the modern chemical industry, especially in petrochemicals and polymers, although less so in pharmaceuticals as yet.

It is also evident that some companies are prepared to take bigger risks than others but do not necessarily have the large resources that may be necessary to fully capitalize on the innovations. This, of course, raises the broadest question of the role of corporate governance in today's world of financial dominance and is the subject of the other sections of this book, which deal with corporate strategy as well (Richards's chapter and the one by Chandler, Hikino, and Mowery).

Indeed, history matters. The view from the top (the aggregate economy) is composed of many such narratives, but the real process of change occurs continuously at the micro level. It then feeds back to the macro economy and provides the engine of growth. Elsewhere in this volume, we have shown that for extensive periods the growth of this industry exceeded the growth of the overall economy and thus helped accelerate its growth. Clearly, these two cases fall into this category, by helping the great post-World War II surge of the U.S. economy. The same thing happened with the dyestuffs industry for Germany in the nineteenth century. In the developed economies of the world, technology ever more has been the driving force for growth since the Industrial Revolution in the eighteenth century—the ultimate feedback of the micro to the macro.

APPENDIX: GRANGER CAUSALITY

In amplification of the appendix in Barry Eichengreen's chapter, this chapter has shown that feedbacks are important. C.W.J. Granger himself makes that point (1987):

> If Y_t causes X_{t+1}, then X_t may, but need not cause Y_{t+1}, so that feedback can occur, but need not. Similarly if Y_t causes X_{t+1}, and X_t causes Z_{t+1}, then Y_t may, but need not cause Z_{t+1}. It has to be remembered when interpreting results based on tests, that missing common causal variables can always alter the interpretation, that causality may be lost if one variable is controlled so as to reduce the strength of the causal link, and that temporal aggregation or using data measured over intervals much wider than actual causal lags can also destroy causal interpretation.

A simple analogy that may help to illustrate the problem is the phenomenon of a thunderstorm. An observer at some distance will see first the lightning and, after an interval, the thunder. Thus, lightning might appear to cause thunder, but of course they occur simultaneously. Since macroeconomics, even in the hands of a sophisticated observer like Eichengreen, cannot get down to the microeconomics of companies and technologies, this chapter serves as an illustration of the true complexities in interpreting the results of such an innovative industry. It is evident, for example, that a new or improved technology requires investment (in R&D and physical capital) ahead of the growth in demand for the products of that investment. This investment, of itself, under many conditions contributes directly to demand growth. Expectations of increase in demand normally fuel such risk-taking investment. The time lag to full production may be (and was, in these case histories) many years, spanning macroeconomic fluctuations of a cyclical character, and thus the supply must lead demand, at least for the products (or services). Demand-side economics assumes instantaneous supplies available, but this does not happen in the world of the chemical industry or that of many other industries. Of course, over a time such as 50 years, the supply curve will track the demand curve closely, and that is what Eichengreen's correlation studies show; but the variability between these curves over shorter periods of time (say 5 to 10 years) may be significant. As Eichengreen says, his causality studies need to be treated with caution.

Stiglitz (1991) gives an excellent analysis of the feedbacks inherent in modern industrial societies and their contribution to multiple equilibria. He also stresses the institutional and socioeconomic factors that determine the relative growth rates of societies, much along the lines of the matrix used in this volume.

REFERENCES

Aftalion, Fred (1991), *A History of the Internatonal Chemical Industry*, University of Pennsylvania Press, Philadelphia, PA.

Aneja, Aran P. (1994), "Wear No One Has Made Before," *Chemtech*, (August): 48–52.

Broadhurst, R. (1979), in *The Dyeing of Synthetic Polymer and Acetate Fibers*, D. M. Nunn (eds.), Dyes Company Publishing, Trent, Bradford, U.K.

Burkinshaw, S. M. (1995), *Chemical Principles of Synthetic Fibre Dyeing*, Blackie Academic & Professional, London, U.K.

Cain, Gordon (1997), *Everybody Wins*, Chemical Heritage Press, Philadelphia, PA.

Carra, Sergio, Federico Pariso, Italo Pasquen, and Piero Pino, (eds.) (1982), *Giulio Natta: Present Significance of his Scientific Contribution*, Editrice di Chimica Srl., Milan, Italy.

Chandler, Alfred D., Jr. (1990), *Scale and Scope*, Harvard University Press, Cambridge, MA.

Chemical Enterprise (1994), Chemical Heritage Foundation Press, Philadelphia, PA.

Cortes, Geoffrey Ward and Robert M. Waymouth (1995), "Oscillating Stereocontrol: A Strategy for the Synthesis of Thermoplastic Elastomeric Polypropylene," *Science* (January) 267(13): 212–19.

Ewen, John A. (1997), "New Chemical Tools to Create Plastics," *Scientific American* (May 1997): 86–91.

Gomory, Ralph (1992), "The Technology-Product Relationship: Early and Late States," in Nathan Rosenberg, Ralph Landau, and David Mowery (eds), *Technology and the Wealth of Nations*, Stanford University Press, Stanford, CA, pp. 383–94.

Granger, C.W.J. (1987), *The New Palgrave Dictionary of Economics*, John Eatwell, Murray Millgate, and Peter Neumann (eds.), Macmillan Press Ltd., London, U.K., pp. 381–82.

Haber, Ludwig F. (1971), *The Chemical Industry 1900–1930*, Clarendon Press, Oxford, U.K.

Hounshell, David A. and John Kenly Smith Jr. (1988), *Science and Corporate Strategy: DuPont R&D, 1902–1980*, Cambridge University Press, Cambridge, MA, pp. 407–22.

Howard, Frank A. (1947), *Buna Rubber: The Birth of an Industry*, van Nostrand, New York, NY.

ICI (1954), *The Lauching of a New Synthetic Fibre—A Historical Survey*, June, London, U.K.

Kline, Stephen J. and Nathan Rosenberg (1986), "An Overview of Innovation," in Ralph Landau and Nathan Rosenberg (eds.), *The Positive Sum Strategy*, National Academy Press, Washington, DC, pp. 275–305.

Landau, Ralph (1956), U.S. Pat. 2 752 362.

Landau, Ralph (1958), "Chemical Engineering in West Germany," *Chemical Engineering Progress* (July) 54(7): 64–68, 115.

Landau, Ralph and David Brown (1965), "Making Research Pay," AIChE-Institution of Chemical Engineers Symposium Series No. 7, London, U.K., pp. 735–43.

Landau, Ralph and Alfred Saffer (1968), "Development of the M-C Process," printed in *Chemical Engineering Progress*, vol. 64, no. 10, October 1968, pp. 20–26.

Landau, Ralph (1979), "The Inventor and the Entrepreneur," Introduction to the General Electric Centennial Symposium on Science, Invention and Social Change, held at the General Electric Company headquarters, Schenectady, NY, September 20–21, 1978. Published under the title, "Innovation for Fun and Profit," in *Chemtech*, January 1979.

Landau, Ralph (1994), *Uncaging Animal Spirits*, The MIT Press, Cambridge, MA.

Landau, Ralph, Timothy Taylor, and Gavin Wright, (eds.) (1996), *The Mosaic of Economic Growth*, Stanford University Press, Stanford, CA.

Lefort, E. (1933), U.S. Pat. 1 998 878.

Malone, Michael S. (1995), *Fortune* (October 30): 119–20.

Menkes, Suzy (1995), "Scorned No Longer: Synthetics Go Modern," *New York Times* (January 1) Section 1, p. 41.

Nunn, D.M., (ed.) (1979), *The Dyeing of Synthetic Polymer and Acetate Fibers*, Dyes Company Publishing, Trent, Bradford, U.K.

Plumpe, Gottfried (1990), *Die I.G. Farbenindustrie AG: Wirtschaft, Technik und Politik 1904–1945*, Duncker & Humblot, Berlin, Germany.

Remsberg, Charles and Hal Higden (1994), *Ideas for Rent—The UOP Story*, UOP, Des Plaines, IL.

Rosenberg, Nathan (1996), "Uncertainties and Technological Change," in Ralph Landau, Timothy Taylor, and Gavin Wright (eds.), *The Mosaic of Economic Growth*, Stanford University Press, Standford, CA.

Rosenberg, Nathan, Ralph Landau, and David Mowery, (eds.) (1992), *Technology and the Wealth of Nations*, Stanford University Press, Stanford, CA, "Successful Commercialization in the Chemical Process Industries," Chapter by Ralph Landau and Nathan Rosenberg, pp. 73–119.

Spitz, Peter H. (1988), *Petrochemicals: The Rise of an Industry*, John Wiley & Sons, New York, NY.

Stiglitz, Joseph (1991), "Social Absorpton Capability and Innovation," *CEPR Publication No. 292*, Stanford, CA.

Teltschik, Walter (1992), *Geschichte der deutschen Großchemie: Entwicklung und Einfluß in Staat und Gesellschaft*, VCH, Weinheim, Germany.

6 The Impact of the Legal System on Innovation

TOM CAMPBELL and RALPH LANDAU

Any economy based on private enterprise and capitalism requires a legal system which, at the very minimum, provides for the protection of private property rights. In this brief essay, we first sketch in general terms the key ingredients of a modern legal system in a capitalist society and then discuss the elements of that system that have particular relevance to chemicals. We make no attempt at a broad survey of comparative legal systems; instead, our focus is on providing our own views, based on our experiences as businessmen and rooted in our academic backgrounds, of the impact of the legal system on competitiveness in the chemical industry.

THE INGREDIENTS OF A MODERN LEGAL SYSTEM

The first modern legal framework arose in Great Britain, and it clearly aided the rise of the Industrial Revolution. Legal historians such as Horowitz (1992) and Rubin and Sugarman (1984) have emphasized that in the past, the common legal standard in most countries had been that employers had an absolute duty to protect an employee from harm—the concept of strict liability. English legal reforms, however, replaced strict liability for employers with a negligence standard, under which the employer was only liable to take "due care." Eventually, the negligence standard gave way to workers' compensation schemes, where the cost of industrial accidents is actuarialized. It is likely that England would not have taken the lead it did in the Industrial Revolution if it hadn't seen these changes in its employer-liability laws.

The development of the common law of England has evolved steadily since the Industrial Revolution, and indeed it inspired much of American law. In Germany, the Emperor drew on the Roman law tradition to frame much of the legal system we know today in Germany, although of course this system also greatly evolved. The Japanese legal system's origins are somewhat less clear, having derived from the Chinese many centuries ago, with sizable borrowings from the German model during the Meiji Restoration. Recently, Japanese law has come to resemble somewhat the systems of the West, particularly due to the constitution inspired by General Douglas MacArthur after the Second World War. This resemblance is especially true in the Japanese unfair competition law, which borrowed extensively from the U.S. Sherman and Robinson-Patman Acts (Iyori and Uesugi 1983).

A modern legal system must include more than the simple protection of property rights. It must include some or all of the following:

First, for property rights, including the enforcement of contracts, it must spell out a judicial system for dealing with disputes, including the specification of penalties. Lawyers to represent the disputing parties must be educated and compensated; judges must be elected or appointed, and so on.

Second, criminal punishments also must be meted out, so that the society can maintain public order. Not only lawyers and judges are required but also methods and places of punishment. Some societies are more crime-prone than others, particularly in the United States where drugs, violence, a heterogeneous population (as compared with Japan and Germany, in particular), and the presence of a more-or-less permanent underclass have created a much larger penal system than in the other three countries. America's need for a more fully developed and more expensive criminal justice system clearly detracts from the growth rate of the American economy.

Third, with the rise of science and technology as the driving force of long-term growth, property rights have been extended to intellectual property rights. These would include a patent system and provisions for the protection of trade secrets.

Fourth, in societies such as in the United States, where justice and equity are ingrained in the populace and where, even in the Constitution, excess power in governments is feared and civil rights respected, a legal system must be created that permits ordinary citizens to redress their grievances through the judicial system. This system has extended to protecting such citizens against injurious acts of other individuals or companies, as well as acts of government. This branch of law deals with torts. All developed economies have some system of dealing with accidents or intentional wrong-doing; the American system has distinguished itself by granting more power to enforce standards of behavior and to collect damages for harms suffered (including attorneys' fees) in private hands. Other legal systems rely much more on government enforcement. The result may be less vigilance, but also less incentive to bring lawsuits driven by attorneys, rather than actually harmed individuals. On rent-seeking by private plaintiffs, see for example, Brown (1973) and Priest (1993).

Fifth, governments also may be given powers to limit the activities of either individual citizens or business entities where it is judged the greater good of the overall population or economy is at stake. This government right has led to the antitrust powers pioneered in the United States in the 1890s as the outgrowth of a long tradition of fearing too much concentration of power, not just by branches of the government but by the private sector.

Sixth, as a further manifestation of this concern over the concentration of power, there has arisen a large body of regulatory laws, which may be enforced by monetary or other means. Citizens and businesses likewise have certain rights to challenge such laws by way of the judicial system. In America, an affected party's ability to challenge administrative regulations far outstrips such a party's opportunity to do so in Europe or in Japan. Private parties may advise in the regulatory process, but nowhere do they have the right to challenge the result of that process, as they do in America, under a standard of "arbitrary, capricious, an abuse of discretion, or otherwise not in accordance with law" for decisions of law or as unsupported by "substantial evidence" for decisions of fact.[1]

---

[1] United States Administrative Procedure Act.

While these six characteristics are by no means a complete account of the nature of a legal system, it is already obvious that the legal system potentially has a great influence on the growth rate and welfare of a country.

THE IMPACT OF LEGAL SYSTEMS ON ECONOMIC GROWTH AND THE CHEMICAL INDUSTRY

The impact of different national legal systems on economic growth is a fascinating subject; unfortunately, there exists a very inadequate scholarly basis for such comparisons among the four countries under review. Even for the United States alone, such studies are rare. The only study to date to measure the effect of various litigation system reforms on industrial productivity and employment is that by Campbell, Kessler and Shepherd (1996). This study measured changes in productivity and employment in each of the 50 states over a period of 20 years, during which various changes in litigation rules were adopted. The results were a marked positive correlation between improved economic productivity and reforms that reduced liability, and a similarly marked positive correlation between heightened employment and reforms that reduced liability. The converse was also shown: Changes in the legal system that increased the likelihood or extent of liability, showed marked decreases in productivity and employment. Such studies seem totally lacking for the other countries in this survey. Thus, this section can deal only with largely anecdotal evidence.

What seems absolutely clear is that the United States is the most legalistic and litigious society of the major economies in the world (Galanter 1983). Regulations overflow each year. Even the tax-collecting system of the Internal Revenue Service, an essential function of any government, is enormous in its regulatory sweep and involves much litigation, both in specially established and more general courts. The unannotated tax code alone runs to 3,052 pages,[2a] and the summaries of relevant court decisions to many more.[2b]

[2a] *United States Code*, 1994 edition, U.S. Government Printing Office, Washington, DC 1995. Title 26, Vol. 1, §§–1000, 1525 pages; Vol. 2, §§1001–9722, 1527 pages.

[2b] The pages in each volume of Title 26 of *U.S. Code Annotated* (1988 and 1989, as indicated) West Publishing Co., St. Paul, MN, total 11,643 in the main volumes and 3,120 in the supplements. The pages per volume are as folows:

		Pp. in main text	Pp. in Pocket Supplement
©1988	§1–100	827	260
1988	101–162	740	227
1988	163–210	729	155
1988	211–350	649	151
1988	351–480	925	380
1988	481–800	882	178
1988	801–1050	875	334
1988	1051–2000	830	201
1989	2001–3100	757	98
1989	3101–5000	919	272
©1989	5001–6000	517	46
1989	6001–6400	801	236
1989	6401–7200	760	206
1989	7201–7440	414	98
1989	7441–end (excl. tables)	918	278

The consequence of this widely known litigiousness is that the United States has many more lawyers than any other major economy, certainly more than the other three countries of this book combined.[3] America's law schools each year produce at the graduate level almost as many attorneys as its colleges graduate engineers, with only a bachelor's degree.[4] The most difficult matter to assess is not only what this system costs in total but also what its effects may be on innovation, on change, on the basic processes that produce growth.[5] Products not made or invented, or ideas not embodied in plants or services, may never see the light for fear of litigation. A concrete example has been in the field of new pharmaceutical products, where many are sold for years in other countries before they are permitted in the United States and some never make it into the United States at all (Goldberg 1995). Certainly, in some cases the U.S. government's failure to permit a product may prove wise, as was the case with the morning-sickness drug thalidomide. The U.S. Food and Drug Administration (FDA) never did approve it. In Germany, where the drug was approved, it became apparent only after some years that thalidomide could, and did, cause birth deformities such as children born without limbs. But this illustration is not typical. The true effect of the FDA's method of operation on American welfare and the growth rate cannot be measured. Yet, it is a fact that the American pharmaceutical industry (a branch of the chemical industry) is the strongest and most innovative in the world! There are many reasons for this state of affairs, including America's system of health-care financing, but this chapter cannot deal with the issue in depth.

Many product litigations occur every day. These include the uniquely American system of class-action suits, along with the uniquely American practice of contingent fees and the wide use of juries for civil cases (not practiced in the United Kingdom). The United States also is the only country that lacks some version of a "loser-pays" rule, so that there is great incentive to file suits against wealthy corporations, hoping for a settlement if not a large jury award. The combination of these features can be disastrous. Attorneys, rather than actually injured parties, initiate and manage litigation. The cases of asbestos and breast-implant devices are fresh in our minds, and the greatest rewards really have gone to the lawyers (Kolata 1995). Unlike in the English system, American plaintiffs can, and often do, operate on a contingency-fee basis, which means they get nothing if the case is lost and often receive one-third of the ultimate awards.

While such a system allows otherwise indigent but harmed citizens to get legal representation and receive compensation, it also provides perverse incentives for the lawyers involved, which is one reason why class-action suits have become so frequent. If the ultimate rewards go mostly to the lawyers, then lawyers will see no reason to simplify the procedures. When all members of the injured class do receive some compensation, there is very little money to go around, compared to the sums allotted to the lawyers on both sides. In a shocking recent example, a U.S. District Court threw out a settlement reached on behalf of the owners of Bronco II automobiles: The attorneys had received more than $2 million in fees, while their clients—the owners—

---

[3] Professor Galanter reported 896,000 attorneys in the United States in 1983 (Galanter 1983, p. 48).

[4] In recent years, the number of law graduates has fluctuated from around 43,000 to 44,000 (Barnes 1995, p. 27), whereas about 66,000 undergraduate degrees are awarded in engineering by America's colleges every year, according to the Engineering Manpower Commission (1991).

[5] Huber's (1988) work is most often cited for the cost estimate of the litigation system, at more than $80 billion per year, with indirect costs added in to total more than $300 billion per year. For a criticism of Huber's estimates as too high, see Hager (1990). The Rand Institute (1988–1989, pp. 43–46) estimates the total cost of the tort system in 1985 at $29 to $36 billion.

received an informational letter from the Ford Motor Company (Geyelin 1995). In rejecting the proposed settlement, the court held that the 680,000 owners received "effectively zero," but their attorneys collected handsomely. What often happens, furthermore, is that the probable costs of pursuing a litigation to its ultimate conclusion may appear so great to the corporate defendant, whose lawyers usually get paid progressively as the case proceeds, that they feel compelled to settle the case before the ultimate outcome. This system lends itself to "blackmail" by unscrupulous lawyers for the plaintiffs. At the very minimum, such a system gives rise to frivolous suits that the courts have generally been reluctant to dismiss. Juries often feel understandably sorry for those who are suffering, even though there may be no evidence that the true cause of their injuries is what is claimed. To the best of our knowledge, this aspect of the American legal system is far more extensive than that of the other countries. The English, for example, do not allow contingency-fee cases, although there is talk of changing this, to a limited extent. A more general comparison of British and American legal systems has been made by Huber (1992), who also considers the relation of insurance to liability and the threat to innovation of the American litigation explosion. Contingency fees are not permitted in Germany (Schlesinger 1988, pp. 352–65 and 702–708; Pfenningstorf 1991).

How much has the distinctive U.S. legal system affected America's chemical industry? The effect is very difficult to state with any precision. However, the U.S. tort system does at least plausibly affect the overall rate of productivity growth, as it seems to do in the states. The study by Campbell, Kessler and Shepherd (1996) found that American states which had reformed their tort systems had a productivity more than 20 percent above that of the states who had not. Perhaps that kind of productivity improvement could be measured for the entire United States, as contrasted with its international trading partners, by adopting such rules across the entire country. One comparison of the effects of the legal systems of several countries found that lawsuits consume 2.3 percent of GDP in the United States, whereas in Germany the figure is 1.3 percent, in England 0.8 percent, and in Japan 0.7 percent (from an editorial in the *Wall Street Journal*, May 3, 1996). While these data can only be approximate at best, they seem consistent with the vast number of lawyers that the United States has compared with the other countries. The chemical industries of other countries seem to flourish just as well, or even better in some instances, without the plethora of legal actions the United States has. Of course, it's important to remember that establishing a framework for economic growth is not the only purpose of a legal system. For example, the openness of the U.S. system does contribute to the American society's sense of fairness, an intangible but valuable good.

The avidity with which American lawyers pursue every conceivable possibility for finding alleged victims and filing suits is documented almost every day in the newspapers. One relatively recent example of this litigation problem is found in alleged securities fraud cases. It has particularly affected younger, and often start-up, companies, whose earnings are likely to be volatile. Even cautious forecasts by management of future results may result in declines in the stock value, and virtually a whole legal industry files suit automatically on behalf of stockholders when such a situation occurs.[6] Since these small companies are especially vulnerable to the costs of extensive litigation, they are under pressure to settle.

---

[6] Recent controversy surrounding this problem led President Clinton to veto legislation designed to reform the securities litigation system and led Congress to override that veto, the first of President Clinton's vetoes to be overridden (Taylor 1995).

Another illustration of America's litigation problem is the case of toxic-waste cleanup in the United States. A law was passed years ago to create a Superfund, funded by a tax on raw materials for the industry; but very few sites have actually been cleaned up, and perhaps one-third of the amount spent has gone to the lawyers.[7] This law expired in 1995, but its tax continues to be collected through the various continuing resolutions,[8] and what will happen to it is uncertain at this writing. Has the law helped or hurt the industry? Perhaps it has done little either way, in economic terms. There is solid evidence that the Superfund law, by making all owners of property encumbered with toxic waste jointly and severally liable, has resulted in such property not turning over and cleanup being delayed (Osborn and Williams 1994). The Rand Corporation's Institute for Civil Justice has been studying the costs of Superfund to date, and finds that transaction costs, including legal expenses, probably exceed one-third of the fund's expenditures so far. The largest legal problem for the chemical industry remains the cleanup of the numerous toxic-waste sites and the conditions surrounding the possible reauthorization of the Superfund legislation. See also the chapter by Esteghamat for further details.

The only published study that deals with the effects of the liability system on the chemical industry specifically is that of Johnson (1991), which points out that recent standard reports on the health of the industry do not cite liability as a major concern. Generally, the industry seems to prefer the relative certainty of the regulatory system to the greater uncertainty of the liability system, because one never knows what should not have been done until after it has been done. A prime example of this incertitude, of course, is the blizzard of silicone-breast-implant liability suits that drove the successful Dow-Corning Company into bankruptcy, although there is still no reliable scientific evidence of an injurious effect from ruptured implants, as demonstrated in a book by Marcia Angell (1996), editor of the highly regarded *New England Journal of Medicine*.

The fundamental unmeasured issue, however, is that of what innovations do not occur because of company fears of tort litigation. In one remarkable case, the DuPont company has decided to cease selling any material that they know would be introduced into the human body. This decision was made in the aftermath of a frivolous suit against a bankrupt manufacturer of jaw-joint replacements made of Teflon. In the search for the proverbial deep pockets, DuPont was drawn in. They won the suit, but at the cost of perhaps $10 million to defend themselves against a sale of a few pounds of material to a company whose business was obscure to them. DuPont has been heavily involved in a much larger case involving its fungicide Benlate. At first, the company was willing to make substantial settlement payments to farmers whose crops were allegedly damaged by impurities in the product. However, after extensive additional testing, DuPont decided that there was no merit in these suits and went to trial on many cases. They won most of these suits but lost some others, despite the absence of credible scientific evidence that their product was to blame. While these episodes may not greatly affect the chemical industry as a whole, they add another example to the

---

[7] Epstein (1995, p. 297) reports, "Of the $30 billion spent on the program in the last thirteen years, less than a third went to cleanup work and nearly the same amount went for lawyers." Epstein cites news accounts, including Bukro (1994).

[8] House Conference Report No. 104-350 (November 16, 1995), section 11131, "Extension of Hazardous Substance Superfund Taxes." See also *Superfund Week* (1996, p. 1).

far-greater burden the pharmaceutical and biomedical industries bear in foregoing innovative and important products because of litigation fear.[9]

The cumulative effect of the scale and scope of litigation in the United States has resulted in some recent political moves to redress the balance. Many states have acted, as Campbell, Kessler, and Shepherd (1996) note. Congress had been moving on various tort-reform measures in the mid-1990s, but little has come of these actions. The one significant improvement that has become law was noted earlier—in late 1995, a securities litigation improvement act became law.[10] The lobbying power of the American Trial Lawyers Association often is regarded with awe. No such efforts at tort reform seem to be alive in Germany or Japan.

One area of regulation that does indeed affect the chemical industry in all four of the countries is that governing the environment. This issue is taken up in detail in the chapter in this book by Esteghamat. One comparision that reflects well on the U.S. system is that of Germany, where the chemical industry has found itself severely regulated in many directions, not just environmentally. As one German executive has said in a private communication, "Other countries may issue regulations, but here in Germany we implement them!" Part of the problem surely arises from the crowding of population in Germany and the proximity of large population centers to much-larger single plants than is customary in the United States or Britain. Japan is an even more crowded center, but the government and industry have worked together in the typical Japanese consensus system to solve the problems that have arisen. Indeed, Japan makes no bones about it—the country is organized to favor producers. Certainly, Japan's extraordinary record of growth rate since the Second World War offers some support for this approach, although Japan's recent faltering economy casts some doubt on the underlying justification of their postwar system in the more recent times.

On the other hand, Germany's 20th-century history of war has made the country wary about any genetic-engineering activities by firms or the government. It is therefore no accident that all three of the large German chemical firms have located their biotechnology activities in the United States. Gottfried Plumpe (1996) has stated that in this area, the United States certainly has an advantage, including, of course, the best molecular biologists and research systems (both private and public) of any country and an entrepreneurial spirit that fosters so many new biotechnology companies. Very recently, Germany has begun to emerge from this war-engendered reluctance to engage in biotechnology, and the German companies may now increase their activities in Germany, which also has a very strong research university tradition. The cost of this research, however, continues to escalate, and it is likely that chemical companies (such as Monsanto) and pharmaceutical companies will buy out the small biotechnology firms. But in the case of biotechnology overall, the U.S. government and industry have worked out together and are careful to observe the necessary precautions and test procedures for approval. There are, of course, still groups and individuals who object

---

[9] Another recent illustration of the direction of tort law, although it does not affect the chemical industry directly, is the June 26, 1996, decision by the U.S. Supreme Court that states lawsuits can still be instituted against alleged medical-device injuries, even if the devices comply with federal regulations. Thus, even when there are no manufacturing flaws and the device has gone through government-monitored testing and regulation, suits can still be filed alleging faulty devices.

[10] Public Law 104-67; 109 U.S. Statues 737 (1995), codified in scattered sections of Title 15 of the *U.S. Code*, authored by Cong. C. Cox.

to such research, and they have tried to use the court system to halt it or slow it down, but with only limited success.[11] Biotechnology is the frontier of the pharmaceutical industry in the 21st century, and biology is its lead science, just as chemistry was in the 19th century and physics in the 20th. It would be folly for the United States to suppress this extraordinary scientific breakthrough; rather it needs to be handled constructively, as seems to be happening. Britain and Japan are also engaged in such work but have neither the scale nor the university or entrepreneurial culture so essential for this field to develop to its full potential.

The patent systems of the world apparently are converging. The United States still has its unique feature of allowing the date of first conception to decide the true inventors, whereas patent applications filed by other countries in the U.S. are limited to their date of filing abroad. The difference here can be illustrated in the history of the lengthy proceedings for determining the true inventors of polypropylene polymers, which are the most rapidly growing of the polymers and already of great commercial importance to many companies. The original idea was conceived by Professor Giulio Natta, in Milan, and assigned to Montecatini. But after more than a 20-year battle within the interference action declared by the U.S. Patent Office, Phillips Petroleum Co. emerged as the victor, and the commercial rewards of collecting royalties on existing capacity built under the Montecatini license are huge! This story is told in the previous chapter by Ralph Landau. However, the new World Trade Organization requires the United States to change its patent system to conform more closely to those of most of the other member countries,[12] and there is a vigorous dispute about these proposed changes, or their effect on competitiveness.

One other aspect of intellectual-property protection appears in the matter of trade secrets. In the chemical industry generally, excluding pharmaceuticals, patent protection has not been dominant, especially as the industry has matured. However, trade secrets are ever present. The question of guarding confidentiality by employees or licensees is a difficult matter to address. In many instances, scientists or technologists have left larger companies to strike out for themselves or to work for a competitor. The courts may enforce properly drawn and executed agreements between company and individual employee, but it has become harder to win such cases, as the right of an employee to use his or her general skills and experience in any job is often supported. In 1996, the U.S. government passed the Economic Espionage Act, which will help the preservation of legitimate trade secrets. Silicon Valley has had many such cases, probably inspired initially by the fact that many companies were formed by just such employees leaving their companies. Intel Corp. is a case in point. Its founders worked for the first semiconductor company set up by William Shockley but found it unsuitable. They then formed the Fairchild Semi-Conductor company, which grew quite large, but they again became frustrated and moved again to form Intel, an enormous success that has given the United States a great competitive advantage in microprocessor manufacture. This pattern created a precedent for such movement and spawned a culture for Silicon Valley.

This entrepreneurial culture hardly exists in the other three countries. Particularly in Japan and Germany, such intercompany mobility simply does not occur. Very few personnel ever leave one chemical company to go to another. This stability has good

---

[11] For a description of one such attempt, involving the "Flavr Savr" biologically engineered tomato, see Leary (1994).

[12] Legislation has been introduced (H.R. 359, 105 Cong. 1 sess., 1997) to return the American patent system to a time limit that commences with the grant of, rather than the application for, the patent.

and bad features. The good facet is that people's loyalty to their firm is conspicuous, and lengthy stays improve the stability and wisdom of the company; investments in "human capital" specific to the firm's own environment are undertaken by employees. (When firm loyalty is less, employees shun this kind of investment.) The bad point is that good ideas and technology are not always undertaken by employers, and fresh ideas may never see the light of day. Which effect prevails? It is difficult to measure. The Japanese system of lifetime employment in some large companies has certainly contributed to political stability, a more equitable distribution of the fruits of growth, and the hiring of the best and the brightest. However, it probably has also suppressed innovation. The British entrepreneurial culture is weaker than America's, although stronger than that of either Germany or Japan. The smaller size of the British economy does not offer such opportunities to the technologists or entrepreneurs, and this has been especially true in chemicals, where there are relatively few companies.

In the matter of antitrust, it is notable how different the paths of the U.S. and German chemical industries are. Just about the time (in the 1890s) when the United States was passing its first act, the Sherman Act, the German government legalized cartels (Hawk 1984). As a result, German industry generally, and the chemical industry in particular, entered an era of pervasive cartel formation that was not ended until the Allied victory in the Second World War. These differences have profoundly influenced the industry and corporate structure in the two countries. The net result in growth does not seem to have been affected unequivocally either way, in economic terms.

Britain also possesses antitrust rules: both those of its own creation, and those it has as part of the European Union. Historically, however, antitrust feeling or action has not been strong in Britain, and the formation of ICI as a national standard bearer had led to no substantial concerns about monopoly power. The same is true in Japan, where the large industrial *keiretsu* seem to flourish without legal hindrance. In Germany, the beneficial effects of competition seem to be provided by the fact that the German chemical industry, from its inception, has depended heavily on exports. In the international area, Germany certainly encountered vigorous competition and thus sharpened its business and technological strengths. The Japanese industry, however, had little direct dependence on exports and thus suffered problems that hampered it from becoming a world-class industry, despite its stature as the second-largest chemical industry in the world. ICI also depended substantially on exports and gained from this necessity.

In general, we have found that companies in the other countries may not always abide by confidentiality agreements and that their legal systems cannot address these deficiencies. Lack of mobility among personnel certainly helps them to keep secrets, but various cases suggest that use of confidential property is not considered unethical. Recent examples of pirating of copyrighted or confidential data have occurred in Asia.[13] By and large, the chemical industry generally respects confidentiality, but for a limited period of time. Companies therefore must keep innovating rapidly, so as to render such lost information relatively harmless. This may have been the situation in the recent case of an engineer who stole microprocessor secrets from Advanced Semiconductor Devices and Intel and, after fleeing to Argentina, sold them to, among others, Communist nations! The two companies state that they do not worry unduly, as their progress makes obsolete most of what they lost, and the recipients do not have the capability to build on the purloined information.

---

[13] One of the greatest scandals of the recent past involved a scheme by Hitachi to steal trade secrets from IBM. See *New York Times* (1983, pp. D1–D15.)

REFERENCES

Angell, Marcia (1996), *Science on Trial: The Clash of Medical Evidence and the Law in the Breast Implant Cases*, W.W. Norton, New York.

Barnes, Patricia (1995), "Cutting Classes," *ABA Journal* (December) 81: 27.

Brown, John P. (1973), "Toward an Economic Theory of Liability," *Journal of Legal Studies* (June) II: 323–49.

Bukro, Casey (1994), "EPA Out to Clean Up 'Ridiculous' Situation," *Chicago Tribune* (11 April) section 4 (North Final Edition), pp. 1–2

Campbell, Thomas J., Daniel P. Kessler, and George B. Shepherd (1996), "Liability Reforms and Economic Performance," in Ralph Landau, Timothy Taylor, and Gavin Wright (eds.), *The Mosaic of Economic Growth*, Stanford University Press, Stanford, CA, pp. 267–80.

Economic Espionage Act of 1997, *Chemical & Engineering News*, December 1, p. 9.

Engineering Manpower Commision (1991), *Modern Plastics* (June) 68.

Epstein, Richard (1995), *Simple Rules for a Complex World*, Harvard University Press, Cambridge, MA.

Galanter, Marc (1983), "Reading the Landscape of Disputes: What We Know and Don't Know (And Think We Know) About our Allegedly Contentious and Litigious Society," *U.C.L.A. Law Review 4* (October) 31(1): 4–70.

Geyelin, Milo (1995), "Settlement Between Bronco II Owners, Ford Is Thrown Out by Federal Judge," *Wall Street Journal* (24 March) section A, p. 3.

Goldberg, Robert M. (1995), "Breaking Up the FDA's Medical Information Monopoly," *Regulation* (2): 40–52.

Hager, Mark M. (1990), "Civil Compensation and its Discontents: A Response to Huber," *Stanford Law Review* (January) 42(2): 539–80.

Hawk, Barry E. (1984), *United States, Common Market and International Antitrust: A Comparative Guide*, 2nd edition, Prentice Hall Law & Business, Clifton, NJ.

Horowitz, M.J. (1992), *The Transformation of American Law, 1870–1960: The Crisis of Legal Orthodoxy*, Oxford University Press, New York, NY.

Huber, Peter W. (1988), *Liability: The Legal Revolution and Its Consequences*, Basic Books, New York, NY.

Huber, Peter W. (1992), "Liability and Insurance Problems in the Commercialization of New Products," in Nathan Rosenberg, Ralph Landau, and David Mowery (eds.), *Technology and the Wealth of Nations*, Stanford University Press, Stanford, CA.

Iyori, Hiroshi and Akinori Uesugi (1983), *The Antimonopoly Laws of Japan*, Federal Legal Publications, New York, NY.

Johnson, Rollin B. (1991), "The Impact of Liability in the Chemical Industry," in Peter W. Huber and Robert E. Litan (eds.), *The Liability Maze: The Impact of Liability Law on Safety in Innovation*, Brookings Institution, Washington, DC, pp. 428–55.

Kolata, G. (1995), "A Case of Justice, or a total Travesty? How the Battle Over Breast Implants Took Dow Corning to Chapter 11," *New York Times* (13 June) section D, p. 1.

Leary, Warren E. (1994), "FDA Approves Altered Tomato That Will Remain Fresh Longer," *New York Times* (19 May) section A, p. 1; section B, p. 7.

Osborn, John E. and Steven N. Williams (1994), "Reauthorizing Superfund: Problems and Prospects," *N.Y. Law Journal* (29 March) 211: 1–4.

New York Times (1983), "Hitachi Guilty in IBM Case" (9 February) section D, pp. 1–15.

Pfenningstorf, Werner, with Donald G. Gifford (1991), *A Comparative Study of Liability Law and Compensation Schemes in Ten Countries and the United States*, Insurance Research Council, Oak Brook, IL.

Plumpe, Gottfried (1996), in Ralph Landau, *Strategy for Economic Growth*; in Ralph Landau, Timothy Taylor, and Gavin Wright (eds.), *The Mosaic of Economic Growth*, Stanford University Press, Stanford, CA.

Priest, George L. (1993), "Lawyers, Liability, and Law Reform: Effects on American Economic Growth and Trade Competitiveness," *Denver University Law Review*, 71(1): 115–49.

Rand Institute, *Annual Report*, 1988–1989, Institute for Civil Justice, Santa Monica, CA.

Rubin, Gerry R. and D. Sugarman (1984), "Towards a New History of Law and Material Society in England 1970–1914," in Gerry R. Rubin and David Sugarman, *Law, Economy and Society, 1750–1914; Essays in the History of English Law*, Professional Books, Abingdon, U.K. pp. 1–123.

Schlesinger, Rudolf B. (1988), *Comparative Law: Cases, Text, Materials*, 5th edition, Foundation Press, Minola, NY.

Superfund Week (1996), "Superfund Gets Hit by Government Shutdown" (5 January) 10(1): 1, Pasha Publications.

Taylor, Jeffrey (1995), "Congress Sends Business a Christmas Gift, Veto is Overridden on Bill Curbing Securities Lawsuits," *Wall Street Journal* (26 December) section A (Eastern Edition), p. 2.

7 Technological Change in Chemicals: The Role of University–Industry Relations

NATHAN ROSENBERG

INTRODUCTION

This chapter examines the role played by universities in generating technological change in the chemical sector. For reasons that will become apparent, it will pay particular attention to the emergence of the discipline of chemical engineering, a discipline that has played a critical role in technological change in the chemical industry over the past century. By general consensus, chemical engineering was a distinctly American achievement. It will be argued here that the catalytic agent has to be found in the unique nature of the American university–industry interface.

Against the backdrop of 19th-century industrialization, American priority in this achievement is surprising. Britain was the acknowledged "workshop of the world" in the mid-19th century, and that claim applied to the chemical sector just as appropriately as to the mechanical and steam-driven realms. Britain's textile industry, which absorbed a large proportion of the output of the chemical industry, was by far the largest in the world. In 1852, for example, Britain possessed well over one-half of the world's cotton spindles (18 million). The United States was second, with 5.5 million spindles; France third, with 4.5 million; and Germany a very distant fourth, with 900,000 (Landes 1969, p. 215). Moreover, Britain's chemical industry, whose output was overwhelmingly inorganic, was also the largest in the world in 1850. This had little to do with the technological progressivity of the British industry. Rather, it was primarily attributable to its role in satisfying the enormous needs of textile manufacturers for such essential inputs as bleaches, mordants, and detergents, as well as the huge requirements of the glass industry for soda and, therefore, sulfuric acid. In the 1850s Britain was by far the largest producer of heavy chemicals: sulfuric acid, soda, bleaching powder, and so on.

Yet in terms of professional engineering associations or the commencement of teaching curricula, chemical engineering in Great Britain is clearly a late arrival. British

I owe a huge debt to Ralph Landau for patiently guiding me through many of the complexities of chemical engineering. He has been a superb teacher dealing with an, at best, indifferent pupil, and I must therefore insist on sole proprietorship of all errors or opacities in this paper. Ashish Arora, Alfonso Gambardella, and Peter Murmann also have been exteemely helpful.

mechanical engineers formed their professional association, the Institution of Mechanical Engineers, in 1847, in the wake of power-driven machinery, and the Institution of Electrical Engineers was founded in 1871. But it was not until 1922 that the Institution of Chemical Engineers was founded, certainly stimulated, at least in part, by serious production difficulties encountered during World War I—e.g., in introducing the important Haber/Bosch process. Also in 1922, a chair in chemical engineering was established in Imperial College, London.

By contrast, the American Institute of Chemical Engineers (AIChE) had already been founded 14 years earlier, in 1908, and the British Institution of Chemical Engineers was closely patterned after the earlier American model (Guedon 1980). Britain's Society of Chemical Industry had established a chemical engineering group within its organization in 1918. Cambridge University established its chemical engineering department only after the Second World War. The discipline subsequently played an influential role in facilitating Britain's entry into the realm of petrochemicals after the Second World War.

It is a further revealing measure of the "arms length" at which the British establishment held the chemical engineer that the Institution of Chemical Engineers failed to receive a Royal Charter until as late as 1957. In spite of strenuous efforts by the Institution of Chemical Engineers, the number of students trained in chemical engineering in British institutions remained extremely modest. By the outbreak of the Second World War in 1939, the cumulative total of such students had barely reached 400, a number that had been exceeded at MIT alone in the four-year period from 1920 to 1924.[1]

It seems natural to ask, Why didn't the discipline of chemical engineering emerge first in Britain? Notice that this question is not the same as one that has been commonly asked: Why did Britain fail to respond to the opportunities opened up by the momentous initial British breakthrough in synthetic dyes—a breakthrough that will always be associated with the 18-year-old William Henry Perkin's great initial foray into the world of synthetic dyes in 1856? The two questions are not entirely unrelated, but they are by no means the same. Chemical engineering deals with a much broader set of problems than does synthetic organic chemistry. As we will see, the products that emerged from the laboratory researches of the German synthetic-dye chemists and, later, German pharmaceutical firms, did not make great demands on the particular expertise of the chemical-engineering profession (Spitz 1988, pp. 58–59).

An initial response to the question of why the discipline of chemical engineering did not emerge first in Britain is that the question is incorrectly formulated, because the basic concept of chemical engineering *was* first articulated by an Englishman in the 1880s. George E. Davis called for the formation of a society of chemical engineers in 1880, reflecting his basic insight that many of the problems confronting the chemical industry were really reducible to a small number of common engineering problems. At the time, the industrial context was dominated not by the colorful laboratory breakthroughs in synthetic organic chemistry but by a preoccupation with heavy chemicals such as sulfuric acid and sodium carbonate. The Society of Chemical Industry was formed, an organization that remains active today in both the United States and the United Kingdom. But the vast majority of those in the society in the 1880s certainly did not think of themselves as engineers. Rather, they regarded themselves as chemists with an expertise in the specific product of a particular industry, and not members of a professional group that shared a broad commonality cutting across the boundary lines of a number of industries (Scriven 1991).[2]

---

[1]Divall (1994) provides useful information concerning the changing content of British chemical education in the interwar and postwar years, as well as the ways in which this content reflected the influence of the Institution vis-a-vis the priorities of private industry and the universities themselves.

Davis delivered a course of lectures on chemical engineering in the Manchester Technical School in 1887. That course grew into a two-volume *Handbook of Chemical Engineering* and was published in Manchester in 1901. But Davis's lectures never became part of an engineering curriculum. In contrast, in 1888 Professor Lewis Mills Norton taught a course in the Department of Chemistry at MIT that was labeled "chemical engineering."[3] This MIT course was "mainly descriptive, and consisted of a series of lectures describing the commercial manufacture of chemicals used in industry. Material was taken in considerable degree from German practice" (Weber 1980, p. 7). In particular, Norton's course offered little treatment of the mechanical-engineering aspects of the design of large-scale chemical-process plants. It was more deserving of the term "industrial chemistry." Industrial chemists at this time focused on the production of each of a large number of chemical products. They described each of the sequences of steps, from beginning to end, in the production of individual products, but they identified few unifying principles among the manufacturing processes of different products.

Davis recognized that "to produce a competent Chemical Engineer the knowledge of chemistry, engineering, and physics must be co-equal." Without using the term "unit operations," which would not be introduced by A. D. Little until 1915, he clearly identified a number of operations that are common to a wide range of manufacturing activities producing different chemical products. In fact, as W. K. Lewis (1953, p. 699) has pointed out, "Few indeed are the really fundamental phases of chemical engineering to which he does not call attention." Nevertheless, there were no departments of chemical engineering in Britain or, indeed, departments of chemical engineering anywhere outside of the United States, until the 1930s.

WHY IN AMERICA?

The intriguing questions, then, are why Davis's initial conceptualization seems to have fallen on such infertile ground in Britain, and why the further development and diffusion of the field occurred first in the United States.

A preliminary historical observation is that, even in the first half of the 20th century, Britain's two preeminent institutions of higher learning, Oxford and Cambridge Universities, continued to be dominated by the preindustrial, indeed by the prebourgeois, values of a landed aristocracy. John Beer offers a telling quotation delivered by Perkin to an audience of British chemists on the occasion of the 50th anniversary of his momentous discovery of mauve, the first of the aniline dyes. Perkin, who had grown rich from the commercial exploitation of his discovery, observed, "I remember the time when it was said that by my example I had done harm to science and diverted the minds of young men from pure to applied science and it is possible that for a short time some were attracted to the study of chemistry by other than truly scientific motives...." Beer then goes on to make the following incisive comment on Perkin's remark: "Notice first his prejudice toward the man who studies chemistry to make money, and second the insinuation that applied chemistry carries with it the danger of attracting such basely motivated men; lastly notice that he was on the defensive" (1959, pp. 45–46).

In Britain at the turn of the century, the notion that universities had an obligation to pursue scientific research, much less engineering research, remained an alien concept.

[2]Scriven (1991) points out, however, that in 1881 14 of the first 297 members of this society were already calling themselves "chemical engineers."

[3]See Tailby (1982) for a discussion of the prehistory and early history of chemical engineering in Britain.

The very notion of a university as an institution committed to scientific research was, historically speaking, a German 19th-century invention. Germany provided the model for an advanced degree in research, the Ph.D., first in physics, then later in chemistry. This model was not imported into Great Britain, but it flourished in the United States, first at the Johns Hopkins University, which was founded in 1876, and which became the American model for a research university in the 1880s. The model was quickly imitated, for example at the University of Chicago, which opened its doors in 1892, and especially at the state universities that had emerged after the passage of the Morrill Act in 1862.

The state universities were initially founded by land grants from the federal government and subsequently depended on state legislatures for financial support. Partly for this reason, they were unapologetically utilitarian in their definition of the goals of higher education and the role of university research. Indeed, the Morrill Act itself had specified that, at these state universities, a "leading object shall be, without excluding other scientific and classical studies,... to teach such branches of learning as are related to agriculture and the mechanic arts."[4] The Morrill Act reaffirmed and institutionalized a cultural feature that had long been pointed to by European visitors: a strong predilection for the useful and practical, including the immediately useful aspects of science. Alexis de Tocqueville (1898, vol. 2, pp. 40, 48, 52–53) had observed years before the passage of the Morrill Act:

> It must be acknowledged, that in few of the civilized nations of our time have the higher sciences made less progress than in the United States.... In America, the purely practical part of science is admirably understood and careful attention is paid to the theoretical portion, which is immediately requisite to application. On this head, the Americans always display a clear, free, original, and inventive power of mind. But hardly any one in the United States devotes himself to the essentially theoretical and abstract portion of human knowledge... every new method which leads by a shorter road to wealth, every machine which spares labor, every instrument which diminishes the cost of production, every discovery which facilitates pleasures or augments them, seems [to such people] to be the grandest effort of the human intellect. It is chiefly from these motives that a democratic people addicts itself to scientific pursuits.... In a community thus organized, it may easily be conceived that the human mind may be led insensibly to the neglect of theory; and that it is urged, on the contrary, with unparalleled energy, to the applications of science, or at least to that portion of theoretical science which is necessary to those who make such applications.

These observations by an astute young French aristocrat deftly capture an attitude of mind that was to characterize American culture for many decades to come.

THE QUESTION OF SCALE

With respect to the question of the emergence of chemical engineering, it is necessary to begin with a specification of the content of that discipline. Just what is it that chemical engineers do? Chemical engineers are fundamentally concerned with the phenomenon of scale, which is central to commercial success in the manufacture of chemical products. More comprehensively, we should say that chemical engineers are concerned with the design, construction, and operation of large-scale chemical-process

[4]For more detail on the Morrill Act, see Ben-David (1971) and Rosenberg and Nelson (1994).

plants (Landau 1966).[5] New products in the chemical industries have, historically, typically emerged out of laboratory research conducted by bench chemists making use of small beakers, test tubes, and retorts. However, laboratory research does not provide the information required for scaling up to commercial production. In the most practical sense, scaling up the original equipment to a size appropriate for commercial production is often physically impossible and hardly ever economically feasible. One cannot, for example, readily scale up with glass containers. Instead, scaling up requires recourse to apparatus of an entirely different sort, involving entirely different materials that can sustain extremely high pressures and temperatures, as well as pumps, compressors, piping, and vats of a very large scale. As Hougen (1965, p. 230) has observed, "Many significant effects appear in plant-scale operation which escape detection on a bench scale, such as changes in flow patterns, fouling of heat exchangers, deactivation of catalysts, and variations in feed composition...."

What constitutes an optimal scale for commercial production requires experimentation of a sort entirely different from that which led to the original development of the product. The use of pilot plants, and inferences drawn from experimental data provided by pilot plants, will determine optimal size, which will differ from one product line to another. The benefits of larger size are so pervasive in the chemical industry that chemical engineers have, in the past, made extensive use of a "six-tenths rule," which states that capital costs increase by only 60% of the increase in a chemical-process plant's rated capacity. The intuitive reason behind these economies of scale is that when the capacity of a pressure vessel is doubled, it is not necessary to double the surface area of the steel of which the vessel is made; and the number (or size) of monitoring instruments may not increase at all as the chemical-processing plant increases in size. The size of the exponent relating investment cost to scale of plant has in fact been shown to vary within the range of 0.4 to 0.8 in different chemical products. In the case of ethylene, which is today one of the two most important petrochemicals (ammonia is the other), the average for ethylene plants has been very close to the six-tenths rule, or 0.58 (Chapman 1991, pp. 125–26; Pigford 1976, pp. 197–98).

It is conceptually important, however, that growth in capacity for a chemical product not be attributed entirely to static economies of scale. Moving to larger scale will often require technological innovations in plant design, equipment manufacture, and process operation to make that larger scale feasible. Thus, the benefits of scale are unattainable until certain facilitating technological conditions have been fulfilled (Landau and Rosenberg 1994). To the extent that this is so, the economies of large-scale production are, in fact, inseparable from technological change, although it is a widespread practice in economics (for example, in the "growth accounting" literature) to treat them as if they were separable.

The critical point for present purposes is that various circumstances drove the American chemical industry to larger volumes of output, and scale, at an earlier point in time than is generally appreciated. This failure in appreciation owes much to the widespread tendency to focus on the novel industrial developments flowing from the dramatic breakthroughs in the synthetic-dye industry after Perkin. Here indeed, German scientific and commercial dominance was, by the 1880s, simply overwhelming. At the beginning of the century, U.S. production of organic chemicals did not remotely compare in their technical sophistication to that of Germany, on whose imports the United States remained heavily dependent. In fact, at the outbreak of World War I, only two significant domestic producers of dyes served America's huge textile industry, and

---

[5]The more comprehensive definition is important because engineers continue to generate smaller but numerous incremental improvements after a new plant has been constructed.

their meager 3,000 tons of annual output amounted to no more than one-eighth of the nation's requirements. In this respect, the termination of German synthetic-dye imports was a decisive event in the emergence of an American organic chemical industry (Aftalion 1991, p. 115–19). Looked at from a German perspective, 8 German firms, along with their foreign subsidiaries, accounted for 140,000 tons a year of a world total of 160,000 tons a year. No less than 80 percent of German output was exported (Aftalion 1991, p. 104).

But in terms of the sheer volume of output of *all* chemicals, as opposed to synthetic dyes alone, the situation was very different. The United States already had a large chemical industry by the beginning of the 20th century, mainly as a result of its rapid growth in inorganic chemicals and explosives in the closing decades of the 19th century. If one employs that once-favorite yardstick of chemical activity, the output of sulfuric acid, the United States in 1914 was producing almost as much as Germany and Great Britain combined (Haynes 1945–54, vol. II, p. xiv; Aftalion 1991, pp. 115–19). The large size of the American market had introduced American firms, at an early stage, to the problems involved in the large-volume production of basic products, such as chlorine, caustic soda, soda ash, sulfuric acid, superphosphates, and so on. This ability to deal with a large volume of output, and eventually to do so with continuous process technology invented and designed by chemical engineers, was to become a central feature of the chemical industry in the 20th century. In this respect, the early American experience with scale was important in accounting for the industry's later success.

The chemical sector, it must be realized, was drawing on skills in mechanical engineering and machine building that had enabled Americans to establish a distinctive competence among industrial nations in the second half of the 19th century. As early as the 1850s, British machinery experts visiting the United States wrote, sometimes in awestruck terms, of the unique "American System of Manufactures," involving the use of sequences of highly specialized tools and precision component parts in producing very large volumes of standardized final products (Rosenberg 1969; 1976, ch. 1; Hounshell 1984). Thus, the American chemical engineer was drawing on a rich accumulation of machine-design and machine-building experience that was well suited to the needs of bigness in the chemical sector.[6] This earlier movement to large-scale operations, not just in one line of chemical production processes but, simultaneously, in many, is a distinctive feature of the American scene. Even before the turn of the century, American industrial growth was creating widespread pressures toward the design or the adoption of chemical-process plants of larger size, especially for the major inorganic products of the day. German chemists were better trained than their American counterparts at the turn of the century, but American engineers were better suited for the design and the delivery of large-scale chemical plants that offered the prospect of great efficiency improvements in the opening decades of the 20th century. One authoritative study, discussing the American situation shortly before its entry into the First World War, referred tellingly to "the American attitude to the size of chemical works, which was, in short, to build a large plant and then find a market for the products" (Haber 1971, p. 176).

The American economy was already being driven toward the high-volume, high-capital-intensity end of the spectrum by a variety of forces, including large domestic markets that were expected to grow rapidly, and the high cost of labor. Moreover, American universities from the 1860s on were as receptive to the teaching of engineering

---

[6]Frank Howard, in his book on the American synthetic rubber industry, published in 1947, made the astute observation that "it is the chemical engineer who has given modern oil and chemical industries the equivalent of the mass-production techniques of our mechanical industries" (Howard 1947, p. 241).

subjects as German universities were ideologically hostile, relegating the subjects to polytechnics and later to technical universities. Whereas German universities became the home of scientific research early in the 19th century, American universities embraced engineering disciplines very early but were slow to open their doors to scientific research.

With respect to the chemical sector, Haber (1971, p. 22) has noted the tendency toward large plant size in the manufacture of sulfuric acid in the United States by the turn of the century, and Trescott (1982) describes an illuminating episode in the pre-World War I years in which an American firm, General Chemical Company, invented around a process for the large-scale production of concentrated sulfuric acid that had been developed and patented by BASF. A patent infringement suit by BASF against General Chemical was settled out of court by an exchange of patents and other agreements, giving each firm access to information concerning the other's process technologies. But after officials from BASF visited the facilities at General Chemicals, they were so impressed by the efficiency of the larger scale of plants that they placed a large order for General Chemical to design and construct four plants at Ludwigshafen. Each of these units was four times the size of the largest units at Ludwigshafen at that time. General Chemical later received a similar order from Bayer for a large plant that, ironically, was to play a major role in supplying the German munitions industry in World War I.[7]

Another striking case was the rapid and extensive adoption of Solvay's ammonia-soda process in the United States. This process, invented in 1872 by the Belgian Ernest Solvay, involved a number of sub-inventions that allowed the product to be produced not only on a large scale but also by a continuous process. "In short, the Belgian with no university education performed as what would come to be called a chemical engineer. Though this was not evident to his contemporaries, his performance did catch some attention in England, and it surely impressed the aggressive Americans. They soon licensed the process, integrating it and its principles into a fast-developing inorganic chemicals industry that would be invading European markets around the turn of the century" (Scriven 1991, pp. 16–17).

THE BEGINNING OF CHEMICAL ENGINEERING

The American impetus toward large-volume production gave rise to a new engineering discipline, pioneered by MIT in the second decade of the 20th century.[8] As mentioned earlier, Lewis Mills Norton had taught a course as early as 1888 that was labeled

---

[7]Trescott's (1982) account is based on Haynes (1945–54, vol. 1, pp. 265–66). Trescott's 1981 book provides much evidence of large-scale plants in U.S. electrochemicals.

[8]Since this is an article and not a book, it cannot pretend to provide a comprehensive coverage of the contributions of numerous universities and firms to the emergence of the discipline of chemical engineering. The emphasis here on the role of MIT reflects not only the remarkable early contribution of that institution to the discipline but also the author's decision to focus on the sources of American strength in the world of petrochemicals. That focus inevitably led back to MIT's undoubted predominance in the pre-World War II development of the technology of petroleum refining.

A more comprehensive treatment of the subject of this paper would have devoted much more attention not only to the role of such universities as those of Minnesota, Wisconsin, Michigan, Illinois, and Delaware but also to the numerous contributions of DuPont. That corporation made major additions to the generalizable knowledge base that chemical engineering eventually acquired in such fundamental subjects as heat and mass transfer and transport. Fortunately, the contributions of DuPont are carefully examined in the splendid book by David Hounshell and John K. Smith, *Science and Corporate Strategy* (1988), especially in chapter 14.

TABLE 7-1. Origins of Chemical Engineering Departments from Non-Chemistry Departments

Department	School
Mechanical Engineering	U. Colorado
	U. Rochester
Petroleum Engineering	U. Tulsa
	U. Wyoming
Ceramics & Mining Engineering	Iowa State
Sugar Engineering	Louisiana State
Paper Engineering	U. Lowell
Electrical Engineering	U. Wisconsin
General Engineering	U. Toledo
	McNeese State
Energy Engineering	U. Illinois, Chicago Circle
	U. Wisconsin, Milwaukee
YMCA Extension	Cleveland State

Source: Westwater (1980, p. 147).

"chemical engineering" but was, in fact, little more than descriptive industrial chemistry. Other universities that were early (pre-20th century) entrants in the teaching of something that was at least labeled "chemical engineering" were the University of Pennsylvania (1892), Tulane University (1894), and the University of Michigan (1898) (Landau 1958, pp. 63–64).

In the course of the 20th century, many American universities established departments of chemical engineering. As one might expect, the majority of these departments trace their origins to departments of chemistry, and a few appear to have emerged full-blown without any obvious connections to preexisting departments. However, a third category has been identified of 13 universities where chemical engineering departments emerged out of departments other than chemistry. The category is revealing because it exhibits the diverse range of productive activities that came to rely on the kinds of skills in the possession of chemical engineers. It also reveals something of the distinctiveness of American universities in accommodating the special requirements of local industry. As will be seen from Table 7-1, these include origins in petroleum engineering in the oil-producing states of Oklahoma and Wyoming, ceramics and mining engineering in Iowa, sugar engineering in Louisiana, and paper engineering in Massachusetts.

At the University of Wisconsin, the chemical engineering department originated out of the interests of a faculty member in the electrical engineering department by the name of Charles F. Burgess. Burgess was actively involved in the electrochemical industry after 1900 and was engaged in electrolytical research. In 1900, he introduced what may have been the first course in applied electrochemistry that was taught at an American college. In addition, Burgess organized in 1910 the Northern Chemical Engineering Laboratories, which was to become the Burgess Battery Company in 1917. The department was originally founded as the Department of Applied Electrochemistry and Chemical Engineering (Trescott 1981, pp. 244, 247, 252, 264).

The central figure who forcefully articulated in 1915 what came to be the unifying concept of this new discipline was Arthur D. Little. The concept is now called "unit operations." Little presented this concept in a report to the Corporation of the Massachusetts Institute of Technology in December 1915, a report that also led to the establishment that year of the School of Chemical Engineering Practice at MIT, an arrangement that allowed students to gain experience in industry, relating theory to applications while working under faculty supervision. In Little's (1933, pp. 7–8) words:

> Any chemical process, on whatever scale conducted, may be resolved into a coordinated series of what may be termed "unit actions," as pulverizing, mixing, heating, roasting, absorbing, condensing, lixiviating, precipitating, crystallizing, filtering, dissolving, electrolyzing, and so on. The number of these basic unit operations is not very large and relatively few of them are involved in any particular process.... Chemical engineering research... is directed toward the improvement, control and better coordination of these unit operations and the selection or development of the equipment in which they are carried out. It is obviously concerned with the testing and the provision of materials of construction which shall function safely, resist corrosion, and withstand the indicated conditions of temperature and pressure. Its ultimate objective is so to provide and organize the means for conducting a chemical process that the plant shall operate safely, efficiently, and profitably.

This statement of responsibilities is surely far broader than would have been acceptable to Great Britain's Society of Chemical Industry, even without the reference to assuring that the process will be carried out "profitably." It is also a conceptualization far removed from that of laboratory chemists, who were unconcerned with the design of plant and equipment on an industrial scale.

Little had run a consulting firm for many years and was therefore especially well positioned to observe manufacturing technologies from a comparative perspective. It is interesting to note that George E. Davis in the United Kingdom also earned his living as an industrial consultant, which also gave him the chance to observe many technologies, to "abstract" the common features or processes of individual firms — and therefore to advise ways in which these processes might be improved. Davis, however, encountered more difficulties than did Little with a concern over the violation of trade secrets (Donnelly 1988). Little's main focus was the paper and pulp industry, where he consulted very widely and where, as a young man, he had worked for the first U.S. mill that employed the new sulfite process for converting wood pulp into paper. That experience was to play an important role in Little's connections with MIT, where he lectured on papermaking from 1893 to 1916 (Kahn 1981, pp. 25, 49). Little's wide-ranging consulting experience, concentrated as it was on industries in which chemical transformations were a central part of the manufacturing process, probably played a formative role in his growing awareness of the extent to which firms in different product lines were making use of similar chemical-process operations.

A critical feature of the concept of unit operations is that it went well beyond the descriptive approach of industrial chemistry by calling attention to a small number of distinctive processes that were common to many product lines. This act of intellectual abstraction laid the foundations for a more rigorous and, eventually, more quantitative discipline. It is tempting to call Little's manifesto an attempt to provide a General Purpose Technology for the chemical sector. Chemical engineering was now in a position to be able to accumulate a set of methodological tools that could be refined and that could provide the basis for a wide range of problem-solving activities connected with the design of chemical-process plants. Not least important, it now had the basis for a curriculum that could be taught (Pigford 1976, p. 191). In 1920, just a few years after Little's formulation, chemical engineering achieved the status of a

TABLE 7-2. **Number of Students Attending Engineering Courses Selected Years: 1900 to 1918**

	1900	1904	1906	1910	1916	1918
Mechanical engineering	4459	6894	7426	6377	7707	5944
Electrical engineering	2555	4389	5696	5450	7098	5683
Chemical engineering	536	694	1234	869	2774	4548

Note: Reynolds (1986, p. 12) states that number of students in chemical engineering in 1920 to 1921 was "almost 6,000."

Source: Haber (1971, p. 63).

separate, independent department at MIT, under the chairmanship of W. K. Lewis. The growing interest in chemical engineering had, of course, been powerfully strengthened by the military requirements of the First World War. Indeed, the termination of German synthetic-dye imports during the war was a decisive event in the emergence of an American organic-chemical industry (Aftalion 1991, pp. 115–19). Table 7-2 shows the number of students taking courses in different fields of engineering in the first 18 years of the century.

THE MATURING OF CHEMICAL ENGINEERING

After the articulation of the concept of unit operations, chemical engineering was in a position to accumulate a set of methodological tools that could provide the basis for a wide range of problem-solving activities connected with the design of chemical-process plants. The awareness that a limited number of similar operations were common to many industries served to identify research priorities and therefore to point to a disciplinary research agenda. This had not been possible so long as technologists remained lost in the particularities of innumerable specific products. For contrast, a widely used British reference book for technical workers in the mid-19th century began its article on "Distillation" as follows: "Distillation means, in the commercial language of this country, the manufacture of intoxicating spirits; under which are comprehended the four processes, of mashing the vegetable materials, cooling the worts, exciting the vinous fermentation, and separating by a peculiar vessel called a still, the alcohol combined with more or less water" (Ure 1853; as quoted in Lewis 1959). Such "tunnel vision" with respect to an operation as common as distillation was obviously incompatible with the notion of a chemical engineer as someone who dealt with a limited number of operations common to a wide range of industries.

The establishment of a curriculum of marketable skills had to await more extensive interaction between universities and private industry. Eventually, however, almost all American chemical engineering departments followed the MIT lead, establishing themselves as independent departments inside the school of engineering, although Cal Tech, the University of California at Berkeley, and the University of Illinois are distinguished exceptions. Figure 7-1, by Olaf Hougen, provides a useful guide to the changing content of an undergraduate education in chemical engineering over the period from 1905 to 1965.

The teaching of chemical engineering was organized around the concept of unit operations for the next few decades, but it must be immediately added that the concept

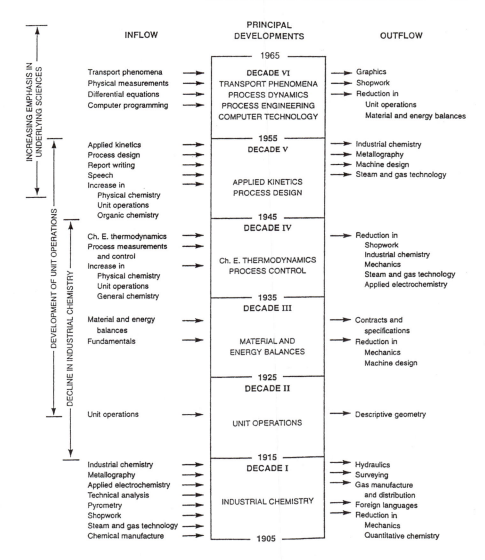

Figure 7-1. Initial Curriculum in 1906 Consisted of Separate Courses in Chemistry and Conventional Engineering. *Source:* Hougen (1965, p. 223).

underwent substantial alteration in its intellectual content, almost from the beginning. An academic environment created strong pressures toward analytical rigor, internal consistency, and generality. One of the first major steps in this direction occurred very quickly, in the form of an immensely influential textbook by Walker, Lewis, and McAdams, which was published in 1923. This book, *Principles of Chemical Engineering*, focused on a limited number of operations that were widely practiced in industry and also provided a conceptual framework that would be useful for addressing a range of unsolved problems common in chemical manufacturing processes. An important feature of the book was its demonstration that many separate unit operations could be compressed into a small number of principles involving momentum, mass, and heat transfer. The book remained extremely influential in the training of chemical engineering students for 25 or 30 years. It amply fulfilled the statement of purpose of the authors (p. vii): "We have selected for treatment basic operations common to all chemical industries, rather than details of specific processes, and so far as is now possible, the treatment is mathematically quantitative as well as qualitatively descriptive."

An engineer trained in terms of unit operations could mix and match these operations as necessary in order to produce a wide variety of distinct final products. Such an engineer was much more flexible and resourceful in his approach to problem solving. Of key importance is that he was now well equipped to take techniques and methods from one branch of industry and to transfer them to other branches. Experience in one place could now be readily transferred to other, apparently unrelated, places (Spitz 1988, pp. 58–59). This capability was especially valuable in the innovation process—particularly as new materials and new products emerged. Thus, research that improved the efficiency of any one process was now likely to be more quickly employed in a large number of places. Putting the point somewhat differently, the identification of a small number of unit operations common to a large number of industries meant that it was now possible to identify specific research topics for which new findings could be confidently expected to experience widespread utilization. Needless to say, this point is of great significance in the growth of an engineering discipline.

The 1930s saw an important textbook innovation by Olaf Hougen and Kenneth Watson of the University of Wisconsin in their *Industrial Chemical Calculations* (1936). Hougen and Watson "showed how thermodynamic principles could be applied effectively to understanding such complex processes as roasting ores, manufacturing synthetic gas, and thermal cracking of petroleum" (Pigford 1976). Chemical engineers at the University of Wisconsin subsequently played a leadership role in reaction engineering.

By the early 1930s, an important distinction had been clarified in the concept of unit operations. Whereas the terms "unit operations" and "unit processes" previously had been used as if they were interchangeable, unit operations now were taken to mean "any mechanical or physical operation of unit nature, used to condition one or more of the reacting materials in a technical process or to control their environment. Unit operations bring about changes of state, of position, or of energy content, but they do not affect chemical constitution or molecular structure." On the other hand, unit processes are now defined "as representing the embodiment of all factors in the technical application of individual reactions. Chemical change, rather than physical or mechanical, is the distinguishing characteristic of unit processes" (Olive 1934, p. 229).

The years after World War II brought significant changes in the discipline of chemical engineering, especially as a result of the federal government emerging as a generous patron of university research. The availability of federal funds reinforced the thrust of university chemical engineers toward greater rigor and more extensive use of scientific concepts, while simultaneously loosening what had been a common dependence on industry for financial support. Indeed, in the postwar period, the recipient of a doctoral degree in chemical engineering would sometimes assume an academic position immediately after receiving the degree. Such a possibility would have been extremely rare before World War II. Part of the reason for this change is that the number of positions available in university chemical engineering departments increased sharply after the war. But another part of the reason is that academic chemical engineering began to take on a life of its own.

In their authoritative study of DuPont, Hounshell and Smith observe that by the 1950s, "[A]cademic chemical engineers began to work on research problems based on increasingly sophisticated mathematical theories. Much of this research had little or no relevance to the problems faced by engineers in industry" (1988, p. 285). There seems little doubt that, at least from industry's perspective, chemical engineering became "too academic" in these years. Speaking of American engineering education in the postwar period, the MIT Commission on Industrial Productivity complained, "Engineering students are taught to analyze systems but not really to design them. Many faculty members have little or no industrial experience, and few of those who do have such experience have worked directly in production-related positions. Engineering graduates of the nation's leading universities thus typically enter industry with little knowledge of manufacturing. Their education has prepared them primarily for careers in research and development" (Dertouzos, Lester, and Solow 1989, p. 79).

But while engineering disciplines did become more remote from the needs of industry as they devoted more effort to preparing students for careers in academia or industrial R&D, chemical engineering did not move as far away as other engineering disciplines did from process design activity, since that is the central focus of the discipline. The MIT Commission also wrote, "Apart from the field of chemical engineering, which in large measure *is* process engineering, university engineering schools have contributed little to the engineering of industrial-production processes" (Dertouzos, Lester, and Solow 1989, p. 78). However, chemical engineering did move toward an examination of the fundamental phenomena underlying various unit operations, such as heat transfer, mass transfer, and fluid flow. The still-unanswered question is how valuable this more fundamental approach will prove to be to the future needs of industry. Will the researches of academic chemical engineers who regard themselves as practitioners of an "engineering science" generate insights that will make major contributions to improved industrial performance some decades down the road?

A major reformulation of chemical engineering occurred at the University of Wisconsin in the late 1950s. "Professors Bird, Stewart, and Lightfoot, collectively known to future generations of students as BSL, prepared a set of notes (in 1957) and eventually a book (in 1960) entitled *Transport Phenomena*, which offered a new approach to the analysis of chemical engineering unit problems. The main lesson of BSL is that there is a strong unifying backbone to apparently different unit operations in the framework of the continuum equations of transport. The necessity for analysis of individual operations or processes does not disappear, but the differential volume and the balance equations become the central theme of this approach" (Peppas 1989, p. 2).

Furthermore, the particular way in which chemical engineering was institutionalized in the United States was of great significance. The university locus of research assured that the focus of intellectual progress would be on general results rather than the ad hoc solution of specific industrial problems. At the same time, the university's strong link with private industry, especially its continued, if partial, dependence on private industry for funding, assured that university research would remain focused on industrial needs. In fact, it seems plausible that the overwhelming American leadership in the discipline of chemical engineering owed much to the close connections between university and industry. A more prominent role of government as a funder of research, as was the case in Europe, led to a very different outcome (Pigford 1976, p. 197).

GRADUATE TRAINING IN CHEMICAL ENGINEERING

The Ph.D. degree came to play a role in chemical engineering much earlier than it did in other engineering disciplines. MIT's chemical engineering department, for example, awarded its first Ph.D. degrees in 1924, before any other engineering departments had yet awarded such a degree. The University of Michigan, the University of Wisconsin, the University of Pennsylvania, and Columbia University each awarded their first doctoral degrees in engineering in the field of chemical engineering (Peppas 1989, p. 157). The percentage of Ph.D. degrees awarded in this field long exceeded all other engineering disciplines. Between 1905 and 1979 more than 9,000 doctoral degrees were granted by U.S. chemical engineering departments. Figure 7-2 shows that the number of such degrees peaked around 1970 and then declined.

Maloney (1982) has analyzed the roughly 2,800 doctoral degrees awarded in chemical engineering between 1905 and 1959. He finds that process studies were the dominant topics until 1930.[9] In the 1930s, unit operations accounted for 30 percent of the thesis topics, and for over 50 percent in the 1940s. Until around 1960, few graduate students were of foreign origin, and most of those who were had Chinese names. But by 1980, Maloney estimates that about one-half of the degree recipients were foreign nationals. Table 7-3 indicates the cumulative number of doctoral degrees in chemical engineering awarded by individual American universities through 1959.

One very striking thing about the data in Table 7-3 is the large number of U.S. institutions that have been major "producers" of doctoral degrees in chemical engineering, a diversity that would stand in sharp contrast with the experience of other countries. This reflects in part the American system of state universities that have played such a major role since the passage of the Morrill Act of 1862. It also reflects the alacrity with which the American higher-education system has historically responded to the emergence of new disciplines of potentially great economic value (Rosenberg and Nelson 1994).

THE ROLE OF SCIENCE

America's early assertion of world leadership in chemical engineering was not built on superiority in the science of chemistry. In this realm, and especially in organic chemistry, Germany had become totally dominant in the second half of the 19th century. The reasons here are deeply intertwined in social and intellectual history.

One useful starting point is the emergence of the German university system in the wake of the trauma of Prussia's defeat by Napoleon in 1806. By the end of the 18th century, theologians had essentially lost control of the German university. Their previous eminence had given way to a roll call of distinguished German philosophers: Christian Wolff, Immanuel Kant, Johann Fichte, Friedrich Schleiermacher, and Georg Hegel. As John Beer (1959, p. 157) succinctly observed: "The influence of these men on the further development of the German university was profound. Whereas former educational theories had maintained the attitude that the body of man's knowledge was largely complete, these philosophers believed exactly the reverse, maintaining that the unknown remained vast and that it was, therefore, the primary task of higher education not to transmit for maximum retention a fixed body of knowledge, but to prepare young minds for research and independent thought."

[9]At one point Maloney says "process and product studies."

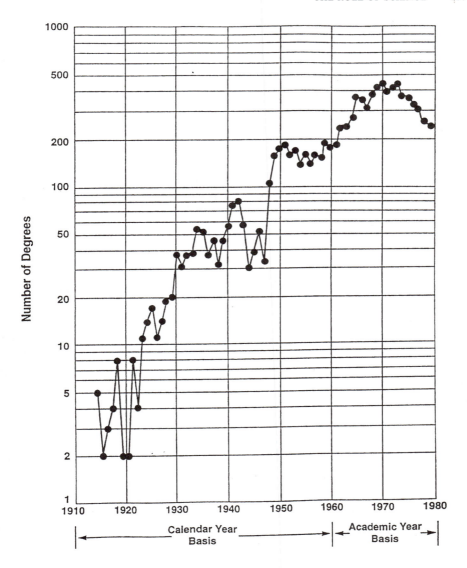

Figure 7-2. Maloney (1982, p. 213).

Wilhelm von Humboldt, scholar and statesman, was placed in charge of the post-Napoleon reorganization of the Prussian educational system, one major element of which was the founding of the University of Berlin in 1809. The University of Berlin placed a strong emphasis on training for the professions, an emphasis that was repeated at universities in Breslau (1811), Bonn (1818), and Munich (1836). A distinctive feature of the German university in the 19th century was that, in making their claim for status and legitimacy on a par with other professionals, natural scientists—among them Wilhelm's brother Alexander von Humboldt—felt that they had to sever their links

TABLE 7-3. Doctoral Degrees, American Universities: from the Beginning through 1959

Brooklyn, Polytechnic Institute of	73	Missouri, University of	5
California Institute of Technology	24	Montana State College	8
California, University of (Berkeley)	23	New York University	23
California, University of (Los Angeles)	13	North Carolina State College	9
Carnegie Institute of Technology	52	Northwestern University	43
Case Institute of Technology	41	Ohio State University	187
Cincinnati, University of	33	Oklahoma State University	7
Colorado School of Mines	1	Oklahoma, University of	7
Colorado, University of	13	Oregon State College	8
Columbia University	175	Pennsylvania State University	45
Cornell University	61	Pennsylvania, University of	32
Delaware, University of	57	Pittsburgh, University of	50
Florida, University of	10	Princeton University	41
Georgia Institute of Technology	28	Purdue University	106
Houston, University of	6	Rensselaer Polytechnic Institute	26
Illinois Institute of Technology	38	Rice University	5
Illinois, University of	140	Rochester, University of	3
Institute of Paper Chemistry	8	Southern California, University of	2
Iowa State Univ. of Science & Technology	95	Syracuse University	20
Iowa, State University of	48	Tennessee, University of	17
Johns Hopkins University	22	Texas, Agricultural & Mechanical College of	9
Kansas, University of	8	Texas, University of	95
Lehigh University	12	Utah, University of	6
Louisiana State University	25	Virginia Polytechnical Institute	30
Louisville, University of	1	Washington University	31
Maryland, University of	23	Washington, University of	53
Massachusetts Institute of Technology	321	West Virginia University	22
Michigan State University	8	Wisconsin, University of	133
Michigan, University of	253	Yale University	82
Minnesota, University of	107		
Total			2824

Source: Maloney (1982, p. 218).

with the realm of the practical and the vocational and, therefore, with the demands of the realm of technology. This break led to the redefinition of a university as a scientific institution whose members were committed to the conduct of research that was to deliver great benefits for society in the long term. The accomplishments of empirical research by Justus von Liebig at the University of Giessen, beginning in the late 1820s, along with the influential impact of his teaching laboratory, represented the beginning of a trajectory of German preeminence in chemical research. Domestic demand for university-trained chemists remained small, however, and many such chemists sought employment abroad, among them August Hofmann, a student of Liebig, who played a leadership role in the establishment of the Great Britain's Royal College of Chemistry in 1845.

The German situation was transformed by the beginnings of a modern textile industry in Germany and a subsequent growing demand for trained chemists in the newly emerging coal-tar dye industry of the 1860s and 1870s. German dominance of organic chemicals came to be based on this symbiotic relationship between the universities and the science-based coal-tar dye industry, powerfully strengthened by huge, expanding export markets to foreign textile industries. The dye companies did much more than simply hire university-trained chemists. Complex webs of connection and cooperation were drawn with the most creative academic laboratories. John Beer (1959, p. 65) pointed out:[10]

> The means by which the competing dye firms tried to outdo each other for the favor of the professors is shrouded in mystery. It required a certain amount of circumspection to secure such cooperation without offending academic dignity and propriety...some of the ways generally used in rewarding professional cooperation are perfectly well known and were considered proper. First, there was the payment of royalties; second, there was the promise of expert technical and legal assistance to bring the discovery under the protection of an effective patent that would safeguard it from legal attack, infringement, or circumvention; third, there were consulting fees that were usually extremely handsome, considering the time and effort involved; and fourth, the dye firm could offer scientific cooperation which not only involved the already mentioned donation of rare and expensive chemicals, and the use of the company's laboratories and staff for tedious confirmatory and analytical tests, but went so far occasionally as to furnish the professor with one or more company chemists who would work in his laboratory on specific projects that were of special interest to the company.

A key feature of these developments in Germany was the drawing of rigid lines demarcating what subjects were, and were not, regarded as appropriate for university teaching and research. The determination to exclude subjects of immediate usefulness led to the exclusion of engineering from the university curriculum. This deficiency led to the establishment of the Technische Hochschulen, institutions of distinctly inferior social status, for the training of engineers. As badges of their social inferiority, the Technische Hochschulen were not allowed to grant doctoral degrees until the very end of the 19th century (1899), when the coveted right was finally conferred on them by Kaiser Wilhelm himself. Even so, it was not until after World War II that the Technische Hochschulen were allowed to call themselves "universities," and to this day the training of engineers in Germany is confined to these institutions.

This history of the subordination of the engineer to the chemist sheds much light on why Germany's eminence in the science of chemistry failed to give birth to the discipline of chemical engineering. But one more critical element needs to be added. The excellence of German research skills in organic chemistry were indispensable to that nation's overwhelming dominance in synthetic dyes, as well as its later outstanding performance

---

[10]Beer's (1959) book remains the best study of its subject. For some valuable perspectives on the subject, see also Ben-David (1971, ch. 7), Jarausch (1990), and Schmookler (1965). Schmookler's hypothesis is that in the past couple of centuries, government support for science has been primarily driven by utilitarian considerations or by some external threat or extrascientific event to which the nation state has felt compelled to respond. A 1990 book, *The Kaiser's Chemists*, by Jeffrey Johnson, supplements Schmookler's argument in the specific context of Germany's establishment of the Kaiser Wilhelm Institutes. Johnson argues that German chemists garnered crucial political support for the establishment of these research institutes by pointing to the economic threat being posed by the American chemical industry, even though the United States was far behind Germany with respect to the quality of its scientific research in chemistry.

in pharmaceuticals and photochemicals.[11] But these industries generated little demand for the unique combination of skills that were, in the United States, embodied in the chemical engineer. For example, although the synthetic dye-stuffs industry relied heavily on the research of well-trained bench chemists to devise a stream of new products for rapidly expanding markets, the manufacturing skills remained those of small-scale and batch production. Once these new products were developed by the chemist, the transition from the laboratory to the manufacturing facility was not so great as to require the skills of a mechanical engineer who also had a considerable professional competence in chemistry.[12]

A division of labor emerged in Germany in which the chemist described his needs to mechanical engineers, and although the engineer had little understanding of the underlying chemical transformations for which he was designing and constructing equipment, the separation worked reasonably well. However, it would not constitute a satisfactory arrangement later in the 20th century, when German industry would turn to petrochemicals. By the same token, American primacy in chemical engineering needs to be understood in terms of the historical fact that the U.S. made large-scale use of petroleum refining well before other industrial countries did so. Indeed, the discipline of chemical engineering largely owed its origins to satisfying the needs of large-scale petroleum refining. The key insight for present purposes is that leadership in the science of chemistry need not, and historically did not, coincide with leadership in chemical engineering.

In fact, scientific leadership in the field of chemistry continued to reside in Germany until World War II, although the U.S. dramatically improved its research and teaching capabilities in the interwar period (see Thackray et al. 1985, pp. 154–60, for quantitative evidence). This was especially apparent in the newly emerging field of polymer chemistry, where the fundamental researches of Staudinger in the 1920s, as well as those of Herman Mark and Kurt Meyer, were responsible for a huge expansion in understanding the structure and behavior of large, long-chained molecules. (One American scientist, Wallace Carothers, made major contributions to polymer science, as well as to the research project that produced neoprene and nylon, after he arrived at DuPont in 1928.) Mark later played a critical role in introducing the science of polymer chemistry to a larger American audience. Mark, a distinguished research chemist at IG Farben, left Germany under the Nazis and was then forced to leave Vienna after the German *Anschluss* in 1938. In 1940, DuPont arranged for Mark to be an adjunct professor at Brooklyn Polytechnic Institute as well as a consultant to DuPont. Mark's consulting commitments were very heavy, but he nevertheless managed, by 1946, to establish an Institute of Polymer Research. Mark was responsible for training a large number of students in polymer chemistry, many of whom eventually went to work for DuPont (Hounshell and Smith 1988, p. 296).

An important aspect of the relationship between the scientific and technological realms should be noted here. Industrial progress has often proceeded in a sequence in which fundamental science first opened new product categories but then had to await

[11]Germany was far and away, the world's largest exporter of pharmaceuticals in 1913, accounting for slightly over 30 percent of the total.

[12]As Ralph Landau (1997, p. 53) has put it, referring to the German scene, "Chemistry was always a queenly science, and engineers were expected to take orders from them in designing plants. Since, in the era of dyestuffs, the scaleups were largely repetitions of batch laboratory operations, and costs of production were not significant in view of the high sales prices for these new products, it is no wonder that engineering was considered a handmaiden to science, and chemists dominated in the upper levels and research organizations of the German chemical industry."

further development until the appropriate methods of manufacturing were worked out. In the case of the modern petrochemical industry, which emerged during and especially after the Second World War, many of the manufacturing methods had already been developed in the interwar years, largely in the petroleum-refining industry. The modern postwar petrochemical industry had its origins in fundamental research initiated by Staudinger in the 1920s, but the methods of producing the necessary chemical feedstocks had already been well established by the end of the Second World War. Thus, when the fruits of the scientific research became available after that war in the form of new products made possible by plastics, synthetic fibers, and synthetic rubber, the basic processing technologies were already at hand as a result of the earlier accomplishments of the chemical engineers.

The discrepancy between the path of American achievement in the chemical industry and the nation's achievement in the underlying chemical science is problematical only if it is taken as axiomatic that the two must move in lockstep. But the axiomatic nature of that association has been much exaggerated in recent years. Japan, for example, had stunning accomplishments in the "high tech" industries in the postwar years, although that nation's status in the underlying sciences was, at best, distinctly modest. Skepticism should also be aroused by America's parallel rise to dominance of the commercial aircraft industry in the interwar years, when the modern discipline of aerodynamics was overwhelmingly a German preserve. The major progress in theoretical fluid dynamics, which was fundamental to dealing with aerodynamic drag in aircraft design, was overwhelmingly the accomplishment of Ludwig Prandtl and his students, at Göttingen University, where Prandtl served as professor of applied mechanics from 1904 until his death in 1953.

It is true that scientific expertise can sometimes lead almost directly to commercial technological success. In Germany, with the exception of the Haber/Bosch process for ammonia synthesis just before World War I, the main thrust of research had led to expansion in sectors where the optimal size of plant was small. Germany's decisive successes in the science of synthetic organic chemistry translated into leadership in dyestuffs and, later, pharmaceuticals and photochemicals. In these product lines the scale of manufacturing operations was small and did not require much drastic redesign from the laboratory apparatus in which the products were first developed, which is very nearly the same as saying that the skills of the chemical engineer were less in demand. The manufacturing technology was still batch operation rather than continuous process (again, with the important exception of the Haber/Bosch process).

However, this German-style division of labor, in which chemists directed mechanical engineers who were placed in charge of designing-process machinery, also had its downside. The mechanical engineers were often ill-equipped to evaluate the numerous trade-offs inherent in the design of efficient chemical-processing equipment. As Warren K. Lewis (1953) pointedly observed of the German scene at this time, "Details of equipment construction were left to mechanical engineers, but these designers were implementing the ideas of the chemists, with little or no understanding of their own of the underlying reasons for how things were done. The result was a divorce of chemical and engineering personnel, not only in German technical industry but also in the universities and engineering schools that supplied that industry with professionally trained men."

Although there are a variety of reasons for which one should not necessarily expect a close association between national excellence in some commercial technological realm and excellence in its underlying science base, two reasons should be emphasized here.

First, scientific knowledge is readily communicable, and the incentives to receive such communication are especially strong when the knowledge is understood to have great potential commercial value. Second, technical progress often takes place without any reliance on research at the current frontiers of science. In the case of the chemical industries, while product innovation has often been intimately tied to scientific research, those ties were far less salient in the early days of chemical engineering, although they did indeed become stronger during the course of the 20th century.

The important role played by pilot plants in the design of new chemical-plant facilities is compelling testimony to the limited assistance offered by a rigorous body of underlying scientific knowledge. Pilot plants are an expensive and time-consuming way to search for reliable design data, but they had to be built because no body of scientific knowledge could answer the necessary questions. Leo H. Baekeland developed Bakelite, the first commercial plastic. It began to be manufactured in 1910 and was a considerable success, licensed abroad in a number of countries. Nevertheless, the product had been developed by empirical methods, and its chemical formula was not even established for fully 20 years. Indeed, the availability of this new product played some role in stimulating research in polymer chemistry (Chapman 1991, pp. 44–45).

As late as the 1960s, Hougen (1965, p. 229) argued: "Nearly all the chemical operations involving the solid state [calcination, centrifugation, crystallization, dissolution, drying, filtration, leaching, mixing, and so on] are of an empirical nature where the art plays a more important role than the present state of science permits.... These are the areas where the engineer must depend upon experience and pilot-plant experimentation for successful application.... The science of the solid state as applied to chemical processing and catalysis is relatively undeveloped."

INSTITUTIONAL DEVELOPMENTS

American educational and professional institutions before World War I had a distinctly provincial status. For example, prior to the First World War and for a decade or more after, the most ambitious and talented young Americans opted to do their graduate work or postdoctoral research in Europe — most especially in Germany, but also France and England. Between 1876 and 1905, more than one-half of the presidents of the American Chemical Society, America's premiere professional organization for chemists, had received their education in Germany, primarily in Göttingen and Leipzig (Thackray et al. 1985, pp. 189–91). The trend in the production of Ph.Ds. in chemical science at American universities was already upward, however, by the turn of the century. Although American universities produced only 418 Ph.D.'s in the two decades from 1881 to 1900, they produced 591 in the next decade, 1901 to 1910 (Skolnick and Reese 1976, p. ii). Between 1863 and 1900, more than four-fifths of American Ph.D.'s in chemistry were awarded by just five institutions: Johns Hopkins, which accounted for 40 percent of the total, followed by Yale, Harvard, Pennsylvania, and Columbia (Thackray et al. 1985, p. 150).

Perhaps more revealing are the fields in which American professors in the 19th century had concentrated their efforts. Most had specialized in fields that offered the most direct payoffs in a mineral-rich country that was undergoing a rapid industrialization experience: mineralogy and analytical chemistry. The majority of chemists employed by American industry around the turn of the century were analytical chemists doing routine testing (Trescott 1981, p. 140), and much of their work was essentially quality control. Only a small number of students of chemistry studied organic chemistry. Whereas it was possible by the 1870s to receive a Ph.D. in organic chemistry at a number of European institutions, there were not many American colleges where one might have taken even a separate single course in organic chemistry. Indeed, until the appearance in 1870 of a new journal, *The American Chemist*, no single journal published in the United States was devoted exclusively to chemistry.

Physical chemistry was an unfamiliar subject in America until students went to Heidelberg in the late 1880s to study with the great Wilhelm Ostwald. In fact, most American chemists were ignorant of the seminal work of their own compatriot, Willard Gibbs. Gibbs was not only by far the greatest of all 19th-century American chemists; his astonishingly complete formulation of the theory of chemical thermodynamics in the 1870s was one of the greatest achievements of world science in the 19th century. But his two-part paper, "On the Equilibrium of Heterogeneous Substances," was published in the obscure *Transactions of the Connecticut Academy of Arts and Sciences* and almost totally ignored in the United States until the 20th century, after it had been translated, late in the 19th century, into German by Ostwald and into French by Le Chatelier. Although an English reprint appeared in London in 1906, an American reprinting was not available until 1928!

The American Chemical Society (ACS) was founded in 1876, at a time when chemistry was just beginning to be identified as both a science and a profession. At the same time, The Johns Hopkins University opened its doors. Its offerings included a graduate program in chemistry, headed by the distinguished chemist Ira Remsen, who subsequently founded and edited the *American Chemical Journal*, and eventually became president of Johns Hopkins (1901 to 1913). Johns Hopkins conducted the most important American Ph.D. program in chemistry for 30 years or so after its founding.

Growth in the ACS was slow in its early years, however, as shown in Table 7-4. Until 1889 it was essentially a New York City organization. "During 1880, six regular meetings lacked even the quorum of 15 required to transact Society business" (Skolnick

TABLE 7-4. American Chemical Society Members vs Census Chemists: 1870– 1970

Year	ACS Members (1)	Chemists (Census Estimates) (2)
1870	—	774
1880	303	1,969
1890	238	4,503
1900	1,715	9,000
1910	5,081	16,000
1920	15,582	28,000
1930	18,206	45,000
1940	25,414	57,000
1950	63,349	77,000
1960	92,193	84,000
1970	114,323	110,000

Source: As reported in Thackray et al., (1985, p. 253).

and Reese 1976, p. 17). By the turn of the 20th century, the membership of the organization began to reflect a growing specialization. This specialization resulted from the rapid growth of industrial research labs in the opening years of the new century and the inevitable cleavage between the society's academic and industrial memberships. But, in addition, the public sector commitment to chemical research increased, both for industrial purposes and out of concern with public health issues involving chemical

TABLE 7.5. Specialty Divisions in the American Chemical Industry

Division	Year Established
Industrial and Engineering Chemistry	1908
Agricultural and Food Chemistry	1908
Fertilizer and Soil Chemistry	1908
Organic Chemistry	1908
Physical Chemistry[a]	1908
Medical Chemistry	1909
Rubber	1909
Biological Chemistry	1913
Environmental Chemistry	1913
Carbohydrate Chemistry	1919
Cellulose, Paper, and Textile	1919
Dye Chemistry[b]	1919
Leather and Gelatin Chemistry[c]	1919
Chemical Education	1921
History of Chemistry	1921
Fuel Chemistry	1922
Petroleum Chemistry	1922
Organic Coatings and Plastic Chemistry	1923
Colloid and Surface Chemistry	1926
Analytical Chemistry	1936
Chemical Information	1948
Polymer Chemistry	1950
Chemical Marketing and Economics	1952
Inorganic Chemistry	1956
Microbial and Biochemical Technology	1961
Fluorine Chemistry	1963
Nuclear Chemistry and Technology	1963
Pesticide Chemistry	1969
Professional Relations	1972
Computers in Chemistry	1974
Chemical Health and Safety	1977
Geochemistry	1978
Small Chemical Businesses	1978
Chemistry and the Law	1982

[a]Originally Physical and Inorganic Chemistry.
[b]Merged with Organic Chemistry in 1935.
[c]Discontinued in 1938.
Source: Thackray et al. (1985, p. 184).

products. The U.S. Department of Agriculture had a Bureau of Chemistry since 1862, whose long-term chief from 1883 to 1912 was also President of the ACS in 1893 to 1894; the Pure Food and Drug Act was established in 1906; and the National Bureau of Standards, established in 1901, had its own research laboratories, as did the U.S. Bureau of Mines, established in 1910 (Skolnick and Reese 1976, p. 12).

The need of the ACS to accommodate the increasing specialization of its membership resulted in the decision, in 1908, to establish the *Journal of Industrial and Engineering Chemistry*. In addition, in 1907 the Society had initiated the publication of *Chemical Abstracts* as a separate entity, rather than including its abstracts in the regular issues of the *Journal of the American Chemical Society*. The ACS also recognized the inevitability of separate divisions of the society. The Division of Industrial Chemists and Chemical Engineers was established in 1908 (changed to the Division of Industrial and Engineering Chemistry in 1919), soon to be followed by four other divisions: Agricultural and Food Chemistry, Fertilizer Chemistry, Organic Chemistry; and Physical and Inorganic Chemistry. The ACS has spawned a number of specialized divisions over the course of the 20th century, as shown in Table 7-5. Indeed, the record of these individual divisions provides a good deal of insight into the emerging concerns of the chemical industry.

One reason the ACS established separate divisions was that the leaders of the ACS believed the establishment of a separate professional organization for chemical engineers would diminish the ability of ACS to speak with authority for the entire chemical community. Indeed, the first chairman of the ACS Division on chemical engineering was A. D. Little, who had earlier been active in the formation of the American Institute of Chemical Engineers (AIChE) and who also served as ACS president in 1912 to 1913. Yet in 1908, a separate, independent AIChE was established. It was the first organization of chemical engineers anywhere in the world. "Many charter members of AIChE and all of its first officers were ACS members. They were aware of the plans for the analogous ACS division and its companion journal, but they persisted in forming the institute for a single reason — to have rigid requirements for membership." Hence the requirement that an applicant for membership in AIChE had to be "not less than 30 years of age... proficient in chemistry and in some branch of engineering as applied to chemical problems, and...engaged actively in work involving the application of chemical principles" (Skolnick and Reese 1976, p. 14). Terry Reynolds (1986) has argued that at the time of its formation, the founders of AIChE were reflecting the status anxieties of production chemists who did not yet have a well-defined territory with unambiguous boundaries and were concerned over incursions into their domain by less-skilled men.

THE IMPACT OF THE AUTOMOBILE INDUSTRY

The situation in the United States, and its distinctiveness among industrial nations in twentieth century chemicals, was altered sharply in the second decade of the 20th century and after by a combination of circumstances involving the growing demands of a rapidly expanding new industry, on the one hand, and the country's supply of certain natural resources, on the other. The new industry was, of course, the automobile. In 1900, the automobile industry did not yet exist, and the *U.S. Census of Manufactures* classified the small number of cars built in that year under "miscellaneous." In less than 30 years, however, automobiles comprised the largest manufacturing industry in the American economy.

TABLE 7-6. Number of Car Registrations in the United States

1900	8,000
1910	458,300
1920	8,131,522
1930	23,034,753
1940	27,165,826
1950	40,339,077
1960	61,671,390
1970	89,243,557
1980	121,600,843
1985	131,864,029
1995	136,066,045

Source: From *Highway Statistics Summary to 1985* (U.S. Dept. of Transportation) and *Historical Statistics of the U.S.*

When the automobile was introduced in the opening years of the 20th century and commenced its spectacular growth, it brought in its wake an almost insatiable demand for a liquid fuel. See Table 7-6.

As a result, petroleum refining, specifically calibrated to accommodate the fuel requirements of the automobile, emerged as an entirely new sector in the first three decades of the 20th century.

America had a high level of car ownership compared to all other countries, and it was by far the best-endowed of all the industrial countries in its known petroleum reserves (as well as unknown reserves that were to be discovered later in the century). These factors were to give the U.S. chemical industry a particular boost. In fact, the United States already had a petroleum-refining industry in the late 19th century, albeit one that was exploiting petroleum sources primarily to extract an illuminant — kerosene. What petroleum refining particularly required to meet the demand for automobile fuel was petroleum cracking techniques that would break large, heavy hydrocarbon molecules into smaller and lighter ones.

Before such cracking was possible, however, it was necessary to establish the exact location of the petroleum deposits. Here the American oil companies, beginning with Humble Oil (an affiliate of Standard Oil of New Jersey), were major beneficiaries of the transfer of sophisticated European exploration techniques. These new techniques were initially operated in the United States by European geologists. Humble was the first to develop its own geophysical staff, and by early 1925 its seismic surveying crews were using a variety of European seismic equipment — torsion balances, magnetometers, and refraction seismographs — in exploratory activities along the Gulf Coast. These new techniques quickly proved their value.

In sixty years before 1920, sixty-eight major fields had been discovered. 'Practical men,' as the old-fashioned unscientific prospectors were called, had made most of the discoveries for several decades, but geologists had gradually risen to considerable importance. The two groups were probably about equally responsible for discoveries made during World War I. In the years 1920 through 1926, geologists had been more productive than practical men; they had found two-thirds of seventy major fields . From 1927 through 1939, of 171 major discoveries, geophysicists found 65, geologists 77, and the old type of prospector found 29.

It is significant also that practical men had only one successful strike out of seventeen wells drilled, as compared to the technologists' one in every 7.5 (Larson, Knowlton, and Popple 1971, p. 75).

A very different instrumentation need also should be mentioned here. The switch from batch production to continuous-process production required the development of new instruments for monitoring and automatic control. Visual and manual techniques that had been adequate before World War I were totally inadequate for continuous processing in which a large number of variables — temperature, pressure, rate of flow, composition of inputs and outputs, and so on — required continuous attention and adjustment. Many of the variables requiring sophisticated monitoring and automatic control capabilities were first encountered in petroleum refineries during the 1920s and 1930s. The smooth flow of operations necessitated the development of instruments for continuous, precise monitoring and adjustment and automatic control that had not been necessary before. Although little scholarly research exists on the subject, it appears that the instrumentation requirements of early petroleum refineries may have been the most important single source of modern automatic control technologies (Perazich, Schimmel, and Rosenberg 1938; Field 1940).

Petroleum refining to meet the needs of the automobile had three important features. First, the demand was great, and it was also widely (and correctly) anticipated at the outset that future growth would be rapid. The output of U.S. petroleum refineries was far more directed toward the production of gasoline than was the case in Europe, where car ownership was much more restricted. The output of gasoline in the United States increased almost four and one-half times over the decade 1919 to 1929, from 99.7 to 441.8 million barrels (Williamson et al. 1963, p. 395). Second, petroleum refining had huge capital requirements. By the 1930s, the industry had become by far the most capital-intensive of all American industries (Chandler 1992, p. 24). Third, productive efficiency required that the small-batch production methods that had characterized the early American petroleum refining, not to mention the German technologies in the synthetic organic-chemical industry, be replaced by large-volume production methods using continuous-process technologies. In achieving this goal, American chemical engineers acquired technical capabilities in designing continuous-process plants that were unmatched in any other country. The discipline of chemical engineering was itself transformed by the effort.

THE RISE OF PETROCHEMICALS AND THE ROLE OF MIT

This transformation of chemical engineering was critical to the technological developments in the chemical industry for the rest of the 20th century because it placed the American chemical industry on a trajectory that was to prove extremely advantageous: The main thrust of technological change in the chemical sector was to be in organic chemicals, and in turn, a large part of the story of the organic chemical industry in the 20th century is the story of changes in feedstocks.

Many organic chemicals can be derived from any of several feedstocks: coal, alcohol, crude oil, or natural gas. The dominant feedstock for organic chemicals, as that industry had emerged in Germany in the last third of the 19th century, had been coal or, more precisely, coke-oven by-products. Since western Europe had very little in the way of petroleum deposits but abundant coal and a huge chemical industry, European chemists concentrated their efforts on synthesizing molecules from coal. While this approach was technically feasible, it also proved considerably higher cost than exploiting petroleum for the same purpose. America's early leadership in petroleum refining provided the knowledge base, and the engineering and designing skills, that turned out to be crucial as the chemical industry's resource base began its historic shift from coal to petroleum hydrocarbons in the years before the Second World War. As petrochemicals came to the fore, the dominant participants in this industrial transformation were Union Carbide, Standard Oil of New Jersey, Shell, and Dow (Spitz 1988, ch. 2). Union Carbide was especially prominent in the development of plastics and intermediates from petrochemical feedstocks.

The dramatic shift in the raw-materials base of the American chemical industry in the interwar years has been described in the following terms: "Between 1921 and 1939, the production of organic chemicals *not derived from coal tar* rose from 21 million pounds valued at \$9.3 million to 3 billion pounds, with a value of \$394 million. Coal-tar chemicals production in 1939... still amounted to only 303 million pounds, valued at \$260 million. The average 1939 price for petrochemicals was, in fact, 13 cents per pound versus 87 cents for coal-tar derived chemicals. The average price of non-coal-tar chemicals had, over the period, been reduced by a factor of three, from a 1921 level of 43 cents per pound" (Spitz 1988, pp. 67–68).

The fact that America alone experienced a rapid growth in its petroleum-refining industry and the beginning of a shift toward petrochemicals in the interwar years helps explain the growth in demand for university-trained chemical engineers. In contrast, Great Britain had no equivalent petroleum-refining industry and experienced only a very small growth in trained chemical engineers in the interwar years. Britain's Institution of Chemical Engineers worked hard to advance the academic training of chemical engineers in the 1920s and 1930s, but with indifferent success. Even ICI, which dominated the British chemical industry after its founding in 1926, showed limited interest in university-trained chemical engineers. University College London had the largest British chemical engineering training program between the wars, yet ICI employed only 25 of their 159 graduates between July 1928 and July 1937 (Divall 1994, p. 264).

This situation was transformed after World War II, largely because of Britain's sharp expansion in the domestic refining of petroleum and the shift (in Britain, partly a consequence of government policy) from coal to oil as the principal feedstock for chemical manufacturing. As British firms confronted the prospect of dealing with large flows of petroleum, the need for chemical engineers forcefully asserted itself. This growing reliance on chemical engineers was strengthened by the entry of some American firms, such as Monsanto, into the British market. By the late 1940s, even ICI was making greater use of chemical engineers in some of its divisions. By 1960, 13 British departments of chemical engineering were producing about 600 graduates per year. This number represented a roughly twelvefold increase over the level for 1939 (Divall 1994, pp. 263–65, 271–72).

From the perspective of the 1990s, the intimate linkage between the petroleum and chemical industries seems natural and, therefore, inevitable. But from the vantage point of 1920 that was emphatically not so. The "natural" connection between the two industries was a human creation, involving the mobilization of vast resources and the sustained exercise of human intelligence in the development of inventions, novel designs, and a wide range of problem-solving activities. Before the invention of the internal combustion engine, petroleum was valued primarily as an illuminant and lubricant, and the more volatile fractions were commonly discarded as waste. Petroleum became a major source of fuel for transportation purposes only as a result of the invention of the automobile. In 1920, however, the state of technological knowledge was such that petroleum, quite correctly, was not regarded as a significant input into the chemical industry. In 1920, the chemical industry thought of its inputs in terms of chemicals in a less-processed state, on the one hand, and feedstocks drawn from coke-oven by-products, on the other. Only after extensive research and the slow accumulation of technical knowledge did the oil companies come to realize that their refining operations could produce not just fuel and lubricants but organic chemical intermediates as well (Spitz 1988, ch. 2). In the 1920s, the offgases of oil refineries, if they were not simply flared, were likely to be employed only as fuel at the refineries themselves. In fact, the oil companies paid hardly any attention to the presence of ethylene, which was eventually to become the primary building block of the petrochemical industry. But of course, the products for which ethylene eventually proved to be such an excellent feedstock, such as plastics and synthetic rubber, did not achieve a real importance until World War II and after.

The transformation of the chemical industry as it existed in 1920 into the petrochemical industry that matured after World War II was in large measure the achievement of the chemical engineering profession. It was a process in which by-products that formerly had been treated as waste materials were converted into sources of great commercial value. Chemical research had achieved a similar outcome in the late 19th century, when a by-product derived from coke ovens, coal tar, provided the raw-material basis for a burgeoning organic-chemical industry. Standard Oil (N.J.) and Shell were induced to undertake the research that brought them into the chemical industry by their growing awareness of the commercial opportunities that might flow from the eventual utilization of the waste products of their refinery operations (Spitz 1988, pp. 68, 116, 514–15).

This achievement involved a uniquely intimate set of interactions between the petroleum-refining industry and the newly formed chemical engineering department at MIT. During the 1920s, that department built on the conceptual platform of unit operations that had been introduced by Arthur D. Little. That concept was enriched and deepened as MIT's chemical engineers confronted the difficulties of satisfying the demand for gasoline. The technique of thermal cracking, which was to remain dominant in petroleum refining until just before the Second World War (Spitz 1988, p. 118), was analyzed in terms of its specific unit operations, including heat transfer, fluid flow, and distillation; and the design process was approached in increasingly quantitative terms rather than crudely empirical ones. In the early 1930s, for example, pressure-volume-temperature relationships for gas mixtures were established that were sufficiently accurate for the design of refining equipment. These approximations would not have satisfied a physical chemist but provided a reliable guide to design in a number of industries other than petroleum refining (Weber 1980, p. 28).

The commitment of MIT chemical engineers to the problems of the petroleum-refining industry, beginning in the 1920s, would be difficult to exaggerate. "The major oil companies quickly hired large numbers of MIT chemical engineers, and for a number of years these men practically dominated the industry. The chemical engineering senior faculty was almost completely hired as consultants or permanent employees by the oil industry" (Weber 1980, p.26; see also pp. 30 and 37). The influence of MIT's chemical engineers in the 1920s was, of course, not confined to petroleum refining. At

DuPont, for example, of the ten chemical engineers in the Chemical Department, more than half were MIT graduates (Hounshell and Smith 1988, p. 279).[13]

The chairman of MIT's chemical engineering department and the leading academic figure was W. K. Lewis, who served as a consultant to Standard Oil Company of New Jersey (later Exxon) for many years. Lewis received his doctorate in organic chemistry at the University of Breslau (at this point, a German city) and then spent a year at a New Hampshire tannery before joining the MIT faculty in 1910. Lewis's first efforts at Standard Oil were to provide precision distillation equipment and to convert batch processing methods to methods that were continuous and automatic, as in thermal cracking (the tube-and-tank process) and continuous vacuum distillation. In doing these things, he was also unknowingly inventing technologies on which the petro-chemical industries of the future were to be based (Scriven 1991, p. 18). By 1924, Lewis had helped increase oil recovery by the use of vacuum stills. From 1914 to 1927, the average yield of gasoline rose from 18 to 36 percent of crude throughput. This work and his earlier "bubble tower" designs became refinery standards. Course work at MIT was quickly altered to embody these new concepts and their underlying design principles.

In the late 1920s, Exxon negotiated a series of agreements with the German chemical giant IG Farben that would provide access to their research on hydrogenation and synthetic substitutes for oil and rubber from coal. It was also anticipated that the German findings might be used to increase gasoline yields and promote Exxon's entry into chemicals. Lewis was consulted in putting together a new research group for this purpose, and he recommended Robert Haslam, head of the Chemical Engineering Practice School. Haslam took leave from MIT and formed a team of 15 MIT staff members and graduates who set up a research organization in Baton Rouge. Many of the members of this group later rose to positions of eminence in petroleum and chemicals (Spitz 1988, ch. 3; Landau and Rosenberg 1992).

In this single step, by exploiting the German technology for American refining use, Exxon overcame much of the gap that had so long existed between applied and innovative research in the petroleum industry, including introducing the use of chemical catalysis in petroleum refining. The Houdry process had first introduced catalytics into petroleum refining in 1937, thus replacing earlier processes that had relied exclusively on thermal cracking. Houdry's catalyst, an acidic compound of silica and alumina, allowed temperature to be reduced to about 500°C and pressure to about one atmosphere, and also raised the yield of gasoline. But the Houdry process, which played an important role in the early years of the Second World War, suffered from the serious deficiency that its operation built up a deposit of coke on the catalyst, the removal of which required periodically shutting down the process. In other words, it was not a truly continuous-process technology. The process was soon replaced by fluid catalytic cracking, which was introduced during the Second World War and became the predominant cracking technology in the postwar years. Tables 7-7 and 7-8 show the changing composition of U.S. cracking technologies between 1927 and 1957. Note

---

[13]DuPont maintained close connections with other universities as well, such as the University of Illinois, where Carl Marvel, a prominent professor of organic chemistry, was a preeminent figure. Marvel was instrumental in identifying Wallace Carothers, then at Harvard, as a talented young organic chemist who might be induced to move to DuPont. Carothers's decision to move constituted, in fact, a major event in American industrial history, since not only did his research there lead to the development of nylon, a singular innovation and the first truly important polymer, but Carothers also made fundamental contributions to polymer chemistry, a field in which industrial research had long dominated research in the university community. See Hounshell and Smith (1988, part III).

TABLE 7-7. U.S. Cracking Capacity by Process: 1927–1942

Year	Total Cracking Capacity (Thermal) +Catalytic is 000 bbls/day)	Share of Major Processes						
		Burton	Dubbs	Tube and Tank	Holmes-Manley	Cross	Other Thermal Processes	Houdry
1927	1,246	0.198	0.128	0.170	0.170	0.135	0.199	—
1928	1,359	0.154	0.132	0.219	0.146	0.168	0.181	—
1929	1,476	0.097	0.124	0.293	0.141	0.164	0.181	—
1930	1,720	0.065	0.133	0.283	0.128	0.159	0.232	—
1931	1,829	0.052	0.157	0.244	0.117	0.140	0.290	—
1932	2,011	0.032	0.110	0.316	0.121	0.131	0.290	—
1933	1,882	0.032	0.120	0.261	0.130	0.145	0.312	—
1934	1,887	0.008	0.120	0.267	0.134	0.132	0.339	—
1935	2,153	0.006	0.166	0.234	0.127	0.097	0.370	—
1936	2,169	0.005	0.176	0.215	0.130	0.087	0.387	—
1937	2,195	—	0.193	0.210	0.125	0.071	0.400	0.001
1938	2,348	—	0.207	0.198	0.117	0.075	0.397	0.006
1939	2,138	—	0.215	0.194	0.109	0.079	0.392	0.011
1940	2,284	—	0.206	0.192	0.102	0.074	0.370	0.056
1941	2,352	—	0.214	0.182	0.100	0.081	0.357	0.066
1942	2,456	—	0.213	0.157	0.099	0.076	0.385	0.070

Source: From *Oil and Gas Journal*, Annual Refining Issues (1927–1942). Total capacity includes shut-down plants. Shares are calculated by dividing total cracking capacity by the capacity of each process. As reported in Enos, p. 286.

particularly the shift away from thermal processes and toward alternate cracking technologies in the years after World War II.

Much of what took place in modern petroleum processing before World War II originated in Baton Rouge from the phalanx of MIT chemical engineers who worked there. With the continuing advice of Lewis, and later of MIT Professor Edwin R. Gilliland, Baton Rouge produced such outstanding process developments as hydroforming, fluid flex coking, and fluid catalytic cracking. This last technology ultimately became the most important raw-material source for propylene and butane feedstocks to the chemical industry.

Thus, MIT's "outreach" program played a crucial role in solving industrial problems. But the development of many of the new processing technologies could not have been carried out entirely in a university context, if only because a university laboratory could not possibly attain the large scale that was essential to meaningful results. In 1938 a research consortium, originally consisting of Jersey Standard, Indiana Standard, Kellogg, and IG Farben, was formed to develop a catalytic cracking method that would not infringe on the Houdry patents. The largest expenditures of any of the member companies by far were by Jersey Standard, which spent almost $30 million on fluid catalytic cracking between 1935 and 1956. The key patent was originally applied for by W. K. Lewis and E. R. Gilliland in January 1940 and assigned to Standard Oil Development Company (Enos 1962, pp. 196–201). Fluid-bed catalytic cracking made use of a powdered catalyst. "After being aerated, this powder was placed in a tall standpipe which produced such a pressure build-up that the catalyst flowed into and

TABLE 7-8. U.S. Cracking Capacity by Process: 1943–1957

Year	Total Cracking Capacity (Thermal + Catalytic is 000 bbls/day)	Share of Major Processes					
		Thermal Processes	Houdry	Houdriflow	T.C.C.	Fluid	Other Catalytic Processes
1943	2,541	0.901	0.079	—	—	0.020	—
1944	2,544	0.786	0.092	—	0.030	0.092	—
1945	3,075	0.715	0.084	—	0.065	0.133	0.003
1946	3,199	0.700	0.087	—	0.069	0.141	0.003
1947	3,483	0.678	0.080	—	0.069	0.168	0.005
1948	3,781	0.660	0.074	—	0.071	0.189	0.006
1949	3,749	0.626	0.069	—	0.079	0.222	0.004
1950	4,130	0.583	0.060	0.005	0.075	0.277	—
1951	4,163	0.562	0.057	0.013	0.085	0.283	—
1952	4,444	0.533	0.051	0.018	0.087	0.310	0.001
1953	4,104	0.388	0.040	0.023	0.124	0.421	0.004
1954	4,190	0.325	0.039	0.025	0.139	0.468	0.004
1955	4,442	0.253	0.033	0.029	0.155	0.527	0.003
1956	4,625	0.207	0.012	0.040	0.160	0.576	0.005
1957	4,754	0.181	0.009	0.038	0.152	0.616	0.004

Source: From Oil and Gas Journal, Annual Refining Issues (1943–1957). Total capacity includes shut-down plants. After 1942, thermal cracking capacity is not broken down by process. From 1953 through 1957, thermal cracking capacity is included in the category "thermal operations." The remainder of the "thermal operations" (vis-breaking, coking, etc.) are excluded from this tabulation. As reported in Enos, p. 286.

through the reactor with the oil vapor. Because the aerated catalyst acted more like a fluid than a powder, the operation came to be known as fluid catalytic cracking" (Larson, Knowlton, and Popple 1971, p. 168).

Lewis's strategy was to focus the discipline of chemical engineering on an overall systems approach to the design of continuous automated processing of a large variety of products, first in petroleum and then in chemicals. America's earlier traversal of the expanding petroleum industry had provided a vitally important learning experience in the design of continuous processing technology that was transferred to the much more diversified canvas of the petrochemical industry in the postwar years. America's later commercial success in petrochemicals would have been a vastly different story had the country not enjoyed the immense learning benefits that flowed from her earlier experience with petroleum. The speed with which American industry exploited the new products emerging from scientific research in polymer chemistry — plastics, synthetic rubber, and synthetic fibers — was largely a function of the chemical engineering skills that had been acquired in petroleum refining in the interwar years.

As mentioned earlier, the rise of the petrochemical industry began with the entry of a few companies, particularly Union Carbide, Shell, Dow, and Exxon, not long before World War II. As the scale of operations forced them to resort to continuous processing, these firms soon encountered the need for chemical engineering skills, as had previously occurred in the petroleum-refining industry. In fact, the techniques and skills developed for petroleum refining could be applied to the problems of the petrochemical industry, as chemical raw materials began the shift from coal-based to petroleum-based feedstocks. To be sure, the shift was not simple. The problems in manufacturing chemicals were different, and in many respects even more challenging, involving corrosion, complex product separations and purifications, diversified markets, toxic wastes and hazards, and more (Landau 1966). But the experience with petroleum refining provided a storehouse of concepts, methodologies, and experience on which to draw—and the keys to that storehouse were largely in the hands of the American chemical engineering profession. Contractors who had been engaged in the design and construction of petroleum refineries could transfer these skills directly to the design and construction of petrochemical plants (Spitz 1988, pp. 518–19).

Many of the chemical engineering skills learned in the interwar period generated an immense payoff during the Second World War. In particular, the achievements in the petroleum-refining sector, especially in fluidized-bed catalytic cracking, which was developed almost at the time of America's entry into the war in December 1941, came both to dominate the refining of petroleum in the postwar world and also, ultimately, to be the most important processing technology for propylene and butane feedstocks in the chemical industry. Furthermore, the fluidized-bed principle was also applied to "the roasting of pyrite, the oxidation of naphthalene into phthalic acid, and the production of olefins, etc." (Schoenemann 1980, p. 257).

The discipline of chemical engineering matured during the Second World War, when the skills of this discipline were employed in several projects critical to the allied war effort. Catalytic cracking plants provided high-octane gasoline that significantly improved allied fighter aircraft performance—perhaps providing the British with their bare margin of superiority in the Battle of Britain. Ironically, the American petroleum industry received considerable assistance in this important strategic achievement from an agreement that Jersey Standard had entered into in 1927 with IG Farben. A concern over the possible exhaustion of American oil reserves at the time had stimulated American interest in high-pressure coal hydrogenation, an activity at which the Germans had already developed considerable expertise. Although the subsequent discovery of rich new oil fields reduced the American interest in the hydrogenation of coal, the knowledge acquired turned out to be valuable for producing the 100-octane gasoline necessary for the Air Corps' high-compression engines. "At the outbreak of the war in 1939, the Baton Rouge hydrogenation plant was producing both the synthetic blending agent and the synthetic base, and was the largest single source of 100-octane fuels in the world" (Gornowski 1980, pp. 306–307).[14] Although the German experiment with coal hydrogenation was ultimately a commercial failure, the knowledge that it generated led to other successes as well, "for instance the increase of gasoline yield by hydrogenation of residues, the aromatization of gasoline, and the production of pure aromatic and unsaturated hydrocarbons" (Schoenemann 1980, p. 257).

Catalytic cracking plants also provided the essential raw materials for the production of synthetic rubber in World War II. This program was second only to the Manhattan Project in terms of rapid and extensive mobilization of human resources to achieve an urgent wartime goal. (Of course, chemical engineers played a vital role in the Manhattan Project as well; they were responsible for the preparation and reprocessing of the uranium fuel and the manufacture of plutonium.) Synthetic rubber, which in recent years has accounted for about two-thirds of U.S. rubber consumption in automobile tires, was the first synthetic polymer to be produced in huge quantities from petroleum-based feedstocks (Spitz 1988, p. 141). By 1945, U.S. rubber consumption was not only about 20 percent greater than that of 1941, but no less than 85 percent of the 1945 total

[14]The performance of the Luftwaffe's aircraft was handicapped by their reliance on Germany's 87-octane fuel.

was accounted for by synthetic rubber. In 1940, before the crash program was initiated, synthetic rubber had accounted for a mere 0.6 percent of the total (Herbert and Bisio 1985; Howard 1947).[15]

Although the episode is not well known, the technical capabilities of the chemical engineer also played a crucial role in the wartime introduction of penicillin. Fleming's brilliant insight that common bread mold was responsible for the bacteridical effect that he had observed in his Petri dish came in 1928, but penicillin remained unavailable at the outbreak of the Second World War. Producing penicillin on a very large commercial scale during that war required a crash program in which the solution came not from the pharmaceutical chemist but from chemical engineers designing and operating a pilot plant. Chemical engineers demonstrated how the technique of aerobic submerged fermentation, which became the dominant production technology, could be made to work by solving the complex problems of heat and mass transfer (Elder 1970).

One aspect of the wartime development of penicillin is worth noting. Although the British pioneered the scientific research leading to the development of penicillin, their main subsequent research interest was in finding new uses for penicillin, as well as in improving its effectiveness in the clinical treatment of infection. In Germany, with its strong tradition and accumulation of skills in chemical synthesis, the synthesis approach was adopted. Only years after the war was it established that this route was inherently much more difficult and costly than obtaining the penicillin directly from the mold (Sheehan 1982).[16] In America, by contrast, the skills of the chemical engineer were enlisted to identify efficient ways of achieving large-scale production methods and increased yields. This joint achievement of the chemical engineer and the microbiologist is the first great success of biochemical engineering.

CONCLUDING OBSERVATIONS

Chemical engineering grew rapidly in America in the first few decades of the 20th century, for several interrelated reasons. The analytical base was provided by the concept of unit operations. The practical necessity was provided by America's early experience with large-scale production, and in particular with the huge demand for large supplies of liquid fuels, a demand that arrived with the automobile. The task of designing and constructing large-scale, continuous-process chemical plants required a new class of professional, whose design and engineering skills were indispensable to the achievement of the benefits of large-scale operations.

The development of chemical engineering offers a vivid example of how the interaction between industry and academia can give shape to both. In chemical engineering, the interface between university and industry was unusually intimate — consider the MIT professors of the 1920s serving as regular consultants over extended periods of time in dealing with the problems of petroleum refining, and in which their students undertook research topics drawn from immediate needs at the technology frontier of the industry. Moreover, the most significant findings drawn from the consulting experiences of the MIT faculty were quickly transmitted to students through

[15]As Howard (1947) has shown, the American wartime achievement in synthetic rubber drew heavily on knowledge of earlier German chemical research on Buna-S rubber.

[16]The Germans were also influenced by their expectation that sulfa would eventually prove to be the drug of choice.

changes in the content of the teaching curriculum. As long as professors maintained an active role in a teaching capacity, they were under a natural pressure to place the knowledge acquired from their problem-solving activities as consultants in a larger and more general context. This meant fitting that knowledge together in an internally consistent way with other knowledge in their discipline. When reverting to their teaching roles, they needed to systematize their knowledge as an essential precondition for writing textbooks and other forms of publication. This systematization had profound implications for the diffusion of new technological knowledge, not just because open universities "naturally" diffuse their knowledge, but because the need to systematize knowledge for teaching purposes meant that they had to spend time and sustained effort in further activities that inevitably facilitated the spread of useful knowledge (Arora and Gambardella 1994).

The prominent role played by an academic institution devoted to teaching as well as research was of great importance for a related reason. Even though students in chemical engineering programs went to work in different firms or organizations, they had all been taught a common language of concepts, theories, and methods. This shared language facilitated the development of a professional community of people who could easily communicate with and learn from one another, and vastly reduced the barriers to the diffusion of technical knowledge across organizational boundary lines. This ability was what made chemical engineers, after the introduction of the concept of unit operations, so different from the earlier industrial chemists, who tended to speak in very distinct, idiosyncratic, industry-specific, or even firm-specific, languages.

The history of the German chemical industry offers an alternative model, since chemical engineering did not emerge as a distinct subject until after the Second World War. In the years after World War II, petroleum feedstocks demonstrated their crushing economic superiority over coal-based feedstocks. Until that time, Germany had remained wedded to a vision of the chemical sector that had been articulated in 1898 by Carl Duisberg, who would later become the first chairman of IG Farben (as quoted in Guedon 1980, p. 67): "In opposition to many of my friends I place myself... on the standpoint... that the chemist does not require [engineering] as a necessity. Nothing, in my opinion, is worse than to make of a chemist an *ingénieur-chimiste* as is done in France, or chemical engineer as is very often done in England. The field of chemistry which the chemist has to master is at the present so enormous that it is practically impossible for him to study at the same time mechanics which is the special field of the engineer. Division of labour is here absolutely necessary. I leave to the engineer and to the chemist their respective sciences, but I desire that both work together." In Duisberg's vision of the strict division of labor between the chemist and the engineer, the engineer was very much the junior partner, who simply carried out the instructions of his scientifically trained superior. Although the chemist's elevation on the organizational hierarchy served the German chemical industry well for many years, it became a serious handicap after the Second World War, when the industry turned to petrochemicals and good plant design came to require sophisticated engineering skills. Then, Germany essentially borrowed the notion of chemical engineering from America.

Before that time, Germany universities and Technische Hochschulen had trained chemistry students to a very high professional standard. These chemists, however, knew no engineering, and required extensive on-the-job training in the different chemical divisions of the firms in which they were eventually employed. This socialization process for newly trained technical workers must have reinforced the inward-looking nature of their perspectives and limited their ability to communicate with, and to learn useful things from, the world outside their firms. In this way, the absence of a discipline of chemical engineering in Germany reinforced a tendency toward secrecy and reduced the level of interfirm communication. After World War I, BASF and IG Farben under Bosch did not allow academic institutions access to the many confidential secrets involved. This stood in sharp contrast with the earlier case of dyestuffs (Landau 1997, p. 56). This policy may have reduced new entry and the possibilities for greater competition in the industry, as well as slowing the diffusion of useful knowledge (Guedon 1980, pp. 67–68; Landau 1958). But the secretiveness of German chemical firms may not have been the decisive factor here. An alternative possibility is that German chemical firms were no more inclined to secretiveness than their American counterparts. The real difference may have been that American petroleum firms, facing a rapidly expanding market in the interwar years, were far more open and transparent than American chemical firms, and that the petroleum firms shaped the nature of the university/industry relationship in the United States.

Of course, Germany and the United States had developed along different trajectories in the first half of the century. Any comparisons between them need to reflect differences in the size and the composition of demand confronting the chemical sectors of the two countries, differences in skills accumulated over the course of previous decades, differences in the resource base of each country, and differences in government policies resulting from wartime experiences. In the case of Germany, a key factor was the policies of self-sufficiency instituted in preparation for possible future wars. Before the First World War, Germany had been heavily dependent on the import of Chilean nitrates. This dependence on imports had rendered Germany twice vulnerable: her supply of fertilizers for agriculture and of nitrates for munitions were both vulnerable to a naval blockade. This vulnerability was responsible for the extensive search process that led to the development, just before the First World War, of the Haber/Bosch process at BASF for manufacturing ammonia by combining atmospheric nitrogen with hydrogen at high pressures. These skills in high-pressure technology, flowing from the Haber/Bosch process, also made it possible to convert coal (and lignite) into both liquid fuels and synthetic rubber. This capability was to prove immensely important to Nazi Germany in World War II, when the Luftwaffe made use of ersatz fuel and the Wehrmacht traveled on ersatz tires. Skills in high-pressure technology also led to Germany's leadership in the technology of hydrogenation.

A final formative factor in the German experience that shaped Germany's chemical industry in the first half of the 20th century, and tilted it away from chemical engineering, was that nation's earlier extraordinary successes in the discovery and manufacture of synthetic dyes. The underlying skills in organic chemistry were subsequently transferred to pharmaceutical products, where Germany has remained consistently successful through the subsequent course of the 20th century. Both the synthetic dyes and the pharmaceuticals involved reliance on small-batch production technologies, not the large-scale continuous production for which chemical engineering expertise is so useful.

Ironically enough, the research breakthroughs in polymer chemistry that eventually gave rise to a wide range of new petrochemical products—plastics, synthetic fibers, synthetic rubber—happened primarily in Germany in the 1920s. Given the technologies that prevailed at the end of the Second World War, and given also certain immutable facts of nature, petroleum-based hydrocarbons were vastly superior to coal: Polymer-based products such as plastics and synthetic fibers required straight-chained organic chemical compounds (aliphatics) that are derived from petroleum-based hydrocarbons rather than the aromatic ring compounds that predominate in coal tar. German scientific leadership in polymer chemistry in no way guaranteed German technological or commercial leadership in the products that were eventually based on the findings of polymer chemistry. The excellent scientific-research capabilities of the institutes of the Kaiser Wilhelm Society and, after 1948, the institutes of the Max Planck Society, could not overcome certain intractable economic facts of the natural world. Many polymer-based products can indeed be manufactured from a coal-tar base, but they cannot be manufactured as cheaply as from petroleum-based feedstocks. German scientific research capabilities, given Germany's (coal) natural resource base, had thus placed Germany at a competitive disadvantage. History is full of such ironies. For example, the British development of the Gilchrist-Thomas process made possible the exploitation of western Europe's immense deposits of high-phosphorus ores, laying the basis for Germany's emergent steel industry in the closing decades of the 19th century, and placing Britain at a serious competitive disadvantage.

The decisive strategic development for the global postwar chemical industry was the discovery and exploitation of the huge, low-cost Middle Eastern oil deposits. The availability of this new oil source substantially reduced Germany's and Europe's resource disadvantage in petroleum vis-a-vis the United States. The new resource base of the petrochemical industry laid the basis for Germany's transition to the large-scale, continuous process technologies required in petrochemicals (Stokes 1994). Germany's long and distinguished industrial tradition required an entirely new approach to the design of large-scale plants, and the transition was not always an easy one. This transition was, however, facilitated by expertise spread through multinational oil companies as well as by the specialized engineering firms that played a critical role in the postwar years in the rapid diffusion of petroleum-based technologies. In the drastically new industrial context of the postwar period, Germany essentially borrowed the findings and methodologies of American chemical engineering (Landau 1958; Schoenemann 1980; Buchholz 1979). Britain similarly made a transition to petroleum sources after the war, a transition that was substantially complete by 1960. The demand for chemical engineers increased sharply in postwar Britain as coal was replaced by oil as the principal feedstock for the chemical industry, along with the increasing prominence of oil companies on the domestic scene. The United Kingdom output of petrochemicals tripled in just six years, from 1953 to 1959.

A final observation is appropriate. The technological history of the German chemical industry in the first half of the 20th century is, in large measure, a story of path dependence: Skills originally acquired in synthetic dyes were later transferred to pharmaceuticals; capabilities acquired in the Haber/Bosch process were later transferred to hydrogenation of coal; the professional division of labor between the chemist and the mechanical engineer in synthetic dyes was carried over into other sectors of the chemical industry; and so on. Historians frequently and correctly call attention to ways in which events can be understood only by reference to earlier events that shaped later ones in path-dependent fashion. But path dependence is hardly ever the entire story. Indeed, if it were, history would be remarkably deterministic—and thoroughly dull! When economic incentives are sufficiently pervasive and powerful, and where there is at least relatively free entry into markets, such incentives may serve as battering rams to replace an inefficient technological system or a division of professional labor that has outlived its usefulness. Whatever handicaps the German chemical industry may have suffered by failing to embrace the discipline of chemical engineering in the first half of the 20th century, they proved not to be enduring. By the end of the 20th century, the chemical industries of Germany and the United States had largely converged, and the different paths taken earlier were far less visible than they had been at the outbreak of the Second World War.

Yet, this episode also underlines that the speed at which such convergence processes take place is to a large extent determined by economic institutions and public policies. Without the emergence of a market for petrochemical technology, as well as firms that were prepared to satisfy the demands of that market, and without U.S.-guaranteed access to cheap middle Eastern oil, Germany might have taken much longer in replicating the American achievements in chemical engineering.

REFERENCES

Aftalion, Fred (1991), A History of the International Chemical Industry, University of Pennsylvania Press, Philadelphia, PA.

Arora, Ashish and Alfonso Gambardella (1994), "The Changing Technology of Technological Change: General and Abstract Knowledge and the Division of Innovative Labour," *Research Policy* (September) 23, 523–32.

Beer, John (1959), *The Emergence of the German Dye Industry,* University of Illinois Press, Urbana, IL.

Ben-David, Joseph (1971), *The Scientist's Role in Society,* Prentice-Hall, Englewood Cliffs, NJ.

Bird, R. Byron, Warren E. Steward, and Edwin N. Lightfoot (1960), *Transport Phenomena,* John Wiley & Sons, New York, NY. (Originally published in 1958 as "Notes on Transport Phenomena.")

Buchholz, Klaus (1979), "Verfahrenstechnik (Chemical Engineering) — Its Development, Present State and Structure," *Social Studies of Science,* pp. 33–82.

Chapman, Keith (1991), *The International Petrochemical Industry*, Blackwell, Cambridge, MA.

Chandler, Alfred D., Jr. (1992), *"Global Enterprises – Big Business and the Wealth of Nations,"* unpublished manuscript, July 29.

Davis, George E. (1901), Handbook of Chemical Engineering, Davis Brothers, Manchester, UK.

Dertouzos, Michael, Richard K. Lester, and Robert M. Solow (1989), *Made In America*, MIT Press, Cambride, MA.

Divall, Colin (1994), "Education for Design and Production: Professional Organization, Employers, and the Study of Chemical Engineering in British Universities, 1922–1976," *Technology and Culture,* 35(2) 265–66.

Donnelly, J. F. (1988), "Chemical Engineering in England, 1880–1922," *Annals of Science*, 45: 555–590.

Elder, Albert Lawrence, ed. (1970), *The History of Penicillin Production*, American Institute of Chemical Engineers, New York, NY.

Enos, John (1962), *Petroleum, Progress and Profits*, MIT Press, Cambridge, MA.

Field, Philip (1940), *Industrial Research and Changing Technology*, National Research Project Report No. M-4, Works Progress Administration, Philadelphia, PA, January.

Gibbs, J. Willard (1874–1878), "On the Equilibrium of Heterogeneous Substances," *Transactions of The Connecticut Academy of Arts and Sciences*, vol. 3, New Haven, CT, pp. 108–248.

Gornowski, Edward J. (1980), "History of Chemical Engineering at Exxon," in William Furter (ed.), *History of Chemical Engineering*, American Chemical Society, Washington, DC.

Guedon, Jean-Claude (1980), "Conceptual and Institutional Obstacles to the Emergence of Unit Operations in Europe," in William Furter (ed.), *History of Chemical Engineering*, American Chemical Society, Washington, DC.

Haber, Ludwig F. (1971), *The Chemical Industry, 1900–1930*, Clarendon Press, Oxford, UK.

Haynes, William (1945–1954), *The American Chemical Industry: A History*, 6 vols., Van Nostrand, New York, NY.

Herbert, Vernon and Attilio Bisio (1985), *Synthetic Rubber*, Greenwood Press, Westport, CT.

Hougen, Olaf (1965), "Chemical Engineering Education in the United States of America," *The Chemical Engineer* (September): 222–31.

Hougen, Olaf and Kenneth Watson (1936), *Industrial Chemical Calculations*, John Wiley & Sons, New York, NY.

Hounshell, David (1984), *From the American System to Mass Production, 1800–1932*, The Johns Hopkins University Press, Baltimore, MD.

Hounshell, David A. and J. Kenly Smith, Jr. (1988), *Science and Corporate Strategy: DuPont and R&D*, Cambridge University Press, New York, NY.

Howard, Frank (1947), *Buna Rubber*, Van Nostrand, New York, NY.

Jarausch, Konrad (1990), *The Unfree Professions: German Lawyers, Teachers, and Engineers, 1900–1950*, Oxford University Press, New York, NY.

Johnson, Jeffrey (1990), *The Kaiser's Chemists*, University of North Carolina Press, Chapel Hill, NC.

Kahn, E. J., Jr. (1981), *The Problem Solvers. A History of Arthur D. Little, Inc.*, Little, Brown, Boston, MA.

Landau, Ralph (1958), "Chemical Engineering in West Germany," *Chemical Engineering Progress* (*July*) 54: 63–68.

Landau, Ralph, ed. (1966), *The Chemical Plant*, Reinhold, New York, NY.

Landau, Ralph and Nathan Rosenberg (1992), "Successful Commercialization in the Chemical Process Industries," in Nathan Rosenberg, Ralph Landau, and David Mowery (eds.), *Technology and the Wealth of Nations*, Stanford University Press, Stanford, CA.

Landau, Ralph and Nathan Rosenberg (1994), "Innovation in the Chemical Processing Industries," in Nathan Rosenberg, *Exploring the Black Box*, Cambridge University Press, New York, NY.

Landau, Ralph (1997), "Education: Moving From Chemistry to Chemical Engineering and Beyond," *Chemical Engineering Progress*, (January) 93(1):52–65.

Landes, David (1969), *The Unbound Prometheus*, Cambridge University Press, New York, NY.

Larson, Henrietta M., Evelyn H. Knowlton, and Charles S. Popple (1971), *New Horizons, 1927–1950*, Harper and Row, New York NY.

Lewis, Warren K. (1953), "Chemical Engineering: A New Science?" in L. Lohr (ed.), *Centennial of Engineering 1852–1952*, Museum of Science and Industry, Chicago, IL.

Lewis, Warren K. (1959), "Evolution of the Unit Operations," in issue on Chemical Engineering Education — Academic and Industrial, *Chemical Engineering Progress Symposium Series*, vol. 55, no. 26, pp. 1–8.

Little, Arthur D. (1933), "Chemical Engineering Research," in Sidney Dale Kirkpatrick (ed.), *Twenty-five Years of Chemical Engineering Progress*, American Institute of Chemical Engineers, New York, NY.

Maloney, J. O. (1982), "Doctoral Thesis Work in Chemical Engineering in the U.S. from the Beginning to 1960," in William Furter (ed.), *A Century of Chemical Engineering*, Plenum Press, New York, NY.

Olive, Theodore R. (1934), "A New Classification of Unit Operations," in *Chemical and Metallurgical Engineering*, 41(5):229–231.

Peppas, Nikolaos (1989), "The Origins of Academic Chemical Engineering," in N. Peppas (ed.), *One Hundred Years of Chemical Engineering*, Kluwer Academic Publishers, Dordrecht, Netherlands.

Perazich, George, Herbert Schimmel, and Benjamin Rosenberg (1938), *Industrial Instruments and Changing Technology, National Research Project. Report No. M-1, Works Progress Administration, Philadelphia, PA, October.*

Pigford, Robert (1976), "The Past 100 Years in Chemical Technology," *Chemical & Engineering News,* (6 April):190–203.

Reynolds, Terry (1986), "Defining Professional Boundaries: Chemical Engineering in the Early 20th Century," *Technology and Culture,* 27(4):694–716.

Rosenberg, Nathan, (ed.) (1969, *The American System of Manufactures,* Edinburgh University Press, Edinburgh, Scotland.

Rosenberg, Nathan (1976), "Technological Change in the Machine Tool Industry, 1840–1910," in Nathan Rosenberg, *Perspectives on Technology,* Cambridge University Press, New York, NY.

Rosenberg, Nathan and Richard Nelson (1994), "American Universities and Technical Advance in Industry," *Research Policy,* 23(3):323–48.

Schmookler, Jacob (1965), "Catastrophe and Utilitarianism in the Development of Basic Science," in Tybout, Richard (ed.), *Economics of Research and Development,* Ohio State University Press, Columbus, OH.

Schoenemann, Karl (1980), "The Separate Development of Chemical Engineering in Germany," in William Furter (ed.), *History of Chemical Engineering,* American Chemical Society, Washington, DC.

Scriven, L. E. (1991), "On the Emergence and Evolution of Chemical Engineering," *Advances in Chemical Engineering, vol. 16, Academic Press, San Diego, CA,* pp. 3–40.

Sheehan, John (1982), *The Enchanted Ring: The Untold Story of Penicillin,* MIT Press, Cambridge, MA.

Skolnick, Herman and Kenneth Reese (1976), *A Century of Chemistry: The Role of Chemists and the American Chemical Society,* American Chemical Society, Washington, DC.

Spitz, Peter (1988), *Petrochemicals: The Rise of an Industry,* John Wiley & Sons, New York, NY.

Stokes, Raymond (1994), *Opting for Oil,* Cambridge University Press, New York, NY.

Tailby, S. R. (1982), "Early Chemical Engineering Education in London and Scotland," in William Furter (ed.), *A Century of Chemical Engineering,* Plenum Press, New York, NY.

Thackray, Arnold, Jeffrey L. Sturchio, P. Thomas Carroll, and Robert F. Bud (1985), *Chemistry in America, 1876–1976,* D. Reidel Publishing, Dordrecht, Netherlands.

Tocqueville, Alexis de (1898), *Democracy in America,* Century Co., New York, NY.

Trescott, Martha M. (1982), "Unit Operations in the Chemical Industry," in William Furter (ed.), *A Century of Chemical Engineering,* Plenum Press, New York, NY.

Trescott, Martha M. (1981), *The Rise of the American Electrochemicals Industry, 1880–1910,* Greenwood Press, Westport, CT.

Ure, Andrew (1853), *Dictionary of the Arts, Manufactures and Mining,* D. Appleton and Company, New York, NY.

Walker William. H., Warren K. Lewis, and William H. McAdams (1923), *Principles of Chemical Engineering,* McGraw-Hill, New York, NY.

Weber, Harold C. (1980), *The Improbable Achievement: Chemical Engineering at MIT,* Chemical Engineering Department, privately printed, Cambridge, MA. (Abridged version appeared in William Furter (ed.), *History of Chemical Engineering,* American Chemical Society, Washington, DC. 1980.)

Westwater, J. W. (1980), "The Beginning of Chemical Engineering Education in the USA," in William Furter (ed.), *History of Chemical Engineering,* American Chemical Society, Washington, DC.

Williamson, Harold F., Ralph L. Andreano, Arnold R. Daum, and Gilbert C. Klose (1959–1963), *The American Petroleum Industry, 1899–1959: The Age of Energy,* Northwestern University Press, Evanston, IL.

PART IV
Surveying the Levels of the Matrix

8 The Industry Evolves within a Political, Social, and Public Policy Content: A brief look at Britain, Germany, Japan, and the United States

MICHELINE HORSTMEYER

Businesses are social institutions, shaped by the society in which they operate. The chemical business is no exception. A matrix of economic, political, and social forces defines the growth path of the industry in each country. This chapter offers a broad-brush, abbreviated review of those changes in government, public policy, and social-political climate especially relevant to an understanding of the chemical industry's development over the last 150 years.[1] It should be emphasized that this chapter is intended to provide a background rather than a scholarly history of the period and the countries.

Britain is discussed first, as befits the birthplace of the modern chemical industry. Germany follows, as the next country to dominate the chemical industry. The discussion of Japan is kept especially brief, since much of the development has taken place after World War II. The chapter then concludes with the United States.

BRITAIN

From Landed Gentry to Rentier Capitalism: The Arrival of the Industrial Revolution

During the late 18th and first half of the 19th century, Great Britain was in the forefront of what we now call the First Industrial Revolution. It was a "revolution" only in hindsight. Its roots ran back to the Enlightenment, and the changes it wrought were gradual. Official England did little to encourage it. Little thought was given to a national industrial policy (Wiener 1981).[2] Capitalism flourished and industrialism took root in a land that largely practiced laissez faire economics, much in the style that Adam Smith had exposited in his landmark *Wealth of Nations* (1776).

[1] A number of books and articles treat the history of these countries; some of these titles are grouped in a separate section in the bibliography.

[2] Also see Halevy (1961), particularly his chapter "Industry," pp. 256–337, and Woodward (1958), pp. 14–15.

233

Britain's governing class throughout most of the 19th century was drawn from the well-established families of the landed aristocracy, many of noble lineage. The career path for this class involved receiving a classical education at a socially prominent "public" (but really private) high school such as Eton or Harrow, and perhaps continuing on to Oxford or Cambridge University (Wiener 1981, Malchow 1991, Rubinstein 1993). Some graduates were offered a lifetime career in the service of the Crown overseas, either in the officer corps, the diplomatic service, or the administrative civil service. Others went into the clergy, the professions, and the financial industry. To cement its empire, Britain maintained the world's most powerful navy. The empire and navy were paid for by a regressive tax structure which disproportionately burdened the lower-income groups and at the same time reduced the risk for capitalists with money to invest.

The government made little effort to strengthen the nation's infrastructure or direct the course of domestic industry, including the chemical industry.[3] The industrial developments of the early 19th century were largely mechanical. The prime skills necessary were an inventive turn of mind and a practical facility to design and craft machinery. Artisans acquired and perfected their abilities through years of apprenticeship and experience. Enterprising foreigners with ideas and energy settled in this "workshop of the world" to exploit opportunities. The elder Marc Isambard Brunel arrived in 1799,[4] and William Siemens came in 1843.[5] Ludwig Mond, who founded his alkali works in 1873, was probably among the last of this type.

The nation's coal mines, iron foundries, textile mills, and factories enabled Britain to achieve its early dominance in the iron, weaponry, railroad, machinery, and shipbuilding industries. The fledgling chemical industry, centered largely in the north of England, rose to prominence during the first half of the 19th century. The industry drew on the talents of such self-taught chemists as Joseph Priestley (1733–1804) and John Dalton (1766–1844), who were working in makeshift, private laboratories. It responded to the escalating demands of rapidly growing textile, glass, soap, and metal industries and flourished in an atmosphere of benign neglect by the state. William Henry Perkin (1848–1907) discovered the first synthetic dye in 1856, generally viewed as the first significant product of the organic-chemical industry.

Late 18th- and early 19th-century English upper classes responded to the economic changes occurring around them by funneling money in new directions. A new class of entrepreneurs borrowed start-up funds from these wealthy investors. Many also relied on their savings and on money from friends and family. Retained earnings from profitable ventures made possible further growth. Some small local banks supplied working capital through short-term loans; others joined in small partnerships to finance larger ventures (Halevy 1961).[6] A symbiotic arrangement between informal and formal financial networks developed to service these growing capital requirements (Dintenfass

---

[3]Canals, steambouts, and railroads were all privately planned and funded. Legislation in Parliament "sanctioned" and "authorized" canals and railroads, but generally after petition by interested parties. See Woodward (1958), pp. 40–47, 575; also Halevy (1961), p. 305.

[4]Brunel was known for his machine for making ship's block and as the engineer for the Thames Tunnel. His famous and only son, Isambard Kingdom Brunel, was a leading railway and marine engineer.

[5]Siemens patented the open-hearth furnance used in glassmaking and steelmaking. It eventually supplanted the Bessemer process. Siemens also laid the first direct electrical telegraph cable between Great Britain and the United States and worked with electric lighting and electric railways.

[6]See Halevy (1961), Part II, Chapter III, "Credit and Taxation," pp. 338–83. Also refer to Marco Da Rin's chapter in this volume.

1992). These partnerships were important in launching early textile, mining, and machinery operations. London bankers organized a money market for private short-term debt, called *bills of exchange*. In this way, the immense fortunes of the landed gentry—shored up with the help of favorable tax policies—moved from farming into railroads (both at home and abroad), canals, mines, urban infrastructure, industrial ventures, and increasing trade. Many wealthy landowners became rentier capitalists during the course of the 19th century (Wiener 1981).

The risk-taking owners-enterpreneurs of this time, who were responsible for building and running the nation's multiplying factories, mines, and foundries, were practical seat-of-the-pants inventors, trial-and-error scientists, and self-taught industrialists. Often long, hard years of work lay behind their firms' financial success. They failed to create their own national political party and thus lacked a clear voice or unified presence in Parliament. The prototypes were men like James Watt (1736–1819)[7] and Henry Cort (1740–1800).[8] Like them, many owner-entrepreneurs were born in the north of England and retained strong personal ties to that region throughout their lives. Many were radical in local politics and nonconformist in religion (Malchow 1991).[9] They were apprenticed to fathers, uncles, or family friends at an early age and remained in the same field until the end of their lives.

By the mid-19th century, with a 50 year head start in industrializing, Britain had become the wealthiest country in Europe on a per capita basis. It was also the most urban European nation, with a fast growing middle class populating London and its spreading suburbs. As individuals migrated into this sprawling hub, they moved from blue-collar positions into white-collar jobs in the growing commercial, financial, service, and consumer sectors (Rubinstein 1993). England also dominated the world's chemical industry in 1850, employing more than 10,000 people, largely making soda by the LeBlanc process, which was then at the cutting edge of technology. Enactment of the highly contested Corn Laws in 1846 sharply reduced the trade barriers to imported corn and marked Britain's passage from an agricultural to a commercial and industrial economy, as well as from a protectionist nation to the world's foremost bastion of free trade. Free trade made eminent sense for the industrial frontrunner trying to penetrate new European markets and supply her empire.

Limitations of the System: The Downward Slide Begins

In the middle of the 19th century, most Englishmen were confident that a combination of Empire and industrial dominance would sustain Great Britain's status and wealth indefinitely. Yet in retrospect, this time marked the beginning of a long, downward slide for Britain's industrial and economic fortunes. The Great Exhibition of 1851 in London was an early indication that England was being nudged aside by her American and German cousins; one historian characterized the international gathering as, a "...pageant of an epoch already passing away" (Woodward 1958, p. 47). The United States and Germany displayed surprisingly sophisticated machines in the glass-and-iron Crystal Palace, suggesting that they were closing the gap with the British frontrunner—though the British won most of the prizes. By the Paris Exhibition of 1867 Britain had lost ground, besting her competitiors in only ten percent of the departments (Ensor 1992,

[7]Watt, a Scottish instrument maker and engineer, was best known for his development of an efficient steam engine.

[8]Born in Lancaster, Cort made possible the large-scale and inexpensive conversion of cast iron into wrought iron.

[9]English "Radicals" of the time were members of the extreme section of the Liberal party.

p. 319). More to the point, Germany's chemical industry would take the lead away from Britain in the field of organic chemistry by the 1870s, even though it was an Englishman who had opened the door to synthetic dyemaking only a few years before.

A cultural argument sometimes put forward for this economic decline is that many successful, early-19th-century industrialists eventually merged into the ranks of the landed gentry. For example, Martin Wiener (1981) argues that the values of the British aristocracy were firmly imprinted on England's first industrial leaders and their descendants. Retired factory owners evolved into a "gentrified bourgeoisie" sitting on the back benches of Parliament, a passive presence echoing the voices of the true bloodline gentry and deferring to their decisions. Industrialists' sons graduated from the elite schools that were dedicated to the crafting of "English gentlemen." The diploma-bearing offspring either turned their backs on their fathers' enterprises or ran them into the ground. In most cases, they manifested a decided distaste for manufacturing industries, technology, and science (Wrigley 1986).

A more recent analysis of a fairly large sample of English and German businessmen working from 1870 through 1914, however, challenges some of these views (Berghoff and Möller 1994). Data reveal that there *was* an ample supply of fresh enterpreneurial talent available to British industry in cities such as Birmingham, Bristol, and Manchester during these years. Indeed, not all among the upper and upper-middle classes disdained production and profit. Sons of landowners, professionals, bankers and merchants accounted for roughly 33 percent of British steelmakers active between 1865 and 1953 (Dintenfass 1992). The German bourgeoisie had little geographical mobility, with a smaller number emigrating elsewhere, while English businessmen were more willing to move out of their villages or towns to set up their companies. Market demand for degreed businessmen was less in Britain than in Germany, but so was supply.[10] Of the British and German businessmen who went on to secondary education, all placed a strong emphasis on classical studies. For both national groups, schools often were chosen for reasons of status or to continue a family tradition, not in practical anticipation of future professional needs. However, the sons of English businessmen did *not* attend the nation's most prestigious high schools: Eton and Harrow schoolboy populations continued to be drawn from the nobility, the landed gentry, financiers, and professionals.

Independent young men didn't necessarily follow in their fathers' footsteps since by the late 19th century even lofty Harrow contributed some graduates (albeit a minority) to industry and commerce, with between one-quarter and one-third taking up careers in business. (Dintenfass 1992).[11] The typical English businessman of this time had an Anglocentric view of the world. He was reluctant to gain business experience in any other part of the world, rarely spoke a language other than English, and generally expected the foreign customer to learn English and adapt his needs and wants to English products (Berghoff and Möller 1994).

As funding for universities increased in both countries during the late 19th and early 20th centuries, a larger percentage of businessmen obtained degrees. The British diploma, however, continued to cost far more than one from Germany's state-supported school. Moreover, in Germany degrees were often in technological or scientific areas,

[10]As late as 1913, there were only 9,000 full-time students in British universities, as opposed to 60,000 in German ones.

[11]The majority of 19th-century industrialists and merchants had long been unwilling to send their sons to the elite public schools (Dintenfass 1992, pp. 62–63. These were generally considered the grooming grounds for the sons of blue bloods, bankers, and professionals.

whereas in Britain a classics degree from Oxbridge was still the degree of choice. For example, many German chemical manufacturers held advanced degrees in chemistry, an expertise rarely found in Britain.

A small minority of successful British businessmen were elevated into the nobility, acquiring a title that, toward the end of the century, depended more and more on wealth than on birth.[12] Some observers viewed this trend with disdain, saying it cheapened the nobility and signaled the "incipient embourgeoisement of the aristocracy" (Berghoff and Möller 1994). However, not a single titled businessman in the sample gave up, or even neglected, the family's business. In some cases the businessmen received their titles during their retirement, long after they had handed the business over to a son and entered a political career. In addition, retired industrialists did not purchase large country estates to which to retreat (Malchow 1991; Dintenfass 1992; Rubinstein 1993). As the 19th century progressed, capital markets offered more lucrative investment opportunities than did land, and successful businessmen were far more apt to choose the better return on investment. Those few who opted to settle in the countryside lived by regular workaday hours and not by the gentryman's schedule of hunts and unpaid public service (Berghoff and Möller 1994). And when one examines business choices and decisions of this period, no direct evidence links them to anti-industrial values. As one historian notes, if there was a "gentry cast" to their (i.e., the businessmen's) minds, there are few traces of it in the records of British enterprises (Dintenfass 1992, p. 64).

By the latter part of the 19th century, the prototypical English businessman spent a far greater amount of his time and energy on philanthropies of every cast than did his German cousin, who tended to concentrate his paternalism on his own workforces. For one thing, the sense that the fortunate shared a responsibility for helping those less fortunate is deeply imbedded in the ideal of a cultivated English gentleman. Moreover, while Bismarck had begun to weave a social safety net for Germany in the 1880s, Britain had no comparable system, underscoring the need for voluntary and private charitable efforts. Participation in the political process was another mark of the honorable English gentleman. In the last decade of the 19th century, commercial and industrial classes actually constituted a majority of the elected MPs in the British parliament, supplanting the once-dominant landed gentry. For many of them, however, Parliament was no more than a diverting sideshow at the end of their useful lives (Malchow 1991).

Attempts to explain the decline of Britain's industrial strength based solely on the distinctive cultural moorings of British businessmen are incomplete at best. An additional set of explanations traces Britain's decline to her underdeveloped scientific and technical-educational system. The emergence of science-based industries such as the synthetic-dye industry drew attention to the dearth of scientifically educated managers and technically trained workers. The problem was well recognized at the time by a string of blue-ribbon commissions (Select Committee, 1868; Royal Commission, 1881; Haldane Commission, 1906). As one historian writes, it is "one of the strangest paradoxes in modern history" that Britain, with its liberal political system, should have retained schools that clung to the outmoded curriculum of classics for the upper classes and practical subjects for the working classes, while Germany, a "far more authoritarian society," should have built a more modern school system (Landes 1969, p. 348).

When the British government did tiptoe into the field of science education—for example, as with the founding of a Department of Science and Art (1853)—the science

---

[12]For example, Edward Guinness and Henry Allsopp, both brewers, were elevated into the nobility, the latter becoming Baron Hindlip. Also, H. F. Eaton, the silk broker, became Lord Cheylesmore (Cannadine 1990).

classes offered by the self-taught, poorly paid teachers were abstract and dry, with little relevance to industrial production. Elite, private universities dragged their feet as well, in keeping with a long-standing, aristocratic disdain for practical activities. They preferred to put their energies and funding into disinterested, amateur research rather than industrial research and development. This class-stratified society attached much status to those engaged in abstract, noncommercial work. Indeed, it wasn't until 1887 that chemists at Cambridge acquired their own laboratory![13] Later on, the course of instruction in British chemistry departments was heavily geared toward teacher training, especially after the Balfour Act of 1902 established a state system of secondary education and many schools were built that needed staffing.

Although 19th-century Britain produced world-reknowned scientists, seldom did they have any desire to run a company. In the chemical industry, W.H. Perkin and Edward Nicholson are noted examples of chemists who left successful firms to pursue independent research. Consequently, historians believe that the lack of scientists willing to apply science for practical ends contributed to Britain's declining competitiveness in technological fields (Wrigley 1986, p. 184).

But the problem went beyond the supply of industrial scientists and included the demand, as well; British industry of that time offered few opportunities for research scientists. British manufacturers claimed that scientific research was simply unaffordable, due to the pressure of foreign competition, and skimped on such spending or neglected it entirely. Few understood why their foreign competition (meaning especially the Germans during the last third of the 19th century) fully supported in-house research, even at the cost of immediate profits. For example, the German firm BASF funded a 17-year effort to commercialize the production of synthetic indigo after the dye was first made in a German laboratory in 1880. It was an effort that a British firm would not have made.

There is some evidence that some British chemists—most likely preferring the pursuit of pure science to applied science—could not find employment in their own country during the early years of the dye business (Beer 1959, p. 47; Wrigley 1986, p. 180). When necessary, British firms commonly relied on foreign chemists and consultants, further suggesting that there was an insufficient number of British chemists, that they were not well trained for industrial work, or both. Those British nationals that were hired were relegated to doing routine analysis rather than research. In fact, a few British scientists emigrated, presumably because of better opportunities abroad. Foreign-born scientists, like the famous German chemist August Hofmann, returned to their native lands. The British dye industry lost a group of talented German chemists, including not only Hofmann but also Heinrich Caro, Carl Martius, and Otto Witt. Britain's loss was Germany's gain. All went on to make important contributions to the German dye industry.

The story about engineers and technicians is similar. Britain was slow to move into technical education, waiting more than 30 years after its leading industrial competitors had a system in place, despite almost 40 years of blue-ribbon governmental commissions (Ensor 1992, p. 318). The starting point was the Technical Instruction Act of 1889, which provided for mostly secondary technical instruction under the jurisdiction of county councils. British universities moved just as hesitantly into engineering education. It wasn't until 1893 that Britain's civic universities, established with the idea of helping

---

[13]A British delegation to Germany and Switzerland in 1872 found that there were more students studying and doing research in chemistry at the University of Munich than in all the universities and colleges of England combined. (Dintenfass 1992, p. 36).

to man British industries, received their first state money. State funding reinforced existing methods of training: The engineering curriculum continued to produce engineers who relied more on craft skills than on scientific knowledge. Moreover, the supply of university-trained engineers soon outstripped demand, since employers still believed that engineers should receive most of their training on the factory floor (Wrigley 1986, p. 169).

British obtuseness with regard to science and technology hit a highwater mark with its LeBlanc soda industry. Production reached its peak at about 1880. Thereafter, competition from the more efficient Solvay process — licensed solely to Brunner Mond within Britain — started cutting into sales. Moreover, U.S. protectionism had largely eliminated that export market. So in 1890, an amalgamation of inefficient LeBlanc producers combined to form a large, publicly traded United Alkali, perhaps hoping to realize with economies of scale what they had failed to achieve with technology. Brunner Mond — then the largest chemical company in the world — and United Alkali together allowed Britain to continue to dominate the world's heavy-inorganic-chemical industry until the early years of the 20th century, but the latter also slowed any movement toward new industrial processes and markets.

British shortsightedness repeats in the story of its dye industry. In the later 19th century, British dyers first stopped making dyes themselves; instead, they exported key inputs, especially coal tars, to German firms and then imported the finished dyes. This killed off the part of the British inorganic-chemical industry that had supplied other intermediates for dyestuffs. By the 1880s, German dye manufacturers were drawing on home supplies of coal tar and intermediates, forcing Britain out of the picture entirely.

Along with a hostility to applied science and technology, several other factors conspired to aid Britain's industrial decline in the latter part of the 19th century. For example, Britain clung to its free trade policy up until the outbreak of World War I. While this policy surely offered some overall benefits to the English economy, the British chemical industry suffered in the face of protectionism by others. In the chemical sector, Germany raised its tariffs against Britain in 1879, and the United States imposed high duties on British exports in 1897.

In the area of patent policy, British scientific research was not as well protected by patent law as was German research. The absence of a unified patent law in the German states prior to 1876 meant that German dye entrepreneurs were able to freely copy British (and French) innovations in dyemaking. Fledgling German companies patented their innovations in Britain, while British firms were unable to patent in Germany. By 1876, when it was in Germany's self-interest to shield its growing R&D investment in the dye industry, the newly unified government passed a strong patent law that offered protection to German innovations. Finally, it wasn't until 1907 that holders of British patents were required to work them in Britain. Until that date German dye manufacturers could make all their dyes in the Reich and export them to Britain and elsewhere.[14]

In the area of finance, British industry lacked the active, long-term banking partners the Germans enjoyed and which proved particularly valuable to high-growth, technologically driven sectors such as chemicals. Toward the end of the 19th century, the British banking sector consolidated into five large banks that supplied short-term loans, competed on interest rates, and made little effort to establish enduring relationships

[14]However, a British high-court ruling in 1909 erected legal hurdles to the law's enforcement that again shifted the advantage toward Germany and away from Britain. See Murmann and Landau chapter in this volume for a fuller discussion.

with firms. The London Stock Exchange shared this aversion to long-term monitoring; it tended to steer investment toward less risky and more profitable foreign utilities, transportation projects such as railroads, and public works rather than British industry. British financiers were also wary of venturing into the relatively uncharted investment waters of science-based firms. These new businesses must have intimidated even the wiliest of city bankers, who were, more likely than not, illiterate in science and technology. Ivan Levinstein — a German chemist who moved to Britain in 1864 and ran one of the most successful dye firms in the country — complained that he was unable to secure funds in 1877 from a financial community that, he believed, lacked any capacity to judge the value of scientific work (Wrigley 1986, p. 175).

During this time, some English historians writing about the gradual spread of the "company system," deplored the disappearance of the owner-entrepreneur. One noted, "Property passed to shareholders concerned only for dividends; control was exercised on the shareholders' behalf by boards of directors, nominally elected by them, but in fact mainly co-opted, often representing only financial, social, or personal 'pulls' and devoid of any specialized understanding of the firm or even of the industry. Thus for the alert individual carrying his business in his head came to be substituted a collectivity finding safety in rules and procedures" (Ensor 1992, p. 112). But in fact, owner-enterpreneurship dominated Britain's business landscape until well into the 20th century. Despite the Companies Act of 1862, conversion of firms into limited companies did not pick up speed until 1872, and the motivation was often to limit liability rather than divorce ownership from management.

Some historians have argued that Britain's industrial decline has been overstated, because the nation was never quite the industrial giant that it was reputed to be. For example, W.D. Rubinstein (1993) contends that the English economy was always fundamentally a commercial, financial, and service-driven one, even at the height of its industrial power at mid-century. Eventually, Rubenstein claims, the growing consumer market overshadowed the industrial, and the nation passed from being a basically agricultural economy to a commercial/financial/consumer one with only a brief — and overemphasized — industrial interlude. As another historian notes, if Britain in 1870 was no longer the world's workshop, she was more than ever its warehouseman, its banker, and its commission agent; and these were relatively better-paid functions (Ensor 1992, p. 507). From this perspective, Englishmen should be earning plaudits — not reproach — for recognizing their true areas of comparative advantage. Back in the third quarter of the 19th century it was hard to see the direction in which the new worlds of chemistry and electricity were heading. Why take imprudent investment risks in enterprises based on brand-new technologies? Far smarter to excel in the industries that had brought the nation to the pinnacle of wealth and power in the first place and to focus new energies into the expanding fields of finance, commerce, and consumer services.

Whatever the merits of this argument, in the dye industry, Britain appeared to have many advantages. Her textile sector guaranteed a large domestic customer base. She had a plentiful supply of the necessary raw material tar distillates, which were a by-product of coke-recovery ovens. The inventor of synthetic dyestuffs, W.H. Perkin, called England his home. British industry in general was remarkably open to talented individuals, wherever they were from. British dye companies, however, employed few trained chemists, relying instead on practical dyers who had learned their skills through long-standing trial-and-error methods. But the new synthetic dye industry was a truly science-based industry. To produce efficiently, to replicate the same dyes over and over, and to discover new dyes, the industry had to have managers thoroughly trained in organic chemistry. Yet British dye manufacturers were reluctant to abandon the practical, shop-floor training and the hit-or-miss procedures that had brought them past success and profits. The aforementioned Ivan Levinstein argued that the nation's very success in earlier industries had handicapped it for developing industries based on sciences such as chemistry (Wrigley 1986).

After World War I: Labour Finds Its Voice

At the conclusion of hostilities in 1918, the British found themselves without a foundation for a competitive, domestic-based industrial economy geared to move into the Second Industrial Revolution. In the chemical field, the war highlighted the defects in Britain's domestic chemical industry, particularly dyestuffs, where there was some concern that the supply of chemicals would be insufficient to dye the troops' khaki uniforms. Education remained oriented toward the classics and the professions. Unlike in Germany and the United States, there was little to no contact between universities and growing, scientifically based industries, such as chemicals. While some British companies went public to broaden access to capital funds, most remained family controlled. Banks did begin to participate more actively in industrial financing, yet most retained a strong aversion to long-term financing and preferred to do business only with larger companies.

Heavy losses of wealth and manpower during World War I weakened the governing classes, leaving a void that was soon filled by the Labour party, representing the middle and working classes. While British trade unionism can be traced back to the early decades of the 19th century, these unions were organized by craft. It wasn't until the 1880s that unions began to organize by industry and make some inroads toward enlisting unskilled and casual labor. At the same time, Socialist and Labour representatives began to secure seats on local councils, where they could address political issues beyond the purview of the unions.

Britain lacked the deep-rooted suspicion of concentrated economic power that had led America to pass various antitrust acts in the late 19th and early 20th century. Instead, the government began to encourage the consolidation of the chemical industry. A first step was the formation of British Dyes, Ltd., in 1915, with a 50 percent subsidy coming from the Treasury. But while government funds went into the support of scientific research, the new company's board of directors continued to exclude chemists, and the commitment to research remained inadequate.

In 1926, with the government's stamp of approval, four leading British chemical companies—United Alkali, Nobel Industries, British Dyestuffs, and Brunner Mond—came together to establish Imperial Chemical Industries (ICI), heralded as Britain's national champion to match the newly formed IG Farben in Germany. At a time when most British companies were still small and family-owned and run, ICI was a large, professionally managed stock company right from the start. While ICI's bosses (Ludwig Mond and Henry McGowan) lacked scientific training, they understood that science was critically important to the success of their business. They oversaw ambitious research efforts in the fields of high-pressure chemistry, dyes, and by the mid-1930s, pharmaceuticals. ICI prospered, partly because cartel agreements allowed it to occupy a monopoly position within the United Kingdom and a preferential position within the Empire, partly because it was successful in carving out markets and focusing on selected products, and partly because its R&D efforts proved very fruitful.

The Great Depression of the 1930s was not as severe in Britain as in other parts of the world, to some degree because Britain abandoned the gold standard in 1931 in favor of keeping credit available and expanding its money supply. As the world turned protectionist, British gave up its cherished free-trade policy in 1932, in favor of the Dominion trade-preference system, which it held onto until 1973. The pain of the Depression and the strain of World War II shaped a Labour-dominated era that sought "fair shares for all" in a modern welfare state. In 1942, at the height of the German blitz, Sir William Beveridge, former director of the London School of Economics and a civil servant, issued the famous *Beveridge Report*, which called for the building of a socialist economy and a comprehensive welfare state. The *Beveridge Report* provides the backdrop to the Conservative-Labour battles that dominated nearly forty years of postwar politics.

Labour clearly won the early rounds of this political war, as Winston Churchill's personal ascendancy as a wartime leader failed to translate into votes for his beleaguered Conservative party (Hennessy 1993). Returning British soldiers had seen liberated countries in Europe where workers enjoyed substantially better housing and public facilities. British industry still clung to First Industrial Revolution enterprises like coal mining and shipbuilding, where outdated mines and shipyards were manned by a dissatisfied working class living in grimy towns and villages. Labour's program of nationalization hit railroads, coal mines, aviation, and the Bank of England in the late 1940s. Road transport, docks, harbors, iron and steel, and electrical power were next.

In the chemical industry, it is interesting to note that Britain began the shift from coal to petroleum as primary feedstock before Germany did—although it took Britain about ten years longer to actually make the transition.[15] Britain had only a small amount of petrochemical production before the war, but BP Chemicals was formed in 1947, and ICI set up its own petrochemical operation in 1951. By 1965, petroleum supplied 70 percent of British feedstock requirements. The industry that energed from 6 years of total war had matured during 80 years of considerable growth. Between 1870 and 1950, the chemical industry increased its share of employment almost fourfold and saw an eightfold increase in its share of British exports (Dintenfass 1992, p. 5).

Labour had promised fair shares for all, but in the postwar decades, the question became fair shares of what? Britain experienced severe industrial strife, slow growth, and rising inflation, although throughout most of these years ICI maintained a strong worldwide presence, particularly in the Dominions, and continued to diversify its product mix. Britain retreated from its Empire in India, the Middle East, and Africa and, after years of ambivalence, finally became a full member of the European Community in 1973. With membership, Britain replaced its Dominion trade-preference system with Common Market trade policies.

Many of the factors that contributed to Britain's lagging industrial performance in the 19th century persisted well into the 20th. Dintenfass (1992) points out that while Great Britain shifted toward a service-based and appliance-oriented corporate economy like that of other Western nations, it continued its record of declining competitiveness.[16] The author points his finger at several reasons. British industry has always been technologically conservative, averse to innovation in the tools and techniques of production. An early illustration is the aforementioned 1890 decision of 45 (out of 50)

[15]See Aftalion (1991, pp. 269–70).

[16]Dintenfass summarizes that "...while the economy has proven ever more successful at generating wealth, the absolute increase in the British production of goods and services has been *relatively* slow compared to that achieved elsewhere" (Dintenfass 1992, p. 2).

chemical companies to stay with the LeBlanc method of producing soda rather than switching to the more efficient Solvay process.[17] Moreover, Britain has suffered a chronic shortage of technologically skilled labor right up through all levels of the corporate organizational pyramid. These postwar shortages of human capital plagued industry, regardless of national employment levels. Dintenfass pins the blame on the nation's utter failure to perceive the economic benefits of a well-articulated, formal system of education and training in science and engineering. In addition, he believes that the country's postwar government and private investment record is poor. More important than the relatively low quantity of investment is the low productive return on investment. Errors of judgment meant monies were misallocated, with funds flowing freely to such defense-related industries as aerospace, while chemicals, vehicles, and machinery claimed comparatively little. But world markets for cars and chemicals proved to be more than three times the size of airplane markets.

The Conservative Rebound

Both British Labour and Conservative postwar governments unsuccessfully sought to increase productivity and ensure labor peace. But it became clear that while Labour's "Fair Shares for All" program had succeeded in weaving a more egalitarian social welfare net, the success came with a price tag measured in unproductive nationalized industries, stultifying government intervention in industry, much labor strife, and steep marginal tax rates.

In the 1980s, British Conservatives returned to power. In a series of dramatic acts, Margaret Thatcher denationalized almost every industry that Labour had tucked under government control during the previous 40 years, reduced marginal tax rates, reined in the unions, and generally attempted to encourage private enterprise. Of course, Thatcher was helped by fortuitous events elsewhere: a cooling off of inflation, North Sea oil and gas production, and the patriotic fervor that followed the Falklands war. But the conservative turn in British politics was nonetheless striking. Within this climate, the opposition British Labour Party struggled to redefine itself, finally peeling off its Marxist-socialist legacy in 1995. By 1997 the voters were ready to return this "New Labour" to power.

GERMANY

Junkers and the German Class Structure

The area now called Germany was little more than a group of miscellaneous political entities during the early decades of the 19th century. The Congress of Vienna, held in the aftermath of France's defeat in the Napoleonic wars, led to a loose German confederation of 39 independent states, each with its own government and coinage. Although Prussia dominated the political landscape as the largest state, a peasant's allegiance was to Saxony or Bavaria, not to a still mythical "Germany."

Over the next half century economic, political, and social forces persuaded states, one by one, that it was in their best interest to join forces with their overbearing Prussian neighbor, the real engine behind the drive toward German unity. A free-trade area, or

---

[17]These companies had substantial investments in the LeBlanc process. Their reluctance to switch to the Solvay procedure may not have been due as much to technophobia as to economics.

Zollverein, was created in 1834. Germany's first written constitution was signed in 1848, a concrete symbol of shared political allegiance. Facing a common foe in Austria-Hungary (1866) and then again in France (1871), the squabbling states were forced together. German nationalism, which many believed was rooted in a "folk spirit" or *Volkgeist*, flowered, and this nationalism, more than anything else, supplied the moral cement needed to hold together the emerging nation-state. In 1871, complete political unification was achieved, and King Wilhelm I of Prussia was crowned German Emperor Kaiser Wilhelm I.

The new constitutional monarchy was quite different from Great Britain's. Britain's Prime Minister is elected by and answers to Parliament. The German Chancellor, on the other hand, answered only to the Kaiser. Even the Iron Chancellor, Otto von Bismarck — the first and most powerful of Germany's Chancellors — was dislodged only two years after strong-willed Kaiser Wilhelm II ascended the throne in 1888. Junkers, the landowning aristocracy, were typically not as wealthy as their English cousins. They controlled local politics and held the plum appointed state and imperial positions; lower-level professional civil servants did their bidding.

During the course of the 19th century, the status-laden bureaucracy and Junker landowning elite consolidated, becoming virtually indistinguishable from one another. The military, particularly the German army, was also closely identified with the Junkers. Scions of the old Junker nobility staffed the officer and reserve officer corps. A strong alliance of blood ties held the monarchy, bureaucracy, and military together. Together, they buttressed themselves against currents of change that threatened agricultural prosperity, the established political order, and the traditional way of doing things.

Junkers played a crucial role both politically and militarily in 19th century Germany, yet their social and economic influence was slowly declining. The passage of time eroded their hereditary feudal rights and privileges. Some Junkers did develop into agrarian industrialists running their own mills and breweries. A very few of the great landowning nobles of Silesia ignored class rigidities and made money in industry. Still, most viewed "money-grubbing" industrial capitalists and bankers with disdain. Moreover, while many Junkers held both science and technology in great esteem, it was scholarly debate and not practical application that intrigued them. Nonetheless, they were prescient enough to support the Reich's reliance on science and technology in the industrial catch-up race with England. Many were among the last champions of agricultural interests, vainly trying to preserve the prosperity of their East Prussian enclave. They fought long and hard within the state bureaucracy and the Reichstag for a tariff against cheap Russian and American wheat during the 1870s. Eventually, however, a booming German industrial sector began to eclipse a declining agricultural sector unable to compete with the vast grain fields of the American midwest, Russia, and Argentina. Thus, although a rank-conscious society continued to defer to the old nobility at court and in state service, nobility was increasingly devoid of economic clout.

A scholarly debate continues over just how porous the German upper classes were to commoners. Otto Pflanze (1990) contends that some enterprising German bourgeoisie entered the Junker ranks, whether through marriage or military promotions. Hardworking bureaucrats were elevated through the upper levels of the state civil service. Very wealthy businessmen were ennobled by the crown, one example being Friedrich Engelhorn, the founder of the largest German chemical company of the 19th century, BASF. Bismarck himself actively supported ennobling the new elites. On the other hand, Martin Wiener (1981) believes that the great majority of German businessmen lacked entrée to the upper reaches of German society, and that the elite, heriditary German class structure was, in fact, rigidly based on birth, education, and profession. The Berghoff and Möller (1994, p. 271) study mentioned earlier reveals that very few businessmen received titles in imperial Germany. Specifically, only 11 percent (or 148) of all titles went to entrepreneurs between 1871 and 1918, and the lion's share of these titles went to one class of businessmen, Frankfurt private bankers. Though they were rarely accepted into the ranks of the aristocracy, the German bourgeoisie readily identified with conservative Junker values. The middle classes—including the great majority of businessmen—staked out a "psychological space" far closer to the proper-tied nobility than to the working classes.

The German entrepreneur of this time was less distracted by national politics than his British counterpart. He left conservative politics to the Junkers and liberal politics to the journalists and lawyers. Compared with the case in Great Britain, very few retired businessmen entered politics as members of the Reichstag, their representation varying between a high of about 15 percent to a low of about 5 percent during the period from 1870 to 1914 (Berghofff and Möller 1994, pp. 280–81). The men who built the German chemical industry, for example, typically remained middle class, apolitical, and vitally engaged in running their businesses for most of their lives. However, German business-men were represented in politics through extraparliamentary lobbying channels. An elaborate system of well-organized and well-financed employers' associations acted as very effective mouthpieces for their interests.

Four Decades To Catch Up: From the 1830s to the 1870s

By mid-century, Germany was about a generation behind Great Britian industrially but catching up rapidly. This period of accelerated industrial growth lasted from the 1830s and 1840s until about 1873. Germany recognized that it was too late to build an Empire that would rival England's, but the time was propitious for erecting an industrial powerhouse. The German government orchestrated economic growth with many substantial policy changes, some of which were later reversed. It eliminated the legal constraints on internal migration and established a common currency, along with a bank for issuing it. The state regulated production and marketing in several industries, and it often stepped in when private initiative and capital were lacking, particularly in banking and transport. Government-subsidized railroad construction that began in the 1830s and peaked in the 1870s led to a steady demand for coal and iron and encouraged growth in the machine-building industry. The effect of railroads in tying together raw materials, power supplies, factories, and markets was more substantial in the larger and more diffusely settled area encompassed by the German Confederation than in the more compact England. The shift from sail to steam power increased mercantile tonnage on the Rhine. Prior to 1870, government policy severely restricted the formation of joint stock companies. An 1870 law abolished this control, resulting in 857 new companies between 1870 and 1873, more than in the previous 70 years combined! The lack of a unified patent law prior to 1876 helped in the transfer of dye technology to German manufacturers. The passage of a uniform patent code in 1877 encouraged German dyestuffs manufacturers to invest in corporate R&D labs and provided protection for scientific advances (Wrigley 1986, p. 172; Beer 1959, p. 103).[18]

---

[18]See the Murmann and Landau chapter in this volume for further detils.

But of the many government actions, scientific and technical education was Germany's primary ticket to industrial success.[19] Germany boasted a strong tradition of higher education; indeed, the roots of the modern research university can be traced to 18th century Germany. Government officials, trained at the Berlin Technical Institute, were often better educated technically than the owners of the mines and foundries. In the late 19th century, state governments grew convinced that polytechnic colleges, later known as *Technische Hochschulen*, gave them the means to promote the diffusion of scientific and technological knowledge. They moved quickly to set up their institutions for training scientists, engineers, and managers, so that their cities would be identified with industrial progress. In a division of labor, the polytechnics concentrated on applied chemistry, while the universities focused on theoretical organic chemistry. Germany's universities and technical institutes were so closely aligned with the state that professors held ranks corresponding to those in the state's civil service bureaucracy. In this case, the state made a conscious effort to confer status and privileges on its intellectual elite (Ben-David 1971).

In the 1870s and 1880s, some in the conservative classes still clung to the classical curriculum. For example, many German university professors, like their Oxbridge counterparts, derided subjects of an applied character and, for a time, successfully resisted the introduction of engineering into the curriculum. However, their efforts did not deter Kaiser Wilhelm II from giving the nine polytechnics the right to grant the Doktor-Ingenieur degree (roughly equivalent to the Ph.D.) in 1899. A few years later, the barrier between pure and applied science crumbled further when the universities began offering chemists some training in engineering and related applied fields (Wrigley 1986).

Nonetheless, the majority of 19th- and early 20th-century German businessmen continued to attend the prestigious local Gymnasiums that had served their families for generations, studying a curriculum that focused heavily on classical studies. Moreover, only a minority (24 percent) of the Berghoff and Möller (1994) sample of German businessmen of this period went on to attend a university or technical college — although this figure far outstripped the 13 percent who sought higher education in the English sample. It is important to note that the majority of businesses at this time did not require a scientific education, or any higher education at all.

The chemical industry, however, was different, even at this early date. Most German chemical manufacturers did hold science degrees. As far back as 1890, Germany, with the world's foremost organic chemical industry, had twice as many academic chemists as did Britain. German superiority in organic chemistry research was particularly striking, with 574 scientific abstracts published in 1882 compared to 59 from Britain (Wrigley 1986, p. 171). German-university chemistry students worked long hours in a laboratory under the close direction of a professor or lab director. This was strong preparation for the hierarchical, and often wearisome, nature of industrial research. The triumph of the dye industry can in part be attributed to the tedious hours that teams of chemists spent on repetitive, specialized tasks. Only a few of the many German chemists who worked in the growing companies were brought into decision-making management positions, where they could help direct the course of research, but these

---

[19]Compulsory primary education arrived in Germany more than a century earlier than in Britain. Right from the start, there was close articulation among primary, secondary, and higher institutions of learning, whereas in Britain there was practically no coordination among the three levels. Moreover, the state contributed very substantially to the cost of teaching and research in Germany. In Britain, government financing long remained minimal (Dintenfass 1992, p. 34).

few were especially skilled at moving from theory to application. Chemists, both academic and industrial, enjoyed high status and commensurate salary. This became a crucial factor in attracting top-notch talent to a field such as dyestuffs, which was based on cutting-edge advances in organic chemistry. Carl Duisberg, head of research at Bayer, said that in 1896 he employed about 100 university-trained chemists and 25 polytechnic-trained engineers. Four thousand chemists worked in the entire industry.

Germany's technology-oriented educational system did a stellar job of promoting economic growth. By the 1870s, German armament and machine-tool factories had reached a level of competence equal or superior to that of foreign competitors. Textile makers went from hand to mechanical spinning. Shipbuilding boomed. Coal, pig iron, and steel production set records. The electrical-equipment industry took off. The nation's chemical industry, with its enormously successful dye component, also soared. In relatively short order, Germany's financial success was so great in the dye market that firms could utilize much of their own retained earnings to finance R&D and future growth in dyes and also to expand into heavy chemicals.

After 1870, when the government began to encourage joint stock companies, managerial capitalism spread quickly in the newly unified nation. While in Britain family ownership continued to dominate firms until well into the 20th century, in Germany, salaried managers, technicians, and commercial specialists held top decision-making posts and received both professional and social recognition.[20] In addition to their broad experience, their formal educational qualifications often exceeded those of owner-businessmen in the sample (Berghoff and Möller 1994).

The Socialist Impulse and Bismarck's Welfare State

By 1871, Germany's population of 41 million had outstripped England's (31.8 million) and was about equal to the U.S. population of 39 million at the time (Ensor 1992, p. 102). Within Germany, an internal migration took place from the rural east to the more urbanized west. Although fewer hands worked the land, intensive farming increased output more rapidly than did population growth. The surplus food was exported. Despite these population shifts, 67 percent of Germans continued to live in small villages. By 1871, England's social transition to a fully urban, industrial economy was complete; in Germany, the transition would be complete by the end of the 19th century; in the United States, that level of urbanization would not be achieved until 1945.

But the 1870s brought substantial economic changes that shifted the focus of the German government from the support of industry to conciliation of labor. In 1873, the halcyon years of growth came to an abrupt end. The stock market collapsed. A depression, which in certain industries was worldwide in scope, lasted until 1879. Small companies, including scores of young dyestuffs companies that had sprung up throughout the German states prior to unification, were especially hard hit. The slow and halting recovery dragged on from 1879 until 1894.

Although German wages had risen steadily during the years of growth, they remained lower than in both England and the United States. Indeed, during the Congressional tariff debates of the 1880s and 1890s Americans complained of the competitive pressures resulting from the low wage levels of German workers. Smaller paychecks and rising layoffs rankled the German masses. Many felt they were not

---

[20]When compared to the United States, however, family controlled corporations played a larger role in the German economy, both before and after WWI (Dintenfass 1992, p. 65).

sharing in the nation's growing wealth. Many blue-collar workers turned to a new political doctrine—socialism. A class consciousness awakened that would leave its imprint on the German political landscape for decades to come. Karl Marx and Friedrich Engels had a profound influence in their native land.

Between 1865 to 1869, German workers had won the right to a secret ballot and a limited right to organize and strike, and strikes grew more frequent and violent. A growing number of socialist representatives were sent to Parliament. German chemical manufacturers were not immune from these sentiments. By and large, workers in this industry earned generous wages and were provided with decent company housing. However, they also toiled for long hours in vaporous rooms filled with vats of dangerous chemicals, and management firmly opposed any attempts to unionize (Haber 1958).

In the face of spreading socialist agitation, Chancellor Otto von Bismarck at first responded to political unrest with ruthless measures: Socialist leaders were jailed, political offices closed, newspapers and pamphlets confiscated, organizations outlawed, and strikes brutally suppressed. These steps were generally supported by the German bourgeoisie. They united with the elite classes to form a common front to counter the socialist foe.

But soon Bismarck was forced to change his tactics from "sticks" to "carrots." To the shock and dismay of many of his oldest allies, Bismarck endorsed the "right-to-work" principle of full employment. He successfully guided the medical, accident, and old-age and disability insurance programs through the Reichstag. Bismarck's social welfare reforms mark a shift in government attention away from creating a fertile environment conducive to industrial growth and toward counteracting the deleterious side effects of rapid industrialization.

Whatever his motivation, the fact is that for his immediate purposes, carrots worked little better than sticks. Political and economic conflict continued to plague society. Some say Bismarck didn't go far enough; he never pursued protection of the worker *inside* the factory and mine *during* working hours. Others believe he overestimated the material benefits the laws conferred and underestimated the capacity of an idealistic political movement like Marxist socialism to woo the hearts and minds of the underclass. Consequently, union membership continued to grow, reaching 3.7 million in 1912, of which more than 60 percent were affiliated with the socialist union. By the 1890 general election, the Social Democrats received more votes than any other party; in 1891, the party platform supported a Marxist program of revolution.

While these years from the 1870s through the 1890s were difficult for the economy as a whole, they were also a time when the roots of today's German chemical industry became firmly established. Three of the earliest pioneers of the German chemical industry were Bayer, BASF, and Hoechst. Having driven the producers of natural dyes and the less-efficient British and French dye firms out of the market, these three survived the profit-destroying, competitive struggle that occurred within the German industry during the 1870s and 1880s. They made the long-run investment in basic research, development, and marketing that ensured their domination of the worldwide market. The inorganic end of the chemical industry came into its own. Prosperity in the dyestuffs industry stimulated the growth of acid and alkali production. Potash deposits were exploited as the fertilizer industry expanded, with exports going to farming-intensive nations such as Holland, Belgium, and Scotland. The Solvay process took root and spread rapidly since, in contrast to British firms, German firms did not have substantial investments in the older LeBlanc process. At the end of the century, 80,000 people were working in the German chemical industry.

The main forms of assistance to industry during the last few decades of the 19th century were trade protection and encouragement of cartels. The protectionism of this era is best viewed as a deal hammered out between declining agricultural and rising industrial interests. The German government erected tariff walls in 1879, 1882, and again in 1887 on such things as iron and grains. Historians generally agree that Bismark's tariffs allied the landed elite, the peasants, the major industrialists, the military, and the higher civil servants, insulating them from the uncertainties of market forces and forestalling the rise of social democracy. On balance, tariff protection aided the growth of the nascent German chemical industry.

By the 1890s, at just about the time the United States was making a first attempt at outlawing cartels with the Sherman Act, German industry in general, and the chemical industry in particular, entered an era of pervasive cartel formation that did not end until 1945. Monopolies and cartels were encouraged as a way to rationalize the market. They arose in the steel, coal, glass, and cement industries. Chemical cartels soon began to appear. The two main competitive blocs, comprised of Hoechst-Cassella-Kalle and Bayer-BASF-Agfa, formed in 1906. Collaborative agreements were written, some international in scope. For example, German and British companies carved up the world's bleaching-powder markets in 1901. However, the German cartels continued to face vigorous competition in international markets.

By the early years of the 20th century, many of the factors discussed here had contributed to a Germany that was ready for war. Working classes continued to move to the left politically. In the last national election of the Second Reich, held in 1912, the Social Democrats scored a great victory, winning about 35 percent of the vote and frightening the opposition Junker, military, university, and bureaucratic elites. The Prussian ideals of restraint and sacrifice that characterized Bismarck's Germany gave way to the "pomp and circumstance," bluff, and bombast of the Wilhelmine court.[21] Some historians have viewed World War I partly as an attempt by the elites to shore up their sagging stature with a stunning victory.

Meanwhile, the German economy had caught and surpassed Britain's. From 1870 to 1913, the German rate of GNP growth per capita averaged 1.8 percent per year, while Britain's was only 1.3 percent. Whereas only one-third of German exports were finished goods in 1873, 63 percent were by 1913. Industrial products accounted for about 60 percent of the 1913 GNP. The chemical industry formed part of this industrial behemoth, to be sure, but only a *relatively small part*. While some organic chemicals — for example, synthetic dyestuffs, pharmaceuticals, photographic chemicals, and solvents — dominated the world markets, dyes ranked only eleventh in a list of German exports. Dyes were overtaken by leather goods and dwarfed by woolen textiles, iron, and iron products. Indeed, despite their expansive factories, their thousands of workers, and their subsidiaries in all major countries, the "big three" dyestuffs firms (BASF, Hoechst, and Bayer) were, in terms of assets, only about two to three times as large as the biggest German brewery. Britain remained the leading inorganic-chemical-producing country, with Brunner Mond the biggest chemical company in the world in terms of sales and Solvay of Belgium coming in a close second (Keck 1993, p. 128; Chandler 1990, appendix C1). Nonetheless, by 1914, when Germany was poised on the brink of war, it was generally acknowledged to be the strongest nation in Europe, both economically and militarily. In 1914, the autocratic Kaiser imposed his will upon his increasingly militaristic nation and led it into World War I.

---

[21]Wilhelm II ascended to the imperial throne in 1888, as noted before.

War and Recovery: Germany in the 20th Century

The Great War had a profound effect on Germany and its chemical business. The industry, financially supported by the government, played a substantial role in producing the weapons of modern warfare. Dyestuffs companies took care of providing the chlorine needed for gas attacks. The need for acetone soared as demand for smokeless powder climbed. The war effort also marked the first large-scale use of synthetic rubber, patented by Bayer in 1910.

The most important technical innovation coming out of the chemical industry during World War I was the Haber-Bosch process, the synthesis of ammonia out of atmospheric nitrogen. The process involved high temperature and pressure. It was large-scale, continuous, and capital-intensive. Haber-Bosch plant requirements wedded chemistry to engineering.[22] In fact, high-pressure chemistry continued to monopolize the chemical industry's center stage for years to come, to the detriment of profits.

If the war effort gave the German chemical industry a technical push, the postwar German economy hobbled further progress. More than five years of fighting spilled the blood of a generation of German men. Defeat in 1919 led to severe postwar economic dislocations triggered by huge war reparations and debt payments. A deflationary postwar recession was followed by the worst hyperinflation in Germany's modern history. A currency reform in 1923 aborted the inflation but essentially wiped out the liquid savings of many German families and businesses. Extreme political groups gained a foothold among the electorate.

The chemical industry suffered severe setbacks in this unstable economic and political climate. The Haber-Bosch patents were among the booty captured by the victors. The industry faced a serious shortage of coal and other raw materials, as well as an excess capacity in dyestuffs. However, "in kind" war reparations payments in dyes and pharmaceuticals proved beneficial to the industry's recovery. In response to these postwar realities, an industry cartel headed by IG Farben was able to organize in 1925 with virtually no objection from the public or the authorities. Its unprecedented size and strength enabled IG to continue to export 60 percent of its output — no small achievement, given the rise of new protectionist barriers in Europe and America. Dyestuffs initially accounted for about one-third of the output, but new products such as synthetic ammonia and nitrogenous fertilizers later took center stage.

The Great Depression hit the already weakened German economy very hard. Chancellor Heinrich Brüning's government made many of the same mistakes as did America's Herbert Hoover; both pursued tax and interest-rate increases combined with a cut in government spending, moves that deflated the economy at a time when incomes were already falling precipitously. Gottfried Plumpe (1995) tells of IG's efforts to persuade the Chancellor to relax these policies, because their chemical customers did not have enough money to pay for IG products! Protectionist barriers increased, and trade throughout the world dropped considerably. The country was rent by social splits and rising unemployment. By 1933, one out of every four employable Germans was looking for a job. Plagued by political infighting within the ranks of its shifting coalitions, the ruling Social Democrats had but a tenuous grip on the government. A young, Austrian rabble rouser named Adolf Hitler was able to take advantage of this

---

[22]It is interesting to note that while in the United States the linking of chemistry to mechanical engineering gave rise to the field of chemical engineering, it had no such effect in Germany. The German chemical industry continued until well into the 20th century to be run by highly valued chemists assisted by well-trained mechanical engineers (Landau 1997).

economic and political turmoil to transform a crew of thugs and ruffians into a movement of national proportions. The Nazis took over in 1933.

Hitler charged into an economic-recovery program characterized by increased government spending and oversight. The state-subsidized autobahn system became the 20th century infrastructure equivalent to the railroads of the 19th century. In 1936, Hitler launched a four-year plan for complete autarchy, a barely concealed preparation for war. IG was a dominant player in this effort. Synthetic gasoline from coal hydrogenation, synthetic "Buna" rubber, and some synthetic fibers such as rayon were all strategic war products coming out of the cartel.

The frightening political context notwithstanding, this was an exciting time for chemical research. Breakthroughs in the field of macromolecular chemistry paved the way for later developments in plastics, fibers, and synthetic rubber. IG eliminated competition in dye production between units; this allowed individual companies to concentrate on a particular dye. As war approached, the government became the largest customer of the chemical industry, buying its burgeoning output of explosives, synthetic rubber, and synthetic gasoline. Hitler's totalitarian regime ended with Germany's unconditional surrender in 1945.

Postwar Germany was divided into four zones and occupied by the four victorious allies. The British, French, and Americans allowed their sectors to unite and form a democratic West Germany, in order to resist the threat of Soviet Communism — the cold war had begun. The Soviet sector, East Germany, developed into a communist satellite. For the second time in this century, Germans faced the task of rebuilding in the face of devastation. Many of Germany's plants had been bombed, all eastern factories had fallen into Soviet hands, overseas interests were sequestered, trademarks were confiscated, and technologies divulged.

As reconstruction began, the political pulse of Germany was anti state control (Nichols 1994). The new constitutional framework maintained political stability and returned Germany to the decentralized structure it had possessed during the Second Reich. In the newly created Federal Republic the few small parties were relatively unimportant, and the two largest parties, the Christian Democrats and the Social Democrats, saw themselves as representing the interests of all people, not just a single class or section. Together they began to build what became known as a "social market" economy (Giersch, Paque, and Schmiedieng 1992).

The market side of this duality was personified by Ludwig Erhard, an economics minister who succeeded in convincing first the Western Allies and then Konrad Adenauer, the first elected Chancellor, that a complete unshackling of the economy from Nazi and wartime controls was needed. The result was an economic environment characterized by few controls and conducive to capital formation.[23] Erhard's policies met with such success that they were dubbed Wirtschaftswunder, or economic miracle. While the government played an active role in such things as the autobahn network, waterways, the postal system, rail system, telecommunications, and subsidizing research and development, it generally drew the line at direct subsidies or an explicit industrial policy — with the exception of subsidies to the coal industry and agriculture.

Currency reform wiped the economic slate clean and paved the way to recovery. Issuance of the new Deutsche Mark in 1948 and the retirement of the old Reichsmark eliminated most liquid savings for the second time in less than 30 years. Owners of real

---

[23]Capital controls, mainly on the inflows of capital, were maintained for several decades to preserve the German ownership of important assets.

assets fared better, receiving nearly the equivalent of their holdings in the new currency. The collective memory of two confiscatory revaluations explains German support for a powerful Bundesbank acting independently of the government to dictate a noninflationary monetary policy. German exports boomed through the 1950s and into the 1960s, thanks to an undervalued mark exchange rate, strong European demand, the help afforded by the Marshall Plan (1948–52), the formation of the EEC (1957), and lower tariffs resulting from the Kennedy Round of the GATT talks. The NATO military shield, obviating the need for a large defense budget, enabled Germany to channel its resources toward its economy.

The social side of "the German model" focused on building a consensus about the institutions and decisions in this market economy. In this system of government by consensus, unions and firms were viewed not as antagonists but as "social partners," who ought to cooperate in the management of companies. As a result, labor unions were given seats on the supervisory boards of corporations equal to those of management, although the chairman, usually drawn from the ranks of former management or the banks, cast the deciding vote. Wages and working conditions were negotiated for whole industries by deals struck between union and management representatives. Change came about by small, incremental steps. The "social market" economy put a human face on capitalism, softening the adverse features of unfettered competition with a wide and deep safety net of welfare benefits. This mix of capitalism, welfarism, and consensus sets postwar Germany apart from many other industrialized countries.

The chemical industry recovered and prospered under these conditions. IG Farben, notorious for its association with concentration-camp gas chambers and slave labor, was broken up during the early 1950s into three successor companies: BASF, Hoechst, and Bayer (Teltschik 1994). Chemists were at the helm of all three. Having experienced the benefits of belonging to a cartel, the companies were careful not to compete with one another, and instead carved out separate markets.

But by the mid-1960s, reconstruction and growth had ceased to be unifying and energizing goals for the German nation. Defeat in two wars had discredited and wiped out large segments of the power structure. The Social Democrats, who had renounced Marxism only in 1959, now came into the government, first in a Grand Coalition with the Christian Democrats in 1966 and then with the Free Democrats in 1969. Tax rates and regulations increased as the government introduced many new social welfare programs. A number of radical student groups rediscovered and embraced Marxism. By the late 1960s, the student protests had snowballed into a cultural revolution, with the universities as centers of radical activism. Many German young joined the Green Party and became ardent environmentalists. Distrustful students shunned science and engineering disciplines, and a vocal part of modern German society grew hostile toward science-based industry, including nuclear energy, biotechnology, and the chemical industry. Legislation, regulation, and bureaucracy grew and contributed to Germany's economic slowdown during the 1960s and 1970s. It was a period of slow growth for virtually all industrialized Western nations. The oil shock of the 1970s jolted the economy and boosted inflation. The "economic miracle" fashioned by Ludwig Erhard had stumbled.

The Berlin Wall tumbled in late 1989, and the long-awaited unification of Germany was completed in 1990 as the Soviet Union collapsed and the cold war ended. But this joyous occasion came with a price tag. East German firms were burdened with dilapidated plants, highly polluted sites, and saddled with debt — leaving them unable to compete in the free market. Unemployment and crime increased. A steady stream of

East Germans migrated West, straining social services and causing resentment. Moreover, increasing numbers of ethnic Germans from the former Soviet Union and its satellites applied for asylum. These sizable demographic shifts, coupled with large numbers of foreign "guest" workers who had been in Germany since, in some cases, the 1950s, led to a nativist and nationalistic reaction in what had long been an ethnically homogeneous nation.

Though the Germans of the 1990s continue to enjoy one of the highest standards of living in the world, the weakening fiscal and industrial situation has prodded the government and employers to begin nibbling away at the formerly sacrosanct welfare system. At the same time, the traditional German work ethic and sense of discipline has metamorphosed into a German leisure ethic and demand for security. German labor spends substantially fewer hours "on the job" per year than do its counterparts in other nations. Non-wage costs to meet generous welfare, pension, social security, and holiday commitments are exceptionally high, adding about 80 percent to wage rates that are already among the top in the world. Taxes are steep. A top German executive today has a marginal tax rate above 60 percent, and most managers receive no stock options. Moreover, since the entry of the Green Party into national politics in 1983, Germany has become one of the most regulated nations in the world, driving the costs of doing business even higher. Many German businessmen, including those in the chemical industry, are choosing to locate their new plants on more economically hospitable soil. German unemployment has risen to more than 11 percent, a level not encountered since the pre-Hitler government, for all the above reasons.

JAPAN

From 1600 to 1868, Japan was a feudal, militaristic society closed to the outside world. A turning point occured in 1868 when the Emperor (Meiji) was reestablished as the source of power, and feudal institutions were abolished in 1872. Japan embraced a brand of capitalism similar to that in Germany, in which a pattern of cooperation between the government and industry was deemed necessary to catch up to the rest of the modern world. Large industrial combines, known as *zaibatsu*, arose from the sale of government enterprises during a deflationary period in the 1880s. The government severely limited foreign investment, thus solidifying the power of the *zaibatsu*. Japan's chemical industry at this time was essentially insignificant. It was small, backward, and dependent on foreign technology.

During the early decades of the 20th century, Japan's policies for industrial catch-up were generally geared to gain a military advantage. This resulted in government measures to stimulate heavy industries, including chemicals, textiles, steel, and machinery. Japan's military gained more influence over the aristocracy, aided by victories in the Korean War (1894–95) and the Russo-Japanese War of 1905.

Japan was a non-belligerent during World War I that benefited economically by supplying all warring nations. When a recession hit during the immediate postwar years, the growth rate fell. The devastating Kanto earthquake of 1923 worsened an already unstable economic situation. Even during the ensuing boom years of the 1920s, the growth rate of per capita income remained below 1.5 per cent per year.

The worldwide depression of the early 1930s knocked the wind out of the Japanese economy. The military grew to become the dominant factor in the governance of Japan. In 1931, Japan began an attempt to build its own Empire, the Greater East Asia

Co-Prosperity Sphere, with the invasion of Manchuria. When hostilities spread throughout China, the financial system had to sustain both the Japanese war machine and a growing public deficit. During World War II, Japan's alliance with the Axis powers tied its future to Germany's. After the defeat of 1945, Japan entered into a period of relatively democratic governance under a U.S.-imposed constitution. Like postwar Germany, Japan abandoned militarism. Instead, it embraced the goal of rapid economic growth through peaceful means. A troika of government, industry, and labor cooperated to establish a high-savings, high-investment economy. As in Germany, the American military shield permitted Japan to escape the costs of military preparedness and concentrate on economic rebirth and growth. War-torn Japan pulled itself out of the ashes of defeat and achieved growth rates as high as 5 percent per year, raising its standard of living to among the highest in the world.

Japan's version of postwar capitalism, however, was very different from Germany's. Germany supported a free market and broke up industrial cartels. In contrast, the Japanese government carefully planned and controlled economic life through its powerful civil service bureaucracy. The government continued to support the large industrial combines — *zaibatsu* before the war and now called *keiretsu*. Cross-ownership linked *keiretsu* members to each other and usually all of them to a particular bank.[24] This arrangement erased the threat of takeover bids and averted most bankruptcies. The Japanese chemical industry served the domestic market and had little direct dependence on exports. Although the industry bought large amounts of western technology, in many respects it lacked the energizing contact with foreign competition that might have shaped it into a world-class industry.

For more than 40 years of the postwar era, one party — the Liberal Democrats — has dominated the Japanese political scene. This party took a significant and direct role in implementing an explicit industrial policy. Both the Ministry for Industry and Trade (MITI) and the Ministry of Finance (MOF) were the government's most powerful instruments for this purpose. For example, during the first few decades after the war, Japan was very short of foreign exchange. Tokyo's controls on capital and foreign exchange were stringent enough to approximate a form of central planning. Government policy spotlighted selected industries but fell short of selecting favored companies.

This policy largely succeeded, until well into the 1980s, in creating a capitalism without costs. Japanese society enjoyed the benefits of an extraordinarily high growth rate, full employment, secure jobs, and a fairly equitable distribution of incomes and assets. A meritocratic bureaucracy channeled the best and the brightest into the highest posts. Savings were substantial, and interest rates were low. Exports and overseas investments propelled Japan to economic stardom. Its economic "co-prosperity sphere" now extends over much of Asia, and its economic influence reaches around the world. True, domestic prices stayed high and wages remained low, so the standard of living did not fully reflect Japan's economic progress; but labor unions and workers were mollified by the lifetime employment policies that were adopted in response to labor shortages.

Carol Gluck (1995) has suggested what she calls "A Grand United Theory of Japanese History." She points to the underlying cultural traditions that bind this cohesive society together. These have remained remarkably stable over the course of history, unraveling only in response to major external shocks. There has been, however, a slow but constant evolutionary change between these periods of "shock." More recently, Japan's economic path seems to be moving away from government planning.

---

[24]For more details see the chapter by Hikino et al. in this volume.

Slowly but unmistakably, markets are becoming more competitive and open to imports. Labor markets are growing more flexible. Deregulation and liberalization of the financial markets has forced changes in the tightly controlled main bank system that once tied together the large *keiretsu*. Japan's bursting real estate bubble in the 1990s has led to nearly a decade of low growth rates and substantial financial woes.

Change is not occurring fast enough for some. Problems such as the rapid aging of the population, a serious budget deficit, and a dearth of foreign investment in Japan have combined to prod industrialists to publicly call for an increase in the rate of deregulation and a streamlining of the central bureaucracy. While Japan's model of consensus and government control was unquestionably successful in generating its remarkable catch-up during the 1960s and 1970s, it now seems that an alternate model will be necessary for Japan to sustain rapid growth into the 21st century.

UNITED STATES

The American system of representative self-government was set forth in a written constitution in 1789, a revolutionary step in a world long used to governance by royal edict. The system is undergirded by a simple principle: Those who are governed have the right to determine the shape of the government. It took a Civil War and a series of smaller political battles, however, before all people, regardless of race, gender, and property ownership, were brought together under this system. In practice, governing has usually involved two major parties, with third parties playing bit parts from time to time.

The nation lacked a hereditary aristocracy, and such class divisions that did exist were much more fluid and porous than those in Europe. The entrepreneurs who founded and grew the chemical industry were from the lower and middle classes; economic success gave them access to the upper classes. The *belief* that there was equality of opportunity (historians differ on just how much there actually was) and the possibility of achieving great wealth provided a social cement missing from the stratified nations of Europe.

The young nation's people multiplied and spread during the 19th century. The population more than quadrupled from 1815 to 1870. Almost 39 million people inhabited the country in 1870, bypassing the population of the United Kingdom and just a few million short of Germany's population. Then the republic's population *trebled* during the period from 1865 to 1914, as waves of eager immigrants arrived. Citizens enjoyed a consistently rising standard of living, with real per capita income more than doubling during this half-century. Pioneers steadily expanded their settlements westward, pushing indigenous tribes out of their way. The frontier seemed to stretch interminably, beckoning, full of resources that were needed for creating an agricultural and industrial giant. Ample deposits of coal, oil, natural gas, and plentiful hydroelectric power were all to prove indispensable to the development of the chemical industry.

In 1815, eight out of ten Americans lived in the original thirteen colonies along the eastern seaboard. By 1865, a majority of Americans lived west of the original thirteen. Not until the 1890s did the westward stream dry to a trickle. The railroads played a central role here; from the 1830s to the 1870s, they transformed what had been little more than a collection of populated hamlets bound together by waterways into a vast, unified market, extending from the Atlantic to the Pacific. By 1914 there were more than 250,000 miles of railroad track in the United States, greater than the rest of the world combined!

The young nation benefited from its close commercial relationship with Great Britain, the home of the first Industrial Revolution. By the time the first shots of the Civil War were fired on Fort Sumter in 1861, the major technologies of that first industrial wave — the steam engine, textile machines, and coal-fired furnaces — had been exported and were firmly planted in the United States. The majority of manufacturing centered in the northeast and mid-Atlantic regions, although agriculture related manufacturing industries settled in the midwest.

Through the first half of the 19th century, the U.S. chemical industry was relatively insignificant. By mid-century, about 1,000 people were employed in America's chemical sector, only one-tenth the size of the British chemical workforce. Unable to compete with cheap British soda, U.S. chemical manufacturers of this time concentrated on fertilizers, explosives, and other inorganics. As early as the 18th century chemists graduated from programs at Columbia, Harvard, Princeton, William and Mary, and Dartmouth. By 1820, 22 different colleges/universities had chemistry in their curricula.

The second installment of the Industrial Revolution, of which the nascent chemical industry was such a vital part, found fertile ground in which to grow during the last half of the 19th century. By the 1870s, the United States enjoyed an expanded transportation and communication network, a thriving commercial life, and a booming agricultural sector. These changes, each a mini-revolution in its own right, worked together to support a tremendous burst of entrepreneurial and technological energy. The United States was clearly among the world's leaders during this second wave, characterized by the growth of capital-intensive "heavy" industries. A hallmark of this stage of industrialization was the methodical application of new scientific, engineering, and management principles to factory processes and products, especially in the chemical and electrical industries, which together employed close to one-half of America's scientific personnel.

In the chemical industry, as in other industries, businesses paid scant attention to developing export markets. It was all they could do to satisfy a domestic market that was already large and seemed ever-increasing. Lower freight rates and ease of transportation promoted a concentration on bulk, commodity-like inorganic chemicals such as acids, caustic soda, and soda ash. Producers focused on chemical agents for the tanning, bleaching, and dyeing trades, as well as fertilizers and explosives. By the end of the century, local factories had increased in size to serve regional markets and then increased in size again to serve national markets.

Chemical firms readily adopted new technologies that promised cost reduction: for instance, the Solvay process for soda, new electrochemical processes for chlorine, and later, the contact process for making sulfuric acid. All involved continuous operations. This constant flow of materials created a need not only for well-trained chemists but also for mechanical engineers. Universities such as MIT, the University of Pennsylvania, and the University of Michigan responded by starting chemical engineering programs during the 1880s and 1890s. Many American chemistry students of the 1880s and 1890s went to German universities to finish off their schooling. The chemistry programs at Johns Hopkins and University of Chicago were reshaped in the German mold with a decided emphasis on basic research. At most other universities and colleges, the course of study was more practical, with an expectation that graduates would either go into teaching or industry. By 1910, about 4,800 Americans were studying chemistry, and an additional 850 were enrolled in chemical engineering programs. (See also the chapter by Rosenberg.)

Trade protection also shaped the U.S. industry at this time. The United States remained the single largest market for British exports of alkalis and bleaching powders for most of the century, until the tariffs of the 1880s and 1890s shut out British imports. The more-efficient Solvay process and new electrolytic technology spread quickly since American manufacturers had not invested in the older LeBlanc methods. By 1901, the United States was completely self-sufficient in alkalis and bleaching powders; but while the inorganic side of the U.S. industry was developing behind tariff walls, the organic side was dying. Nine firms produced dyestuffs in the 1870s and early 1880s. When the tariff on imported dyes was cut in 1883, Germans drove five of the nine companies out of business. By 1899, the United States was completely dependent on imported dyes. Haber (1958) suggests that this was because American businessmen were partial to large-tonnage products that promised a quick return on investment but had little interest in the meticulous research and technical service required by dyestuffs.

There is a half-full, half-empty quality to the debate about the extent to which the government aided and abetted the momentous economic changes of the 19th century. Some economic historians[25] emphasize that enterprising Americans drew upon government credit and resources to make rivers and harbors navigable; explore and map the land; build dams, tunnels, bridges, canals, roads, and railroads. The government was involved in banking, beginning with the charter for the First Bank of the United States in 1791. The government intervened in land policy with the passage of the Homestead Act of 1862. It established land-grant agricultural and technical colleges throughout the new western states with the passage of the Morrill Act of 1862, and graduates of these schools supplied industry with talented well-educated engineers, and agriculture with scientifically trained farmers. Other economic historians argue that these steps fall far short of any systematic industrial policy; they characterize governmental interference in the business sector as minimal prior to the 1880s.

During the last two decades of the 19th century, however, government policy clearly became more activist. State, and eventually federal, regulation of the rails (through the Interstate Commerce Act of 1887) was an early step. Simultaneous with cartel formation in Europe, the American antitrust movement pushed the Sherman Antitrust Act (1890) and the Clayton Act (1914) through Congress. Toward the beginning of the 20th century, many Americans believed that the growth of giant corporations and the emergence of a fabulously wealthy group of individuals posed a threat to their personal economic well being and also offended the American sense of "fair play." For others, of course, worldly success proved the virtue of men like Carnegie, Mellon, and Rockefeller.

World War I and the Government-Industry Relationship

On the eve of World War I, most of the American chemical industry involved small, private companies owned by their founders. Aggregate output was quite large, with production midway between England's and Germany's. The maturing nation was self-sufficient in fertilizer, acids, and alkalis. During these formative years, the industry was steadily shaped by the emergence of chemical engineering as a discipline in its own right. The American reliance on bulk, continuous production of a few widely used commodity chemicals meant that the mechanical elements of chemical production were complex because of the large volume of production and the large number of interrelated

[25]See Coats (1980) for a review of how economic historians line up on this issue.

products. Conversely, production of dyes and pharmaceuticals in Germany involved complex chemistry but relatively small volumes of production. The developing field of chemical engineering answered the U.S. industry's practical need to solve production problems that were based as much on mechanical engineering as on chemistry. Thus, by the 1920s the United States was already on the way to having a well- established program of chemical engineering at its best universities.[26]

Europe had been at war for almost three years when the U.S. entered World War I, in April of 1917. For about six months the government relied on voluntary efforts in mobilization, and then it moved to establish more control over the nation's economic life. A year later, the vast engine of the U.S. economy was harnessed for total war. Despite the waste, shortages, and inflation, the war experience demonstrated that the government could effectively organize certain kinds of output through emergency agencies. Ties were forged among the military, the civilian government, and the private sector. In the chemical industry, cut off from German dyes, the United States was forced to breathe life into its all-but-dead dyestuffs sector. By 1918, the dyestuffs firms were able to supply the needs of the warring nation, with DuPont and National Aniline emerging as leaders.

Following the end of World War I, America was swept by strong isolationist, pro-business sentiment. The nation closed itself off to immigrants with the 1924 National Origins Act, and it put up tariff walls around a number of industries, such as the recently regenerated dyestuffs industry. The prosperous 1920s brought most Americans an increased level of comfort. With a consciousness of impermanence and loss born in the pain of World War I, Americans were impatient to live in the here and now. Instead of practicing thrift and saving, they used installment credit to buy new mass-produced consumer goods. Movies and ball games became national entertainments. More and more, small investors went into the stock market. The American automobile industry took off during this decade, marking Detroit as a new hub of American economic growth. In the 1920s, the number of farms and farmers actually declined; urban areas expanded, growing at five times the rate of small towns. Both Christian fundamentalism and Prohibition were efforts to preserve and protect small-town and rural American values in an increasingly urban, multiethnic nation.

These were formative times for the U.S. chemical industry. When oil refiners discovered that petroleum and natural gas could be the sources for a wide array of new chemicals, the petrochemical industry was born. The new feedstocks were plentiful and cheap. The first petrochemical unit was started in 1920 by Standard Oil. The expanding auto industry pulled the chemical industry along its phenomenal growth path; the two worked hand in hand to deliver mass produced vehicles to an increasing number of households. New petrochemical plants sprouted in the shadows of Gulf refineries.[26]

The chemical industry was prevented, both by the Sherman Act and the Federal Trade Commission (1914), from outright participation in any of the cartel arrangements proliferating in Europe. However, there was still a process of upstream and downstream consolidation and diversification, which led to the formation of Union Carbide and Carbon (1917) and Allied Chemicals and Dyes (1920). The majority of the other American mergers took place between small companies. In addition, bilateral and trilateral technical "agreements" with European cartels skirted the letter of the law and allowed American companies to join forces with ICI and IG in various sectors, one

---

[26]For more details, see the Rosenberg chapter in this volume.

[27]For more details see the chapter by Arora and Rosenberg in this volume.

good example being a DuPont-ICI agreement in 1929. But no American company quite rivaled the national market positions of the European giants.

Following a speculative mania that saw the trading values of stock climb to dizzying heights, the Wall Street crash of October 1929 abruptly closed the door on the prosperous 1920s and ushered in a depression the likes of which the nation had never known. The Great Depression of the 1930s staggered the U.S. economy. Industrial production plummeted, giving rise to the worst unemployment the country had ever seen. Investments were wiped out, savings were gone, debts mounted. Both salaries and hourly wages fell; many of those who did have jobs worked for a pittance. Tax collections dropped, forcing government to cut back essential services. With the passage of the Smoot Hawley Tariff in 1930, the United States became the most protectionist of the industrial powers, and world trade shrank; some believed this move helped to precipitate the ensuing worldwide depression.

By 1932, when Franklin Delano Roosevelt was elected president, business interests that had ridden high during the heady years of the 1920s found themselves saddled with blame for the hard times. For most citizens, it mattered less what Roosevelt did than that he acted boldly. With the electorate enthusiastically behind him, FDR piloted through Congress a raft of legislation, collectively known as the New Deal. Many of these laws emerged out of a spirit of business-government cooperation. A dizzying array of program acronyms became standard currency in everyday conversation.

The chemical industry was not as hard hit as some other industries during these Depression years. The industry was relocating south, near the abundant oil and gas deposits, and evolving into the petrochemical industry. The south also offered new markets for expanding sales to paper, cotton, and synthetic-textile companies. Moreover, the industry was heavily dependent on scientific and technological research, and these were very fruitful years in the laboratory, even though there was little sales growth and prices remained low. DuPont focused on R&D in its Central Research Lab, leading to an explosive burst of innovation that ran from around 1935 up until 1950. Discoveries included a variety of synthetic fibers, plastics, drugs, agrochemicals, Freon gas, Teflon, and nylon. By 1937, 40 percent of DuPont's sales involved products that did not even exist ten years prior.

After 1935, the New Deal shifted away from recovery and toward reform. In retrospect, the single most important legislative step of this period was the Social Security Act of 1935, which proved to be the cornerstone of a permanent government welfare system; its purview was steadily expanded with almost every subsequent Congress. The second most significant change was the Wagner Act, which gave workers a clear right to form unions and organize. When the Depression struck, only 5 percent of the workforce was organized. In 1935, the AFL's Committee for Industrial Organization broke off to establish itself as the Congress of Industrial Organizations (CIO). With its newfound tactic—the sit-down strike—and the support of the Wagner Act, the CIO began to unionize mass production in such labor-intensive industries as automobiles and steel. By 1941, more than 10 million workers were unionized; that's more than three times as many as a decade before. But in non-labor-intensive industries such as chemicals, unions hardly left a mark. Large-scale petrochemical plants were highly automated and became more so after World War II.

The New Deal ended not so much in defeat as in a stalemate. America's antistatist tradition—the belief that big government was inherently evil—confronted the New Deal with rising frequency. However, the New Deal left a lasting legacy: It established the principle that the government had ultimate responsibility for the well-being of its citizenry. To rescue capitalism from its own excesses, New Dealers socialized the economic life of the nation, as the government assumed a partnership with private efforts in everything from banking to utilities.

It was left to the exigencies of war to lift the nation completely out of the depression. As in World War I, much of the country hoped to avoid another European quagmire. But isolationism died at Pearl Harbor in 1941. Again, the economy was rapidly converted to a war footing. After initial snarls, production figures for such things as fabricated metals, artificial rubber, machine tools, planes, and merchant ships reached astronomical totals; by the beginning of 1944 the value of U.S. manufacturing was double that of all the enemies combined.[28]

Scientists played a more important role in the war effort than in any previous war, making gains in rocketry, radar, and sonar. The prima donna of government-funded secret scientific work, the Manhattan Project, created the atomic bomb. Chemical companies such as DuPont, Union Carbide, Harshaw Chemical, and Hooker Electro Chemical all cooperated on this project. Moreover, the chemical industry achieved record production rates for synthetic rubber, aviation fuel, and such specialty metals as magnesium and aluminum, all required in huge quantities by the military. New materials earmarked for the war effort emerged: polyethylene for electrical insulation, silicones for lubricants and greases, nylon for parachutes and tires, PVC film for protection of equipment, and penicillin for tropical diseases. Arthur D. Little, working with Harvard, came up with napalm for incendiary bombs. The war dried up the pool of unemployed; in fact, toward the end of the fighting there was a labor shortage. As a result, both women and blacks found that "help-wanted" signs applied to them, too.

After hostilities ceased, demobilization was rapid. Servicemen returned home, eager to resume normal lives, and the birthrate rose precipitously. Indeed, the nation's population increased by almost 20 percent during the 1950s. The GI Bill of Rights helped an increasing proportion of young people attend college. Young and growing middle-class families emptied out of cities and took up residence in tract homes in the spreading suburbs. Mobility and speed, always valued by Americans, became nearly an obsession during these years. Thousands of miles of high speed, multilane highways were thrown across the land, 90 percent financed by the federal government.

For three decades after World War II, perhaps the most important government-industry interactions happened through the defense and space budgets, which included sizable allocations for research and development. During the 1950s and 1960s, business grew increasingly global in scope, and the chemical industry was one of the first to become truly international. Such companies as Dow and Monsanto achieved spectacular growth by setting up production units overseas. The United States dominated the world's economy during the late 1940s, '50s, and '60s, and the U.S. chemical sector dominated the world's chemical industry, exporting petrochemical plants, as well as homegrown American know-how and technology. The Kennedy Round of GATT talks during the early 1960s reduced tariff walls for everyone at just about the time that the European chemical industry was getting back on its feet. The Europeans seized the opportunity to regain a foothold in the American market. German dye manufacturers undercut the prices of U.S. companies. Synthetic-dye manufacturers were also squeezed between higher prices for imported intermediates and lower prices for finished textiles. As a result, DuPont and other American companies were forced out of the dye business.

---

[28]See Morison (1980, pp. 556–67) and Schlesinger (1957, p. 436) for a review of these times.

United States firms also faced increasingly vigilant antitrust watchdogs in Washington. DuPont, a target of this rediscovered zeal, was forced to sell off licenses, its share of General Motors, its agreements with ICI, and several joint ventures (Aftalion 1991, pp. 250–51).

The 1960s and 1970s saw a general loss of public confidence in traditional institutions and systems of governance. During the 1960s, young people, particularly those attending colleges and universities, rebelled against what they viewed as the repressed, conformist society of their parents. The New Left opposed military involvement in Vietnam, materialism, and capitalism. These concerns helped spur a "War On Poverty" and a "Great Society." Programs such as Medicare, Medicaid, and various other forms of assistance constituted the greatest expansion of social welfare since the New Deal. The public image of chemical companies took a beating. Some firms were involved in producing munitions for the Vietnam war — Dow Chemical, for example, supplied both napalm and Agent Orange. Also, chemical manufacturers were challenged by the growing environmental movement.

By the mid-1970s, the future of the American economy faced greater uncertainty than at any other time since the Great Depression. The disaffected questioned whether economic growth was possible, or even desirable. Two major increases in the price of oil sparked inflation, and when the Federal Reserve fought that inflation, it brought on deep recessions. The petrochemical industry was especially hard hit by price increases for their key input. Excess capacities had been building up for years, and the industry was now faced with overcapacity in petrochemicals, thermoplastics, synthetic fibers, and fertilizers. Petrochemicals and thermoplastics presented the most difficult problems, because certain processes cannot operate efficiently at under 60 percent capacity. The result was plant closings and redirected production into such areas as specialty chemicals. Companies divested themselves of their petrochemical units and divisions such as thermoplastics, polymers, and chemical fibers.

Although the United States had never adopted an explicit industrial policy outside of wartime, the role of government in the economy had grown steadily over the course of the 20th century. By the 1970s, environmental regulations, consumer legislation, subsidies, taxation, and direct intervention into market relationships formed a complex network of government activity. Each part was in some sense justifiable, but there were plenty of examples of overlapping bureaucracies and policies working at cross-purposes. Systemic rigidities made change difficult and the results often unpredictable. None of these programs particularly favored the chemical industry, except for some tariff barriers that were eventually abolished. On the other hand, environmental regulations raised costs for the chemical industry. For example, regulations forced DuPont to give up its tetraethyl lead business and restructure its Freon gas division.

When Ronald Reagan was elected president in 1980, his campaign promised a tax cut, less government interference in people's lives, and a strong U.S. presence abroad. During his presidency, marginal tax rates dove to the lowest level among industrial societies, and a military buildup firmly established the United States as the world's foremost superpower. The collapse of the Soviet Union in 1991, however, has made foreign policy in the 1990s harder to articulate. People favor tax cuts but also want lower budget deficits. Citizens favor less government, but they do not want to see reductions in expensive entitlement programs like Social Security or Medicare. The rise of global markets and competition has brought both economic benefits and anxieties. The social and economic ills of the day — wage stagnation, growing gaps between rich and poor, the rise of crime and drugs, the inadequacies of public education, and the decline of the nuclear family — have public policy solutions that are either very costly, or perhaps ineffective, or both. Yet, at present the United States is the most economically successful of the industrial nations and is enjoying an enviously long, inflation-free period of growth with low unemployment.

REFERENCES

Aftalion, Fred (1991), *A History of the International Chemical Industry*, University of Pennsylvania Press, Philadelphia, PA.

Beer, J. J. (1959), *The Emergence of the German Dye Industry*, University of Illinois Press, Urbana, IL.

Ben-David, J. (1971), *The Scientist's Role in Society*, Prentice Hall, Englewood Cliffs, NJ.

Berghoff, H. and R. Möller (1994), "Tired Pioneers and Dynamic Newcomers? A Comparative Essay on English and German Entrepreneurial History, 1870–1914," *Economic History Review*, XLVII(2):262–87.

Cannadine, David (1990), *The Decline and Fall of the British Aristocracy*, Yale University Press, New Haven, CT.

Chandler, Alfred D., Jr. (1990), *Scale and Scope: The Dynamics of Industrial Capitalism*, The Belknap Press of Harvard University Press, Cambridge, MA.

Coats, A. W. (1980), "Economic Thought," in *Encyclopedia of American Economic History: Studies of Principal Movements and Ideas*, vol. 1, Glenn Porter (ed.), Scribner's Sons, New York, NY, pp. 468–83.

Dintenfass, Michael (1992), *The Decline of Industrial Britain, 1870–1980*, Historical Connections Series, Routledge, London, U.K.

Ensor, Robert (1992), *England: 1870–1914*, Oxford University Press, Oxford, U.K.

Giersch, Herbert, Karl-Heinz Paque, and Holger Schmiedieng (1992), *The Fading Miracle*, Cambridge University Press, New York, NY.

Gluck, Carol (1995), "Patterns of Change: a Grand Unified Theory of Japanese History," *Bulletin of the American Academy of Arts and Sciences* (March) XLVII(6).

Haber, L. F. (1958), *The Chemical Industry During the Nineteenth Century: A Study of the Economic Aspect of Applied Chemistry in Europe and North America*, Oxford University Press, Oxford, U.K.

Halevy, Elie (1961), *A History of the English People in the Nineteenth Century. England in 1815*, Barnes and Noble, New York, NY.

Hennessy, Peter (1993), *Never Again: Britain 1945–1951*, Pantheon, New York, NY.

Keck, Otto (1993), "The National System for Technical Innovation in Germany," *National Innovation Systems. A Comparative Analysis*, Richard Nelson (ed.), Oxford University Press, Oxford, U.K. pp. 115–57.

Landau, Ralph (1997), "Education: Moving from Chemistry to Chemical Engineering and Beyond," *Chemical Engineering Progress* (January) 52–65.

Landes, D. S. (1969), *The Unbounded Prometheus: Technological Change and Industrial Development in Western Europe from 1750 to the Present*, Cambridge University Press, London, U.K.

Malchow, H. L. (1991), *Gentlemen Capitalists: The Social and Political World of the Victorian Businessman*, Stanford University Press, Stanford, CA.

Morison, Samuel Eliot, Henry Steele Commager, and William E. Leuchtenburg (1980), *The Growth of the American Republic*, vol. 2, Oxford University Press, New York, NY.

Nichols, A. J. (1994), *Freedom with Responsibility. The Social Market Economy in Germany 1918–1963*, Clarendon Press, Oxford, U.K.

Pflanze, Otto (1990), *Bismarck and the Development of Germany, Volume I: The Period of Unification, 1815–1871; Volume II: The Period of Consolidation, 1871–1880; Volume III: The Period of Fortification, 1880–1898*, Princeton University Press, Princeton, NJ.

Plumpe, Gottfried (1990), *Die I.G. Farbenindustrie AG: Wirtschaft, Technik und Politik 1904–1945*, Duncker & Humblot, Berlin, Germany.

Plumpe, Gottfried (1995), private conversions.

Rubinstein, W. D. (1993), *Capitalism, Culture and Decline in Britain, 1750–1990*, Routledge, New York, NY.

Schlesinger, Arthur Meier (1957), *The Rise of Modern America 1865–1951*, Macmillan, New York, NY.

Teltschik, Walter (1994), *History of the Large German Chemical Industry; Development and Influence in State and Company* (tr. Ralph Landau), VCH, Weinheim, Germany.

Wiener, M. J. (1981), *English Culture and the Decline of the Industrial Spirit: 1850–1980*, Cambridge University Press, Cambridge, U.K.

Woodward, E. L. (1958), *Oxford History of England. The Age of Reform, 1815–1870*, Clarendon Press, Oxford, U.K.

Wrigley, Julia (1986), "Technical Education and Industry in the Nineteenth Century," in Bernard Elbaum and William Lazonick, *The Decline of the British Economy*, Clarendon Press, Oxford, U.K.

SELECT BIBLIOGRAPHY FOR FURTHER READING

Ashby, Eric (1963), *Technology and the Academics. An Essay on Universities and the Scientific Revolution*, Macmillan, London, U.K.

Bark, Dennis L. and David R. Gress (1993), *A History of West Germany. Volume 1: From Shadow to Substance: 1945–1963; Volume 2: Democracy and Its Discontents: 1963–1991*, Blackwell, Cambridge, MA.

Berghahn, Volker R. (1994), *Imperial Germany, Economy, Society, Culture and Politics*, Berghahn Books, Providence, RI.

Crafts, N. F. R. (1995), "The Golden Age of Economic Growth in Western Europe, 1950–1973," *Economic History Review*, XLVIII(3):429–47.

Daunton, M. J. (1989), "Gentlemanly Capitalism and British Industry 1820–1914," *Past and Present*, (122):119–58.

David, Lance E. and Robert A. Huttenback (1988), *Mammon and the Pursuit of Empire: The Economics of British Imperialism*, Cambridge University Press, Cambridge, U.K.

Dornseifer, Bernd and Jurgen Kocka (1993), "The Impact of the Preindustrial Heritage. Reconsiderations on the German Pattern of Corporate Development in the Late 19th and Early 20th Centuries," *Industrial and Corporate Change*, 2(2):223–48.

Fischer, Wolfram (1994), "The Role of Science and Technology in the Economic Development of Modern Germany," *Science, Technology, and Economic Development*, pp. 71–112.

Freidel, Frank (ed.) (1959), *The Golden Age of American History*. George Braziller Inc., New York.

Gispen, Kees (1989), *New Profession, Old Order: Engineers and German Society, 1815–1914*, Cambridge University Press, Cambridge, U.K.

Grant, Wyn, William Paterson, and Colin Whitston (1988), *Government and the Chemical Industry: A Comparative Study of Britain and West Germany*, Claredon Press, Oxford, U.K.

Haber, L. F. (1971), *The Chemical Industry, 1900–1930. International Growth and Technological Change*, Clarendon Press, Oxford, U.K.

Nash, Roderick (1967), *Wilderness and the American Mind*. Yale University Press, New Haven, CT.

Nevins, Allan and Henry Steel Commager (1966), *A Short History of the United States*, 5th ed., Knopf, New York, NY.

Nipperdey, Thomas (1983), *Germany from Napoleon to Bismarck, 1800–1866*, Gill and MacMillan, Dublin, Ireland.

Porter, Michael E. (1990), *The Competitive Advantage of Nations*, The Free Press, New York, NY.

Reischauer, Edwin O. (1977), *The Japanese*. Belknap Press of Harvard University Press. Cambridge, MA.

Reynolds, Terry S., ed. (1991), *The Engineer in America, An historical Anthology from Technology and Culture*, University of Chicago Press, Chicago, IL.

Rosenberg, Nathan (1969), *The American System of Manufactures*, University of Edinburgh Press, Edinburgh, U.K.

Sansom, G. B. (1965), *The Western World and Japan: A Study in the Interaction of European and Asiatic Cultures*. Alfred A. Knopf, New York.

Schlesinger, Arthur M. Jr. (1986). *The Cycles of American History*. Houghton Mifflin Co., Boston.

Spitz, P. H. (1988), *Petrochemicals: The Rise of an Industry*, Wiley & Sons, New York, NY.

Supple, Barry (1994), "Fear of Failing: Economic History and Decline of Britain," *Economic History Review*, XLVII(3):441–58.

Weaver, R. Kent and Bert A. Rockman, eds. (1993), *Do Institutions Matter? Government Capabilities in the United States and Abroad*, The Brookings Institution, Washington, DC.

9 Monetary, Fiscal, and Trade Policies in the Development of the Chemical Industry

BARRY EICHENGREEN

There are a number of grounds for believing that macroeconomic and trade policies can matter for a nation's output and productivity growth. The stability and buoyancy of the macroeconomic environment can matter critically for growth. Investment plans are difficult to formulate when the economy is volatile. Firms delay investing, foregoing first-mover advantages. If growth at home drives the demand for the products of domestic industry (perhaps because of constraints on the range or volume of products that can be sold on export markets), then policy to ensure adequate domestic demand will be indispensible.[1]

Although trade policy can surely matter, there are opposing arguments over what policy is most useful to a nation's growth. Some emphasize the importance of a liberal trade regime (Easterly et al. 1993). The implication is that governments which expose producers to the chill winds of foreign competition promote efficiency and discourage rent seeking. However, in the presence of market imperfections, departures from free trade have at least a theoretical justification. For example, if production is only profitable after a period of learning by doing, firms will first need to endure losses in order to gain the know-how required to reap profits. If firms are liquidity constrained as a consequence of capital market imperfections, they may be unable to undertake projects that would be profitable in the long term. This provides a rationale for infant-industry protection. Another possible market failure that is relevant here occurs if the price mechanism is unable to coordinate the initiatives needed to initiate industrial growth. For example, if production costs are a declining function of the range of non-traded inputs available to final-goods producers, then it may be in the interest of no potential supplier of inputs to commence operations unless others do so as well—and producers of final goods have no incentive to start up unless input suppliers

For helpful comments I am grateful to Ralph Landau, Marco da Rin, and other participants in the Stanford University project on the development of the chemical industry. This paper owes much to the capable research assistance of Chang Tai Hseih, Darren Lubotsky, Kris Mitchener, and Martin Petri and to the financial support of the Center for Economic Policy Research at Stanford University.

[1] Fischer (1991) provides evidence consistent with this link for a cross section of developed and developing economies.

also do so. In this case, a temporary import tariff may protect final-goods producers against low-cost foreign competition and encourage them to commence operations despite an otherwise inadequate range of non-traded inputs; in turn, this will promote the emergence of a network of domestic suppliers. East Asian economic growth after World War II has been prominently interpreted in this light (Rodrik 1995).[2]

Of course, neither of these arguments for trade protection should be swallowed too quickly. In the given case for infant-industry protection, intervention to correct the capital market imperfection is the first, best solution, while subsidies to domestic producers (which distort trade) are second best. The argument for a "big push" through trade protection does not hold up if the required inputs are traded internationally. If firms are not constrained financially and there are no seriously diminishing returns to scale and scope, they can solve the coordination problem by internalizing it—that is, by producing the requisite inputs within the firm. But since the topic of this chapter is macroeconomic and trade policy, we leave the analysis of alternate microlevel policies toward industry or the financial sector to other chapters.

While various authors have argued that macroeconomic and commercial policies can affect the pace and pattern of growth, for every proponent there is a skeptic. Econometric evidence from macroeconomic cross sections is often fragile, yielding few robust correlations (Levine and Renelt 1992). This skepticism provides the motivation for focusing on a particular sector, the chemical industry in the present instance. Channels whose operation is hard to discern in the aggregate may be easier to distinguish at the industry level. We focus on four countries—Britain, the United States, Germany and Japan—that are leading suppliers of chemical products. Their histories can be divided into three phases: the pre-World War I period, the interwar period (more precisely, the "transwar" period, since we include the wars themselves), and the post-World War II era. The phenomena to be explained are summarized in Tables 9-1 and 9-2: Between 1913 and 1989, the share of world chemical production accounted for by Japan rose from 2% to 14%, while that accounted for by Britain, Germany, and the United States declined from 12% to 4%, 25% to 7%, and 36% to 24%, respectively.

Our approach is to see how far we can push the argument that two channels through which macroeconomic and trade policies could have operated—providing a favorable macroeconomic environment and redressing problems of market failure—account for variations in the industry's growth. We find some support for the importance of both macroeconomic policies and import protection. Interest rates, exchange rates, and inflation rates—all standard indicators of the macroeconomic environment and of the stance of policy—have consistently influenced the development of these four national industries. However, it appears that the influence of trade policy was more prominent in the 19th century, when the emergence of each national industry (aside from Britain's, whose early growth reflected the special circumstances of the first industrial nation) was promoted by tariff protection, than after World War II, when international trade was increasingly free. Moreover, although protection encouraged firms to initiate the production of new outputs, usually that protection did not lead to self-sustaining growth; instead, it had to be maintained to sustain the new sectors. Conversely, macroeconomic policy matters more in the 20th century, when the macroeconomic environment varies internationally for reasons related to policy, than in the 19th, when

---

[2] Other recent work (Young 1992; Krugman 1994) has emphasized the predominant influence of physical and human capital accumulation in East Asian growth. This perspective is compatible with emphasis on coordination failure à la Rodrik, whose point is that coordination problems had to be solved before it was worthwhile to undertake high levels of investment in East Asian countries.

TABLE 9-1. Chemical Production in the Advanced Industrial Countries: 1913–1994 (Percentage of Combined Production, Based on Total Sales)

	1913	1927	1938	1954	1989	1994
Industrial Countries	94.2	93.4	91.4	85.7	65.3	62.1
United States	35.6	44.4	33.8	52.0	24.2	25.1
Germany[a]	25.2	16.9	23.6[b]	7.0	7.4	7.9
United Kingdom	11.5	10.8	9.7	8.9	3.7	3.9
France	8.9	7.1	6.3	4.7	4.7	5.3
Japan	1.6	2.5	6.3	4.0	14.0	18.0
Others[c]	11.4	11.7	11.7	9.1	11.3	12.9

[a] West Germany in 1954 and 1989.
[b] Of which about 8.5% represents the production in Eastern Europe.
[c] Belgium-Luxembourg, Canada, Italy, Netherlands, Switzerland, and Sweden.

TABLE 9-2. Exports of Chemicals by Country of Origin: 1899–1993 (Percentage of World Exports)

Exports from	1899	1913	1929	1937	1950	1955	1959	1993
United Kingdom	19.6	20.0	17.5	16.0	17.9	16.7	15.0	5.2
France	13.1	13.1	13.5	9.9	10.1	9.2	8.6	6.2
Germany[b]	35.0	40.2	30.9	31.6[a]	10.4	17.4	20.2	12.7
Other Western Europe[c]	13.1	13.1	15.3	19.4	20.5	19.6	21.1	13.1
United States	14.2	11.2	18.1	16.9	34.6	28.3	27.4	13.0
Canada	0.4	0.9	2.5	2.9	5.2	6.3	4.4	3.4
Japan	0.4	1.0	1.8	3.0	0.8	2.4	3.1	13.0
Other	4.2	0.3	0.4	0.3	0.5	0.1	0.2	33.4
Total	100.0	100.0	100.0	100.0	100.0	100.0	100.0	100.0
Total in $ billion U.S.	0.26	0.59	1.04	0.98	2.17	3.91	5.48	309.2

[a] West Germany in 1950–93.
[b] For post-World War II West Germany, the 1937 percentage would be between 26 and 27.
[c] Belgium-Luxembourg, Italy, Netherlands (except in 1899 and 1913). Sweden, and Switzerland.
[d] 1993: West Germany is used.
Source: Maizels (1963) and *U.S. Chemical Manufacturers Statistical Handbook* (1995).

gold-standard ideology and institutions more closely harmonized macroeconomic conditions internationally. German and U.S. economic growth outstripped British growth over the four decades leading up to World War I. In the post-World War II period, there is a strong correlation between GNP growth rates and chemical-output growth rates across countries: both were lowest in Britain and highest in Japan. This correlation suggests that economy-wide considerations, including macroeconomic policy, were important for the performance of the chemical industry. In the conclusion to this chapter, we step back and assess the evidence for and against the null hypothesis that macroeconomic and commercial policies mattered little to the growth of the industry.

THE PRE-WORLD WAR I PERIOD

The British chemical industry was the world leader until the final decades of the 19th century, when it was overtaken by its German and U.S. rivals. From a macroeconomic perspective, it is useful to distinguish four explanations for 19th-century industrial performance. First is the macroeconomic environment, namely the operation of the gold standard and the late-19th century "Great Depression" with which it was associated. Second is trade policy, which may have provided a more hospitable environment for industry in some countries than in others. Third is the shifting pattern of comparative advantage that encouraged the reallocation of resources from industry to finance, insurance, and trade in more mature economies such as Britain's. The fourth is the institutional inheritance. We will review developments in the chemical industries of our four countries before attempting an assessment of the role of these factors.

Great Britain

The first-mover advantage enjoyed by the first industrial nation makes the development of the British chemical industry distinctive. But while neither macroeconomic policy nor import protection figures prominently in the industry's early history, that history is necessary background for understanding subsequent performance.

There was little need for coordination among suppliers of inputs and demanders of outputs in the early British chemical industry, given its simple structure. The industry relied on materials produced from the waste products of animal husbandry (for instance, bone ash, glue, and grease) used in trades such as pottery, painting, printing, and fertilizer and on by-products of forestry (wood tar, bark, and charcoal) used in the manufacture of paint, varnish, and gunpowder. By mid-century, British industry had come to rely on the LeBlanc process and the chemicals it produced: sulfuric acid, chlorine bleach, and bleaching powder. By the 1860s, the British LeBlanc soda industry, dominated by small firms, was easily the largest in the world, producing more than one-quarter of a million tons in sales annually.

For several decades, firms in Britain and France experimented with techniques for producing soda (sodium bicarbonate) from ammonium carbonate and saturated brine. Ultimately, this led to the soda that Ludwig Mond began producing in Britain in 1872, using techniques pioneered by the Belgian Ernest Solvay, and which after a time proved cheaper and purer than that manufactured using the LeBlanc process. Soda production by the partnership of Brunner-Mond began in 1874; within 20 years, Brunner-Mond had built up the world's largest alkali business, supplying soda ash and other chemicals to industries such as glass, paper, soap, and textiles. Brunner-Mond was a member of the Solvay group, a network of companies that served markets throughout Europe and Russia. The associated firms controlled most continental and American markets until Germany raised its tariffs in the 1880s to develop an alkali industry of its own. Thereafter, Brunner-Mond continued to dominate British production and exports to the United States. When the American company Solvay Process of New York began to cut into its American sales, Brunner-Mond negotiated a share of its American rival's profits in return for withdrawing from the U.S. market (Kennedy 1993, pp. 9–11).[3]

---

[3] Another important early firm, British Dynamite Co., was founded by Alfred Nobel in 1871. By 1900, its plant at Ardeer was producing around 5,000 tons of explosives per year, roughly one-tenth of world consumption.

In the final third of the 19th century, Britain still possessed the world's largest chemical industry, centered at Merseyside for the newer producers of heavy chemicals and in South Lancashire and Tyneside for the older LeBlanc producers. Workers in the chemical industry increased from 1,700 in 1870 to 6,000 in 1894. Heavy chemicals—soda ash, chlorine, sulfuric acid and bleach—still dominated production, although new technologies had already begun to change the industrial landscape. The coal-tar dye industry founded by William Perkin in 1856 began to capitalize on the ability to control a wide range of organic-chemical reactions. The demand for coal-tar products, in turn, had consequences for the coking industry and for technical education. Coke ovens were engineered to capture by-products such as benzene, naphthalene, and anthracene, providing new inputs into synthetic-dye production. A research school at the Royal College of Chemistry began turning out graduates trained in the new techniques. As the industry grew in scale, technological sophistication, and capital intensity, it became necessary for producers to coordinate a growing range of activities—such as coking, the production of raw materials, and the supply of professional chemists.

By the 1880s, the LeBlanc soda producers were encountering difficulties. Trade depression and competition from firms utilizing the ammonia soda process transformed their profits into losses. They modernized their equipment, but to no avail. They attempted to regulate the soda trade but failed. A voluntary scheme to limit the production of caustic soda and bleaching powder collapsed in 1889.[4]

This left amalgamation. In 1891, the principal British LeBlanc producers merged to form the United Alkali Company, which became the largest chemical business in the world. United Alkali incorporated 48 firms, all but 3 of which were chemical works, with an annual production of 700,000 tons of acid; 150,000 tons of bleaching powder; 180,000 tons of caustic soda; and 280,000 tons of crystal soda, soda ash, and bicarbonate. The company employed 12,000 workers and 50 chemists. Virtually all the alkali was produced using the LeBlanc method. Initially, the prices of alkali and bleaching powder rose above those that had preceded the merger, dragging up profits, but soon exports to North America began to fall. By 1900, it was clear that the company had failed in revitalizing the LeBlanc soda industry.

Germany

In the four decades preceding World War I, German producers emerged as the British industry's leading rivals. German chemical output grew at about $4\frac{1}{2}\%$ per annum from 1875 to 1895 and 4% from 1895 to 1913. To put it another way, the value of German chemical production rose from 480 million marks to more than 1.15 billion marks between 1875 and 1895, a time when the value of global chemical production only increased from 3 to 3.5 billion marks (Greiling 1952, p. 23). On the eve of World War I, German chemical production and sales accounted for fully one-quarter of the world total, more than twice that of Great Britain, as shown in Table 9-1.

Historians cite technical expertise, business acumen, and the ample supply of university-trained chemists as factors in the rise of the German industry. These chemists often advanced to managerial positions, facilitating communication between manage-

---

[4] In July 1890, talks begun to unite the soda manufacturers in a "Chemical Ring" or "Chemical Union" met with some success. With the increase in the prices of bleaching powder and caustic soda, however, *The Times* began editorializing against the scheme, and it collapsed.

ment and technical staff (Haber 1958, p. 129; Hohenberg 1967, p. 77; Greiling 1952, p. 24). But macroeconomic factors may have also played a role. German GDP rose by about 3% per year, high by historical standards and half again as fast as British GDP (Sommariva and Tullio 1987, p. 15), stimulating the demand for chemical products.

Prussia had joined other German states in 1834 to form the Zollverein, a customs union with a common external tariff, and duties on most imported chemicals were moderate during the Zollverein years from 1834 to 1870. With the completion of rail links between Dutch ports and the interior of Germany, producers found it difficult to compete with imports from Britain.[5] German firms were competitive in the production of dyes, however, and the country's share of world chemical production rose from 10% in 1850 to 16% in 1875 (Greiling 1952, p. 23). Following German unification in 1871, import duties were raised. Tariffs on alkalis were raised in 1879 and again after 1882, with the intent of promoting the development of a German alkali industry, although protection remained lower than in the United States, France, and Russia. In fact, since the dyestuffs industry was globally competitive from the start, it opposed the tariffs of 1879 in the hope that German trade liberalization could be used to bargain down foreign trade barriers. However, German alkali imports, which averaged 70,000 metric tons per annum for most of the 1870s, fell to virtually zero in the course of the 1880s.

The United States

The early U.S. chemical industry was based on the needs of an agricultural population. Production was mainly for domestic consumption and was primarily for tanning, bleaching, and dyeing. At the time of the 1879 Census of Production, the first to distinguish chemicals, the industry was small in terms of productive capacity. American firms produced a mere 40 million pounds of soda salts, used primarily in the manufacture of paper, textiles, glass, and soap. Ten times this amount was imported to meet demand. For the most part, American manufacturers found it unprofitable to compete with British exporters and did not adopt the LeBlanc process.

After the Civil War, chemical production expanded dramatically, reflecting the effects of a favorable business climate and the consolidation of the national market. Trade protection also played a role: Specific and ad valorem duties were placed on dyes in the 1870s and then rolled back in the 1880s. Haber (1958) suggests that the insulation these duties provided from foreign competition cultivated the development of an indigenous dyestuffs industry. The 1897 Dingley tariff then again raised duties on soda ash and caustic soda by 50%. Although the Payne-Aldrich tariff raised many import levies in 1909, those on most chemical imports were lowered (the rate on finished organic chemicals was reduced by one-half, for example). The Underwood Tariff Act of 1913 then reduced many chemical duties still further—down to zero in the case of sodium phosphate and potassium cyanide.[6]

It is interesting to note the arguments advanced during the tariff hearings in the 1880s and 1890s. Calls for protection were based primarily on the competitive pressure resulting from cheap European labor and the need to protect American workers' living

---

[5] Soda duties were reduced in 1865 when Britain was extended most-favored-nation status, leaving German soda manufacturers unable to meet the price of British imports (Haber 1958, p. 48).

[6] Although the general trend of these laws was toward less protection of the chemical industries, as a result of the 1913 Act, it is true that duties on cellulose products were raised and a 5% ad valorem duty was established for benzene, toluene, xylene, and napthalene.

standards from low-wage German competition. The infant-industry argument was invoked only infrequently.

By 1913, U.S. firms accounted for more than one-third of global sales, nearly matching the combined size of the British and German industries, as shown in Table 9-1. This surge reflected the rapid growth of the U.S. economy, itself a function of capital deepening, productivity advance, and an immigrant-fed population. Although U.S. per capita incomes were only marginally (if at all) higher than in Britain, per capita consumption of chemicals in the United States was nearly half again as large (Maizels 1963, p. 290). American producers made advances both by importing technology and by developing processes of their own.[7] But the prewar U.S. chemical industry focused on domestic sales: In 1913, U.S. chemical exports were less than one-fifth those of Britain and Germany combined (see Table 9-2). The prewar industry remained specialized in the production of inorganic chemicals such as acids and sodium compounds (especially caustic soda and soda ash). Dyes were produced by only a few firms which imported intermediates from Germany; as in the early days of soda salt production, domestic producers supplied only 10% of the market.

Japan

The first modern Japanese chemical manufacturers produced sulfuric acid and soda ash (Molony 1990, pp. 19–21). The demand for sulfuric acid and soda derived from their use in the manufacture of money: Caustic soda and bleach were used to manufacture paper currency, while sulfuric acid was used to refine silver and gold. The production facilities were established by the State in the first decade of the Meiji era, in the 1870s, as part of a comprehensive industrialization program. In addition to chemicals, the State invested in coal mines, railroads, telegraph lines, iron foundries, shipyards, and machine shops. However, the effort to promote domestic chemical production was short-lived and high cost, and the government disposed of its chemical facilities in 1879.[8] As late as 1905 to 1910, domestic production of sulfuric acid was still only one-tenth of what Japan imported from Britain (Molony 1990, p. 326).

Superphosphate, produced from phosphate rock and sulfuric acid, was the first chemical fertilizer produced in Japan. The earliest production facility was established by Osaka Alkali in 1886. Output expanded slowly because of British exports of chemical fertilizers, Chinese exports of soybean cake fertilizer, and resistance from farmers unfamiliar with chemical fertilizers. Although the government did not establish a pilot plant as it had for sulfuric acid and soda, it assembled a group of private investors to form Tokyo Artificial Fertilizer. This group sold surplus sulfuric acid (an input into the production of superphosphates) at low prices and organized an educational campaign to encourage the use of chemical fertilizers. Superphosphates, however, did not become popular until the Sino-Japanese war of 1894–1895 disrupted Chinese exports of soybean cake fertilizer.

In general, from the 1880s to World War I, the Japanese chemical industry received few direct subsidies and little tariff protection Japan signed treaties in 1858 and 1866

---

[7] The Solvay process for soda production was adopted in Syracuse in the 1870s; coal-tar distillation began in Philadelphia in the 1880s; and electrochemical production took hold in the 1890s. Cylindrical tanks loaded onto flatcars provided a reliable form of transport, overcoming internal barriers that had been a hindrance to the development of the domestic market. The number of firms grew from around 3,300 in 1869 to as many as 11,000 by 1914. Employees increased from 42,000 in 1869 to 300,000 in 1914.

[8] Johnson (1982, p. 84) argues that direct state investment in industrial facilities was a source of "inflation, trade deficits, corruption, and looming bankruptcy."

preventing it from imposing tariffs; these remained in place until 1899. And even after 1899, treaty bindings kept most tariffs below 10% to 15% (Lockwood 1954, p. 539). The state did encourage the chemical industry, though indirectly: It provided fertilizer producers with electricity at favorable prices; it invested in a network of vocational schools and scientific training in universities; it sponsored the establishment of research facilities such as the Tokyo Industrial Experimental Laboratory. However, given the small size of Japan's chemical industry in this time period—even by 1913, it represented less than 2% of global production and 1% of global exports (see Tables 9-1 and 9-2)—we will not discuss Japan's experience further in the context of this time period.

The Role of Macroeconomic and Trade Policies

The most striking feature of the late-19th-century chemical industry was Britain's relative decline. The British industry's difficulties during this time were compounded by a series of trade depressions. Trade was depressed in the wake of the Franco-Prussian war in the early 1880s and again in the first half of the 1890s. These periods of stagnation reflected the downward trend in commodity prices that characterized the quarter century from 1870. Beginning in 1871, one nation after another, starting with Germany, joined Britain on the gold standard. As additional countries sought to accumulate gold reserves, the increase in the demand for gold put downward pressure on the price level, with depressing effects on industry and trade. The slow growth of the economy posed particular difficulties for a chemical industry that supplied basic goods to producers of glass, paper, soap, and textiles.

Yet macroeconomic policy, as modern readers conceive of it, probably accounted for little of the difference in industrial performance across countries. The same fiscal framework of broadly balanced budgets and low public debts prevailed in Britain, Germany, and the United States.[9] The same gold-standard statutes governed the conduct of monetary policy.[10] Nor did taxes, which were modest in all three countries, significantly affect the incentive to save and invest. Britain's savings rate was no lower than that of Germany or the United States; rates of domestic investment however, were lower, reflecting the country's unusually high propensity to export capital (Eichengreen 1991). Contemporaries complained that this propensity reflected a bias in the capital market, with British financial institutions preferring overseas railway securities to domestic industrial investments, especially those of small and newly established firms (Lavington 1921). Recent research lends little support to this view, suggesting rather that British financial markets in fact equalized risk-adjusted domestic and foreign returns on publicly traded securities (Edelstein 1981). If so, Britain's low rate of domestic investment reflected the low profitability of domestic projects, explanations for which must lie elsewhere.

[9] If anything, U.S. performance should have suffered from the uncertain status of gold convertibility prior to the Gold Standard Act of 1900; in the 1890s in particular, doubts about the government's commitment to the maintenance of convertibility led to a period of high interest rates that could not have been good for economic performance. Similarly, the United States, and not Britain, suffered through a series of financial crises prior to 1913 that led to disruptive interest-rate movements and sharp macroeconomic swings; see Grossman (1993).

[10] Germany was on the gold standard from 1871. While the United States only reinstituted de jure gold convertibility in 1879, fully a decade and a half following the conclusion of its Civil War, it had effectively restored de facto convertibility at the previous rate in 1873. And while the U.S. Treasury retained some capacity (and obligation, under the provisions of the Bland-Allison and Sherman Silver Purchase Acts) to purchase and coin silver until the passage of the Gold Standard Act of 1900, such purchases were limited and had only a minor impact on the U.S. price level.

More important at this time than monetary and fiscal policies were the commercial policies of the three governments.[11] Britain, the dominant exporter of chemicals and steadfast proponent of free trade, imposed no duties on imports. But Germany raised its tariffs in 1879 and again after 1882, and imports of British alkalis fell. The United States, the single most important market for British soda exports, increased its duties on soda ash, caustic soda, and bicarbonate soda by 50% in 1897 and placed a tariff on formerly duty-free bleaching powder (Haber 1958, p. 216). These measures virtually halted British alkali exports to the United States. All the while, Britain adhered to free trade, although British chemical producers did ponder openly whether tariff protection might offer a solution to their problems.[12]

There is no question that these policies shifted chemical production toward the United States and Germany at the expense of Great Britain, in the short run. At issue is whether these policies launched the German and American chemical industries onto longer-term trajectories of self-sustained growth. The German experience with tariffs on alkalis does appear to have been a case in which the increased competitiveness of domestic producers proved permanent, rendering their newfound foothold stable even in the absence of continued protection. German alkali exports, which drew no direct benefit from protection, rose from negligible levels at the start of the 1880s to more than 30,000 metric tons in 1890 and 60,000 metric tons in 1899.

Were these developments contingent on the imposition of tariffs, or would they also have occurred in its absence? The infant-industry argument requires a capital market failure that, in the absence of protection, would have prevented aspiring producers of alkalis from obtaining the funds needed to defray their start-up costs. In fact, there is little evidence that finance was a binding constraint. Some companies were profitable from the outset; others were readily able to obtain external finance from Germany's highly developed system of universal banks (Hohenberg 1967, p. 45), thereby relaxing the finance constraint when internal funds did not suffice. As to the possibility of a "big push" through trade protection, it is not obvious that protection from imports resolved insurmountable coordination problems on the input side. The alkali-producing sector did not require coordinating a wide range of activities to supply inputs. Insofar as it was required, the absence of binding financial constraints allowed German firms to solve this problem by integrating vertically. And, insofar as coordination among firms was required, this could be accomplished through government-sanctioned industry agreements. The founding in 1877 of an industry organization, the *Verein zur Wahrung der Interessen der Chemischen Industrie Deutschlands*, facilitated cooperation. Between 1904 and 1907 several German chemical cartels were formed; for example, the *Dreibund* involved cooperation among BASF, Bayer, and Agfa, while the *Dreiverband* was an agreement between Hoechst, Casella, and Kalle.[13] Insofar as the true underlying constraint was an adequate supply of chemists, the government could and did relax it by supporting the university system, but for the purposes of the present argument, this is not a macroeconomic or trade policy.

[11] One can argue, of course, that the choice of trade policy was not independent of the monetary regime. Britain, which could stabilize its balance of payments through adjustments in the Bank of England's powerful discount rate, may have found it relatively easy to operate a system of unrestricted international trade. Germany and the United States, in contrast, had less ability to control the direction of international capital flows and consequently relied on restrictions on commercial transactions.

[12] Not by coincidence was this the moment when Joseph Chamberlain launched his tariff-reform campaign, advocating tariffs to force other countries to lower their barriers to British goods.

[13] The question is whether these cooperative agreements functioned mainly to internalize externalities that were not adequately addressed by the price system or to restrain competition in a collusive fashion. The natural hypothesis is that they served both functions.

In the U.S. case, specific and ad valorem duties were placed on dyes in the 1870s. An 1882 report of the Tariff Commission indicated employment and capital growth of over 200% and production growth of almost 150% for the dye industry in the 1870s. Predictably, the dyestuffs-consuming textile industry soon began to demand that domestic producers face the test of international competition. While the Tariff Commission appealed in general to the infant-industry argument for exceptions to the case for trade liberalization, it acknowledged that the chemical industry, as a thriving adolescent, obviously did not deserve infant-industry protection. As a result, the specific duty on dyes was eliminated in 1883. Almost immediately, German competition drove out of business the dozen or so dyestuffs companies that had been founded starting in 1877. This episode shows that while generous protection from import competition could support the development of an indigenous dyestuffs sector, it did not allow dyestuffs firms to become established in such a way that they could retain their viability in the absence of protection. It does not suggest a role for government intervention in support of the industry on grounds of market failure.

The 1897 Dingley Tariff, which increased import duties on caustic soda and soda ash by 50%, also offers little support for the argument that protection of the U.S. market helped to launch U.S. producers on a trajectory of sustained growth. True, British alkali exports to the United States fell sharply starting in 1897 and did not recover subsequently, and the U.S. industry became all but self-sufficient by the turn of the century. The value and volume of British alkali exports to the United States, however, had already been trending downward in the first half of the 1890s, unlike British exports to other countries, which remained stable; the contrast plausibly reflects increasing competitiveness of the U.S. industry in the preceding period. In other words, there is no strong correlation between the level of protection received by U.S. producers and the rate of growth of domestic production.

Hence, if tariff protection was critical to the successful overtaking of the British chemical industry by its German and American competitors, its role must have been indirect. That role would have been to help put in place structural conditions conducive to the emergence of large, science-based enterprises capable of exploiting economies of scale and scope.[14] The argument runs as follows. Exploiting opportunities for rapid growth in the final decades of the 19th century required the development of large, high-speed throughput, continuous-process industries capable of exploiting economies of scale in the production of intermediate inputs and branded, packaged consumer goods; a financial system with the capacity to meet the capital requirements of large enterprises and a willingness to commit funds to firms investing in unproven technologies; a flexible system of labor relations adaptable to the technological and organizational imperatives of modern mass production; and a mass market. The United States possessed the requisite large enterprises, which could draw on the financial support of private investment banks (DeLong 1991).[15] In Germany, these prerequisites took the form of industrial cartels and a system of universal banks (Gerschenkron 1966). In neither country did extensive unionization impede the adoption of new technologies.

Britain, on the other hand, acquired a very different set of institutions as a legacy of her early industrialization. Since the capital requirements of early industrialization had been modest, Britain inherited a banking system that specialized in trade credit rather

---

[14]Chandler (1990) argues that scale and scope were important for the development of the German and U.S. chemical industries in the final decades of the 19th century.

[15]Admittedly, investment banks may have been less intimately connected to the chemical industry than to other sectors of the U.S. economy. For further details, see the chapter by Da Rin in this volume.

than industrial finance. As a result, such innovative firms as Brunner-Mond found it difficult to raise external finance in their early years. Since early industrial technology had offered limited economies of scale and scope, Britain inherited an industrial structure dominated by atomistic, single-plant firms. She also inherited a system of fragmented, craft-based unionism that reflected the autonomy on which the skilled craftsman could insist during the long, mid-Victorian boom.

The question regarding Britain is why, if these arrangements were poorly suited to the technological and organizational imperatives of the late 19th century, they did not succumb to institutional competition. If there was a market for investment banks, why did no one start such a bank to fill the niche? If there existed unexploited economies of scale and scope, why didn't clever entrepreneurs merge firms of suboptimal size?

One answer is that a coordination problem existed; it was in no one's interest to deviate from the institutional norms unless others did likewise.[16] It was in no one's interest to start an investment bank to lend to and monitor the activities of large enterprises in the absence of manufacturing firms of the requisite size. It was in no one's interest to create large enterprises in the absence of investment banks to meet their financial needs. It was in no one's interest to challenge the prevailing system of industrial relations in the absence of firms capable of exploiting the technologies of modern mass production. The implication is that two equilibria could coexist: the dynamically efficient German-American equilibrium, characterized by large firms, industrial banks, and flexible labor relations; and the inefficient British equilibrium, featuring small firms, clearing banks, and craft-based unions. Given the prevailing British ideology of laissez faire, there was no mechanism for coordinating a shift from one equilibrium to the other.

It is important to recognize that government policy could play a role in creating the preconditions for reaching one equilibrium or the other. The existence of a mass market was one of the conditions that created an incentive for entrepreneurs to invest in the creation of large enterprises, for investors to develop innovative technologies suitable for modern mass production, and for financiers to invest in the creation of industrial banks that could meet those enterprises' financial needs. Policy could create that mass market through the imposition of tariffs against imported products. In this way, protecting key sectors such as steel and chemicals at the point in time when the decision was made whether to sink the costs of investing in the financial and technological requirements of producing for a mass market could have permanent effects.

How applicable is this story to the rise of the German and American chemical industries? Other chapters suggest that economies of scale implying significant capital requirements and the need for a mass market to amortize that investment were more relevant to some segments of the industry than to others. They were pronounced in the electrochemical segment, which constituted the fastest growing segment of the industry during this period. Early U.S. electrochemical companies, which clustered around Niagara Falls to make use of its electricity-generating capacity, were large by the standards of the day. Their high fixed costs had to be amortized over long production runs. These observations are consistent with the notion that tariff policy had sustained effects, at least in such dynamic market segments as electrochemicals.

Even this favorable interpretation of the role played by import protection in the development of the chemical industry in countries such as the United States, however, does not suggest that tariffs had a favorable impact on industrial development more

---

[16]Here I paraphrase Eichengreen (1996).

generally. The United States possessed a domestic market sufficiently large to support a number of electrochemical producers. The abundant supply of natural resources— electricity-generating capacity in particular—with which the nation was endowed positioned the industry to exploit that latent demand. The existence of a backlog of still largely unexploited technologies facilitated the emergence of an entirely new segment of the industry. This exceptional constellation of preconditions, on which the favorable effects of import protection relied, were not as prevalent elsewhere.

THE TRANSWAR PERIOD

World War I and the decades that followed transformed the chemical industry. The war interrupted trade in chemical products, leading governments to initiate programs of import substitution to meet their military needs. The consequence was worldwide overcapacity, which led in the 1920s to a consolidation movement whose purpose was to restrain competition and rationalize production. Cartels rapidly became an international phenomenon; for example, an international dyestuffs cartel operated after 1927 (Schroter 1992). Unlike prewar cartels, which had tried to prevent overcapacity from developing, their interwar successors were formed to deal with the overcapacity that resulted from wartime stimulus to demand and from policies promoting national self-sufficiency. The influence of macroeconomic and commercial policies was especially prominent in these years. One consequence of overcapacity was trade protection, which had particularly devastating effects on export-oriented chemical industries. Generally, however, the output of chemical industries moved in tandem with overall industrial production, declining with the onset of the Great Depression and recovering earliest in countries that were quick to leave the gold standard and adopt reflationary monetary and exchange rate policies.

Great Britain

World War I had a dramatic impact on the British chemical industry. It stimulated the demand for certain products and interrupted imported supplies of others. It highlighted the fact that the industry was now smaller and less diversified than its German rival. It revealed the British economy's dependence on Germany for organic dyes. It had a negative impact on the synthetic-dye industry, one of whose principal inputs, picric acid, was needed to fill shells for the army. In response, the government took steps to encourage the production of dyestuffs, organizing a Chemical Supplies Committee to coordinate a national production program and allocate raw materials.[17]

The period also lent impetus to consolidation. In 1915, to relieve the plight of textile firms threatened by the dye famine, the government organized and financed the conversion of Read. Halliday & Sons into British Dyes Ltd. and encouraged the rival Manchester firm of Ivan Levinstein to take over the Hoechst indigo factory. These companies were then combined to form the British Dyestuffs Corporation. As noted earlier, the LeBlanc alkali manufacturers had combined in 1890 to 1891 as the United Alkali Company. Their rival Brunner-Mond used war profits to acquire several smaller companies, notably the Castner-Kellner alkali concern. Other chemical companies followed suit. In 1920, Nobel Explosives of Scotland acquired most of the British explosives industry and formed Nobel Industries. Lever Brothers took over the entire

---

[17]We return below to the implications of these measures for the emergence of a viable dyestuffs sector.

soap industry, but for two producers. Small gas companies, coal owners, and coke manufacturers engaged in tar distillation combined regionally to form companies such as South Western Tar Distillers, Midland Tar Distillers, Lancashire Tar Distillers and Scottish Tar Distillers.

But the single most significant merger for the chemical industry occurred in 1925, when Nobel, Brunner-Mond, British Dyestuffs Corporation, and the United Alkali Company combined to form Imperial Chemical Industries (ICI). This was a response to German dye firms forming the cartel that in 1925 became IG Farbenindustrie AG. Like Farben, ICI produced basic goods consumed by other industries. Its main businesses were alkalis, nonferrous metals, explosives, and dyestuffs. Its only major new product was synthetic ammonia, which by the end of the 1930s accounted for 20% of sales. ICI found itself on the verge of bankruptcy by 1929, as the problem of overvaluation created by Britain's return to the gold standard in 1925 at the prewar rate of exchange hampered exports and rendered imports artificially cheap. Then came the Great Depression, and profits plummeted. However, ICI's prospects recovered along with the economy: Between 1927 and 1937, its sales grew at an annual average rate of 7% (Smith 1992, p. 145). The company sought to further strengthen its position by negotiating cartels and market-sharing and mutual-assistance agreements. In 1929, ICI joined an international cartel that already included the principal producers of France, Switzerland, and Germany. A nitrogen cartel in which IG Farben and ICI were the principal participants was formed in 1929. ICI also concluded a technical-aid agreement covering sharing patents, processes, and R&D results with DuPont of the United States. By basing the agreement primarily on patents, DuPont maintained that stipulating sales territories was legal under the patent monopoly (Smith p. 148). In 1938, nearly two-thirds of worldwide dyestuffs sales were made by members of the European dye cartels.

Britain tiptoed away from its long-standing policy of free trade with the Dyestuffs Act of 1920, which imposed tariffs on imports of a range of chemicals, and the Safeguarding of Industries Act in 1921, in which 6,500 "strategic goods," including dyestuffs, were accorded duties of $33\frac{1}{3}$% ad valorem. Obviously, the measures taken during World War I to promote the development of dyestuffs capacity had not created a sector capable of withstanding international competition. But Britain did not truly convert to protectionism for ten more years, until the Abnormal Importations Act of 1931 and the General Tariff of 1932. Its abandonment of free trade reflected lobbying by sectors (including chemicals) under competitive pressure caused by the Great Depression and the hope that protection might reduce unemployment. Capie (1983) suggests that Britain applied the highest tariffs to industries facing the stiffest foreign competition, as measured by import penetration. Of the 20 industries he examines, chemicals had one of the highest rates of effective protection (63% in 1935).

Germany

The German chemical industry expanded significantly during World War I. Large sums were spent researching the Haber-Bosch process for the fixation of atmospheric nitrogen, which resulted in production of nitrogen (for explosives) that had been imported previously from Chile. Dyestuffs companies became involved in the development of poison gases.

Aside from the dyestuffs sector, the industry was still dominated by small- and medium-size firms, but the government invited collaboration and cartelization. By 1923, Germany had 93 chemical cartel agreements in place. Two years later, the eight largest companies merged to form IG Farbenindustrie AG, the gigantic conglomerate that accounted for about one-third of the turnover of the German chemical industry and more than one-half of the industry's exports. By the end of the interwar period it accounted for 8.7% of all German exports.

Germany lost important chemical factories because of the postwar redrawing of Europe's borders and important patents because of the activities of entities such as the U.S. Alien Property Custodian (discussed below). Industry capacity, however, remained above prewar levels because of the industry's wartime expansion—not only in Germany, but in all the leading national producers. The postwar problem was not producing, but selling. Global production of chemicals more than doubled between 1913 and 1927. Germany's share of that total fell from 25% to 16%.

Weimar Germany's tariff policy was determined initially by the Allies, who granted most-favored-nation status to all signatories of the Versailles Treaty. The occupied area west of the Rhine, and hence the rest of Germany to which the Rhineland was linked, became a free-trade zone. Germany did impose one new tariff, which averaged the same height as its prewar predecessor, in 1925, and another new tariff was adopted in 1929. Overall, however, Germany's tariffs in the 1920s were moderate by the standards of other countries. In particular, most chemical products other than bleaching powder and soda ash had zero tariffs. Reflecting their export orientation, German chemical producers lobbied for reductions in German tariffs designed to elicit reciprocal reductions in the rest of the world.

Such an export-oriented industry was, of course, devastated by the Great Depression. But between 1932 and 1938 German chemical output then more than doubled, led by export sales. The country negotiated bilateral trade treaties and clearing arrangements with its allies and dependencies in central and eastern Europe, importing foodstuffs and raw materials in exchange for industrial goods. Investment in the chemical industry rose from 15% of all investment in 1932 to over 25% of total investment in 1939 and continued at high levels during World War II.[18]

The United States

World War I stimulated the production of chemicals that had not been manufactured previously in the United States; for instance, the coal-tar dyes that were were no longer available from Germany. Production of compressed gases, especially oxygen, increased as well. The wartime interruption of German exports of dyes and other chemicals prompted Congress to pass the Emergency Tariff of 1916, which imposed duties of 30% ad valorem plus 5 cents per pound for finished coal-tar products, and 15% plus $2\frac{1}{2}$ cents per pound on coal-tar intermediates; coal-tar crudes remained duty free. The Emergency Tariff was limited to a five-year period, after which tariffs were to be reduced by 20% annually. Moreover, if at the end of five years American companies did not produce at least 60% of domestic consumption of any dye or medicinal, the tariff on that product would be eliminated. United States industry grew rapidly under the shelter of the Emergency Tariff, with 128 companies producing coal-tar intermediates, 78 producing dyes, and 31 making synthetic medicinals.

The industry received few direct wartime subsidies, although mention should be made of the Alien Property Custodian (APC). The APC took over from foreign interests a number of patents and processes, including some for coal-tar processing and dyes. Thirty chemical companies underwrote the Chemical Foundation to make these patents

[18]Data on production during World War II are not available, although the number of workers (including forced labor) increased steadily through 1944 (Stratmann 1985, p. 47).

available to American producers, and President Wilson transferred the patents to it by executive order.

With the resumption of German competition after the war, Congress extended the industry's protection under the veil of national security. The Dye and Chemical Control Act of 1921 imposed a six-month embargo on imports of chemical products that were produced domestically. The 1922 Fordney-McCumber Tariff protected the industry far more generously than had been the case prior to World War I. While the Hawley-Smoot Tariff of 1930 further raised duties on many imports, it did not change most duties on chemicals, dyes, and coal-tar intermediates.

The U.S. industry in the interwar period, like that of its British and German counterparts, underwent considerable consolidation. Some 500 mergers were completed in the 1920s, achieving economies of scale and managerial reorganization.[19]

The 1920s represented a period of buoyant growth of the market for consumer durables: Automobiles, household appliances, and other durables accounted for 11% of GDP, up from 7% before the war (Olney 1991). The economy expanded vigorously, fueling the demand for the products of the chemical industry; the only downturn between 1921 and 1929—when Henry Ford retooled his Detroit auto works from the production of the Model A to the Model T—was short and mild. Much of the demand for chemical products at this time derived from the growth of these industries. Production of synthetic rubber, stimulated by the popularity of the automobile, became increasingly important. Union Carbide began developing processes for manufacturing plastics from petroleum products. Production of rayon mushroomed. Compressed gases extended their wartime gains, while the output of coal-tar products, after declining to prewar levels, experienced a sharp rise in the 1930s. Rayon, compressed gases, and plastics, which had been relatively unimportant before World War I, accounted for fully 30% of the value of industry output by the late 1930s.

Between 1929 and 1933, industry sales fell by 39%. In April of 1933, the newly inaugurated president, Franklin Delano Roosevelt, took the country off the gold standard, allowing the dollar to depreciate. For the next nine months industry recovered vigorously, led by automobiles and steel. With the chemical industry possessing strong links to these sectors, it too recovered (although evidence discussed below suggests that, by international standards, it derived less stimulus than might be expected from the change in the policy regime).

The decade of the 1930s was one of rapid technological change. The development of polymer science led to breakthroughs in chemical products, although the mass production of new chemicals did not occur immediately. By World War II, the United States was largely self-sufficient in chemicals. In dyestuffs in particular, domestic production displaced imports; by 1939, only 11% of apparent consumption was imported. World War II provided a huge military demand for products of the chemical industry and fueled the further expansion of investment in production facilities.

Japan

Before World War I, the dyes used by large Japanese textile firms were almost entirely imported. Once imports were curtailed by the war, the government extended subsidies to encourage domestic production of dyes and soda-ash (Kobayashi 1976, pp. 247–48;

---

[19] As Smith (1992, p. 143) has written of the synthetic ammonia market, "In the United States two plants, those of Allied and DuPont at 210,000 tons/year and 70,000 tons/year, respectively, supplied the domestic market with three-quarters of its capacity. Synthetic ammonia was the most dramatic example of economies of scale; for most technologies the American market was large enough to absorb the output of several large plants."

Lockwood 1954, p. 528; Uchida 1980, pp. 166–67). The postwar resumption of trade created difficulties for dyestuffs and soda ash producers still seeking to get on their feet, and they lobbied for protection. Japan's relatively open trading regime did not survive the 1920s, a decade characterized by global overcapacity in industries such as dyestuffs and persistent macroeconomic difficulties in a country devastated by an earthquake in 1923. Tariffs on dyes were raised to more than 100% ad valorem in 1926.

Japan returned to the gold standard in January 1930, just as the world economy was entering the Great Depression. Prices dropped precipitously. Domestic manufacturers of chemical fertilizers pressured the government to raise import duties still further; fertilizer users strongly opposed such a step. The compromise was the Important Industries Control Law of 1931, which legalized cartel agreements to fix levels of production, establish prices, and limit entry. But then, following Japan's departure from the gold standard and the depreciation of the yen in December 1931, tariffs on all imported goods, including items that had previously been allowed into the country duty free, were raised by 35% (Schumpeter 1940, pp. 737–38; Lockwood 1954, pp. 540–41).

With Japan's invasion of Manchuria in 1932, the country shifted to a wartime footing. To overcome the Depression, Finance Minister Takahashi launched a policy of deficit financing (earning him his sobriquet "the Keynes of Japan"). Military expenditure rose from 28% of the budget in 1930 to 43% in 1935, a period over which the budget deficit averaged 6.7% of GNP. The foreign exchange value of the yen was cut drastically; it depreciated more rapidly than the currency of any other country. In late 1935, Takahashi attempted to control the resulting inflation by curtailing military spending. The military assassinated him, and Takahashi's successor gave the military a free hand. Since the military did not want to restrain spending, it resorted to controls and rationing. After the outbreak of war with China in 1937, state control of the economy tightened, including a new trade law that authorized the government to restrict or prohibit the export and import of any commodity and to control the distribution and use of all imported raw materials.

The Role of Macroeconomic and Trade Policies

Different national chemical producers emerged from World War I with their economies in varying degrees of disarray. Germany, where conflicts over reparations and domestic taxation produced the largest budget deficits and most rapid rates of inflation, experienced the greatest dislocations. For Germany's chemical industry, the hyperinflation was a mixed blessing: positive before 1923 and negative thereafter.

Feldman (1993, p. 843) writes that "the chemical industry is a prime example of a 'winner' in the inflation."[20] For example, lags between when taxes were levied and collected reduced the real tax burden on industry almost to zero. While the hyperinflation wiped out financial savings and largely eliminated the burden of debts, it did not affect the real assets of companies in the chemical industry and elsewhere; it thus increased their capacity to undertake acquisitions. In the early stages of the inflation, interest rates rose by less than inflation; this made borrowing attractive and investment profitable for firms such as those in the chemical industry, with good financial connections. Chemical concerns used this low-cost finance to undertake expansion, plant improvement, and inventory accumuation, improving their international competitiveness. Initially, inflation reduced production costs, as the wages of both skilled

---

[20] He continues, however, for reasons explained below, that "its leaders were not overly enthusiastic about the conditions of German industry or their industry in 1924–25."

and unskilled chemical industry workers failed to keep up with rising product prices (Feldman 1993, pp. 613, 678). But by early 1923, German unions succeeded in indexing wages to prices. After that, the primary effect of the inflation on German industry operated through demand, rather than cost-related, channels. Bresciani-Turonni (1937), in a conclusion echoed by subsequent authors, suggests that the demand-side effects were positive on balance through 1922, after which the chaos associated with uncontrolled inflation tended to depress output.

The 1924–25 stabilization recession was severe. The remaining demand was mainly a demand for consumer, not producer, goods, a fact that redounded unfavorably on the chemical industry. For the balance of the 1920s, interest rates remained high, reflecting the slow return of confidence in the currency and lingering uncertainty about Germany's commitment to shouldering its reparations burden. Voth (1995) shows that high interest rates significantly depressed German investment up through the onset of the Great Depression. Having repeatedly raised tax rates to compensate for the inflation-induced erosion in the real value of receipts, the Reich, state government, and municipalities were now able to extract significant amounts of revenue from industry, to the point that they could grant generous salary increases to public employees.[21] That said, the chemical industry navigated this recession relatively well; whereas dividends as a share of capital in 1924 to 1925 were a mere 1.16% in the coal industry and 1.93% in the iron and steel industries, they remained a healthy 5.52% in chemicals. Of the 17 industries for which the *Statistical Abstract for Germany* provides these data, only brewing, textiles, lignite, banking, insurance, and public utilities reported higher dividends (Bresciani-Turonni 1937, p. 377).

Once the international gold standard was restored in the second half of the 1920s, the British chemical industry lagged behind. Between 1925 and 1929, the output of chemical products rose by only 16% in the United Kingdom, compared to 40% in Germany, 50% in Japan, and 51% in the United States.[22] The difficulties of British industry are commonly ascribed to overvaluation of sterling associated with the country's return to the gold standard at the unrealistic prewar rate. Of course, Britain's poor performance may well have reflected long-term structural problems as well as these macroeconomic factors.

The depression that engulfed the world in 1929 provides an example of a macroeconomic shock with different effects across countries. The impact of the depression varied among countries as a function of proximity to the United States (whose increasingly restrictive policies, starting in 1928, helped precipitate the slump), as a function of economic size and openness (smaller, more open economies suffered disproportionately from the collapse of international trade), and as a function of economic structure (producers of primary products, external debtors, and countries with fragile banking systems were in the weakest positions) (Eichengreen 1992). But for each nation, the depth and duration of the slump depended first and foremost on the policy strategy adopted in response. Countries whose governments and central banks intervened in financial markets to stem the collapse of prices also succeeded in stemming the collapse of output. Countries whose central banks intervened as lenders of last resort limited the prevalence of bank failures, the spread of financial panics,

---

[21] In addition, the hyperinflation wiped out the value of nominally denominated grants to science departments, setting back research and development. To recapitalize its research activities, the industry funded three societies for the development of chemical research, education, and literature in the post-hyperinflation years (Haber 1971, p. 368).

[22] The figure for Japan is actually for 1926–29, data for 1925 not being readily available.

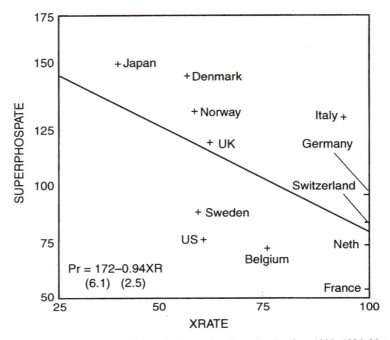

Figure 9-1. Exchange Rates and Change in Superphosphate Production, 1928–1936. Note: 1936 production − 1928 production. *Source:* Svennilson (1954).

the collapse of the money supply, and the worsening of deflation. Where central bank discount rates were cut and money stocks were stabilized, the Depression was largely halted as early as 1932, allowing recovery to commence and growth to resume.

The single most important policy decision, which in turn dictated the others, was whether to defend or abandon the gold standard. In countries that maintained gold convertibility, money supplies were tied to gold reserves, and gold reserves dwindled as these were sucked toward the United States by monetary contraction there.[23] In countries such as France and Belgium, which remained tied to gold until the mid-1930s, policy was immobilized. They could not reduce interest rates to stabilize demand, because this threatened gold losses and jeopardized the convertibility of gold and currency. Under the gold standard, their central banks had no capacity to stabilize the money supply or to intervene in support of the banking system. In such countries, only when a series of financial crises forced them off the gold standard was policy freed up and could recovery commence.

Countries that abandoned the gold standard were able to recover more quickly. Currency depreciation enhanced the competitiveness of their exports—no great advantage in a period of high tariff barriers, but beneficial nonetheless. More important, it

[23] Under a gold standard, a negative monetary shock, as in the United States, will reduce money supply relative to money demand. The excess demand for money will reduce absorption relative to output, move the balance of payments into surplus, and draw gold from other countries (Schwartz 1981).

allowed central banks to cut interest rates, steady the price level, and stabilize the banking system. Eichengreen (1992) emphasizes the close connection between the timing of devaluation and the timing of recovery. The United Kingdom and Japan devalued in 1931 and their recoveries commenced in 1932. The United States devalued in 1933; its recovery commenced in 1934. Belgium devalued in 1934; its recovery commenced in 1935. France, the Netherlands, and Switzerland devalued in 1936, and their recoveries commenced in 1937. If the advantages of abandoning gold convertibility were so apparent, why did countries such as Germany, Austria, Belgium, and France resist it so strenuously? The simplest explanation is that countries which suffered double-digit inflations in the 1920s were unwilling to abandon the gold standard lest it lead to another outbreak of inflation in the 1930s.

There is a clear correlation between the magnitude of a nation's currency devaluation and its change in industrial production between 1929 and 1935 (Eichengreen and Sachs 1985). Figures 9-1 and 9-2 provide parallel analyses for two chemical products (chosen on the basis of data availability): superphosphates and dyestuffs. On the horizontal axis is the change in the exchange rate between 1929 and 1935; a movement to the left indicates devaluation. On the vertical axis is the production of superphosphates or dyestuffs in 1936 or 1937 (with the 1928 or 1929 level indexed to 100). These plots suggest that the chemical industries of countries which devalued by larger amounts recovered more successfully from the shock of the Great Depression, at least as

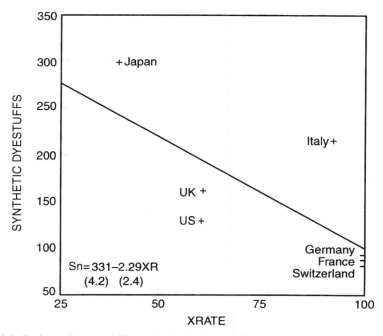

Figure 9-2. Exchange Rates and Change in Synthetic Dyestuffs Production, 1927–1937. Note: 1937 production −1928 production. *Source:* Svennilson (1954).

measured by the output of these two products.[24] In fact, the relationship is strikingly similar to that for overall industrial production.

Macroeconomic policy is not the entire explanation for industry performance in the 1930s, of course. The two plots also point to outliers: Japanese (and Italian) production appears to have recovered more quickly from the Great Depression than macroeconomic factors alone can explain, while the opposite is true of U.S. production.[25] But for chemicals, as for other sectors, macroeconomic policy appears to have exercised powerful effects in the 1930s.[26] Had World War II not intervened, these very different recovery experiences and different rates of investment in capacity could have had a sustained impact on the different national chemical industries. However, World War II transformed the industrial landscape yet again.

THE POST-WORLD WAR II PERIOD

World War II stimulated another wave of technological change in chemicals and a vast increase in production in the United States. In the immediate postwar years, total world production of chemicals increased by 20% between 1950 and 1953 (OECD 1954, pp. 67, 126). The development of plastics and synthetic fibers accelerated in the 1930s with the introduction of polystyrene by IG Farben, nylon by DuPont, and polyethylene by ICI. The organic-chemical industry shifted from coal-based materials such as acetylene and carbon monoxide to petroleum-based olefins, notably ethylene, propylene, and butylene. Factories changed, with vats, towers, and furnaces giving way to the pipes and columns of the petroleum-refining industry. Petrochemical plants began producing vast quantities of organic chemicals to be converted into resins, fibers, pesticides, solvents, dyes, and pharmaceuticals. Chemical production typically grew more rapidly than the overall rate of economic growth. During this time, there is a strong correlation between the rate of growth of chemical production and the rate of growth of national product: of our countries, both were slowest in Britain, fastest in Japan. This could mean either that the chemical industry was a catalyst for growth or that the same economy-wide factors that led to overall growth were also especially conducive to the growth of the chemical industry.

Great Britain

In the early 1950s, British chemical output grew more rapidly than industrial production. Although prices fell with the expansion of capacity, new plants came onstream in 1953, producing silicone compounds, polyvinyl chloride, and phosphates; and other facilities were built to address shortages of phthalic anhydride, titanium dioxide, and

---

[24] In each case, the relationship is statistically significant at standard confidence levels, as indicated by the regression equation and t-statistics in the lower left-hand corner of the scatter plot.

[25] Previous authors have sought to account for America's somewhat anomalous recovery in terms of output-restricting and cost-increasing policies associated with the National Industrial Recovery Act, which may have attenuated the stimulative effects of the dollar's depreciation (Weinstein 1981).

[26] Compared to the effects of monetary policy and currency depreciation, the impact of tariff protection on the development of national chemical industries is more difficult to assess. Reader (1979), for example, considers the impact on British petroleum production of the British Hydrocarbon Oils Production Act of 1935, which protected ICI from low-cost, imported petroleum products. He shows that this Act stimulated production via very generous rates of effective protection, but only temporarily; it did not produce a British petroleum-refining sector capable of moving down a learning curve and standing on its own feet in the absence of protection.

pentaerythritol (OECD 1954, p. 45). This rapid expansion persisted. By 1966, chemicals accounted for 9.7% of value added in British manufacturing, up from 9.1% in 1958. Production declined with the 1973–74 oil shock, recovered from 1975 to 1980, and then fell off again in the recession of 1980.

Despite outperforming the economy as a whole over this time, the British industry did not keep pace with its foreign rivals. The country's share of world chemical exports, which had already fallen from 23% at the turn of the century to 16% by 1954, had dropped to 12% by 1965. This decline followed the trend for Britain's exports of all manufactures, which fell from 33% of the world total at the turn of the century to 13% by 1965. In 1958 to 1965, for example, the growth of U.K. industrial production averaged 4% per annum, chemical output 6.8%, and chemical exports 7.5%. All these figures were lower than those for Germany, the United States, and Japan.

Macroeconomic analyses of the British economy's slow growth indict "stop-go" policies, failure to embrace European integration, and high tax rates (Cairncross 1992; Bacon and Eltis 1976). "Stop-go" involved cutting central bank discount rates, inflating consumer demand, and allowing incomes to rise, especially just before elections, followed by a sudden rise in rates to restrict demand, often too late to head off a balance-of-payments crisis. The consequent instability of interest rates and of macroeconomic policy generally hindered the growth of investment, incomes, and productivity.

During the 1950s, Britain shunned membership in the European Coal and Steel Community at the beginning of the decade and in the European Economic Community subsequently. Instead, Britain adopted a strategy of encouraging exports to the sterling area. The growth of the sterling area, however, lagged behind that of continental Europe. Favorable access to sterling area markets may also have sheltered producers from the rigors of competition, although recent authors have observed that sterling area markets were exposed to significant international competition; the sterling area feather-bed was actually a "rather hard mattress," as Schenk (1994) puts it. The point would seem to apply to the chemical industry.

Although high taxes are often cited as an exceptional burden on the chemical industry and the British economy, the case is not clear-cut, since taxation of corporate profits was generally declining over this time. From January 1, 1947, companies were required to pay the standard rate of income tax on all profits (around 45%) plus a profits tax, which was higher on distributed profits than on retained earnings. But after reforms in 1965 and 1973, and then a number of allowances and exemptions progressively appended to the 1973 tax law, the corporation tax existing in the late 1970s was far removed from a tax on company profits. For example, before paying taxes, a company could deduct interest on loans taken out to finance purchases of fixed and working capital and investments in a wide range of assets. For many firms, the effect of these provisions was essentially to abolish the corporation tax. Of the 20 leading U.K. industrial companies in 1977 (based on capital employed) only 8 made tax payments. ICI, one of the eight with a tax burden, paid £23 million on £540 in profits, an effective rate of 4.3% (Kay and King 1978, p. 199). While it is hard to argue that high corporate tax rates provided a significant disinducement to industrial investment, it is possible that personal marginal tax rates, which in some periods reached very high levels, had negative incentive effects for managers and enterpreneurs.

An alternative set of explanations for Britain's slow growth, rather than focusing on macroeconomic factors, argues that the same problem of ill-suited institutions and difficulty coordinating their reform that hampered growth in the late Victorian and

Edwardian periods also have held back the British economy after World War II (Elbaum and Lazonick 1986). Unlike defeated Germany and Japan, where the war discredited existing institutions and policymakers, Britain enjoyed no exceptional postwar window for change. The inheritance of undersized firms remained. The dominance of the stock market and lack of banks interested in industrial lending encouraged managers to think in short horizons and heightened their sensitivity to financial-market conditions; long-term investment was thereby made more difficult. The fragmentation of the union movement made it impossible to secure agreements to coordinate the changes in work rules necessary for the adoption of new technologies; instead, ICI and others suffered a flood of strikes over pay and work rules. Lack of cooperation between unions, employers associations, and government made it impossible to adopt a German-style system of apprenticeship training.

A lengthy period of poor economic performance inevitably strains existing institutions, a factor that helped lead in the 1980s to the Thatcher revolution and a weakening of Britain's trade-union movement. The country now has a U.S.-style industrial-relations system to complement its U.S.-style financial markets. Only time will tell whether these two sets of arrangements are better suited to the technological and organizational imperatives of the 21st century than those of their German or Japanese counterparts.

Germany

When Germany regained de facto trade sovereignty in November 1949, its exports had already begun to revive. Monetary stabilization ignited a surge of economic growth. Trade liberalization prompted a restructuring of the German economy, with Germany relying on imports for goods in which it had no comparative advantage and shifting those resources toward export-oriented activities. As a founding member of the European Payments Union, Germany subscribed to the Code of Liberalization of the Organization of European Economic Cooperation, which established a schedule for the removal of quantitative barriers to trade. Germany joined the GATT in October 1951.

Under the new tariff adopted in 1951, duties averaged about 6% of the value of all imports and 18% on dutiable goods—low by historical standards. The chemical industry, wary of foreign competition given the disorganized state of the German economy, initially advocated import protection but reversed its stance once Germany's postwar recovery was underway. Nominal and effective rates of protection for chemicals circa 1958 were 13.0% and 11.4%, which is roughly where they remained into the early 1970s (Weiss 1992, p. 147). Thus, the chemical industry was less protected than other sectors.

Although a significant portion of Germany's prewar industrial capacity was lost to Poland and the creation of East Germany, those factories that remained had suffered little damage. Production in 1948 was only 58% of 1938 levels—reflecting rationing, binding price controls, and the effects of generalized shortage—but rose to 234% of the prewar level by 1958. Between 1953 and 1967, the industry grew at an average annual rate of more than 10%, more rapidly than German industry as a whole. While the 1970s were less favorable because of the two OPEC oil shocks, growth resumed vigorously thereafter.

Following the war, the Allies split IG Farben into four companies: Hoechst, Bayer, BASF, and Casella. The successor companies, however, maintained their prewar patterns of specialization and developed an informal understanding to avoid the duplication of expertise and development of excess capacity (Adams 1994, p. 106). The relationship between the chemical industry and government remained excellent until the 1970s, when environmentalism led to the adoption of stringent antipollution laws and industry complaints about regulatory burdens.

The United States

During and after World War II, the growth of the U.S. chemical industry was extraordinarily fast. The production of plastics and synthetic rubber expanded rapidly in response to World War II shortages and the interruption of foreign supplies. Petroleum and natural gas increasingly replaced coal, cellulose, fats, and oils as chemical-industry inputs, and the development of a petrochemical industry encouraged the entry of large petroleum companies into the production of chemicals, dramatically affecting competition. Real output of the chemical industry reached two and one-half times its 1937 level by 1950 and five times that level by 1959. The chemical industry came to occupy an increasingly prominent position in the American economy. While real GNP rose 221% between 1947 and 1968, shipments of chemicals increased ninefold. The annual rate of growth was 10% for the chemical industry versus $4\frac{1}{2}$% for the economy as a whole.

United States tariffs on chemical imports were cut as a result of the Reciprocal Trade Agreements Act, which authorized the president to reduce duties by up to 50%. By 1945 a 50% reduction for most products had been achieved, and Congress adopted legislation allowing for a further 50% cut. After a slow start in the first three GATT rounds (Geneva in 1947, Annecy in 1949, and Torquay in 1950–51), subsequent GATT negotiations cut U.S. tariffs substantially. As Backman (1970, p. 9) writes, "That tariffs have not permitted American producers to avoid intense competition is evident from the experience with plastics, resins, and synthetic fibers where appreciable foreign competition has been felt. For many products (chlorinated solvents) tariffs have not been restrictive and imports have affected both prices and volume of domestic producers." This is not to say that protection was inconsequential. A study cited by Backman estimates that without tariffs, imports of organic chemicals would have captured as much as 10% of the market instead of the 2% actually achieved prior to the Kennedy Round in 1962.

It does not appear that much preferential tax treatment was afforded the chemical industry. Firms extracting oil, gas, and other minerals were allowed special depletion allowances, which were justified by analogy with the depreciation allowance for capital consumption allowed to manufacturing firms. These allowances had first been permitted in 1918.

Japan

Following World War II, the U.S. Occupation Authorities gave top priority to the resumption of fertilizer production. Ammonium sulfate production matched prewar levels in 1949 and doubled those levels by 1955. The output of soda ash and dyestuffs recovered more slowly, matching prewar levels in 1953 to 1954. A major development in the post-World War II Japanese chemical industry was the emergence of new organic chemicals, particularly synthetic resins and polymers that served as inputs into the rapidly expanding plastics and synthetic-textile industries. The growth of the organic-chemical industry benefited from the development of oil refining, from which production of organic chemicals obtained inputs. By 1961, organics accounted for over one-half the value of the output of Japanese chemical firms.

The Japanese forged the institutions of their "high-growth system" in the years after World War II. Their technique consisted of measures to nurture selected industries, commercialize the products of the favored sectors, and regulate competition. Tools at the disposal of the bureaucrats included control of foreign exchange and imports of technology, preferential financing, tax breaks, protection from foreign competition, and authority to order the creation of cartels and bank-based industrial conglomerates (World Bank 1992).

The petrochemical industry developed in this context. A "Petrochemical Nurturing Policy" was adopted at an MITI ministerial conference in 1955. Foreign currency allocations were authorized, and funding was provided by the Japan Development Bank. Licenses were granted for imports of new chemical technologies. The petrochemical industry was deemed "strategic," allowing investors accelerated depreciation on their investments. It was allocated land to build its installations at a nominal cost, exempted from custom duties on imported catalytic agents and special machinery, and refunded duties on refined petroleum products used as raw materials. MITI created a Petrochemical Cooperation Discussion Group to regulate competition and coordinate investment.

However, Japan's tax policy did not particularly favor the chemical industry. A 1949 commission headed by Carl Shoup, an economics professor at Columbia University, was organized to reform Japan's tax system. The commisson recommended converting the system from one based on consumption taxes to one based on income taxes. The most important measures affecting the chemical industry were special deductions allowed firms in strategic sectors (including chemicals), under which up to 50% of a firm's export income (raised to 80% in 1955) was exempt from taxes. Royalties paid to license foreign technology were tax deductible. In addition, the government promoted a system of tax-free reserves for losses caused by price fluctuations, breach of contract, abnormal hazards, bad debts, pollution control, and special repairs. In 1967, tax credits were adopted for R&D expenditures. In 1970, special depreciation allowances accounted for 8.4% of total depreciation in Japan's chemical industry. In contrast, special measures accounted for 29.1% and 22% of the depreciation taken by the construction and steel industries, respectively (Pechman and Kaizuka 1976, table 4-9).

The Role of Macroeconomic and Trade Policies

The most striking aspect of the growth of the chemical industry since 1970 has been the shift in production away from the four countries that are the focus of this book and toward producers in Asia and Latin America. In 1970, the eight leading national producers of chemicals were the four countries discussed throughout this chapter, together with Belgium, France, the Netherlands, and Italy. In the quarter century after 1970, in none of these eight nations did chemical production more than double. Over that time, chemical industry output has grown tenfold outside the nine countries that started the period as the leading producers. The output growth in the industrialized countries has been achieved without any net increase in employment—indeed, with a net reduction.

Figure 9-3 provides a crude measure of productivity growth for workers in the chemical industry for three subperiods: the 1970s, the 1980s, and the first half of the 1990s. One feature that stands out is the relatively low rates of productivity growth in

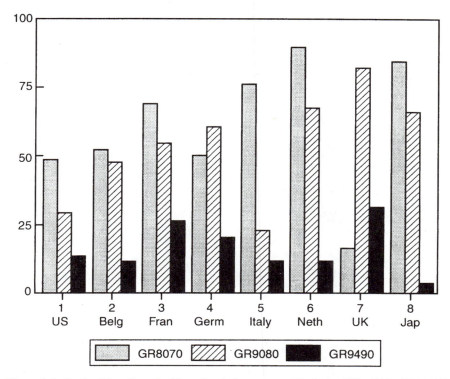

Figure 9-3. Productivity Growth (Growth of Output per Worker), 1970–1980, 1980–1990, 1990–1994.

the United States in each of the three subperiods. A second interesting feature is that two countries show faster productivity growth in the 1980s than in the 1970s. In the United Kingdom, the effects of the Thatcher revolution in removing restrictive labor practices helped the British chemical industry move from the slowest productivity growth in the 1970s to the fastest in the 1980s. The other case—Germany—is one of unexceptional productivity growth in the 1980s following a disappointing record in the 1970s.

The competitiveness of the chemical industry can be viewed from two angles: It can be compared to other domestic sectors that produce substitutes for chemical products (the products of the steel industry, which can be substituted for plastic in the construction of automobile bodies); or to foreign producers of chemical products. In the United States, for example, unit labor costs have risen more rapidly in chemicals than in U.S. manufacturing as a whole. While unit labor costs in all manufacturing increased by 60% between 1970 and 1994, those in chemicals rose by 84%. Much of the difference was concentrated in the first half of the 1980s. This was a period when productivity growth in chemicals did not keep pace with the rest of the manufacturing sector; between 1980 and 1985, labor productivity rose by 24% in all manufacturing but only one-half as rapidly in chemicals and allied products. However, over this same time, unit labor costs in all manufacturing rose by 6% but those in chemicals increased four times

TABLE 9-3. Unit Labor Cost Levels in Chemical and Allied Products and in Total Manufacturing: 1970–1990 (United States = 100)

Country Pair	1970	1975	1980	1985	1990
Germany-United States					
Chemicals	59.6	96.4	106.9	80.2	160.0
All Manufacturing	59.7	95.2	112.3	70.1	146.1
Japan-United States					
Chemicals	52.1	77.4	87.1	81.6	142.5
All Manufacturing	48.1	79.5	78.6	65.5	99.5
United Kingdom-United States					
Chemicals	na	na	115.3	72.4	111.4
All Manufacturing	75.3	99.8	146.1	87.6	137.1
France-United States					
Chemicals	68.8	115.3	122.2	85.3	129.3
All Manufacturing	66.0	110.6	123.1	77.8	129.1

Note: *na* = not available.
Source: van Ark (1995, p. 69).

as rapidly. While product innovation and quality improvement may be relatively rapid in the chemical industry, thus causing conventional indices of unit labor costs to overstate the cost disadvantage of the sector, the data still suggest a secular problem of cost competitiveness on the domestic side.

By comparison, the international terms of trade—chemical export prices relative to import prices—show little trend from the 1980s into the first half of the 1990s. In a way, this is surprising, since the real and nominal effective exchange rates of the dollar strengthened dramatically from 1983 to 1985 before weakening from 1985 to 1987. Only a mild (2%) rise occurred in the relative price of U.S. chemical exports over the first period and no decline in the second. While the relative prices of exports and imports fluctuated more widely for particular chemical products, here too there is little evidence that this variation was driven by movements in the exchange rate.

Table 9-3 presents evidence comparing unit labor costs across industries, for manufacturing as a whole and for the chemical industry in particular. For total manufacturing, there is a secular improvement in U.S. cost competitivness, interrupted by the dollar's appreciation in the first half of the 1980s. Interesting divergences between total manufacturing and chemicals are evident: For example, after moving with total manufacturing through 1980, unit labor costs in German and Japanese chemical production rise relatively rapidly thereafter. The opposite trend is evident in the United Kingdom, whose chemical industry improved its competitive position even more rapidly than did British manufacturing as a whole in the 1980s.

Currency swings can readily be shown to have significant effects on the level of production, employment, and profits in the chemical industry.[27] For the American chemical industry over the period from 1975 to 1994, for example, a 10% rise in the real

---

[27] Specific statistical results are reported in the appendix to this paper. These are a series of regression results in which U.S. chemical-industry output (overall and in various sectors), employment, and profits are the dependent variable, to be explained by a set of variables including industrial production (as a measure of cyclical conditions), interest rates (Moody's AAA rate), the Baa-Aaa spread (a measure of credit availability), the real exchange rate (proxied by the real effective exchange rate measured in terms of relative unit labor costs), and time.

effective exchange rate, measured in terms of relative unit labor costs, leads to a 17% decline in chemical-industry output. Within the sectors of the chemical industry basic chemicals, industrial organic chemicals, drugs and medicines, and inorganics as a whole are even more strongly affected by the exchange rate. Synthetic materials and plastics, on the other hand, show relatively little responsiveness. In terms of employment, a 10% rise in the real exchange rate leads to a decline of 0.12% in chemical-production employment, essentially the same percentage decline in employment in manufacturing production, and about three times as large as the decline in total manufacturing employment. In terms of industry profitability, chemical-industry profits hovered in the range of 12% to 14% of total profits for all manufacturing industries from 1959 to 1973, fell off to between 8% and 10% of all manufacturing industry profits between 1980 and 1985, and then bounced back to between 16% and 18% of all industry profits between 1989 and 1994. This pattern matches the pattern of the dollar exchange rate; the fall in profits with the strong dollar of the early 1980s and the rebound in profits once the dollar began to fall in 1985. Interestingly, it appears that real currency appreciation cuts U.S. chemical industry profits more dramatically than manufacturing profits as a whole. Similarly, interest rates have a stronger impact on profits in the chemical industry than in U.S. manufacturing as a whole.

In terms of overall levels of investment, of the four countries considered here, Japan has clearly invested at the highest rate: between about 28% and 32% of GDP in the years since 1975. Investment rates in Germany, the United States, and the United Kingdom cluster together, ranging from 16% to 21% of GDP over the last few decades, with Germany at the top of that range, and America and Britain trading places at the bottom. Figure 9-4 shows the share of investment in each country accounted for by their respective chemical industries. Taken together, the figures suggest that the same factors that account for high overall levels of investment in some countries (Japan) and low ones in others (the United Kingdom) can also account for differing investment rates in the different national chemical industries.

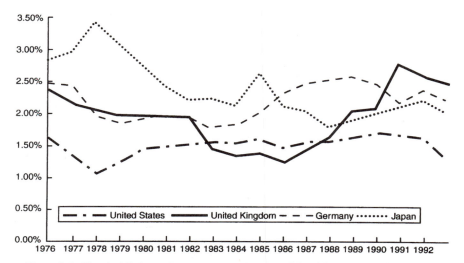

Figure 9-4. Chemical Industry Investment as a Percent of Total Investment, 1976–1993.

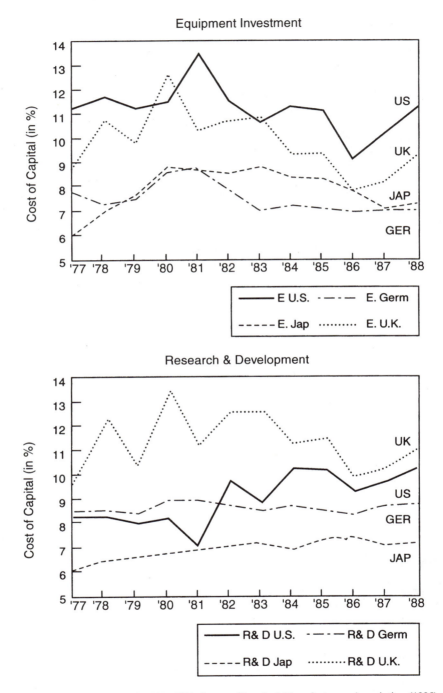

Figure 9-5. Cost of Capital, 1977–1988. *Source:* Chemical Manufacturers Association (1995).

An obvious place to start in analyzing the determinants of investment rates is with the cost of capital. The cost of capital is a composite that depends on, among other things, the cost of external funds, the taxation of retained earnings, and tax deductions for depreciation and other investment-related expenses. Because depreciation and tax rules differ by type of project, so too will the cost of capital. Figure 9-5 therefore distinguishes investment in equipment with a 20-year physical life from R&D with a 10-year payoff lag for each of the four countries. For both categories of investment, the cost of capital is consistently higher in the United States and the United Kingdom than in Germany and Japan, which is consistent with secular differences in national investment rates. These differentials are relatively stable, although Britain's cost disadvantage diminishes over the years from 1977 to 1988 (which coincides with the Thatcher revolution), while the cost of investing in R&D in the United States appears to have risen over time. Of the two relatively high-cost-of-capital countries, the United States followed more favorable policies toward R&D, while U.K. policy has been biased toward physical investment. A similar comparison emerges for the two low-cost-of-capital countries: German policy has been more favorable toward R&D, Japanese policy toward equipment and machinery.[28]

On what do these variations in the cost of capital depend? The cost of capital is a weighted sum of the cost of debt and equity. McCauley and Zimmer (1989), from whom the time series in Figure 9-5 are drawn, find that the cost of debt varies only modestly across countries; indeed, it was no lower in Japan than in the United States for much of the 1980s.[29] The cost of equity finance is higher and varies more across countries; hence, those countries relying most heavily on debt finance have historically had a cost-of-capital advantage. In the 1980s, the ratios of debt to equity finance were highest in Germany, somewhat lower in Japan, and quite significantly lower in the United States and the United Kingdom. These ratios reflect well-known differences in financial structure in the English speaking countries, with firms in the United States and the United Kingdom relying relatively heavily on stock markets for external finance, while German and Japanese firms raise debt finance from their countries' banks. Thus, it would appear that the financial structures alluded to in previous sections continued to cast a shadow over industrial performance in the post-World War II period.[30] Despite the fact that Japanese firms relied somewhat more on equity finance than did their German counterparts in the 1980s, the former enjoyed a cost-of-capital advantage owing to Japan's extraordinarily low cost of equity finance.

These estimates measure the cost of investing but do not explain it. To explain differences in the cost, it is necessary to account for international variations in the

---

[28]This idea is consistent with studies of postwar Japanese growth emphasizing the importance of factor accumulation, in comparison with the total-factor-productivity-driven increase in output in the postwar United States.

[29]The authors start with a weighted average of rates on bank and bond debt, adjusted for differing propensities for firms to hold liquid assets, and subtract the actual inflation rate to produce a standard real interest-rate measure. By this measure, the United States faced the highest real rates in the first half of the 1980s (the period of the Volcker disinflation), West Germany the lowest. Correcting for tax deductions on corporate interest payments to place costs on a real after-tax basis, eliminates much of Germany's advantage: The cost of debt is lower than in the United Kingdom in the second half of the 1980s but not the first. These adjustments also tend to confirm the picture of relatively favorable tax treatment in the United States.

[30]Recent literature reflecting economic problems in both Germany and Japan has suggested that bank-centered financial systems are a mixed blessing. While the efficiency of their monitoring works to reduce the cost of external funds, the degree of competition in the provision of finance is less than in stock-market-centered systems, making it difficult for start-up firms without a track record to obtain external finance.

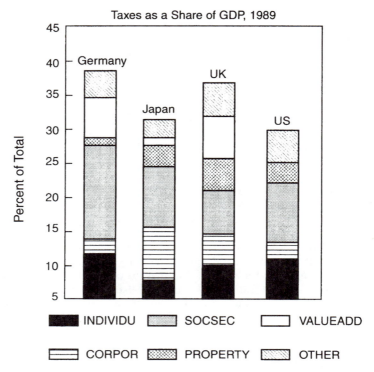

Figure 9-6. Taxes as a Share of GDP, 1989.

interest rates, price/earnings ratios, depreciation rules, and inventory holding rates that enter into the calculation. Tax policy, while a logical place to start, does not take us very far. Figure 9-6 shows that Japan, which has the lowest cost of capital, in fact raises the largest share of national income via corporate taxes. Germany, the other low cost-of-capital country, has the highest tax burden overall. McCauley and Zimmer (1989) similarly conclude that tax rates on interest, dividends, and capital gains vary too little across countries to explain much of observed differences in the cost of capital.

Since taxes and accounting conventions account for little of observed international differences in the cost of capital, it follows that the relatively low cost of capital in Japan and Germany has as its counterpart low rates of return to domestic savers in those countries. Conversely, the high cost of capital in the United States means higher rates of return to savers in the United States. To put the same point another way, because Japanese corporations pay so little to borrow, Japanese savers, who do the lending, necessarily receive low rates of return. The conventional argument is that relatively high savings rates in Japan and Germany drive down the rate of return. From the 1960s through the 1980s, conventionally measured household savings rates were three times as high in Japan as in the United States (and twice as high in Germany).[31]

---

[31] Boskin and Roberts (1988), among others, suggest that such measures overstate international differences, because they treat spending on consumer durables as consumption rather than savings. Adjustment for the relatively high propensity of U.S. households to spend on consumer durables attenuates, but does not eliminate, the gap.

For high domestic savings to drive down interest rates (and provide corresponding stimulus to domestic investment), capital cannot flow freely across national borders; otherwise, funds would flow from Japan to the United States to the point that interest rates in the two economies were equalized (and domestic savings rates had no impact on domestic investment). While international capital mobility was indeed an economic fact of life in the 1980s, the degree to which this mechanism in fact operated was quite modest.

There remains a strong correlation between domestic saving and investment, indicating that foreign lending and borrowing was not sufficient in magnitude to neutralize the effects of domestic-savings rates (operating through the cost of capital) on capital formation.

Explaining these savings-investment correlations and observed differences in national savings rates is beyond the scope of this chapter.[32] The literature on savings-investment correlations indicates that while international capital mobility is rising, it is far from complete. Moreover, governments pursue policies to limit the magnitude of net capital flows; for example, Japan's net export of capital is arithmetically equivalent to its current account surplus, a delicate variable politically. For explaining higher savings rates in Japan and Germany than in the United States in the postwar period, the evidence suggests that the opportunities for rapid growth afforded by catch-up, Japan and Germany's relatively strong fiscal positions, and—in the case of Japan—demographic structure have been particularly important (Masson, Bayoumi, and Semiei 1995). Masson, Bayoumi, and Semiei find that an increase in government saving crowds out only one-half as much private saving; thus, a stronger fiscal balance can have a significant effect on national savings rates. However, budget deficits in our countries have rarely exceeded a few percentage points of GDP in the postwar period, and assuming that their impact on national savings is at most one-half their size, fiscal policies can account for only a fraction of observed differences in savings rates across countries. Masson et al. similarly find that the real interest rate has a positive and significant impact on savings in the industrial countries. However, its overall effect is small, and policymakers' ability to manipulate real rates via monetary policy is limited.

These conclusions are not meant to dismiss macroeconomic and tax policies as determinants of savings rates. Rather, the point is that their impact is small by the standards of the international variations in savings ratios. The conclusion to which the literature points is that other institutional and demographic characteristics of countries beyond their monetary, fiscal, and tax policies must be invoked to provide a full explanation of savings rates.

CONCLUSIONS

This paper has considered the roles of monetary, fiscal, and trade policies in the development of the chemical industries of the United States, the United Kingdom, Germany, and Japan. The contrasting experiences of the four nations' chemical industries are explicable, first and foremost, in terms of the economy-wide determinants of productivity and output growth. The rapid growth of the Japanese industry reflects the operation of the same factors that account for the fast growth of the Japanese economy as a whole; the same statement applies to Britain, substituting the word "slow"

[32] On savings-investment correlations, see Feldstein and Horioka (1980) and Bayoumi (1990). On Japanese savings rates in particular, see Horioka (1993).

for "fast." Monetary, fiscal, and trade policies are only a few of the many influences constituting the framework for long-term growth.

Prior to 1913, a similar monetary and fiscal framework—that associated with the gold standard—prevailed in each of the countries with which we are concerned; it is hard to see how that common framework could have had very different effects on the different national chemical industries involved. A more important source of differences was commercial policy, which was more restrictive in Germany and the United States than in the United Kingdom and Japan. Protection against imports was important for helping the late-starting U.S. and German chemical industries get on their feet. In some cases, there is reason to think that temporary import protection had sustained effects: By insulating large domestic markets from import competition, it encouraged investors to sink the costs of establishing modern-mass-production facilities capable of exploiting economies of scale and scope and of competing internationally. The point is not that protection is good policy generally; it stimulated the development of the chemical industry in countries like the United States only because electrochemical companies were well positioned to exploit new technologies, reflecting the influence of, inter alia, their country's exceptional resource availability—in particular, abundant electrical power.

Between the wars, macroeconomic policies diverged sharply. Policy was particularly unstable in Germany, for political reasons. The German chemical industry was neither hurt nor helped on balance; the cheap credit it enjoyed in the first half of the 1920s was offset by the burden of high interest rates that were a subsequent legacy of hyperinflation. The different national industries fared very differently in the 1930s. Those countries which were quickest to abandon the gold standard and monetary links to a deflationary global environment most smoothly recovered from the Depression and restored the vitality of domestic industry. Thus, the two interwar decades vividly illustrate the potential for macroeconomic policies to shape the development of industry.

After World War II, the growth of national chemical industries again moved in tandem with rates of growth of national economies. It is difficult to isolate the role of macroeconomic policy in these secular rates of growth. This paper has highlighted the roles of exchange rates and investment rates in the competitiveness of national industries. In particular, it has presented evidence that the U.S. industry has seen its position enhanced by the competitive level at which U.S. monetary and fiscal policies have kept the dollar exchange rate for much of the last two decades. It is harder to link international differences in investment rates, which favor the German and Japanese industries, to monetary and fiscal policies. The analysis here points to financial structure (firms' reliance on banks, versus equity markets, for external funds) and factors affecting national savings propensities as determinants of those variations. While a sound and stable policy framework can encourage saving and help the financial system do its job, neither variable is easily manipulated by the kind of macroeconomic policies that are the focus of this chapter.

APPENDIX: STATISTICAL FINDINGS ON THE CHEMICAL INDUSTRY

This appendix presents two sets of statistical findings on the chemical industry. The first three tables show the impact of currency movements, as well as other factors, on output, employment, and profits in the chemical industry. The second three tables analyze the time-series interaction of chemical output and gross domestic product in our four countries and three periods.

TABLE 9-4. Impact of the Real Exchange Rate on Chemical Production: 1975–1994 (Annual Data)

Variable	Total	Basic Chemicals	Industrial Organics	Drugs and Medicines	Inorganics	Synthetics	Plastics
Constant	65.03	133.74	85.56	86.61	187.18	11.71	−20.00
	(9.43)	(28.13)	(34.57)	(15.72)	(39.75)	(24.27)	(21.23)
Industrial production	0.36	−0.27	0.38	−0.24	−1.03	0.76	0.91
	(0.14)	(0.40)	(0.50)	(0.23)	(0.57)	(0.35)	(0.30)
Interest rate	1.66	2.76	3.11	1.08	3.91	0.90	−0.68
	(0.33)	(0.99)	(1.22)	(0.55)	(1.40)	(0.85)	(0.75)
Interest-rate spread	−9.34	−16.13	−19.95	−5.27	−23.62	−10.27	−3.06
	(1.73)	(5.15)	(6.33)	(2.88)	(7.28)	(4.45)	(3.88)
Real exchange rate	−0.17	−0.23	−0.30	−0.12	−0.29	−0.05	−0.06
	(0.02)	(0.07)	(0.08)	(0.04)	(0.10)	(0.06)	(0.05)
Time	0.55	−0.01	1.13	1.35	1.87	1.71	4.09
	(0.51)	(1.52)	(1.86)	(0.85)	(2.14)	(1.31)	(1.14)
Time squared	0.02	0.05	−0.03	0.09	0.04	−0.03	−0.09
	(0.01)	(0.04)	(0.05)	(0.02)	(0.05)	(0.03)	(0.03)
R^2	0.99	0.94	0.94	0.99	0.89	0.98	0.99
DW	2.30	2.94	1.78	1.75	2.00	2.61	2.60
Number of observations	20	20	20	20	20	20	20

Notes: Standard errors are in parentheses.

Source: See text.

TABLE 9-5. Regression Results: Determinants of Employment (Quarterly Data), 1977: 2–1994: 4

			Independent Variables (Length of Distributed Lag)					
	Constant	Trend	Unemployment (4)	Real Exchange Rate (6)	Real Energy Price (4)	Real Hourly Earnings (8)	ρ	R^2
Total manufacturing	9.88 (81.21)	−0.0007 (−0.90)	−0.03 (78.90)	−0.0004 (2.25)	0.25 (7.14)	0.01 (1.74)	0.53	0.95
Manufacturing production	9.47 (12.09)	−0.0018 (−1.10)	−0.03 (39.90)	−0.0013 (5.29)	0.24 (3.53)	0.03 (0.19)	0.73	0.88
Chemical production	6.89 (42.77)	−0.0014 (−6.60)	−0.03 (27.08)	−0.0012 (17.62)	0.19 (7.40)	−0.03 (3.25)	0.59	0.92

Note: Dependent variable is the log of the number of employees. Coefficients on the real exchange rate are multiplied by 1,000 to better display significant digits. Reported are the sums of the coefficients of the current and lagged variables. The number of lags is indicated in parentheses under the variable name. The current value of real hourly earnings is not included. In parentheses under the coefficient sum is the F statistic for the null hypothesis that the sum is equal to zero. The 5% critical value is 4.0. The t-statistic is reported for the constant and trend variables. With 44 degrees of freedom, the 5% critical value is 1.68. All equations are estimated using a Cohrane-Orcutt correction.

Source: See text.

Table 9-4 presents regressions, showing the impact of currency fluctuations and other key variables on output of the U.S. chemical industry, together with impact on sectors of the industry. The explanatory variables include industrial production (as a measure of cyclical conditions), interest rates (Moody's AAA rate), the Baa-Aaa spread (a measure of credit availability), the real exchange rate (proxied by the real effective exchange rate measured in terms of relative unit labor costs), and time. All variables enter with their expected coefficients, as discussed earlier in the text, except for the level of the interest rate, which enters with an anomalous positive sign.

Table 9-5 presents a set of three regressions. The dependent variables are employment in chemical production, in manufacturing production, and in total manufacturing. The explanatory variables are a constant, the overall unemployment rate (as a measure of cyclical conditions), the real exchange rate, real energy prices, real hourly earnings, and a time trend. In Table 9-6, the dependent variable is chemical-industry profits as a share of manufacturing industry profits.

The remainder of this appendix analyzes the time-series interaction of chemical output and gross domestic product in our four countries and three periods. The goal is to marshall evidence on how much of the growth of chemical production derived from the growth of the economy as a whole, and in which countries and periods chemical production acted as a leading sector, stimulating growth economy-wide.

For the post-World War I period, we have time series for all four of our countries. For GDP, these come from Maddison (1982) and the Penn World Tables (Summers

TABLE 9-6. The Effect of Exchange Rates and Interest Rates on Profits: 1975–1993 (Annual Data)

Variable	Total Manufacturing	Chemical Industry Relative to Total Manufacturing
Constant	−110.45	34.05
	(74.27)	(19.71)
Industrial production	3.26	−0.21
	(1.09)	(0.28)
Interest rates	6.38	1.27
	(2.21)	(0.58)
Baa/Aaa spread	−15.64	2.14
	(11.91)	(3.16)
Real exchange rate	−0.13	−0.09
	(0.17)	(0.04)
Time	−15.03	2.06
	(4.97)	(1.32)
Time squared	0.34	−0.04
	(0.13)	(0.03)
R^2	0.89	0.84
DW	1.53	1.61
Number of observations	19	19

Note: Standard errors are in parentheses. For clarity, coefficients and standard errors in the second column have been multiplied by 100.

Source: See text.

and Heston). For chemical output they are obtained from Mitchell (1989) for the United Kingdom, Stanford Research Institute (various issues) for the United States, Dresner Bank Statistiche Reihen (1994) and Hoffmann (1965) for Germany, and Japan Statistical Association (1987) for Japan. For 1870 to 1913, annual time series are limited to Britain and Germany because of the negligible importance of U.S. and Japanese chemical production toward the beginning of the period.

We tested first for the cointegration of chemical output and GDP. (Intuitively, the null hypothesis is that there is a long-run tendency for the two variables to move together.) For most countries and periods it is impossible to reject the null of non-cointegration. The exceptions are the United Kingdom for 1920 to 1990 and

TABLE 9-7. Pre-war Chemical Industry (1870–1913)

A. UNIT ROOT TESTS

	t-*statistic*	p-*value*
Germany GDP	0.587	0.998
U.K. GDP	−2.170	0.530
Germany Chemical Production	3.025	0.999
U.K. Chemical Production	−1.265	0.898

B. COINTEGRATION TESTS

	t-*statistic*	p-*value*
Germany	−1.745	0.877
U.K.	−1.551	0.917

C. ERROR CORRECTION MODEL (*t*-statistic in parentheses)

i. Dependent Variable: Chemical Industry Growth

	Lagged GNP Growth	*Lagged Ratio*
Germany	0.593 (.975)	0.098 (4.561)
U.K.	0.153 (.154)	0.081 (2.608)

ii. Dependent Variable: GNP Growth

	Lagged Chemical Industry Growth	*Lagged Ratio*
Germany	0.0717 (1.529)	0.0089 (1.637)
U.K.	0.0131 (0.226)	0.0154 (2.143)

D. GRANGER-CAUSALITY TESTS (*t*-statistic in parentheses)

i. Dependent Variable: Chemical Industry Growth

	Lagged GNP Growth
Germany	1.500 (2.115)
U.K.	0.514 (0.488)

ii. Dependent Variable: GNP Growth

	Lagged Chemical Industry Growth
Germany	0.1315 (4.369)
U.K.	0.0469 (0.804)

Source: See text.

TABLE 9-8. Interwar Chemical Industry (1920–1938)*

A. UNIT ROOT TESTS

	t-*statistic*	p-*value*
Germany GDP	-0.764	0.919
U.K. GDP	-2.171	0.462
Germany Chemical	-0.163	0.946
U.K. Chemical	-3.995	0.028
U.S. GDP	-1.456	0.762
U.S. Chemical	-1.491	0.683
Japan GDP	-0.795	0.915
Japan Chemical	0.052	0.953

B. COINTEGRATION TESTS

	t-*statistic*	p-*value*
Germany	-4.755	0.039
U.K.	-3.603	0.156
U.S.	-3.253	0.264
Japan	-2.485	0.510

C. ERROR CORRECTION MODEL (*t*-statistic in parentheses)

i. Dependent Variable: Chemical Industry Growth

	Lagged GNP Growth	*Lagged Ratio*
Germany	1.471 (1.69)	-0.124 (1.01)
U.K.	-0.318 (0.96)	-0.037 (-2.54)
U.S.	0.172 (0.77)	-0.015 (-0.96)
Japan	0.176 (0.55)	-0.036 (-1.03)

ii. Dependent Variable: GNP Growth

	Lagged Chemical Industry Growth	*Lagged Ratio*
Germany	0.178 (0.69)	-0.047 (-0.75)
U.K.	0.324 (0.84)	-0.048 (-2.71)
U.S.	-0.073 (-0.03)	0.098 (0.22)
Japan	0.406 (1.08)	-0.021 (-0.50)

D. GRANGER-CAUSALITY TESTS (*t*-statistic in parentheses)

i. Dependent Variable: Chemical Industry Growth

	Lagged GNP Growth
Germany	1.732 (2.047)
U.K.	0.050 (0.144)
U.S.	0.040 (0.236)
Japan	0.223 (0.700)

ii. Dependent Variable: GNP Growth

	Lagged Chemical Industry Growth
Germany	0.150 (0.601)
U.K.	0.087 (0.195)
U.S.	-0.461 (0.326)
Japan	0.516 (1.770)

Note: Figures for Japan from 1925–38.

Source: See text.

TABLE 9-9. Postwar Chemical Industry (1946–1990*)

A. UNIT ROOT TESTS

	t-*statistic*	p-*value*
Germany GDP	−3.108	0.117
U.K. GDP	−1.719	0.762
Germany Chemical	−2.234	0.492
U.K. Chemical	−2.387	0.400
U.S. GDP	−4.748	0.002
U.S. Chemical	−2.049	0.596
Japan GDP	−0.722	0.969
Japan Chemical	−1.814	0.712

B. COINTEGRATION TESTS

	t-*statistic*	p-*value*
Germany	−2.920	0.573
U.K.	−3.101	0.320
U.S.	−1.061	0.969
Japan	−4.545	0.015

C. ERROR CORRECTION MODEL (*t*-statistic in parentheses)

i. Dependent Variable: Chemical Industry Growth

	Lagged GNP Growth	*Lagged Ratio*
Germany	2.047 (3.18)	0.039 (1.56)
U.K.	1.351 (3.09)	−0.018 (0.36)
U.S.	0.928 (1.40)	0.076 (3.20)
Japan	1.514 (3.89)	0.017 (0.44)

ii. Dependent Variable: GNP Growth

	Lagged Chemical Industry Growth	*Lagged Ratio*
Germany	−0.124 (1.85)	0.016 (1.89)
U.K.	−0.027 (0.26)	0.017 (0.72)
U.S.	−0.022 (0.26)	0.015 (1.80)
Japan	0.045 (0.57)	0.027 (1.64)

D. GRANGER-CAUSALITY TESTS (*t*-statistic in parentheses)

i. Dependent Variable: Chemical Industry Growth

	Lagged GNP Growth
Germany	2.275 (3.564)
U.K.	1.326 (3.120)
U.S.	0.741 (1.009)
Japan	1.572 (4.344)

ii. Dependent Variable: GNP Growth

	Lagged Chemical Industry Growth
Germany	−0.099 (1.448)
U.K.	−0.023 (0.209)
U.S.	0.036 (0.455)
Japan	0.058 (0.716)

Note: Figures for Japan from 1946–85.

Source: See text.

Germany for the interwar period. That the series are not cointegrated is unsurprising: Chemical output, rather than moving with GDP, has tended to grow more quickly and to follow a different pattern throughout the sample period.

The lack of cointegration confirms the appropriateness of the Granger causality tests. We implement them using vector autoregressions with one lag. Since these tests have low power to confirm the robustness of our results, we also estimate the error-correction model that is appropriate in the case of cointegration. The Granger tests are reported in Tables 9-7 to 9-9. There is strong evidence, for the post-World War II period, that GDP growth "causes" the growth of chemical output in the sense of Granger. (For further discussion of the interpretation of Granger causality, see the chapter by Ralph Landau in this work.) This suggests that much of the rapid growth of the industry in the quarter century following World War II, and presumably also its slower growth subsequently, are attributable to economy-wide trends. The link from GDP growth to the growth of chemical output is weaker in the interwar years: Only for Germany is the relationship statistically significant at standard confidence levels. This is consistent with the view that the course of chemical production in the interwar period was unique in that it was driven by sector-specific technological progress and that the industry received special stimulus from government subsidies and protection. The results are similar for the 45 years before World War I: Again, there is causality running from GDP growth to the growth of the chemical industry in Germany, but not the United Kingdom. An interpretation is that the German chemical industry, which produced heavily for the domestic market, benefited from the rapid growth of the German economy, but that the British industry, which produced heavily for export markets, was harmed only marginally by the slow growth of domestic demand.

Overall, these results are consistent with the view that the effects on the chemical industry of fluctuations in aggregate economic growth grew stronger with time.

The evidence of causality running from chemical production to GDP—that is, for the chemical industry "acting as a leading sector"—is different. There is no indication of this leading-sector effect for either Germany or the United Kingdom prior to World War I (plausibly, given that the chemical industry was small relative to the economy as a whole). For the interwar and post-World War II periods, only for Japan is there evidence that chemical production had strong multiplier effects—that it significantly stimulated the growth of GDP. For Japan, we reject the null of no effect at the 90% level for the interwar years and at the 95% level after World War II.

The results of estimating the error-correction models are broadly consistent with these findings. We focus on the results for Germany for the interwar period and the United Kingdom for the interwar and post-World War II years, where cointegration tests suggest that this may be the more appropriate specification.[33] The Granger tests suggest a significant effect of GDP growth on German chemical output between the wars; the error-correction model confirms this result at the 90% confidence level. Again, only for Germany is there evidence that movements in GDP led to subsequent movements in chemical production in the interwar years. For the United Kingdom, the results confirm the absence of causality running in either direction between the wars (although the error-correction term suggests some short-term response of the industry to fluctuations in the ratio of chemical output to GDP). For the postwar period the

---

[33] Prior to World War I, when we decisively reject cointegration, the error-correction term (the coefficient on the ratio of chemical output to GDP) often enters with the "wrong" sign (positively when chemical output is the dependent variable), confirming that this is an inappropriate specification, given that the two series are not cointegrated.

results suggest that the predominant direction of causality was from GDP growth to chemical-industry growth, with minimal feedback, consistent with the results of the Granger tests.

REFERENCES

Adams, R. C. (1994), "Industry Culture, Public Policy and Competiveness: the U.S. and German Chemical Industries," *Science and Public Policy*, 21: 309–13.

Backman, Jules (1970), *The Economics of the Chemical Industry*, Manufacturing Chemists Association, New York, NY.

Bacon, Robert William and Walter Eltis (1976), *Britain's Economic Problem: Too Few Producers*, Macmillan, London, U.K.

Bayoumi, Tamim (1990), "Saving-Investment Correlations: Immobile Capital, Government Policy, or Endogenous Behavior?" *Staff Papers*, 37: 360–87.

Boskin, Michael and John M. Roberts (1988), "A Closer Look at Saving Rates in the United States and Japan," in John B. Shoven (ed.), *Government Policy Towards Industry in the United States and Japan*, Cambridge University Press, Cambridge, U.K., pp.121–43.

Bresciani-Turonni, Constantino (1937), *The Economics of Inflation*, Allen & Unwin, London, U.K.

Cairncross, Alec (1992), *The British Economy Since 1945*, Blackwell, Oxford, U.K.

Capie, Forrest (1983), *Depresssion and Protectionism: Britain Between the Wars*, George Allen & Unwin, London, U.K.

Chandler, Alfred D., Jr. (1990), *Scale and Scope: The Dynamics of Industrial Capitalism*, Harvard University Press, Cambridge, MA.

DeLong, J. Bradford (1991), "Did Morgan's Men Add Value?" in Peter Temin (ed.), *Inside the Business Enterprise: Historical Perspectives on the Use of Information*, University of Chicago Press, Chicago, IL, pp. 205-36.

Dresner Bank (1994), *Statistiche Reihen*, Dresner Bank, Frankfurt, Germany.

Easterly, William R., Michael Kremer, Lant Pritchett, and Lawrence Summers (1993), "Good Policy or Good Luck? Country Growth Performance and Temporary Shocks," unpublished manuscript, Country Economics Department, The World Bank.

Edelstein, Michael (1981), *Overseas Investment in the Age of High Imperialism*, Columbia University Press, New York, NY.

Eichengreen, Barry (1991), "Trends and Cycles in Foreign Lending," in Horst Siebert (ed.), *Capital Flows in the World Economy*, Mohr, Tubingen, Germany, pp. 2–28.

Eichengreen, Barry (1992), *Golden Fetters: The Gold Standard and the Great Depreison, 1919–1939*, Oxford University Press, New York, NY.

Eichengreen, Barry (1996), "Explaining Britain's Economic Performance: A Critical Note," *Economic Journal*, 106: 213–18.

Eichengreen, Barry and Jeffrey Sachs (1985), "Exchange Rates and Economic Recovery in the 1930s," *Journal of Economic History*, XLV, 925–46.

Elbaum, Bernard and William Lazonick, eds. (1986), *The Decline of the British Economy*, Clarendon Press, Oxford, U.K.

Feldman, Gerald D. (1993), *The Great Disorder: Politics, Economics, and Society in the German Inflation, 1914–1924*, Oxford University Press, New York, NY.

Feldstein, Martin and Charles Horioka (1980), "Domestic Saving and International Capital Flows," *Economi Journal*, 90: 314–28.

Fischer, Stanley (1991), "Growth, Macroeconomics and Development," *NBER Macroeconomics Annual*, pp. 329–363.

Gerschenkron, Alexander (1966), *Economic Backwardness in Historical Perspective,* Harvard University Press, Cambridge, MA.

Greiling, Walter (1952), *Funfundsiebzig Jahre Chemieverband: ein Beitrag zur Industriegeschichte und wirtschaftspolitischen Meinungsbildung in einer erzählenden Darstellung mit ausgewählten Dokumentenzitaten,* Verband der Chemischen Industrie, Frankfurt am Main, F.R.G.

Grossman, Richard S. (1993), "The Macroeconomic Consequences of Bank Failures under the National Banking System," *Explorations in Economic History:* 125–140.

Haber, Ludwig F. (1958), *The Chemical Industry During the Nineteenth Century: A Study of the Economic Aspects of Applied Chemistry in Europe and North America,* Clarendon Press, Oxford, U.K.

Haber, Ludwig F. (1971), *The Chemical Industry, 1900–1930: International Growth and Technological Change,* Clarendon Press, Oxford, U.K.

Hoffmann, Walther G. (1965), *Das Wachstum der deutschen Wirtschaft seit der Mitte des 19. Jahrhunderts,* Springer Verlag, Berlin, F.R.G.

Hohenberg, Paul M. (1967), *Chemicals in Western Europe, 1850–1914. An Economic Study of Technical Change,* Rand McNally, Chicago, IL.

Horioka, Charles (1993), "Saving in Japan," in Arnold Heertje (ed.), *World Savings: An International Survey,* Blackwell, Oxford, U.K., pp. 238–78.

Japan Statistical Association (1987), *Historical Statistics of Japan,* Nihon Tokei Kyokan, Tokyo, Japan.

Johnson, Chalmers (1982), *MITI and the Japanese Miracle: The Growth of Industrial Policy, 1925–1975,* Stanford University Press, Stanford, CT.

Kay, J.A. and M.A. King (1978), *The British Tax System,* Oxford University Press, Oxford, U.K.

Kennedy, Carol (1993), *ICI: The Company that Changed Our Lives,* PCP Publishing, London, U.K.

Kobayashi, Masaaki (1976), *Nihon Keieishi o Manabu* (The Study of the History of Japanese Enterprise Management), Yuhikaku, Tokyo, Japan.

Krugman, Paul (1994), "The Myth of Asia's Miracle," *Foreign Affairs* 73, 6, 62–78.

Lavington, Frederick (1921), *The English Capital Market,* Methuen, London, U.K.

Levine, Ross and David Renelt (1992), "A Sensitivity Analysis of Cross-Country Growth Regressions," *American Economic Review,* 82: 942–63.

Lockwood, William (1954), *Economic Development of Japan,* Princeton University Press, Princeton, NJ.

McCauley, Robert and Steven Zimmer (1989), "Explaining International Differences in the Cost of Capital," *Federal Reserve Bank of New York Quarterly Review,* 13 (summer): 7–28.

Maddison, Angus (1982), *Phases in Capitalist Development,* Oxford University Press, Oxford, U.K.

Maizels, Alfred (1963), *Industrial Growth and World Trade: An Empirical Study of Trends in Production, Consumption and Trade in Manufactures from 1899–1959, with a Discussion of Probable Future Trends,* vol. 21, Economic and Social Studies, ed. National Institute of Economic and Social Research, University Press, Cambridge, U.K.

Masson, Paul R., Tamim Bayoumi, and Hossein Semiei (1995), "International Evidence on the Determinants of Private Saving," unpublished manuscript, International Monetary Fund.

Mitchell, B.R. (1989), *British Historical Statistics,* 2nd edition, Cambridge University Press, Cambridge, U.K.

Molony, Barbara (1990), *Technology and Investment: The Prewar Japanese Chemical Industry,* Council for East Asian Studies, Cambridge, U.K.

Olney, Martha (1991), *Buy Now, Pay Later,* University of North Carolina Press, Chapel Hill, NC.

Organization for Economic Cooperation and Development (OECD) (1954), *The Chemical Industry in Europe,* OECD, Paris, France.

Pechman, Joseph and Keimei Kaizuka (1976), "Taxation," in Hugh Patrick and Henry Rosovsky (eds.), *Asia's New Giant: How the Japanese Economy Works*, The Brookings Institution, Washington, DC, pp. 317–82.

Reader, W. J. (1979), "The Chemical Industry," in Neil K. Buxton and Derek H. Aldcroft (eds.), *British Industry Between the Wars*, Scolar Press, London, U.K.

Rodrik, Dani (1995), "Getting Interventions Right: How South Korea and Taiwan Grew Rich," *Economic Policy*, 20: 53–107.

Schenk, Catherine R. (1994), *Britain and the Sterling Area: From Devaluation to Convertibility in the 1950s*, Routledge, London, U.K.

Schroter, Harm G. (1992), "The International Dyestuffs Cartel, 1927–39, with Special Reference to the Developing Areas of Europe and Japan," *International Cartels in Business History*, University of Tokyo Press, Tokyo, Japan.

Schumpeter, E. B. (1940), *The Industrialization of Japan and Manchukuo, 1930–1940*, Macmillan, New York, NY.

Schwartz, Anna J. (1981), "Understanding 1929–31," in Karl Brunner (ed.), *The Great Depression Revisited*, Kluwer Nijhoff, Boston, MA, pp. 5–48.

Smith, John Kenly, Jr. (1992), "National Goals, Industry Structure, and Corporate Strategies: Chemical Cartels between the Wars," in Akira Kudo and Terushi Hara (eds.), *International Cartels in Business History*, University of Tokyo Press, Tokyo, Japan.

Sommariva, Andrea and Giuseppe Tullio (1987), *German Macroeconomic History, 1880–1979: A Study of the Effects of Economic Policy on Inflation, Currency Depreciation and Growth*, MacMillan Press, Houndsmill and London,

Stanford Research Institute (various issues), *Chemical Economics Handbook*, SRI, Stanford, CT.

Stratmann, Friedrich (1985), *Chemische Industrie unter Zwang? Staatliche Einflussnahme am Beispiel der chemischen Industrie Deutschlands, 1933–1949, Vol. 43, Zeitschrift für Unternehmensgeschichte*, F. Steiner Verlag Wiesbaden, Beiheft, Stuttgart.

Summers, Robert and Alan Heston (1991), "The Penn World Table (Mark 5): An Expanded Set of International Comparisons, 1950–1988," *Quarterly Journal of Economics*, CVI: 327–68.

Svennilson, Ingmar (1954), *Growth and Stagnation in the European Economy*, United Nations: New York, NY.

Uchida, H. (1980), "Western Big Business and the Adoption of New Technology in Japan: The Electrical Equipment and Chemical Industries, 1890–1920," in Akio Okochi and Hoshimi Uchida (eds.), *Development and Diffusion of Technology*, University of Tokyo Press, Tokyo, Japan.

van Ark, Bart (1995), "Manufacturing Prices, Productivity and Labor Costs in Five Economies," *Monthly Labor Review* (July): 56–72.

Voth, Hans-Joachim (1995), "Did High Wages or High Interest Rates Bring Down the Weimar Republic? A Cointegration Model of Investment in Germany, 1925–1930," *Journal of Economic History*, 55: 801–21.

Weinstein, Michael (1981), "Some Macroeconomic Consequences of the National Industrial Recovery Act, 1933–1935," in Karl Brunner (ed.), *The Great Depression Revisited*, Martinus Nijhoff, Boston, MA, pp. 262–81.

Weiss, Frank D. (1992), "Trade Policies in Germany," in Dominick Salvatore (ed.), *National Trade Policies*, Greenwood Press, Westport, CT and London, pp. 131–53.

World Bank (1992), *The East Asian Economic Miracle*, The World Bank, Washington, DC.

Young, Alwyn (1992), "A Tale of Two Cities: Factor Accumulation and Technical Change in Hong Kong and Singapore," *NBER Macroeconomics Annual*: 13–54.

10 Finance and the Chemical Industry

MARCO DA RIN

The chemical industry was born with a long-term horizon; that is, with a need to finance large and risky projects with a far-off payback horizon. This is because the industry's discoveries and advances have been the result of substantial research and development efforts, while the industry's production has been highly capital intensive. Thus, the availability of financial resources, and the conditions under which these resources are provided, has had a powerful influence on the features and outcomes of the chemical industry. It is then not surprising that the characteristics of a nation's financial system have often been a key factor in determining the competitiveness of individual firms and national systems.

Financial systems affect the development of industry through allocation and governance. Allocation consists of channeling funds from lenders to borrowers, which means choosing which projects and firms are worth financing and under what conditions. There are different decision-makers for various sources of finance. Equities and bonds can be bought directly by final investors or purchased by fund managers. A loan may be granted by a commercial bank officer or syndicated by an investment banker. The decision about the share of earnings to retain is made by a firm's own board of directors. These different sources of funds put different constraints on borrowers and offer them different opportunities. Thus, each possibility implies for the firm different costs, obligations, and expectations of behavior, all shaped by the regulatory context of the financial system.

The financial system also provides a mechanism to govern firms, which is commonly referred to as "corporate governance." Ownership rights are the source of rights and duties which, again, differ for holders of different titles. Shareholders are the main claimants on their firm's decision-making process, although their influence is mediated through a board of directors. Bondholders and banks, as lenders, may also have some voice in how a corporation is governed. Managers, as the insiders who make key decisions, set agendas, and gather and distribute information, also shape the process. The way ownership is structured affects how managers view a firm's goals, strategies, and performance. A private firm with strong family ownership will differ from a

I would like to thank Ashish Arora and Johann Peter Murmann for their comments, and Ralph Landau for his encouragement and contagious enthusiasm. Ned Brandt and Kathy Thomas at Dow Chemical's Post Street Archives kindly supplied very useful historical sources. Wim Deblauwe provided expert research assistance.

307

company whose ownership is dispersed among thousands of investors. Moreover, managers enjoy different degrees of autonomy and set different goals for themselves, according to how the interest of shareholders, bondholders, and lenders are voiced.

This chapter discusses how allocation and governance, shaped by the economic and regulatory history of nations, affected the availability and conditions of external finance, thus shaping the evolution of the chemical industry. I approach these issues from an angle that is both historical and comparative. When looking at single firms, I explore how cash flow, capital expenditure, R&D intensity, and corporate strategies at large have been affected by the availability of funds. At the industry level, the key questions are how coordination, consolidation, and restructuring have been influenced by financial markets. The first sections of this paper compare directly the evolution of the chemical industry in Britain, Germany, and the United States at particular periods in time. Japan is then treated in a separate section, to account more properly for its idiosyncratic pattern of development.

FINANCING THE ENTREPRENEURIAL AGE

The British Decline

Synthetic dyes, the first science-based chemical industry, were discovered in Britain by Perkin in 1856. In 1862, there were nine British dye firms and seven (smaller) German ones (Beer 1959, p. 30). For the rest of the decade the British firms enjoyed a first-mover advantage and dominated this rapidly growing market. British firms, however, showed little interest in developing technical capabilities; most never went beyond simple refinements of Perkin's methods. The British industry was thus soon overtaken by German competitors in the 1870s. For an overall discussion of the British and German chemical industries since 1850, see the chapter by Murmann and Landau in this volume. Here, I explore the particular question of how Britain's financial sector contributed to its falling behind in the chemical industry.

The British system of industrial finance developed in the first half of the 19th century, during the Industrial Revolution.[1] Several of its initial traits proved long lasting: minimal regulation, strong reliance on retained earnings for the finance of capital accumulation, "distant" borrower-lender relationships based on short-term loans and tradeable notes, and limited access to the stock markets for new firms. The first intermediaries to provide external finance for industrial ventures were small partnerships of wealthy merchants known as *country bankers*. They financed firms through short-term loans and by discounting "bills of exchange," trade-originated securities. This pattern of industrial finance required entrepreneurs to fund fixed capital themselves, or with kin and friends. Banks were constrained by law to be small. Until 1826, they could not have more than six partners, and limited liability was introduced only in the second half of the century, when joint-stock banks developed as centrally coordinated networks of local banks. These networks maintained the arm's-length attitude of country bankers to lending, to which they added an emphasis on risk diversification.

By the middle of the 19th century, firms in railroads and utilities were able to raise large sums of external finance through British merchant bankers. Industrial ventures, however, raised much less. Industrial securities were regarded as speculative, and

---

[1]On the history of the British financial system, see Collins (1988), Cottrell (1980, 1985), Kennedy (1987), Kindleberger (1984), Michie (1987), Nevin and Davis (1970), and Thomas (1978).

promoters and bankers were not equipped to appraise the idiosyncratic risks involved in screening firms and monitoring their management. City institutions developed their skills by floating public securities but stayed away from giving financial support to new industrial firms. In Germany and the United States, specialized entities (*Kreditbanken* and investment banks) emerged for managing the processes of assisting firms in issuing securities. No such entities emerged in Britain. Instead, informal capital markets remained the main source for industrial finance in Britain for a long time. Thus, markets for industrial securities were slow to develop; indeed, no British chemical firm issued equity until the end of the century. Between 1882 and 1913, innovative industries such as chemistry, engineering, and electricity accounted for only 21% of domestic industrial and transport new-securities issues, while breweries and railways accounted for 55%. Typically, incorporation was used by closely held companies to achieve limited liability rather than to raise funds: As late as 1904, almost 90% of Britain's 3,477 new joint-stock companies were private.

There is a longstanding debate on why British financial markets failed to provide capital to sustain domestic industrialization.[2] City financial institutions have been accused of irrationally preferring investment in utilities and foreign public bonds and turning down domestic industrial securities. But, as Edelstein (1971) shows, British financial markets equalized risk-adjusted returns across different securities and thus were investing rationally and efficiently. This sharpens the question of why Britain's financial markets did not offer much support to industrial development.

In my view, the explanation can be seen as a self-reinforcing circle. Given banks' distant attitude, it was rational for firms to invest in low-risk projects. In turn, given the entrepreneurial focus on relatively simple projects, it was rational for banks to avoid involvement in long-term finance. If existing financial institutions had been able to select promising projects and skilled entrepreneurs and to provide them with long-term support, it would have been profitable for industry to invest in more advanced technological capabilities. This failure to coordinate was at the heart of the British decline. Financial intermediaries have often served as an important catalyst for the early stages of industrialization. But investment in the necessary information-processing capacity and coordination skills is costly and risky. When firms adopt simple technologies, there is little that a bank can gain from developing close ties with such firms, and it does not pay for a bank to undergo such investment. Da Rin (1997a) offers a formal presentation of this argument, which draws out some additional implications.

This argument helps explain the pattern of Britain's chemical industry and broader development in the 19th century. British firms successfully adopted a pattern of production that made use of simple technologies, and they invested in easily redeployable assets. Banks engaged in distant, arm's-length intermediation, circulating securities in impersonal markets. Once selected, the compatibility of these industry and bank choices made them self-sustaining and persistent. The earliest products of the British chemical industry, for example, included basic chemicals, acids, and bleaching powder (alkalis). These products used simple technologies, which required little specific physical and human capital. Thus, British technology was complementary to British-style, arm's-length financing.

---

[2] Collins (1991) surveys in detail the literature on the British industrial decline during Victorian times. Kennedy (1987, ch. 5) argues for a bias of British markets toward "safe" securities. He argues that capital market imperfections made it rational for investors to avoid securities, whose riskiness could not be properly understood. However, he is less convincing in providing an explanation for why such a system persisted and did not evolve into a more efficient one; nor does he provide explicit evidence of credit rationing.

The British financial system discouraged chemical firms from making innovative and risky product and process-specific investments, so that they failed to develop original and innovative technology and fell behind when innovative competitors arrived. Consider the evolution of the soda industry. Its firms started declining when they failed to switch to the Solvay process, retaining the obsolescent LeBlanc process. In 1891 many LeBlanc alkali producers merged into United Alkali, in an attempt to sustain prices by concentration. Predictably, United Alkali soon suffered competition from more efficient producers of electrolytic alkalis and stopped paying dividends as early as 1896. By 1913, United Alkali had reduced its capital by 40% (Haber 1971, p. 140). Its low rate of return on capital, around 2%, contrasted with the profitability of Solvay producers Brunner Mond, then around 15%. The main reason for the surprising decline in Britain's chemical industry, after it had seemed a secure industry leader in the 1850s and 1860s, was the reluctance to invest in research, chemists, and experimentation. This pattern was not exclusive to the chemical industry. British mining, foundry, and ship-building also developed with relatively little interest in technological research. Innovative industries such as the electrical and electromechanical ones remained very small and more dependent on imports. Of course, the financial system was not the exclusive reason. British chemical firms were used to exploiting the large markets provided by the Empire, without the need to conquer new ones, which allowed them to thrive without cutting-edge technical excellence (Landes 1969). Germany, on the other hand, had no such dominions (Kocka 1988). Perhaps as a result, German firms were much more aggressive in setting up marketing branches in many countries. German firms also offered technical assistance to customers, hosting them for training on new methods of fixing dyes.

The only British success stories for this time period come from cartelized industries or consumer chemicals. For example, Brunner, Mond & Co., a producer of alkali by the Solvay process, was part of an international cartel of Solvay producers. Brunner-Mond was founded as a partnership in 1873, and it faced severe financing problems for a number of years.[3] The initial funds were provided by the two partners and by a friend. Banks, unable to understand its technology, were not attracted by such a risky venture and only lent money against a mortgage secured by the Winnington property on which this plant was located. The partnership also got short-term credit for small amounts from Parr's Bank in Warrington, whose manager was a friend of Brunner. Ernst Solvay himself provided credit at a critical time and then became a shareholder. Most of Brunner, Mond's technology came from abroad, and cartel protection meant that the company felt little pressure to invest in research. In fact, Brunner, Mond still employed only four chemists in 1914 (Reader 1970, p. 176). Nobel Industries, which produced explosives, is another example of succeeding through cartelization. The Nobel-Dyna-mite Trust, an international holding company, acted as an internal capital market that exerted strong control over its members (Reader 1970, pp. 179–81). Necessary technology was developed abroad. Nobel Industries borrowed more from firms in the cartel than from banks, and never needed to issue debt or equity. In both alkalis and explosives, science played a lesser role, as in many inorganic chemicals. Given their protected status, it is not surprising that Brunner, Mond and Nobel lost little ground to innovative German newcomers.

Consumer chemicals are another area in which research and technical investments mattered relatively less, and in which even Britain's chemical-success stories reveal the

---

[3] Lischka (1985) and Reader (1970, pp. 48–55) reconstruct the firm's financial history.

shortcomings of its financial system. Courtaulds successfully shifted from leadership in natural silk to artificial fibers (Coleman 1969) when it experienced a decline in profitability at the beginning of the century. In 1904, the company incorporated and issued preference shares and debentures. However, Courtaulds achieved its success largely because of marketing and efficient distribution to consumers, two areas in which British firms traditionally excelled, rather than because of technical skill. Two other successful consumer-chemical firms, Lever Bros. and Crosfield, relied on internal finance for a long time before offering equity to the public (Musson 1965; Wilson 1968). Lever Bros. listed its preference shares on the stock exchange only in 1929. Crosfield did not issue securities until 1896, after the firm had relied on personal savings and retained earnings for more than 70 years.

The picture which emerges is that British chemical firms were discouraged by the financial system from investing in projects with a long-term payoff horizon. This lack of investment meant little and unsystematic research activity, suboptimal plant size, and lack of diversification. This held true across industry sectors, from the alkali companies (Haber 1958, p. 154) to the producers of inorganic chemicals (Chandler 1990, p. 357) and the dye makers (Haber 1958, pp. 165–67). There was little interest in the systematic pursuit of science to improve productivity. It is telling that the most active British experimenters in alkalis were William Gossage, a druggist; Charles Tennant, a linen bleacher; and Walter Weldon, a journalist of fashion magazines (Lischka 1985, pp. 63–65). Research and capital-intensive fields such as pharmaceuticals, intermediates, and electrochemicals hardly developed.

Britain's financial system contributed to the underinvestment in technology. In the absence of an effective corporate governance system able to assess the technologial value of projects and managers, simpler technologies and reliance on established markets were the best response by entrepreneurs.[4] Absent a way to coordinate their choices toward riskier but more productive investments, both firms and banks had little incentive to pay the costs and take the risks of changing their strategies.

The German Rise to Leadership

The relative decline of Britain's chemical industry was based on complementarities between industrial and financial strategies. A similar story, told in reverse, helps explain the rise of Germany late in the 19th century as a power in the chemical industry.

Several features of the German financial system date back to its formation, around the middle of the last century: a close relationship between borrowers and lenders, the development of monitoring and control capabilities by large banks, and the dependence of capital markets on a few large banks. Industrial finance appeared with *Privatbankiers*, partnerships of private bankers with good knowledge of business opportunities.[5] *Privatbankiers* established close ties with their clients and engaged in long-term as well as short-term finance. They were superseded by larger and better capitalized *Kreditbanken*, which pioneered "universal banking" by offering both commercial and invest-

---

[4] The size of the financial intermediaries may also have been relevant. In this respect, De Long (1991) argues that the large market share of J. P. Morgan in the United States made maintaining its reputation extremely valuable. The converse was true in Britain, where financial promoters were many and small, and had a high incentive to grab the money and run, a behavior that was frequently observed.

[5] On the development of German industrial finance, see Balderston (1991), Da Rin (1996), Feldenkirchen (1991), Francke and Hudson (1976), Hardach (1984), Kindleberger (1984), Pohl (1984), Reisser (1911), Tilly (1967), and Wellhöner (1989).

ment banking services. In so doing, they developed the skills necessary for evaluating firms' long-term prospects and exerting control over management, and so were willing to provide long-term loans.[6]

Banks helped firms issue securities. Between 1882 and 1913, the four fastest-growing industries—chemicals, mining, engineering, and electricity—accounted for nearly 70% of new issues (Tilly 1993). By the turn of the century, five big banking groups (*Konzerne*) had developed around the Deutsche, Dresdner, Disconto, Schaaffhausen, and Darmstädter banks. Yet the system remained rather competitive, with each group striving to enlarge its market share by attracting new customers in new regions and industries. These German banks became involved with emerging firms in such "new technology" industries as chemicals, electricity, machinery, and transportation.

The choices of German banks and firms had key complementarities. Banks upheld the specific, risky investments that German firms made at the beginning of industrialization. Banks used their skills for control and resolution of financial distress and supported their clients. They lobbied government officials when necessary or would even become directors or managers of firms.[7] They also assisted firms in their export activities. Given this attitude of banks, it was rational for firms to choose risky but potentially very productive projects. In fact, the potential gains from such projects justified the cost to firms of developing close ties with banks. Tilly (1986) employs modern asset-pricing analysis to show that returns to investment in different branches were on the efficient frontier.

To understand how the German system worked in practice, consider banks' influence on some well-known firms at early stages. BASF incorporated in 1865 with the contribution of the bank W. H. Ladenburg and Söhne (Kirchgäesser 1988, pp. 68–69). In 1862, Ladenburg had financed the Chemische Fabrik Sonntag, Engelhorn, Dyckerhoff und Clemm, predecessor of BASF, and the bank also financed other chemical start-ups. Ladenburg retained 10% of BASF shares, and his partner chaired its board. The bank was also active in later capital increases (1873 and 1889), and in 1901 it floated the first large BASF bond issue. Credit from the bank was strategically important. In 1868, it allowed the ailing BASF to hire Heinrich Caro as head chemist; in 1869, Caro (with others) received a patent for alizarin, which made the fortune of BASF. Another chemical firm, AGFA, was founded in 1867 with help from the banker Mendelssohn-Bartholdy, who became a partner. Its parent company, Kunheim, had also been founded with the help of a banker (Haber 1958, p. 48). DEGUSSA, later a leader in electrochemicals, received credit from Frankfurt financial houses to establish its own research laboratories (Beer 1959, ch. 8).

Kreditbanken were not old-time venture capitalists; they preferred to deal with firms that had already shown some potential for growth. Several firms for instance, Hoechst and Bayer, started being financed and assisted by *Kreditbanken* only when incorporating. Since Kreditbanken extended long-term credit as well as floating firms' bonds and shares, they effectively offered bridge loans that allowed firms to wait for low interest rates before issuing bonds. This encouraged German firms to incorporate and access capital markets early in their life, allowing them to raise very large sums of equity from the public and so to afford investing in capital- and research-intensive production

---

[6]Regulations were important in this respect. The Bank of England refused to extend refinancing to banks experiencing liquidity problems. On the contrary, the Prussian central bank adopted a liberal discounting policy, which helped German banks to offer long-term, illiquid loans.

[7]Jeidels (1905, p. 170) reports data for 1903. The 8 largest *Kreditbanken* had a total of 55 corporate directorships. Not surprisingly, seats were typically proportional to the funds borrowed from banks or raised as securities issues.

techniques. The "universal" nature of these banks allowed firms to tap either debt or equity finance using the same financial intermediary.

Hoechst, for instance, incorporated in 1881 with a capital of 10 million marks, of which 5.5 million were sold through two Frankfurt banks, Hauck & Son and J. I. Weller Sons.[8] Between 1863 and 1908, Hoechst raised 36 million marks of capital on the stock exchange, always with bankers as underwriters and providers of bridge loans. Together with 29 million marks of retained earnings, this allowed Hoechst an impressive cumulative investment of 65 million marks. Forty-two German chemical firms incorporated in the early 1870s with bank help, and the trend continued in the following decades (Riesser 1911, p. 125). Of the seventeen chemical companies that ranked among the largest hundred German firms in 1907, fifteen had been founded early as partnerships and were later incorporated (Kocka and Siegrist 1979, p. 111). Grabower (1910) provides detailed statistics showing how relations with banks differed across the branches of the chemical industry.[9] Early in a company's life, equity capital often accounted for more than 95% of financing. As firms matured, loans and bonds became more important. The involvement of bankers was substantial. Only about one-third of the 223 joint stock companies Grabower considers were founded without a banker. A striking difference with Britain was that capital increases were rarely restricted to existing shareholders, especially for larger and listed firms. When shares were offered to the public the involvement of bankers was essential, partly as brokers and underwriters and partly to assure investors that they would get their money back.

The effects of the German financial system on firms' risk-taking and systematic scientific research are clear. It was in Germany that industrial-research departments were invented, and the chemical industry was most active in this, especially dye-makers. For instance, in 1885 Bayer employed 24 chemists, which grew to 100 by 1896, plus 25 engineers, all at least graduates of institutions of higher learning (Wrigley 1987, pp. 171–73). German chemists perfected the knowledge and exploitation of organic chemistry, laid the basis of modern pharmacology, competed with the Americans in electrochemical processes, and created high-pressure processes for synthetic fertilizers. The immense capital outlays required for capital investment and research far outgrew retained earnings. Both the initial support from banks and the subsequent access to capital markets were fundamental in building industry leadership. The combination of the complex scientific nature of organic chemistry and the need for huge financial sums presents a powerful barrier to entry by new firms. In fact, the few chemical firms that emerged from the dye-making industry in the 1870s are still prominent. For instance, the search for synthetic indigo cost BASF 17 years of research with a cost of at least 18 million marks (Wrigley 1987, p. 177). Such an investment required very patient, daring, and forward-looking finance. Hoechst, which joined BASF in the race to the discovery of synthetic indigo, built five experimental plants to develop industrial production, spending about 15 million marks (Haber 1958, p. 175). Similarly, the exploitation of the Haber process for fixation of nitrogen required two decades of costly research. Mittasch, in developing it with Bosch, performed over 6,500 experiments in three years, trying over 2,500 substances for the catalyst (Haber 1971, pp. 85–95). Bayer's profit rate until 1914 ranged between 20% and 30%, which led to the accumulation of 145 million marks. Yet its balance sheets show that it had to raise an additional 45 million marks to finance its huge capital outlays.

[8] Baumler (1968) has a detailed account of the financial history of Hoechst.

[9] From 1850 to 1906, the German chemical industry raised about 750 million marks of capital for joint-stock companies. They were evenly divided among about 100 firms in heavy chemicals, coal-tar derivatives, dyes, and fertilizers and about 100 firms in rubber, explosives, organic colors, and specialty chemicals.

The complementarities between long-term finance and research and development also allowed the chemical industry and the dye-makers to use profits to open new, lucrative areas of business and thus to sustain their high profit rates. Indeed, the persistent success of Bayer, BASF, AGFA, and Hoechst was due to their ability to keep ahead of competition in innovating and diversifying. BASF pioneered high-pressure ammonia; Hoechst, pharmaceuticals; and Bayer, rubber—all very costly fields to enter. The contrast with the sparing research attitude of British pharmaceutical firms is striking (Liebenau 1984). Hoechst established close links with the founders of chemotherapy, Knorr, Koch, and Ehrlich (Baumler 1968, pp. 33–36). Chemotherapy took several years to be recognized by the medical profession and was a very risky investment. The profits from revolutionary drugs such as antipyrin (1883), pyramidon (1896), and novocain (1905) repaid Hoechst well. Bayer soon followed suit, introducing phenacetin (1887) and aspirin (1897). Specialized pharmaceutical firms, for instance Merck, developed more slowly. For these firms, diversification into new products also helped dampen cyclical fluctuations.

Most products did not share the fantastic profitability of dyes. Haber (1971) illustrates in detail the differences between dyes and other branches of the industry (heavy chemicals, fertilizers), which had more and smaller firms and never reached such impressive profit and investment rates. But the general connection between German finance and science reached beyond chemicals to firms in other sectors as well. In electrochemicals, Chemische Fabrik Griesheim, Kunheim, and Alkaliwerke became world leaders at the end of the century, competing with American enterprises. Although the Germans failed to overtake the British in consumer chemicals, Henkel did challenge Lever by investing in advanced research and developing new electrochemical technology with DEGUSSA, a pioneer in the field. The story of DEGUSSA again illustrates how banks helped with restructuring and refinancing, a role repeatedly illustrated by Grabower (1910). In 1899, DEGUSSA formed Cyanid Gesellschaft to exploit the cyanamide process for the fixation of nitrogen. The project was backed by the Deutsche Bank, which helped DEGUSSA promote its subsidiary, Bayerische Stickstoffwerke. Consolidierte Alkaliwerke set up a competing firm in 1904, with financing from the Metallbank, which also rescued it in 1911.

The strength of *Kreditbanken* was not positive for German industry in every way. For example, the *Kreditbanken* were also responsible for fostering the cartelization of the German industry, since that too, along with innovation, would support the financial stability of their borrowers. However, the benefits of the German financial system—help in accessing capital markets, bridge loans for securities issues, and effective corporate governance in case of distress—certainly had an overall positive impact. Calomiris (1995) has argued that this might have offered German firms a lower cost of capital than in Britain or the United States, although his conclusion remains tentative for lack of systematic data.

The American Awakening

At the start of the 19th century, the United States was a relatively less-developed economy. Through most of this century, there was little production of advanced chemicals in the United States; instead, inorganic chemicals were mostly imported from Britain and organic chemicals from Germany. But the potential of the American economy soon became apparent. In terms of the U.S. chemical industry, the strongest force was the possibility of exploiting economies of scale in producing and selling to America's huge internal market (see Arora and Rosenberg, this volume). Large consumer-chemical producers such as Procter & Gamble and Colgate emerged through consolidation only during the last two decades of the 19th century. Heavy and intermediate chemicals took longer. According to Roy (1990), in 1880 there were 1,324 chemical works, with 25,000 workers. By 1900 they numbered 1,331, with 41,000 workers, and by 1914 they had become 1,987, with 78,000 workers. Their capitalization increased tenfold. More concentration occurred in drugs: the 1,805 drug-makers of 1890, with 2,000 workers, became 416 by 1914, with 9,000 workers. Drug capitalization also increased tenfold. However, U.S. production of organic chemicals would not replace imports until after after World War I.

In the 19th century, America's financial sector offered little support to its budding chemical industry. Even more than in Britain, the fragmentation of the banking system, and its short-term horizon, deprived firms from outside sources of funds. America has always caged its financial intermediaries in a web of regulations that constrained their development.[10] The result has been to fragment banking, to favor securities markets, and to separate commercial and investment banking, as Roe (1994) meticulously documents. The 1864 National Banking Act created a dual system of national- and state-chartered banks, in which banks were subject to capital requirements and federal supervision and were prevented from equity ownership and branching.[11] The Banking Act introduced another lasting feature, preventing banks from lending more than 10% of their assets to a single borrower. Together with the absence of a central bank (the Federal Reserve system was not created until 1913) and a national money market, this Act forced banks to reduce their risk by restricting their lending to liquid short-term loans.

One consequence of the lack of available finance in the U.S. economy was that the consolidation of the chemical industry in the later part of the 19th century took different routes in the United States than in Germany. In Germany, the hundreds of firms making organic chemicals in the 1860s and 1870s (Teltschik 1994) were reduced to a few by competition but also by bank-managed mergers during the 1880s. While such concentration was most acute in dyes, it was common to all sectors (Grabower 1910). In the United States, less reliance on external finance delayed greater concentration until the drop in transportation costs made it inevitable.

The similarity between the situation of British and American firms and banks at this time was strong, but with one major difference: British firms were largely reconciled with the way the capital market allocated capital; for example, there are no records of young British chemical firms trying to persuade banks to lend to them for long-term research and capital investment. American firms, however, often applied for such financing and were refused. The small U.S. banks had little ability to understand technical processes— especially when they were so innovative and risky, as in chemicals. The banks preferred lending to safer and more established industries. Capital markets also provided very little support for science-intensive industries such as chemicals, unlike the case for railways or other heavy industries (Michie 1987). The result was to restrain technical development and investment in research.

Thus, the American chemical industry was in a situation where external finance was very limited, but it seemed clear to entrepreneurs that demand was rising and substantial profits could be made in this industry by firms which could endure. Their

---

[10] On the American financial system, see for example, Carosso (1970), Fischer (1968), Goldsmith (1958), Nevin and Sears (1953), and Sylla (1975).

[11] In 1900 there were 3,731 national banks and 4,369 state banks (White 1983, p. 13), but only 87 had branches (Fischer 1968, p. 35).

strategy was a frenetic wave of tinkering with plants and processes to find empirically viable solutions to immediate production problems. Little effort could be spent on long-term, sophisticated basic R&D, but resources were put into adaptive process innovation, often based on foreign patents (see Arora and Rosenberg, this volume). The main virtue of American inventors at this stage was indeed their ability to adapt and respond to market demand (Khan and Sokoloff 1993).

Electrochemicals are a good example in this respect, as they represented the first big success of innovation in American chemicals (Trescott 1981). Aware of the country's enormous electrical-power resources, for instance in the area around Niagara Falls, several chemists started researching in the 1890s: La Sueur, Castner, Hall, and Dow. They adopted and improved discoveries made mainly in Germany and France, eventually producing processes of their own. From 14 firms in 1899, they grew to 34 by 1909, producing metals (aluminum), alkalis, and calcium carbide. In the 1890s they engaged in feverish activity, each racing to stay ahead. Virtually no external finance was raised. Only when some of these firms reached financial stability did they start engaging in systematic R&D.

The monumental work of Haynes (1954) illustrates the history of 219 chemical companies and provides evidence on how firms were financed at this stage. Firms received little help from banks. Industrial securities became accepted only very slowly, and chemicals were regarded as very risky. Thus, most companies were financed out of the accumulated wealth of their founders and relied almost solely on earnings to expand and survive distress. Chemical companies such as Eastman Kodak, Dow Chemical, Monsanto, and Firestone Rubber survived because their tinkering and process improvements were successful. Interestingly, the most successful pioneers were rarely chemists (Haynes, 1939; Haynes, 1954). Also, the typical American drug companies—Abbott Laboratories, Ely Lilly, Upjohn Co., and Squibb & Sons—all started as small-scale family start-ups, unlike in Germany, where pharmaceuticals were financed by the profits from dyes (Haynes 1954).

Toward the beginning of the 20th century, U.S. chemical companies were still unable to rely on bank financing, but they began to be able to draw on securities markets. The reliance on securities for external finance became a distinctive American trait, especially after the spectacular boom of railroads bonds (Michie 1987). Industrial securities took time to be accepted on national stock exchanges, as they were considered too risky and speculative; by 1898 only 20 industrials were listed on the New York Stock Exchange (NYSE), but they grew to 173 by 1915 (Carosso 1979, p. 79). Roy (1990) provides data on chemical firms listed on the New York Stock Exchange from 1899 to 1913. There were only three companies listed in 1899, with a capitalization of $48 million. Of these, DuPont was the first, and by far the largest, issuer. By 1913, seventeen companies were listed, with a capitalization of $367 million. Of those that would become major chemical companies, only four were listed in this period: DuPont, Union Carbide, General Chemical, and Sherwin Williams. Fewer chemical companies were listed than companies from other industries. Other large chemical firms issued only bonds, usually when incorporating. Detroit Chemical Works, Dow Chemical, Merck & Co., Minnesota Mining and Manufacturing (better known as 3M), and U.S. Industrial Chemicals followed this pattern (Haynes 1954).

Trading of securities became concentrated in New York. American investment banks, less hampered than commercial banks by regulations, established close links with both issuers and wealthy investors (Nevin and Sears 1953). They became investors' portfolio managers and with the merger wave around the turn of the century started acting as directors in many firms (De Long 1991). Houses such as J. P. Morgan, Drexel, Kuhn Loeb, Kidder Peabody, First National Bank of New York, National City Bank, Lee Higginson, and later Goldman Sachs and Lehman Bros. became extremely influential (Carosso 1979). However, chemicals participated in the merger movement relatively less than did other capital-intensive industries; firms and plants remained highly dispersed (see Arora and Gambardella, this volume). For that reason, investment banks played a lesser role in engineering chemical deals, a point made by both Haynes (1954) and Chandler (1990).

To get a feel for the impact of the U.S. financial system on the early development of the chemical industry, it is perhaps useful to look at the financial experience of some of today's major chemical firms. The case of Dow Chemical is most telling (Whitehead 1968).[12] Herbert Dow was an inspired improver of processes but had difficult relations with his financiers. In 1892, he founded Midland Chemical, financed by Cleveland manufacturer J. H. Osborne, but had to dissolve it for financial reasons two years later. Dow then founded Midland Chemical Co., again financed by Osborne and other Cleveland businessmen. This time the impatience of the investors, who had no clue of the technical problems Dow was facing, stopped him right on the verge of success. In 1895, he involved the faculty of the Case School of Applied Sciences in founding the Dow Process Company. In 1897, it was incorporated, receiving small credit from a local bank. In 1902, Dow first issued bonds, secured by a mortgage on the plants, which was the start of Dow's traditional policy of high leverage. After 1910 dividends rose steadily, and common stock paid ever-increasing dividends.

John Queeny had to save from his salary to gather enough funds for founding Monsanto Chemical Company in 1901, since he was turned down by several banks (Forrestal 1977). He started producing saccharin but suffered from ruthless dumping by German firms and had to diversify into vanillin and caffeine. His difficulties in obtaining capital were extreme. In 1903, the only financier he could convince was an old acquaintance. Other troubled capital increases occurred in 1907 and 1913. Success only came with World War I, when Monsanto achieved huge profits. These allowed Monsanto large investments in R&D, exploiting patents confiscated from the Germans. Monsanto became large enough to start exploiting economies of scale. At this point, finance became easier. Monsanto went public in 1927, selling about 30% of its capital.

DuPont, founded in 1802, is a rare case of a venture financed by a substantial family wealth (Chandler and Salsbury 1971). Its growth was attributable to the expansion of the explosives business: between 1870 and 1900, sales grew 250 times. The firm remained a partnership until 1899, when it was incorporated but still family owned. In the next three years, DuPont acquired 54 companies and by 1905 controlled 75% of the U.S. explosive-powder market. At this stage, its products were developed through a trial-and-error process, but its Eastern Laboratory (1902) and Experimental Station (1903) were among the earliest U.S. industrial laboratories. An antitrust ruling then forced DuPont to divest part of its traditional business. It spun off two companies, Hercules Powder and Atlas Powder.[13] It then initiated a policy of vigorous diversification, which accelerated after the war. DuPont's most profitable acquisition was the 23% stake in ailing General Motors, which it bailed out and reorganized (Haber 1971, p. 183).[14] As

---

[12] The following analysis is based also on the financial historical files conserved at Dow's Post Street Archives.

[13] Hercules was formed in 1912 with an extremely high (1:1) capital-debt ratio. Debt was owned by DuPont and was reabsorbed in two years by issuing new equity. From then on, the quick growth of Hercules was financed mostly from retained earnings (Dyer and Sicilia 1990, pp. 76–7).

[14] Hounshell and Smith (1988, appendix III) have a complete list of Du Pont's acquisitions.

already mentioned, at the beginning of the century DuPont pioneered the issue of industrial equity to finance its large investments. For example, of the $3.5 million DuPont invested in 1905–1906, about $1.0 million came from the sale of securities. Bank loans were not even taken into consideration.

The early U.S. chemical industry resembled the British in many respects: simple technology, small plants, little support from banks, and capital markets. But the American experience was unique in several ways. At the turn of the century, U.S. investment banks were selling the industrial securities of large firms. Although the chemical industry was not at the forefront of this trend, securities finance was available to a number of leading U.S. chemical companies before World War I. Moreover, the growing American domestic market allowed firms to escape the "British trap" by generating internal funds sufficient to engage in technological innovation and sustain their effort over time. Astounding growth rates in certain sectors (explosives, alkalis, and electrochemicals) made retained earnings a sufficient source of finance for continued growth. This allowed the U.S. producers to enter a German-style cycle, in which large earnings supported huge research programs, which in turn secured future growth through a combination of increased sales, acquisitions, and diversification. The self-sustaining success of the U.S. chemical industry, however, came only after the accumulation of profits had supported extensive research. The development of the industry was indeed held back by the the financial system. For many American chemical companies, the combination of large profits and large research laboratories could only be achieved after World War I.

FROM ENTREPRENEURIAL TO MANAGERIAL FIRMS IN THE CONCENTRATION ERA

Needs for explosives, intermediates, and drugs led to increased production of chemicals in all countries during World War I. Moreover, the disruption of foreign supplies led many chemical companies to a search for replacements—in some cases, new synthetic materials. A number of the today's largest chemical companies became firmly established during World War I. For example, from 1912 and 1918 the real assets of BASF rose by 70%; of Brunner, Mond, by 120%; Dow, by 210%; and Monsanto, by an extraordinary 1500%.

Wartime profits were generally employed toward two goals: increasing R&D expenditures and diversifing away from glutted markets to preserve profitability. The renewed research effort was pursued in large research laboratories, copied from those of the Germans. New fundamental discoveries soon were made: nitrogen fertilizers, nylon, polythene, synthetic materials (fibers, alcohol, and rubber), new drugs (sulfo-aminides, penicillin) and paints, and many new intermediates. These inventions were less serendipitous than many earlier discoveries; instead, they were the fruit of a coordinated effort of applying science to industry. They required resources that could be provided only by financially strong firms. Indeed, since the 1920s no chemical-industry start-up has become a lasting major diversified player in the industry, which shows a remarkable pattern of persistence in its leaders. The firms that profited most from the war and the boom of the 1920s gained a lasting advantage.

Another lasting consequence of the war was to increase industry capacity for existing products, which resulted in overcapacity after the war. By 1924, production of heavy chemicals in Britain, Germany, and the United States was at 50% of industry capacity.

Consolidation swept the industry on a global scale, as companies sought to achieve financial stability and protect national industries in an era of protectionism. The two leading examples were the mergers leading to the creation of ICI in Britain and IG Farben in Germany, both of which will be discussed further below. As entrepreneurial firms evolved and merged into managerial firms, they required a broader supply of external funds and needed a different governance support, which meant changes in the interface between financial markets and industry.

The character of the British financial system, and its relation with industry, did not change substantially over this period. Consolidation of joint-stock banks proceeded steadily, and by World War I the "Big Five"—Barclays, Lloyds, Midland, National Provincial, and Westminster—dominated the banking system. The interwar period saw further consolidation of commercial banking and only a slightly wider use of industrial securities. The scarce support offered by financial intermediaries and capital markets continued. Bank credit was not very lavish, and only a very small, even decreasing, share of bank advances went to chemicals (Thomas 1978, ch. 3). Underwriting procedures and practices slowly improved with time, and by the 1930s some merchant bankers had specialized in underwriting and marketing large domestic industrial issues. However, speculation in new industries, such as films and the gramophone, first deceived and then disillusioned investors, preventing the long-term development of venture-capital institutions.

A trend toward consolidation and larger firms was observable in most branches of Britain's chemical industry in the interwar years. The war had been profitable for British chemical producers. Nobel, for example, more than tripled its capital from 1913 to 1919 and used these resources to acquire 30 firms right after the war. Brunner, Mond also acquired several competitors in this period. This consolidated their industry positions. Courtaulds concentrated on fibers; Lever took over the entire soap industry, and Glaxo, Beecham, and Boots emerged as leaders in drugs. Chemicals, however, remained a relatively small and fragmented industry, with the exception of ICI. In 1905, there were only four chemical firms among the largest 50; in 1919, there were six; and in 1930, four (Hannah 1976, pp. 102, 187, 189).[15] What remained distinctive in Britain was a pattern of closed family ownership (Chandler 1990). Such coziness had an adverse impact, because the lack of financial intermediaries with strong capacity for corporate governance allowed controlling families to divert funds for private use. Historical research has also stressed that Britain's industry leaders of the time viewed entrepreneurism as a way to achieve social status—to spend their time at Ascot and Windsor—not as a reward in itself. Since investors could still put their money into foreign public securities in the City, British firms were left without much funding or outside governance. According to the Macmillan Committee, the fastest-growing firms began to express frustration with the lack of capital available to them, a feeling that had not been widespread before.

The creation of Imperial Chemical Industries (ICI) in 1926 illustrates many of these trends (Reader 1970). ICI was the government-induced merger of two strong leaders— Brunner, Mond and Nobel—with two obsolete producers—United Alkali and the British Dye Corporation—which were saved from a very uncertain future. The merger allowed a remarkable concentration of financial resources, and capital markets responded positively by providing large sums for equity issues. Centralization provided more resources for research, whose budget quadrupled within four years from the merger,

---

[15] By comparison, at these dates, food and drink had 17, 9, and 17 firms among the largest 50, and textiles 12, 9, and 6 firms, at the respective dates.

reaching 1 million pounds—nearly 2% of sales—in 1930 (Hannah 1976, p.113). In 1933, ICI discovered polyethylene, which sustained the company financially for many years. ICI was the largest spender on R&D in the British chemical industry. Courtaulds, for example, spent only around 1% of its sales in R&D (Coleman 1969). Other firms in pharmaceuticals, for instance Glaxo, mainly kept improving marginally on foreign discoveries.

Yet despite these positive signs, ICI suffered from the fact that it was created with a diffused ownership, which failed to provide effective corporate governance and to integrate the constituent firms into one. Its inefficient central financial department did a poor job of planning capital investment, as well as investment in research. Similarly, professional staff in R&D at ICI numbered 510 in 1941, about one-half that at DuPont or IG Farben. The R&D to sales ratio at ICI was 1.3% in 1935 and remained at this level throughout World War II, lower than that at DuPont or IG Farben (Edgerton and Horrocks 1994; see also Reader 1970). And when sales halved during the depression, so did the research budget. By contrast, Germany's IG Farben did not cut its R&D significantly, even during the depression. The depression also had a negative and lasting effect on ICI's capital expenditure, which dropped from about 30% of sales in the late 1920s to 2.6% in 1932 and remained below 10% throughout the war (Reader 1975, appendix). ICI showed only moderate profitability until World War II, and its return on capital fluctuated between 5% and 9%. Other than its discovery of polyethylene, ICI fell behind in developing new products and diversifying its product lines. Much of its revenue was indeed earned thanks to its protection from competition and easy access to large imperial markets.

The German chemical industry during the interwar period was very keen to embark on cooperation and cartelization; remember that *Kreditbanken* had long promoted a highly cooperative attitude among industry leaders (Da Rin 1996). The 1905 cartel inquiry of the German Interior Ministry located 47 cartels in chemicals (Haber 1971, pp. 267–69). In their role as creditors, banks earn their profits mostly from the interest charged on loans. Therefore, their main interest is to keep their clients alive, rather than highly profitable. This is one reason why we tend to observe less competition in economies where large banks are major lenders to industrial firms, as in Germany or Japan.

Germany's *Kreditbanken* were badly hurt by the German hyperinflation of the early 1920s, which shrank their financial assets. By 1927, no bank remained among the ten largest firms. The banking industry was swept by a wave of mergers, which laid the foundations for the present system centered on the big three: Deutsche, Dresdner, and Commerzbank. However, even with the weakening of banks, an anticompetitive attitude persisted in Germany in the interwar period. Concentration of the German chemical industry reached its apex in 1926, when the big producers of organic chemicals consolidated into IG Farben, which took the form of "community of interests" (*Interessengemeinschaften*) (Plumpe 1990; Haber 1971). Concentration was a response of large dye makers to mounting competition from American firms. The capital of the firms that merged into IG Farben had roughly doubled during World War I. Hyperinflation benefited them, since their stocks of materials were increasing in value. For instance, in 1920 BASF stocks (in good part, fertilizers) amounted to 80% of its assets. However, given the depressed state of the German economy, the liquidity of their assets shrank, so that another decrease in sales would trigger a financial crisis. Forming IG Farben was a protection against that risk. Moreover, the assets of IG Farben grew at an average of 3.5% per year in its first decade, more rapidly than those of any other

German chemical firm. Farben dwarfed its internal competition; it was ten times as large as the second-ranking firm, Deutsche Solvay.

The capitalization of IG Farben was similar to that of ICI. One important advantage over ICI was the high depreciation rates allowed by the German commercial law, which could be very important to a capital-intensive industry. About one-half its initial capital was indeed funded this way. Between its enormous size and favorable depreciation treatment, IG Farben was completely free from bank influence, and its base of investors was dispersed enough so that capital markets exerted little oversight. IG Farben engaged in a centralized and meticulous financial planning, however, which rationalized production and commercialization. Even more importantly, the firm engaged in a centrally planned effort of diversification through R&D and capital investment. IG Farben was far more dependent on exports than was ICI, and apparently this dependence provided enough pressure on management that supplementary governance of financial intermediaries and capital markets was unnecessary. In the decade after their formation, the R&D-to-sales ratio was much higher at IG Farben than at ICI: 9.9% and 2.4%, respectively (Haber 1971). While in 1913 three product categories (heavy chemicals, pharmaceuticals, and dyestuffs) accounted for 96% of the sales of the firms that were later to constitute IG Farben, by 1938 they accounted for only 51%. New products such as synthetic rubber, metals, rayon, fibers, new fertilizers, and gasoline were introduced, and their development absorbed half of the capital investments from 1930 to the end of the war (Plumpe 1990, p. 264). At this time, the German industrial-research laboratories reached their fullest expression. Growing earnings sustained systematic diversification into new products, which in turn sustained further growth of earnings. As Herbert Levinstein wrote, ICI "depends for success on making comparatively few products extremely well and selling them at prices usually fixed by international agreement," whereas IG Farben "relies on making a large variety of substances constantly changing in the range as new products appear in the laboratory and are taken up for large scale manufacture because they seem to have profit-making capacity" (Haber 1971, p. 300).

It is revealing to compare the financial record of IG Farben with that of its leading competitors, ICI in Britain and DuPont in the United States. The three companies were of very similar size; in 1929, the capitalization of IG Farben was $404 million, compared to $464 million at DuPont and $502 million at ICI. This ranking almost corresponded to the stock of capital equipment of the three: $203 million at IG Farben, $443 million at DuPont, and $240 million at ICI (Plumpe 1990, pp. 184–93). DuPont and IG Farben suffered from the depression in similar ways, losing about 11% of their sales per year in 1929 to 1932, much more than ICI's loss of 1%. Later in the decade, they grew at about 11% a year, while ICI grew only 6%. What is most notable is that DuPont could exploit a large and growing internal market in an era of increasing trade barriers, whereas IG Farben still depended on exports for most of its sales. The creation of IG Farben, then, was a successful concentration of financial resources in a period of high uncertainty. Moreover, Farben then showed an ability to generate high earnings, which allowed a larger outlay of R&D.

By the mid-1930s, however, the pattern of research at IG Farben was increasingly dictated by the government. Investments were directed toward commercially unrewarding research, especially on high-pressure chemistry for hydrogenation and for synthetic rubber. While the government subsidized the immediate costs of this research, the main loss was the shift in research focus away from fundamental research, a move that retarded the development of polymer science, and the shift to oil as feedstock.

The American chemical industry established itself in the interwar period as the most dynamic in the world. More chemical firms entered the list of top 200 U.S. manufacturers in this period than firms from any other sector (Chandler 1990). Like the British and German chemical industries, the U.S. industry also profited from World War I. Sales at DuPont, for example, increased tenfold during the war. As in Europe, firms used these resources to expand through acquisitions, plant expansion and research, which induced a self-sustaining process. Acquisitions and new research allowed diversification. Diversification provided new profits. Profits upheld further growth, expansion, and research. The stability of profits earned the industry the reputation of being "recession-proof." Indeed, 38 of the largest 45 chemical companies listed on the New York Stock Exchange continued paying dividends even during the Depression, and their R&D activities were not cut back substantially.

In fact, an increased R&D effort was the basis for the growth of American chemical firms. In 1930 they employed 30% of the scientific personnel in all of U.S. industry (Chandler 1990, p. 171). Dow, DuPont, and Monsanto all spent between 2% and 3% of sales in R&D, increasing such sums with time. Each major firm developed fundamentally new products in the 1920s and 1930s, some of which, like polymer chemistry, revealed their full potential only after World War II. DuPont developed the commercial potential of polyethylene (which had been discovered by ICI), and Dow commercialized polystyrene. Both became huge earners.

The American chemical industry also saw a wave of mergers and consolidations in the 1920s. Haynes (1954) estimates that 500 mergers occurred in the U.S. chemical industry in the 1920s, which gave rise to at least two large firms, Allied Chemical and Union Carbide and Carbon. Dow employed its wartime profits to acquire smaller competitors who could help in exploiting its research on by-products of ethylene and benzene. As a result of this successful policy, Dow's assets quadrupled from 1918 to 1929. American Cyanamid and Monsanto also made a number of acquisitions during the 1920s.

DuPont provides a vivid illustration of the cycle of profits, innovation, and diversification. DuPont grew by exploiting the wartime profits of its explosives division, systematically diversifying into new fields (Chandler and Salsbury 1971; Haynes 1954). In 1913, it had $75 million in assets, and 97% of its income came from explosives. By 1940, it had $935 million in assets, and less then 10% of its income came from explosives. DuPont developed a powerful R&D capability. In the 1920s, its nitrocellulose know-how led to viscose and acetate rayon, cellophane, synthetic ammonia, tetraethyl lead, and lacquers. Synthetic ammonia required an investment of $27 million and 10 years of research. The 1930s saw major discoveries such as nylon and neoprene, whose cost of development was respectively about $4.3 million and $2.5 million.[16] Between 1920 and 1939, DuPont spent $40 million in R&D, and in the 1930s it feasted on the fruits of its 1920s research effort. DuPont also bought 23% of General Motors with wartime profits, which brought the company both an incredible stream of dividends and a crucial strategic alliance.

However, no concentration took place in the United States on a scale equal to that of ICI in Britain or IG Farben in Germany. America had passed antitrust laws to discourage or block such combinations. In addition, the sheer size of the market made

---

[16]Hounshell and Smith (1988) provide a detailed study of DuPont R&D strategy over time.

it difficult for any firm to become hegemonic. DuPont, the largest American chemical firm in 1930, ranked twelfth in size among U.S. companies, and there were 4 chemical companies among the largest 50 firms. ICI, ranking fourth, was the only industrial chemical company among the first 50 in Britain. In Germany, IG Farben was the second largest company. Both ICI and IG Farben dwarfed their national competitors, unlike DuPont, which faced competitors such as Monsanto, Dow, Allied, and American Cyanamid.

The American financial system contributed to these developments of the chemical industry. Between 1927 and 1934, U.S. financial regulation was completely reshaped, opening the way for impersonal market-based finance. Three new laws were particularly important (Perkins 1971). The MacFadden Act (1927) de facto stopped the trend toward a wider use of branch banking. The Glass-Steagall Act (1933) barred commercial banks from dealing in securities and introduced deposit insurance, a measure that helped smaller banks, which would otherwise have been riskier. The Securities Exchange Act (1934) instituted the Securities and Exchange Commission to regulate the securities markets and their intermediaries. The first two changes limited the influence of banks and the last encouraged the use of equity and bond markets.

The chemical companies thus began to raise money more aggressively in equity and bond markets. This reliance of U.S. firms on non-bank financial markets dates back to before World War I, but for chemical companies, its force became apparent in the interwar years. DuPont was the first chemical firm to raise large sums on the stock exchange; it raised $60 million in 1915, a huge sum for the time. Dow and Monsanto issued bonds and raised equity during the 1920s. The high profitability of chemical companies had always been known, but with the consolidation of the industry in the 1920s, its riskiness was lowered and firms could find the support of outside investors. Investors had two other sources of comfort, as well. The chemical companies still remained closely held by the "founders" and their families; as they made the transition to being controlled by professional managers (Chandler 1990), this control continued to assure close supervision and dedication to long-term company growth. In addition, the large and powerful investment banks played an important role in brokering securities issues, and their reputations reassured outside investors, as well.

A capsule financial history of Dow gives a flavor of much of what was happening in the U.S. chemical industry at this time.[17] During World War I, Dow accumulated hefty profits. It distributed extra dividends (between 10% and 65%) between 1915 and 1919. Its investment in technology was sustained as the firm diversified on the basis of its electrochemical know-how. By 1932, Dow employed 600 engineers and 100 chemists in its works and laboratories. Over the 1920s, the closely held equity was spread over a wider public through issue of new equity and a keen effort in publicizing the company outside its home territory. In 1929, Dow distributed 4-for-1 shares to capitalize reserves, and in 1931 it distributed to shareholders $900,000 of surplus with an extra dividend, an act that was repeated in 1934 (50% extra dividend). Such lavish returns were instrumental in attracting the attention of investors toward this company. By 1939, Dow's shareholders numbered 4,211; but Dow, like DuPont, still remained under the control of the founders.

The picture that emerges from these three countries during the interwar period is one of deep change in the nature of the industry, with the ascent of managerial companies, greater sophistication of technology, and fundamental scientific discoveries. However,

---

[17] The following analysis is based on the financial historical files conserved at Dow's Post Street Archives.

European financial systems seemed to provide less oversight for management, although in some cases, as in that of IG Farben, the pressure of global competition provided some substitute for oversight from the financial markets.

THE MATURATION OF THE CHEMICAL INDUSTRY

For the chemical industry, the period since World War II can be divided into two very different phases. During the first phase, beginning with wartime government subsidies that supported sustained capital investment, the chemical industry grew strongly up to the oil shocks of the 1970s. Then growth plummeted, and the industry suffered from accumulated excess capacity. Moroever, the industry's rate of innovation fell during the 1960s and 1970s. By one count, of the 63 major innovations in the chemical in industry since 1930, only 20 were introduced in the 1950s and 1960s and a mere 3 in the 1970s and 1980s (Bozdogan 1989, p. 20). New products such as dyes, rayon, nylon, or plastics, which created fortunes, no longer appeared. This drop in innovation in part reflected industrial maturity, but also a slack in research intensity and, consequently, profits. This trend, however, seems to be reversing itself since the late 1980s.

Financing Growth after World War II

For the chemical industry, two decades of intense and widespread growth followed World War II, sustained by seemingly endless economic growth. Petrochemicals, pioneered by American firms, induced a revolution in both processes and products (see Arora and Rosenberg, this volume). No new major firm has emerged in the chemical industry since the 1920s; today's large diversified competitors are those who came out of the Depression, and firms that were created during this period—whether start-ups or firms specializing in commodity chemicals or new specialties—have played only a marginal role. One reason for this lack of large, new entrants has been the capital intensity and research intensity of the industry, which constitute formidable barriers to entry.

Naturally, the financial sector plays an especially important role in an industry that is capital- and research-intensive. Over this time, steady growth loosened the constraints on capital allocation. This change coincided with a relaxation of corporate governance in the Anglo-Saxon world, as the increase in the size of companies was not matched by increased oversight skills by financial intermediaries; share ownership became ever more dispersed in the United States, Germany, and Britain. The reduced pressure from shareholders likely contributed to the slowdown in research effort and its associated capital investment, which are the basic determinants of long-term growth. However, these forces were somewhat different across countries.

In the United States, the Bank Holding Companies Act (1956) reaffirmed the American penchant for preventing banks from becoming large and powerful. External financing for firms came mostly from issue of bonds and stock, as the securities markets increased their popularity. Institutional investors such as mutual and pension funds or insurance companies appeared, but only a very small part of savings passed through them. Compared to others, chemical firms tended to rely more on retained earnings than on external finance, with the notable exception of Dow. Dow proceeded in its traditional policy of high leverage and continuous expansion of capacity. Its debt-to-equity ratio increased with time; it was about 0.7 in the late 1940s, exceeded 1.2 in the early 1950s, then declined to about 0.4 in the early 1960s, when income was very high.

By the time the oil shock struck, vigorous expansion had brought the ratio back to more than 1. High debt was used by Dow to expand capacity and achieve low production cost in commodity chemicals. The low interest rates of this period made this a profitable strategy, and the later increase in inflation reduced the real value of debt. Dow also pursued a policy of relatively high dividends; for example, its dividends doubled between 1963 and 1970.[18] Other firms were more conservative. Monsanto and DuPont, for instance, rarely maintained a debt-to-equity ratio higher than 0.5.

This period also saw a shift in emphasis from fundamental research to product development and capital expenditure, mostly due to the high reward in perfecting petrochemical processes. For instance, at Dow the ratio between R&D and sales reached about 7% in the 1930s, but after World War II it slowed down to between 4% and 6% and eventually descended to 3% in the 1960s and 1970s. At DuPont, the ratio was 7.8% in 1960 and fell to 6.5% in 1970 and 4.3% in 1980. The lack of innovations at this time is exemplified by the dismal experience of DuPont and Monsanto in setting up "new ventures" programs in the 1950s and 1960s. These programs aimed at nurturing entrepreneurial ventures within the company (DuPont) or at spotting them outside and providing finance (Monsanto). Both projects, however, failed to deliver satisfactory results and were repealed. The primary strategic reason for this failure was a lack of connection between new ventures and firms' operational divisions (as Chandler, Hikino, and Mowery maintain in their chapter).

In Britain, ICI found itself facing both low productivity and the loss of its virtual monopoly over Commonwealth markets. Its profitability was correspondingly lower than that of U.S. firms: trade margin over sales at ICI averaged 13.6% in the 1960s and fell to 11.7% in the 1970s, compared to 13.0% at DuPont and 17.1% at Dow (Pettigrew 1985, table 8). ICI was slow at upgrading its product portfolio and at building a petrochemical base (Pettigrew 1985, p. 217). The long-term research effort of ICI was lower than in the United States and Germany, as its R&D-to-sales ratio had remained below 3.5% from the 1950s throughout the 1970s.[19]

ICI was slow in reacting to the challenges of increased competition. One reason for its sluggishness is the mild scrutiny that this widely held company received from its shareholders. In this, Britain and the United States were similar. At this time, ICI's chairmen served a fixed term and were not held accountable for performance. Also, hostile takeovers were rare until the 1960s, when diffusion of ownership and increased transparency of accounting requirements made it possible for bidders to bypass a company's directors and appeal directly to its shareholders (Hannah 1976).[20] ICI increased its capital by retaining reserves rather than by issuing equity or debt. Only in the 1970s, when inflation made debt very attractive, did ICI make more use of capital markets: in 1969, its debt-to-equity ratio was less than 0.1.

The lack of discipline induced by British capital markets can also be seen in the research-intensive field of pharmaceuticals, which currently accounts for nearly one-

---

[18]Dow Chemical, PSA File 00624.

[19]In a recent analysis (Department of Trade and Industry 1996), Richard Freeman of ICI writes that U.K. companies and sectors of industry continue to underinvest in R&D compared with other industrial companies. The new ICI, for example, spent only 1.8% of sales on R&D, compared with what would normally be in the range of 3% to 4% for heavy-chemical companies of this size in other nations. This has been historically true, going back to the days of ICI and IG Farben.

[20]Ironically, ICI launched one of the first (unsuccessful) hostile bids in 1961, for Courtaulds. The bid alone made Courtaulds' shares appreciate by about 70%. However, the move was criticized as "un-British" by both firms' financial advisors (Morgan Grenfell and Barings) and by the Church Commissioners, a large shareholder of both companies.

third of total R&D investment in the United Kingdom. Glaxo was the largest British spender in R&D in the industry after the war. In 1957, its R&D-to-sales ratio was 3.3%, compared to an average of 6.3% for seven U.S. large pharmaceutical firms. The trading margin at Glaxo was 39.9%, against 64.9% for the U.S. firms. The U.S. firms also paid higher dividends: an average of 26.3% versus 9.2% at Glaxo (Davenport-Hines and Slinn 1992, p. 169). A similar pattern continued well into the 1960s.

In Germany, the Allied powers reorganized the German industrial system after World War II. In the chemical industry, they broke up IG Farben into its main constituent firms: Bayer, BASF, and Hoechst. These were among the largest German firms, and their stock listing in 1953 was an important milestone for the German equities market. However, in the years after World War II the German financial system continued to focus around banks. The "big three"—Deutsche, Dresdner, and Commerzbank—suffered increasing competition from commercial, regional, and savings banks, which entered into industrial lending (Edwards and Fischer 1994). Although stock ownership became widespread in Germany, banks retained power through proxy voting and still had close ties with their clients. In addition, they still controlled capital markets, issuing most bonds themselves and using the proceeds for long-term credit.

The corporate governance structure of chemical firms has been no exception to the general German pattern. When German companies raised new capital on the stock market, they did so with support from banks. For instance, Hoechst increased its capital several times in the 1950s, raising 1 billion marks (Baumler 1968, pp. 355–69). Funds set aside to pay future pensions were another important source of funds. At BASF and Bayer, they accounted for between 25% and 45% of equity, at times being larger as a source of funds than bonds and loans combined. Notice, however, that this form of finance is managed directly by the company, thus making management less accountable to outsiders.

German banks have been the controlling proxy votes for the major chemical firms, constantly representing about two-thirds of shareholders and the vast majority of the votes. Banks have thus shielded chemical companies from hostile takeovers. Moreover, Bayer and BASF have a 5% statutory limit on voting rights, also meant to discourage takeovers. However, in Germany, the lack of a market for corporate control does not seem to have supported managerial entrenchment at this time. Research and profits were both strong (the R&D-to-sales ratio was between 4% and 5%), though less then for American firms. Banks seem to have acted as reasonably efficient delegates of shareholders. Several studies of the German chemical industry have found a positive effect of bank closeness to firms during the postwar period (for example, Cable 1985; Gorton and Schmidt 1996).

By the 1960s, the pattern of finance started shifting somewhat in Germany, away from domestic bank loans and toward alternate forms of finance. BASF and Bayer exemplify the changing pattern. BASF started using bonds in the 1960s. Now their ratio of bonds to bank credit is nearly 1:1. When loans are received, they are now much more likely to be from foreign sources or even in foreign currencies.[21] While monitoring by banks worked fairly well for German chemical companies during the 1950s and 1960s, that success does not prove that this mechanism is superior at all times and places. Bank governance may be particularly effective in times of stable macroeconomic growth, when it is possible to push management without a dramatic restructuring of the

---

[21] Loans show the increasing internationalization of BASF, which by 1994 had obtained seven syndicated loans for 1.9 million marks, one loan for 60 million Swiss francs, and six loans for $880 million. Ten loans were raised for $1,320 million, one loan for $40 million Canadian dollars, three loans for 300 million Swiss Francs, one loan for 100 million French francs, and three loans for 240 million marks.

corporation; however, if restructuring measures are needed, the presence of banks may make adjustment less quick and thorough than would an active stock market.

Restructuring, Stagnation, and Recovery

The 1973 oil shock hit the chemical industry with dramatic force and induced a restructuring whose pattern differed in each country (as described in the chapter by Arora and Gambardella). At about the same time, financial markets in all countries were going through regulatory changes, generally in the direction of liberalization and deregulation. The interplay of the economic shock and finance had two important effects: a reduction in cash flow and a change in corporate governance.

The sudden squeeze on cash flow was caused by several factors, from increased competition to reduced growth and lower operating margins. This meant that research activities were temporarily reduced and redirected toward the incremental perfecting of existing processes rather than the development of new products (Bozdogan 1989). While firms could partially restore profitability in this way, it was at the cost of long-term competitiveness. The major players reacted by leaving less-profitable commodity markets and diversifying into high value-added sectors: specialties, pharmaceuticals, and advanced materials. Financing for these changes came largely from the sales of existing plants for commodity chemicals. This evolution of the industry, separating commodity operations from more diversified firms, made strategic sense. Commodity chemical operations resemble mature manufacturing business in that their profitability depends more on careful management than on substantial new research and capital expenditure. Indeed, their management typically has a financial, not an industrial, upbringing. This contrasts with the research and planning effort necessary to make diversified firms successful in the long run and with the specialized knowledge of niche markets (see Arora and Gambardella, this volume). Here is where the national stories start to differ.

American firms restructured more quickly and thoroughly than did firms in Europe and Japan (Lane 1993; Lieberman 1990). One reason can be found in the more active and liquid American capital markets. One major tool was the leveraged, or management, buyout, which grew from a combination of deregulation, increasing power of investment banks, and depth of financial markets. Using leveraged buyouts, newly formed companies can issue large amounts of bonds and buy plants. The resultant extremely high debt-to-equity ratios provide management with powerful incentives to perform in the interest of shareholders (Jensen 1989), as the menace of default leaves little room for managerial slack. Companies such as Cain Chemical, Huntsman, Sterling, Vista Chemical, Aristech, and others were instrumental in turning over capacity in the chemical industry along these lines. Huntsman, for example, bought polystyrene plants from Shell and Hoechst, and Cain bought polyethylene plants from DuPont. About 50 high-leverage restructurings took place in the 1980s, mostly in commodity chemicals such as ethylene, methanol, polyethylene, PVC, propylene, and styrene (Lane 1993). In the first half of the 1980s, Monsanto, Dow, Union Carbide, and DuPont sold plants and businesses. In 1985 alone, Monsanto sold $900 million of plants and wrote off a similar amount, and Dow divested $1.8 billion. These companies also issued large amounts of equity to finance part of the acquisitions and new plants that completed the move into new lines of business. Dow, for instance, doubled its equity between 1975 and 1980 and again between 1985 and 1990 (also capitalizing reserves).

Leveraged buyouts are one main instrument in the market for corporate control. The other main instrument, takeovers, had less of an impact on chemical firms. There was an unsuccessful attempt to take over Union Carbide after the Bhopal tragedy (Eccles and Crane 1988). Allied Chemical has also been involved in an attempted takeover which, while ultimately unsuccessful, caused the company to divest itself of many of its most profitable activities. Fear of a takeover bid prompted American Cyanamid and Celanese, for instance, to buy back shares as a way of keeping stock prices high and rewarding shareholders (Derdak 1988).

The restructuring of the U.S. chemical industry has been very successful. Profitability has bounced back. In the late 1980s, the return on equity was about 5 points higher for the major chemical firms than for the average large manufacturing corporation (Bozdogan 1989). Moreover, R&D intensity has increased steadily, reaching historical records. At both Dow and Monsanto, for instance, R&D is up to about 7% of sales in the early 1990s. Since 1986, Monsanto is spending more in research than in capital equipment, which is unprecedented.

The monitoring power of America's capital markets has been fostered since the 1970s by the increasing strength of institutional investors. Until the late 1970s, the so-called Wall Street rule held that institutional investors would vote for the incumbent management and, if not, would sell their shares, thus "voting with their feet." Pension funds, mutual funds, and life-insurance companies all operated under rules that required a very passive portfolio policy as a way of diversifying risk. A basic principle of American commercial law, the legal doctrine of "equitable subordination," has also contributed greatly to keeping intermediaries away from active stances in corporate matters. This doctrine implies that a creditor who becomes active when his debtor becomes financially distressed loses the seniority of its claims. Regulation of institutional investors underpinned this attitude. The 1974 Employee Retirement Income Security Act had a major impact in this respect.

The 1980s, however, brought a large amount of deregulation to the U.S. financial industry (Khoury 1985). True, the McFadden, Glass-Steagall, and Bank Holding Company Acts continue to prevent banks from direct involvement in corporate governance and thus help to preserve the autonomy of professional management. However, trading of securities has become ever cheaper and more intense. Several changes in legislation have favored a more active representation of shareholders and bondholders (Roe 1994), which has started weakening the traditional power of American managers. Investment and pension funds, trusts, and insurance companies now intermediate the bulk of savings. Greater activism on the part of ever-larger institutional investors is transforming the nature of the U.S. financial system, although the outcome remains uncertain. All large chemical companies are widely held (except for DuPont). It appears that America's active financial markets are working fairly well to provide managerial oversight.

In Europe, reorganization of the chemical industry has been much slower. Here, besides government intervention and labor market rigidities, an important factor has been the lack of accountability of firms to their shareholders. In Britain, this problem has been endemic. In recent years, however, Britain's financial system has seen some degree of change. Concentration in commercial banking has proceeded further. Clearing banks have been reduced to six and have retained their specialization in short- and medium-term loans and the focus on traditional industries, which have become sectors with relatively low growth. Domestic British banks have gradually given way to other intermediaries such as building societies, pension funds, and life-insurance companies. Takeovers have become more common, as active intervention of fund managers on behalf of investors. Deregulation has made equity markets more similar to the American model, and the importance of institutional investors has increased.

London became the host of the Euro-currency markets, and its stock exchange greatly increased its importance at the European level.

The impact of these changes can be seen in the recent experience of ICI. ICI accumulated losses until the early 1980s, when a far-reaching reorganization finally made it suitable for a more competitive world. ICI exited petrochemicals by swapping its polyethylene plants with British Petroleum PVC facilities in 1982; however, it remained slow in developing new product lines. Its R&D-to-sales ratio hovered at about 3% throughout the 1980s, with no change over the two previous decades. In the early 1990s, ICI split its activities between life science and chemical units, spinning off the former into a new company, Zeneca. This separated the dynamic, research and growth-oriented activities from the more traditional and capital-intensive chemical ones. (Pharmaceutical companies spend around 15% of sales in R&D, against the 5% spent by heavy-chemicals companies.) The change was induced by a takeover threat posed by the Hanson Trust in 1991. From June 1992, when the split of ICI was announced, to December 1994, the equity of the two units appreciated by 57%, more than double the stock market index (Owen and Harrison 1995), and by 1996 the appreciation had climbed to 70%. This does not cover the more recent development where in ICI bought the Unilever specialty chemicals and started selling off its commodities.

Germany's bank-centered system, which excelled at sustaining growth until the 1970s, has been less effective at helping the chemical industry adapt and maintain a high level of competitiveness in the 1980s and 1990s (Cable 1985; Gorton and Schmidt 1996). Among German chemical companies, BASF, Bayer, and Hoechst are all widely held companies with a majority of active voting rights exercised by banks and other institutional investors. Hoechst and BASF each have a strong investor (the Kuwait Investment Office and Allianz, respectively). Chemical firms have then suffered the increasing loss of accountability allowed by the system, especially in the case of BASF (see the chapter by Richards). The transition out of commodity chemicals has been longer and less thorough in Germany than in the United States. Research expenditure has fallen and remained lower for longer. At Bayer, for instance, the ratio of R&D to sales remained around 5.5% throughout the 1980s, recovering to over 7% since 1991. At BASF it has yet to recover. The profitability of German chemical firms has not increased since falling in the 1970s.

There are signs of evolution, even in Germany's financial markets, however. Pension provisions and reserves are starting to play a major role. Pension provisions rose from 5% of liabilities in 1970 to 25% in 1980 and 16% in 1990. Together with securities, which now account for about 10% of liabilities, this rise in pension provisions lowered the importance of bank credit, which has considerably fallen from the postwar levels. At BASF, for example, pension provisions have greatly increased over the years: From 7% of liabilities in 1970, they have climbed to 15% in 1980 and to 18% in 1990. Retained earnings have also become quite large, now about one-half the value of equity.

German institutional investors are still underdeveloped, but new financial intermediaries have appeared and thrived. Mutual funds have been recently introduced. By 1993, 62% of the equity in the German market was held by institutional investors and only 29% by individuals. Nearly one-half of the holdings were abroad. At BASF in particular, 60% of stock was owned by institutional investors in 1995. Given the increasingly international ownership of the large German chemical firms, foreign-fund managers have started voicing their discontent. Some changes in management style have appeared. Kaplan (1994a) shows that in the 1980s, the turnover at the largest German firms was no less sensitive to performance measures than in the United States or Japan. The more substantive issues—in particular, the need to follow ICI and the American firms in splitting the bioscience and chemical businesses—are still pending. In early 1996, mere speculation that Hoechst and Bayer are planning such steps caused an increase of 40% to 80% of their stock, which shows how much behind they had fallen. These steps have yet to emerge, and the German tax system makes them difficult. Nevertheless, it is fair to note that the German system still does much to shield firms from takeovers, and a market for corporate control has not yet developed.

In America, Britain, and Germany, the nature of corporate governance has evolved toward more attention to shareholder value. As the chemical industry has been reshaped in recent years—particularly through the separation of commodity from specialty chemicals—activism of investors has become extremely important. Globalization has affected both the competition among products and the financial markets, where large diversified companies compete for capital. In a shifting competitive environment, financial markets may prove even more important in shaping the chemical industry than they have been in the past.

THE JAPANESE PUZZLE

The Japanese chemical industry was very slow to develop; in fact, Japan relied almost exclusively on imports of chemicals until early in the 20th century. Only fertilizers achieved some importance before World War I. There were many small firms. They stayed small, at least in part, because of Japan's inefficiently regulated financial system. In the late 19th century, Japan's financial system consisted of a large number of small private banks. Securities markets were virtually nonexistent. The existing banks financed mainly agriculture and trade. Their endemic weakness led to various attempts to create a system of commercial banks along the U.S. pattern.[22] The most notable of these attempts was the 1890 Banking Act, which brought about a dual system of unit local banks and few (branched) national banks. At about this time, the Industrial Bank of Japan (IBJ) was created to engage in long-term industrial loans and to serve as a conduit for attracting foreign capital, but for some years, its activity was purely nominal. Probably the most powerful source of finance was available through the *zaibatsu*, which were powerful industrial conglomerates centered around trading companies that had evolved into industrial concerns (Da Rin 1997b).

The story of the first major success story in the Japanese chemical industry illustrates the characteristics of Japan's financial system. Two entrepreneurs, Noguchi Jun and Fujiyama Tsuneichi, pioneered the adoption of electrochemical processes and founded a company called Nichitsu. However, as the growth of Japan's chemical industry was slow, retained earnings were insufficient to allow a powerful accumulation of resources. So, like the American chemical pioneers, Jun and Tsuneichi experimented and tinkered, raising finance in informal capital markets and among friends, acquaintances, and rich individuals. Banks and capital markets turned them down repeatedly. To keep their independence, they also refused finance from Mitsui, one of the most powerful *zaibatsu*. Only when the firm had become rather established, in 1910, was it lent substantial sums

---

[22] For an in-depth discussion of the development of the Japanese chemical industry, see the chapter by Hikino et al. in this volume. In addition, the early development of chemicals in Japan has been carefully reconstructed by Molony (1990). Pressnell (1973) and Goldsmith (1983) provide a good account of the development of Japanese finance.

by the Industrial Bank of Japan (IBJ). Ownership of their firm was very concentrated: the ten largest shareholders controlled 70% of it, and Jun himself about one-quarter. In its broad outlines, this story closely resembles that of Dow or Monsanto.

During World War I—in which Japan did not take part—the development of Japanese heavy industry required large financial sums. The powerful *zaibatsu* had recourse to internal capital markets, borrowing little from outsiders. They were among the few large issuers of securities and accounted for the greatest share of industrial-capital formation. Banks also helped Japanese industry as a whole by recapitalizing and merging. However, the Japanese financial system remained fragmented and unique. It was different from the British in that credit was often given for the long term. It differed from the American in that it made little use of securities. It differed from the German in that banks were not universal. World War I was very profitable for Nichitsu and the smaller Japanese chemical firms. Profit rates ranged between 10% and 30%. During the war, Japanese government help allowed Nichitsu to issue bonds. After the war, Nichitsu—still the only large chemical firm in Japan—was powerful enough to become the center of a new *zaibatsu*. By the mid-1920s, Nichitsu had assumed a holding-company structure that allowed an internal capital market to substitute for external credit. In this period, large profits allowed sustained growth of the group. In 1930, Nihon Chisso, Nichitsu's most profitable subsidiary, ranked sixth among industrial firms. At about this time, two other Japanese companies also came to engage in electrochemicals: Dai Nihon Fertilizers and Showa Fertilizers, both parts of industrial *zaibatsu*. Both were financed by their holding companies, with little involvement of banks.

Apart from Nichitsu, which could depend on retained earnings, and Dai Nihon Fertilizers and Showa Fertilizers, which were part of *zaibatsu*, most Japanese chemical firms in the interwar period were short of resources to invest in systematic research. The few resources spent on research went to understanding imported technology. Indeed, Japan's self-sufficiency in dyes, soda ash, and ammonia remained below 50% until the late 1920s. Profits were generally low.[23] Without access to capital, such firms as Asahi Glass or Nihon Soda were unable to expand to an optimal scale, which only reinforced their competitive difficulties and financial weakness.

In the 1930s, the rise of new industrial *zaibatsu* brought the older conglomerates to respond and invest in other chemicals, mostly fertilizers. The new firms were dependent on the mining parts of the conglomerates, as they exploited coal derivatives. They also catered mostly to clients within their holding group, a factor that kept their size relatively small. Mitsui, Sumitomo, and Mitsubishi opened ammonia plants. An advantage of operating within a *zaibatsu* was that not only did it provide group finance but it also opened the way to issuing securities on the stock exchange, although this was possible for only small amounts. In the late 1930s, with mounting profits, most *zaibatsu* started purchasing technology to enter into explosives, coal-tar derivatives, and dyes. The largest chemical firm, Nihon Chisso, now ranked sixth among all industrial firms. There were 5 chemical firms among the largest 50, and 9 among the largest 100 (Fruin 1992, appendix). As part of the Sino-Japanese war effort, the government played an important role in fostering investment in chemicals for war purposes. This greatly stimulated the capital outlay of the industry, and by 1940, there were 18 chemical firms among the 100 largest.

Japanese financial markets also evolved in several other ways during the 1930s. The *zaibatsu* firms began to use cross-shareholding to avoid diluting group ties and to secure

---

[23] Miyajima (undated) reports the large losses of several firms—especially in dyes—during the 1920s. Dependence on exports was common to most sectors.

stable ownership. The IBJ started developing its skills as an industrial investment bank, becoming the main long-term lender to modern industries. A money market was also created after the Depression, easing the liquidity needs of banks. The banking system consolidated substantially; the number of banks in Japan fell from 1,900 in 1921 to 418 by 1936 and to 61 by 1945.

Thus, the main traits of the struggling Japanese chemical industry until World War II were the dependence on *zaibatsu* and the difficulty of raising enough capital—whether through retained earnings or financial markets—to engage in fundamental research. It is worth stressing that belonging to a *zaibatsu* was a mixed blessing for a chemical company: It provided access to finance and oversight of managerial performance; however, intergroup trading also restrained growth, as it pressured chemical firms to stay within the span of the *zaibatsu*.

The end of World War II brought major changes to Japanese finance and industry. After the war, the American occupation tried to dismantle *zaibatsu* and reshape the financial system along American lines. Other than installing an American-style separation of the banking and securities industries, they had little success. However, their efforts did induce a change from industrial *zaibatsu* to financial *keiretsu*. Where *zaibatsu* were centered on a trading company, *keiretsu* have their core element in a bank and are more horizontal in structure. The financial and credit markets at this time were put under tight control of the Bank of Japan, which regulated them heavily (Ueda 1994). Securities markets were repressed. For a Japanese chemical firm, being connected to a *keiretsu* proved very important, as the repressed securities markets in the postwar period meant that bank financing was of foremost importance.

The financial *keiretsu* system, which features cross-shareholdings of large numbers of firms clustered together, is parallel to, but distinct from, what has come to be known as the established "main bank" system of corporate finance and governance.[24] The main bank system had its heyday in the 1950s and 1960s. This system rests on close relationships between a firm and its borrowers and shareholders, which are typically financial intermediaries and other firms. The relationships give rise to a complex corporate governance structure, whereby a firm borrows from several banks, one of which—the main bank—assumes responsibility for monitoring the firm's financial situation and disciplining its managers. The main bank takes the largest share of the firm's loans and assumes the burden of losses, should it default. Participation of many banks in many such pooled-lending agreements assures that none is tempted to disregard its obligations and save on the cost of monitoring firms. These incentives are reinforced by supervision and enforcement on the part of the Ministry of Finance and the Bank of Japan. The main bank structure of the 1950s and 1960s rested on regulation, which restricted firms' access to securities markets and thus forced them to obtain finance through banks. Thus, the Japanese system of corporate governance was a balance among several elements, including the main bank relationships between firms and their main sources of external finance, and the cross-ownership within *keiretsu*.

Belonging to a *keiretsu* was a mixed blessing, just as belonging to a *zaibatsu* had been. Membership helped to assure stable growth, by providing an internal market for a firm's products and shielding it from competition. In general, a *keiretsu* often seeks to develop industries that can absorb production of others in the group; for instance, Mitsubishi-Monsanto Chemicals was established to provide an intra group market for the carbide produced by Mitsubishi Chemicals. On the other hand, *keiretsu* groups pose

[24] A complete and thorough analysis can be found in the collection of studies edited by Aoki and Patrick (1994). See also the chapter by Hikino et al.

constraints on the growth of their firms. A successful firm cannot easily displace others in the same group, which can hamper diversification and limit growth. In times of recession and contraction of production, belonging to a *keiretsu* limits the possibility to consolidate or terminate businesses. Cross-group merging of complementary firms is unlikely. Intra-group mergers are also made difficult by intra-group politics. The troubled mergers of Mitsubishi Petrochemicals with Mitsubishi Kasei and of Mitsui Toatsu with Mitsui Petrochemicals after the oil shock are two instances of this sluggishness.

The cooperative nature of Japanese industrial organization has also influenced the growth of firms. Nakatani (1984) examines a sample of 317 manufacturing companies, including 33 *keiretsu* chemical firms and 5 independent chemical firms. He finds that over the period from 1971 to 1982, *keiretsu* companies had lower profits and grew more slowly, but with less variability, than independent firms. *Keiretsu* firms were also more indebted, a sign of easier access to bank credit. Belonging to a group also hampers the possibility to diversify. Japan has only a few diversified chemical companies—Mitsubishi Chemicals, Sumitomo Chemicals, Mitsui Toatsu, Showa Denko, and the smaller Ube Industries—and even these have a much narrower span than that of DuPont, Dow, Bayer, and BASF. The *keiretsu* structure has prevented firms from integrating downstream, where other firms in the group were already active, so Japanese "all rounders" are mostly petrochemical businesses.

These pressures have led Japanese firms to being smaller than their competitors. In 1954, the largest chemical firm ranked twenty-first among industrials, and there were 8 chemical producers among the largest 100 firms. In 1973, the largest chemical firm ranked still only eighteenth, and there were 10 among the largest 100 (Fruin 1992, appendix). In 1995 the firm with the largest capitalization was Mitsubishi Chemicals, which ranked sixty-fifth among all corporations, and only 3 firms ranked in the first 100 (*Business Week* 1995). This fragmentation into separate groups contributed to the low profitability of chemical firms in several respects: First, it constrained their freedom to compete in the global marketplace; second, it directed their technical capabilities away from the exploitation of scale economies and toward high customization of products (see the chapter on Japan by Hikino et al.).

It is interesting to notice that in both Germany and Japan, the two countries with bank-centered, stable ownership, the system has performed much better in times of steady growth. In Japan, for example, the 1950s saw the entrance into petrochemicals of the first four companies—Sumitomo Chemicals, Mitsui Petrochemicals, Mitsubishi Petrochemicals, and Showa Denko—with five more firms entering in the 1960s. The backing of strong banks and the guidance of MITI (Ministry of International Trade and Industry) achieved a formidable result and helped coordinate the growth of these firms. No independent firm entered this business. By the time of the first oil shock in the 1970s, Japan had 13 major producers.

Much less effective has been the attempt by group banks and MITI to restructure after the oil shock. The protective role of groups in Japan meant that rival *keiretsu* sought to uphold their respective companies. Within the same *keiretsu*, rivalry between chemical and petrochemical companies made combining them difficult. The Anglo-Saxon, market-based systems appear to ensure quicker managerial action when the strategic decision is about restructuring or downsizing. The need to subordinate the strategy of single firms to the whole *keiretsu* group has slowed the response of the major Japanese chemical producers. Whether this should be considered a damage to the single firm or an efficient arrangement for the group is a difficult question.

The *keiretsu* ownership structure has some other features in common with the German pattern, as well. In both Japan and Germany, ownership is in part in the hands of other enterprises, unlike in America. In Japan, the limit for bank share in the ownership of an industrial company was set at 10% for most of the 1970s, reduced to 5% only in 1987. By putting together a few such participations, banks and insurance companies can easily exercise control over management. Indeed, such ownership patterns can be found in most industrial companies. At the same time, since few financial intermediaries own sizable portions of equity, firms are typically shielded from hostile takeovers.[25] The situation contrasts starkly with the more dispersed and less stable ownership of Anglo-Saxon firms, whose shareholders are both individuals and intermediaries with small holdings.

That Japanese firms are insulated from a market for corporate control, however, does not mean by itself that their managers are exempt from discipline. The willingness of Japanese banks to discipline under-performing managers has been convincingly documented. In particular, Sheard (1994) describes in detail the structure of bank interventions in case of financial distress and gives examples for some chemical companies: Mitsui Chemical (1966), Nitto Chemical (1965), Maruzen Oil (1963–94). In these cases, replacement of incumbent management with bank officials and revision of corporate strategies were enacted. Banks remain extremely important in the Japanese economy; for example, the ratio of bank assets to GNP in 1990 was 0.39 in Japan, 0.36 in Germany, 0.08 in the United States (*Fortune*, August 26, 1991, pp. 174–76). However, Japan also saw in the 1980s a marked trend toward deregulation of financial markets. Powerful banks now coexist with developed securities markets. The four large securities houses that emerged after the war—Nomura, Daiwa, Nikko, and Yamahichi—have increased their activities and power with time. This has weakened the main bank system, as large firms have gained access to new sources of finance, both domestic and foreign. As the deep financial crisis of the latter part of the 1990s is showing, such changes are proving far reaching but also painful.

Unlike in the Anglo-Saxon countries, though, the deregulation of securities markets has not resulted in an increased contestability of corporate governance. Operations such as takeovers or leveraged buyouts have been virtually absent from the Japanese economy. This combination of more decentralized ownership and little apparent contestability of corporate control raises a concern that the accountability of corporate management may be diminished. However, Kaplan (1994b) shows that, although the turnover and pay of Japanese managers are less responsive to stock prices than in America, they are more sensitive to negative earnings. Comparing Japanese and American firms, Prowse (1990) finds support for the idea that the main bank system may solve agency problems better than the American stock market does. Thus, in both Japan and Germany it seems that the steps taken toward deregulation had little deleterious impact on managerial behavior in the the 1980s, but the issue remains one to watch in the 1990s.

CONCLUSIONS

Finance matters. It is especially important in the case of chemicals, where the two main sources of success—research and capital expenditure—both require an up-front investment and the passage of time to generate profits. The way finance exerts its

[25]The average of the top five holders for the fourteen Japanese largest firms between 1967 and 1992 was 19% (Roe 1994, table 3).

influence, however, is less straightforward. Several points emerge from the present analysis.

First, the ability to tap diverse sources of external finance has always been important in the chemical industry—and not only at the early stages of a firm's life. Firms that could rely on large cash flows have found it easier to retain their competitive edge. The profits made during World War I promoted the leadership of several American firms such as DuPont. Firms that find it difficult to obtain funds may see their development slowed. Early U.S. firms developed more slowly than did American domestic demand, partly because they could not find external sources of finance and so had to wait to make large investments in technology. Early Japanese firms had a similar experience.

Second, availability of capital is not a guarantee of success. The effectiveness of corporate governance is equally important, as the failure of United Alkali in Britain illustrates. Another example is the experience, during the post-oil-shock period, of the large German chemical firms who enjoyed stable earnings but apparently felt little discipline from their investors. The source of finance does not matter much for the effectiveness of corporate governance. Bank loans, bonds, or equity may all entail good governance but need not do so. German *Kreditbanken* helped the rise of the German industry, whereas British banks did not. The governance features of a financial system also change with time, mostly due to changes in regulations. The deregulated Anglo-Saxon equity markets, for instance, have become much more watchful of shareholder value in the last two decades than they were in the postwar decades. This change has pushed American firms to pursue high-return strategies more relentlessly than have their European or Japanese competitors. Public policy plays an important role in this respect.

Third, the most appropriate type of corporate governance may differ for firms at different stages of maturity. The early German dye-makers benefited from the support of German banks, which allowed them to bear the enormous risks of basic research. Similarly, U.S. biotechnology firms have benefited enormously from finance by venture capitalists. However, more mature firms require less of an infusion of financial support and more discipline for their management. The relative decline of German versus American chemical companies seems to reflect that the bank-centered German financial system may be less effective than the equity-centered American financial system at this stage of the chemical industry's development.

Fourth, the hallmarks of a country's financial system tend to persist for a long time. Some examples include the arm's-length attitude of British banks and the involvement in corporate governance of German banks. The American and Japanese systems have undergone more frequent changes because of sudden regulatory shake-outs. Nonetheless, one can point in America to a continuing desire to restrain banks from a role in corporate governance and to depend on securities markets instead. In Japan, the persistent theme has been the importance of a group industrial structure, whether in the form of *zaibatsu*, *keiretsu*, or main bank. However, in all four countries considered here, the last two decades have witnessed a trend toward deregulation and integration of world securities markets. We should then expect some degree of convergence toward a greater emphasis on shareholder value (see Richards, this volume).

Fifth, finance must be considered in combination with other features of each economic system, especially other regulations that affect competitiveness. This is most clear in the case of Japan, where the early financing of chemical firms responded to the logic of a financial system dominated by integrated *zaibatsu* groups. Key characteristics of this financial system have persisted in the postwar era of financial *keiretsu*. Until World War II, the German economy was hostile to the concept of competition and fostered collusion, which was cherished by banks. In contrast, the American financial environment, where the individuality of firms was embedded in antitrust legislation from the start of the 20th century, developed quite differently.

Financial systems change continuously, along with regulations, technical change, and overall economic development. Since no financial system is well suited for all places and situations, it is important for a country to encourage flexibility in the design of financial markets, so that they can evolve to support the changing needs of the economy.

REFERENCES

Aoki, Masahiko, and Hugh Patrick (eds.) (1994), "The Japanese Main Bank System," Oxford University Press, Oxford, UK; chapter by Masahiko Aoki, "Monitoring Characteristics of the Main Bank System: An Analytical and Developmental View".

Balderston, Theo (1991), "German Banking between the Wars: The Crisis of the Credit Banks," *Business History Review*, 65(3):554–605.

Baumler, Ernst (1968), *A Century of Chemistry*, Econ Verlag, Dusseldorf, F.R.G.

Beer, John (1959), *The Emergence of the German Dye Industry*, University of Illinois Press, Urbana, IL.

Bozdogan, Kirkor (1989), *The Transformation of the U.S. Chemical Industry*, working paper, MIT Commission on Industrial Productivity.

Business Week (1995), "The Global 1000" (10 July):58–80.

Cable, John (1985), "Capital Market Information and Industrial Performance: The Role of Western German Banks," *Economic Journal*, 95(1):118–32.

Calomiris, Charles (1995), "The Cost of Rejecting Universal Banking: American Finance in the German Mirror, 1870–1914," in Naomi Lamoreax and David Raff (eds.), *Information and Coordination*, Cambridge University Press, Cambridge, U.K.

Carosso, Vincent (1970), *Investment banking in America*, Cambridge University Press, Cambridge, U.K.

Carosso, Vincent (1979), *More Than a Century of Investment Banking*, McGraw-Hill, New York, NY.

Chandler, Alfred (1990), *Scale and Scope: The Dynamics of Industrial Capitalism*, Harvard University Press, Cambridge, MA.

Chandler, Alfred and Stephen Salsbury (1971), *Pierre S. DuPont and the Making of the Modern Corporation*, Harper & Row, New York, NY.

Coleman, D. C. (1969), *Courtaulds–Vol. II*, Clarendon Press, Oxford, U.K.

Collins, Michael (1988), *Money and Banking in the U.K.: A History*, Croom Helm, London, U.K.

Collins, Michael (1991), *Banks and Industrial Finance in Britain, 1800–1939*, MacMillan, London, U.K.

Cottrell, Philip (1980), *Industrial Finance 1830–1914*, Methuen, London, U.K.

Cottrell, Philip (1985), *Investment Banking in England, 1856–1914*, Garland, New York, NY.

Da Rin, Marco (1996), "German Kreditbanken 1850–1914: An Informational Approach," *Financial History Review*, 3(2):29–47.

Da Rin, Marco (1997a), "Finance and Technology in Early Industrial Economies: The Role of Economic Integration," *Ricerche Economiche*, (51) 171–200.

Da Rin, Marco (1997b), "Catalysts for Industrialization: Banks vs Conglomerates," mimeo, IGIER. W.P.U. 131.

Davenport-Hines, R. P. and Judy Slinn (1992), *Glaxo: A History to 1962*, Cambridge University Press, Cambridge, U.K.

De Long, Bradford (1991), "Did J. P. Morgan's Men Add Value? An Economist's Perspective on Financial Capitalism," in Peter Temin (ed.), *Inside the Business Enterprise*, Chicago University Press, Chicago, IL.

Department of Trade and Industry (1996), "The R&D Scoreboard 1996," London and Edinburgh.

Derdak, Thomas (1988), *International Directory of Company Histories*, St. James Press. Chicago, IL.

Dow Chemical Co. (Post Street Archives), Historical Records.

Dyer, David and David Sicilia (1990), *Labors of a Modern Hercules*, Harvard Business School Press, Boston, MA.

Eccles, Robert and David Crain (1988), *Doing Deals*, Harvard Business School Press, Cambridge, MA.

Edelstein, Michael (1971), "Rigidity and Bias in the British Capital Markets, 1870–1913," in Donald McCloskey (ed.), *Essays on a Mature Economy: Britain after 1840*, Methuen, London, U.K.

Edgerton, D. and S. Horrocks (1994), "British Industrial Research and Development before 1945," *Economic History Review*, 47(2):213–38.

Edwards, John and Karl Fischer (1994), *Banks, Finance and Investment in West Germany Since 1970*, Cambridge University Press, Cambridge, U.K.

Feldenkirchen, Wilfred (1991), "Banking and Economic Growth: Banks and Industry in Germany in the Nineteenth Century and Their changing Relationship during Industrialization," in Wang Lee (ed.), *German Industry and German Industrialization*, Routledge, London, U.K.

Fischer, Gerald (1968), *American Banking Structure*, Columbia University Press, New York, NY.

Forrestal, Dan (1977), *Faith, Hope and $5,000*, Simon and Schuster, New York, NY.

Francke, Hans and Eric Hudson (1976), *Banking and Finance in West Germany*, St Martin's Press, New York, NY.

Fruin, Mark (1992), *The Japanese Enterprise System*, Clarendon Press, Oxford, U.K.

Goldsmith, Raymond (1958), *Financial Intermediaries in the American Economy Since 1900*, Princeton University Press, Princeton, NJ.

Goldsmith, Raymond (1983), *The Financial Development of Japan, 1868–1977*, Yale University Press, New Haven, CT.

Gorton, Gary and Frank Schmidt (1996), "Universal Banking and the Performance of German Firms," working paper 96–11, Wharton Financial Institutions Center.

Grabower, Rolf (1910), *Die finanzielle Entwicklung der Aktiengesellschaften der deutschen chemischen Industrie*, Duncker & Humblot, Leipzig, G.D.R.

Haber, Ludwig Fritz (1958), *The Chemical Industry During the Nineteenth Century*, Clarendon Press, Oxford, U.K.

Haber, L. F. (1971), *The Chemical Industry, 1900–1930*, Clarendon Press, Oxford, U.K.

Hannah, Leslie (1976), *The Rise of the Corporate Economy*, Methuen, London, U.K.

Hardach, Gerd (1984), "Banking and Industry in Germany in the Interwar Period 1919–39," *Journal of European Economic History*, 13(S):203–34.

Haynes, Williams (1954), *American Chemical Industry*, Van Nostrand, New York, NY.

Haynes, Williams [1970 (1939)], *Chemical Pioneers*, Books for Libraries, Freeport, ME.

Hounshell, David and John Smith (1988), *Science and Corporate Strategy*, Cambridge University Press, Cambridge, U.K.

Jeidels, Otto (1905), *Das Vernhaltnis der Deutschen Grossbanken zur Industrie*, Duncker & Humblot, Leipzig, G.D.R.

Jensen, Michael (1989), "Eclipse of the Public Corporation," *Harvard Business Review* (September-October):61–74.

Kaplan, Steven (1994a), "Top Executives, Turnover and Firm Performance in Germany," *Journal of Law, Economics and Organization*, 10(1):142–59.

Kaplan, Steven (1994b), "Top Executive Rewards and Firm Performance: A Comparison of Japan and the United States," *Journal of Political Economy*, 102(3):510–46.

Kennedy, William (1987), *Industrial Structure, Capital Markets and the Origins of British Economic Decline*, Cambridge University Press, Cambridge, U.K.

Khan, Zorina and Kenneth Sokoloff (1993), " 'Schemes of Practical Utility': Entrepreneurship and Innovation Among 'Great Inventors' in the United States, 1790–1865," *Journal of Economic History*, 53(2):289–307.

Khoury, Sarkis (1985), *The Deregulation of the World Financial Markets*, Quorum Books, New York, NY.

Kindleberger, Charles P. (1984), *A Financial History of Western Europe*, Allen & Unwin, London, U.K.

Kirchgäesser, Bernhard (1988), "The History of the Deutsche Bank Mannheim and its Predecessors," in *Deutsche Bank, Studies on Economic and Monetary Problems and on Banking History*, vol. 23, Hase & Köhler, Mainz, F.R.G.

Kocka, Jürgen and Hans Siegriest (1979), "Die Hundert Grösten Deutschen Industrieunternehmen in Späten 19. und 20. Jahrhundert," in Norbert Horn and Jührgen Kocka (eds.), *Recht und Entwicklung der Grossunternehmen im 19. und 20. Frhhen Jahrhundert*, Vanderhoecht, Göttingen, F.R.G.

Kocka, Jürgen (1988), "The Rise of the Modern Industrial Enterprise in Germany," in Alfred Chandler and Herman Daems (eds.), *Managerial Hierarchies: Comparative Perspectives on the Rise of the Modern Industrial Enterprise*, Harvard University Press, Cambridge, MA.

Landes, David (1969), *Prometheus Unbound*, Cambridge University Press, Cambridge, U.K.

Lane, Sarah (1993), "Corporate Restructuring in the Chemical Industry," in Margaret Blair (ed.) *The Deal Decade*, The Brookings Institution, Washington, DC.

Liebenau, Jonathan (1984), "Industrial R&D in Pharmaceutical Firms in the Early Twentieth Century," *Business History*, 26(2):329–46.

Lieberman, Marvin (1990), "Exit from Declining Industries: 'Shakeout' or 'Stakeout'?" *Rand Journal*, 21(4):538–54.

Lischka, J. R. (1985), *Ludwig Mond and the British Alkali Industry*, Garland, New York, NY.

Michie, R. C. (1987), *The London and New York Stock Exchanges*, Allen & Unwin, London, U.K.

Miyajima, Hideaki (undated), "Strategic Intervention against International Competition in the Interwar Japan. The Case of Chemical Industries," mimeo, Waseda University.

Molony, Barbara (1990), *Technology and Investment: The Prewar Japanese Chemical Industry*, Harvard East Asian Monographs, Cambridge, MA.

Musson, A. E. (1965), *Enterprise in Soap and Chemicals: Joseph Crosfield and Sons Limited, 1815–1965*, Manchester University Press, Manchester, U.K.

Nakatani, Iwao (1984), "The Economic Role of Financial Corporate Grouping," in Masahiko Aoki (ed.), *The Economic Analysis of the Japanese Firm*, North-Holland, Amsterdam, Netherlands.

Nevin, Edward and Edward Davis (1970), *The London Clearing Banks*, Elek Books, London, U.K.

Nevin, Thomas and Marian Sears (1953), "The Rise of a Market for Industrial Securities, 1887–1902," *Business History Review*, 27(1):105–38.

Owen, Geoffrey and Trevor Harrison (1995), "Why ICI Chose to Demerge," *Harvard Business Review* (March-April), pp. 133–142.

Perkins, Edwin (1971), "The Divorce of Commercial and Investment Banking: A History," *The Banking Law Journal*, 88(6):483–530.

Pettigrew, Andrew (1985), *The Awakening Giant. Continuity and Change at ICI*, Blackwell, London, U.K.

Plumpe, Gottfried (1990), *Die I.G. Farbenindustrie A.G.*, Duncker & Humblot, Berlin, Germany.

Pohl, Hans (1984), "Forms and Phases of Industry Finance up to the Second World War," *German Yearbook on Business History 1984*, SAVER, Muenchen, Germany, pp. 75–95.

Pressnell, L. (ed.) (1973), *Money and Banking in Japan*, Macmillan, London, U.K.

Prowse, Stephen (1990), "Institutional Investment Patterns and Corporate Financial Behavior in the United States and Japan," *Journal of Financial Economics*, 27(1):43–66.

Reader, William (1970), *Imperial Chemical Industries: A History. Vol I: The Forerunners*, Oxford University Press, London, U.K.

Reader, William (1975), *Imperial Chemical Industries: A History. Vol II: The First Quarter Century*, Oxford University Press, London, U.K.

Riesser, Jacob (1911), *The Great German Banks*, U.S. National Monetary Commission, Washington, DC.

Roe, Mark (1994), *Strong Managers, Weak Owners*, Princeton University Press, Princeton, NJ.

Roy, William (1990), "Rise of American Industrial Corporations, 1880–1914," computer file, Department of Sociology, UCLA.

Sheard, Paul (1994), "Main Banks and the Governance of Financial Distress," in M. Aoki and H. Patrick (eds.), *The Japanese Main Bank System*, Clarendon Press, Oxford, U.K.

Sylla, Richard (1975), *The American Capital Market*, Arno Press, New York, NY.

Teltschik, Walter (1994), *History of the Large German Chemical Industry; Development and Influence in State and Company*, Weinheim, VCH.

Tilly, Richard (1967), "Germany: 1815–70," in Rondo Cameron (ed.), *Banking in the Early Stages of Industrialization*, Oxford University Press, New York, NY.

Tilly, Richard (1986), "German banking, 1850–1914: Development Assistance for the Strong," *Journal of European Economic History*, 15(1):113–52.

Tilly, Richard (1993), "On the Development of German Big Banks as Universal Banks in the 19th and 20th Century. Engine of Growth or Power Block?" *German Yearbook on Business History 1993*, SAVER, Muenchen, Germany, pp. 109–30.

Thomas, W. A. (1978), *The Finance of British Industy*, Methuen, London, U.K.

Trescott, Martha M. (1981), *The Rise of the American Electrochemicals Industry, 1880–1910*, Greenwood Press, Westport, CT.

Ueda, Kazuo (1994), "Institutional and Regulatory Frameworks for the Main Bank System," in Masahiko Aoki and Hugh Patrick (eds.), *The Japanese Main Bank System*, Oxford University Press, Oxford, U.K.

Wellhöner, Volker (1989), *Grossbanken and Grossindustrie im Kaiserreich*, Vandenhoeck & Ruprecht, Göttingen, F.R.G.

White, Eugene (1983), *The Regulation and Reform of the American Banking System*, Princeton University Press, Princeton, NJ.

Whitehead, Don (1968), *The Dow History*, McGraw-Hill, New York, NY.

Wilson, Charles (1968), *Unilever 1945–65*, Cassell, London, U.K.

Wrigley, Julia (1987), "Technical Education and Industry in the Nineteenth Century," in Bernard Elbaum and William Lazonick (eds.), *The Decline of the British Economy*, Clarendon Press, Oxford, U.K., pp. 162–88.

11 Structure and Performance of the Chemical Industry under Regulation

KIAN ESTEGHAMAT

The industrial development of the 19th century brought environmental damage. But with few exceptions, industries and governments of that time paid minimal attention to the increasing problems of air, water, and soil pollution. The strategic and economic significance of the chemical and allied industries purchased the trust and cooperation of governments in these industrializing countries; even when environmental laws were passed, they were often weak and seldom enforced. But in the latter half of the 20th century, this state of affairs changed dramatically. This chapter examines the extent to which environmental, health, and safety regulation accounts for differences in the structure and performance of chemical-processing industries across the United States, the United Kingdom, Japan, and Germany.[1]

Health-endangering environmental problems reached new heights in the years following World War II, which led to significant changes in thinking about the environment. These changes were motivated by concern over unprecedented environmental problems, perhaps especially those due to dangers of atomic radiation and pesticides. Raymond Dominick (1992, p. 214) describes the change in mindset that took place during the 1950s in Germany.

> Although air, noise, and water pollution had been around for decades, they reached unprecedented levels in the postwar period. Tall smokestacks dispersed airborne poisons over the entire country, and even the wealthy in their country homes now shared the consequences with the poorer Germans who huddled near the factories. Airport expansion and noise from jet airplanes, which rendered substantial residential areas throughout West Germany uncomfortable or even unfit for human habitation, likewise struck at both rich and poor. Surface and groundwater seldom carried the old and familiar contamination that produced epidemic diseases like cholera and typhus. Instead they now carried chemical residues that were linked to the most fearsome disease of the twentieth century, cancer. Evidence mounted that the millions of pounds of synthetic biocides that were dumped into the environment by modern agriculture also carried life-threatening dangers for all consumers, regardless of social station.

I am thankful to Ralph Landau, Ashish Arora, Takashi Hikino, and Johann Peter Murmann for guidance, encouragement, and many helpful comments.

[1]For historical perspective, see, for example, Grant, Paterson, and Whitston (1988), Teltschik (1992), and Dominick (1992).

TABLE 11-1. Selected Environmental Measures Affecting the Production, Use, and Disposal of Chemicals

Year	Germany	United Kingdom	European Community	Japan	United States
1956		Clean Air Act			
1957		Pesticide Safety Precautions Scheme			
1961	Detergent Law				
1967			Directive on Dangerous Substances	Basic Law for Environmental Pollution Control	
1968	Pesticides Law	Clean Air Act (amendment)		Air Pollution Control Law	
1970				Basic Law (amend.); Water Pollution Control Law	Clean Air Act
1971	Environmental Program	Comprehensive policy statement.			
1972	Waste Disposal Law	Deposit of Poisonous Waste Act			Clean Water Act; Federal Insecticide, Fungicide and Rodenticide Act
1973				Examination and Regulation of Chemicals Law	
1974		Control of Pollution Act		Air Pollution Control Law (amendment)	

Year					
1976	Chemicals Law				RCRA; TSCA
1979			Directive on PCB 6th Amendment		
1980					CERCLA
1982			Directive on Prevention of Major Accidents		
1984		Control of Industrial Major Accident Hazards			Amendments to RCRA
1985			Directive on Waste		
1986				Examination and Regulation of Chemicals Law (amendment)	SARA (Superfund amendment)
1988		Control of Substances Hazardous to Health Regulation		Law Concerning the Protection of Ozone Layer	
1989		Water Act		Basic Law for Environmental Pollution Control (amendment)	
1990		Environmental Protection Act			Clean Air Act (amendment)
1991		Water Resources Act			
1993				Basic Law on the Environment	

Parallel statements about the alarming condition of the environment can be made about the situation in other industrializing countries such as the United States, Japan, and the United Kingdom in the late 1950s and 1960s. Air and water pollution had reached dangerous levels. Use of agricultural chemicals, such as the pesticide DDT, threatened drinking water and the food chain. Around the Jinstsu River near Toyama Bay in Japan, cases of cadmium poisoning ("*itai-itai*" disease) appeared with increasing frequency. Other catastrophes involving mercury poisoning and pulmonary maladies induced by air pollution affected scores of Japanese citizens and were highly publicized. In the United States, damage to the Cuyahoga River in Ohio was so severe that it caught fire. Foam from use of detergents endangered drinking water and suffocated the biota of the waterways. On the Rhine, these suds even interfered with commercial traffic. Well-publicized environmental disasters did much to affect public awareness of pollution. In the late 1960s, the oil tanker *Torrey Canyon* went aground off the coast of Britain, resulting in the release of over 60,000 tons of crude oil. Also at that time, a major oil spill near Santa Barbara in the United States captured public attention. These examples could be multiplied many times over.

Many potential threats to the environment, both new and old, were widely discussed, including dangers of organic pesticides and the phenomenon of resistance among the intended pests, nuclear contamination and fallout, noise pollution, dwindling resources, and the population explosion. Rachel Carson's (1962) *Silent Spring* warned against the harmful effects of agricultural chemicals. This and other writings represented a new kind of environmental thinking that was usually accompanied by a plea for a new ethic in interactions with the environment. Others followed with similar themes. A report published for the Club of Rome expressed concern about environmental damage caused by growth, warning of depleting natural resources (Meadows et al. 1972).

Regulation, however, remained minimal and incremental until the late 1960s and the 1970s, when many industrial countries introduced new and strengthened pollution-control measures. In 1967, Japan enacted the Basic Law for Environmental Pollution Control; in 1969, the United States adopted the National Environmental Policy Act; in 1970, the Federal Republic of Germany passed an Emergency Program for Environmental Protection; and in 1974, the United Kingdom enacted the Control of Pollution Act. Other measures addressed more specific problems.

The chemical industry was not immune to the sweep of this legislation. At times, after events such as the discovery of toxic-waste disposal at Love Canal in New York or the improper release of dioxins in 1976 in Seveso, Italy, the chemical industry even seemed a primary target of such legislation. The most significant piece of legislation dealing with the handling and production of chemicals was the United States Toxic Substances Control Act of 1976 (TSCA). Since then, other laws have been introduced affecting air and water quality, pesticides, toxic substances and hazardous wastes. Over time, the European Community (EC) has played a greater role in regulation of chemicals across Europe. For example, in 1979, the EC strengthened its 1967 Dangerous Substances Directive with the passage of the 6th Amendment, requiring the testing and state notification of chemicals before introduction to markets. The 1982 Seveso Directive prescribed guidelines for prevention of accidents.

Table 11-1 lists some of the major laws pertaining to the chemical industry in the United States, Japan, Germany, the United Kingdom, and the European Community that have been enacted over the last few decades. Government policy and industry response to environment, health, and safety issues have often been re-evaluated and transformed after catastrophes, such as the 1986 fire at the Sandoz plant in Switzerland that resulted in contamination of the river Rhine and the 1984 release of toxic chemicals at a Union Carbide plant in Bhopal, India, that caused 2,800 deaths and 200,000 injuries. Pronounced differences have developed across the four countries in regulatory policy, reflecting each country's national circumstances and priorities. The following section provides a brief comparison of the regulatory arenas in the United States, Japan, Germany, and the United Kingdom, highlighting only areas of difference pertinent to the chemical industry. The chapter then examines the potential impact of regulatory measures at the national level and moves to the industry level to gauge the effects of environmental, health, and safety regulations on the performance and structure of the chemical industry. Finally, the discussion considers the influence of regulations on the international location of chemical-industry production and investment.

A survey of evidence illustrates why national regulatory differences may not have yielded substantial economic advantage across industrialized countries, despite significant effects on the cost of capital and reduced investment. However, there is some evidence that differences in national and regional approaches to regulation have affected the performance of individual sectors in the four national chemical industries, as in the case of biotechnology.

REGULATORY SETTING

The United States, the United Kingdom, Japan, and Germany have approached industrial regulation differently. Variations are particularly marked in the areas of industry-government interaction, standards and policy implementation, handling and disposal of hazardous wastes, and responsiveness of the regulatory process to public concerns.

Interaction between the Regulator and the Regulated

Governments in Japan, to a large extent in Germany, and especially in the United Kingdom have maintained a relatively cooperative relationship with industry while implementing environmental, health, and safety regulations since the late 1960s. The interaction of the chemical industry with governments in these countries has been based largely on mutual trust, with industry experts and regulatory authorities acting as mediators. Industry and government interact through networks of private, informal, personal, and largely unpoliticized contacts among members of the public and the private sectors, and consensus is achieved among middle-level officials over a period of time. In the United Kingdom, environmental, health, and safety regulations (including specific pollution-control standards) are based on discussions that take place in informal, private settings among professional and technical experts from industry and senior civil servants from government. The United Kingdom responded early to pollution concerns through industry-government cooperation and has continued in this way. The German pattern is similar, although somewhat more formal and increasingly challenged by environmental and political groups. In Japan, industry and government cooperate directly. Government authorities act as regulatory advisers and achieve consensus through informal meetings and exchanges of information.

The cooperative pattern of interaction between industry and government has led to a climate of mutual reliance and trust in these countries, particularly in the United Kingdom. In contrast, the relationship between the regulator and the regulated has been particularly adversarial in the United States. Major U.S. regulatory decisions on health and environment have generally been made by government officials and courts, with industry and other concerned parties having only indirect influence. Formulation of policy has often taken place in highly politicized settings, with far greater involvement of lawyers and far less involvement of technical experts from industry. The U.S. approach tends to produce regulatory decisions that are less flexible and more often sets standards that are not achievable with current technology.

The process of regulating the safe handling of vinyl chloride (VC) is a lucid illustration of this contrast in industry-government interaction. Vinyl chloride is an invisible, potentially lethal, carcinogenic gas used in the making of plastic products. The health hazard posed by vinyl chloride received wide media attention in the early 1970s, and public concern directed intense pressure on government and industry to control the exposure of workers in plants, nearby residents, and consumers. In Germany, the United Kingdom, and Japan, representatives from industry, trade unions, and government deliberated in small, informal groups and relied on consensus to set VC exposure limits. In contrast, the key VC regulatory decisions in the United States were made in private consultation between legislative and regulatory agents. Relationships were adversarial, and attorneys were the sole intermediaries among the parties. In a cross-country study of the VC problem, Badaracco (1985, p. 159) describes how the network of interdependence linking government to industry and trade unions in Britain, West Germany, Japan, and France, may have favored cooperation.

> Within this network, government officials chose to move patiently, privately, gently, and unobtrusively. They thereby secured several advantages. Government officials gained full access to industry expertise in arcane areas of science and engineering; they shared the high political risks that pervade modern industrial problems; they resolved a difficult, dangerous problem with minimal frustration, resentment, and hostility for all concerned; their VC standard was more likely to be effective because it was based on expert industry judgment and actual, shop-floor efforts at VC control; and they set a precedent likely to secure similar advantages to future problems. Government officials ran few risks in taking this approach: if cooperation failed, they could have regulated VC on their own terms.

The minimum level of vinyl chloride exposure in the United Kingdom was set above that possible with available technology. Germany and Japan also set exposure limits mindful of technology and cost. Regulation in the United States required the chemical industry to achieve a very strict standard, one that exceeded its technical capability at the time. But by 1977, the exposure of workers in the four countries was reduced to roughly equivalent levels.

Regulatory Standards and Policy Implementation

Implementation of environmental regulation in the United Kingdom has been remarkably more flexible than in Germany and the United States, and somewhat more flexible than in Japan. While other countries have opted for more centralized administration, responsibility for implementation of British environmental regulation has remained relatively decentralized — largely left to the discretion of local authorities. British standards are less complex and impose fewer administrative and legal costs on business. The British have made extensive use of industry self-regulation to implement regulatory policy. Noncompliance penalties have remained relatively mild, and prosecutions, regarded as failure on the part of the regulator as well as the regulated, have been rare. In administering its environmental policy, the United Kingdom makes less use of environmental quality or ambient and source discharge standards than other members of the Organization for Economic Cooperation and Development (OECD) and the European Community (EC). As Vogel (1986, p. 75) notes: "The British make less use of legally enforceable environmental-quality or emission standards than does any other industrial society. They attempt to tailor pollution-control requirements to meet the particular circumstances of each individual polluter and the surrounding environment."

In contrast, the American, German, and Japanese approaches to regulatory policy place greater reliance on uniform standards and formal procedures. The "command-and-control" style of regulatory policy in the United States has often obliged regulatory officials to enforce rigid standards that both the regulator and the regulated have perceived to be arbitrary. The U.S. Congress has traditionally allowed little flexibility in implementing policy objectives. In the words of the first Environmental Protection Agency (EPA) administrator, William D. Ruckelshaus, "Congress wants to do the job for the agency. Politically, Congress finds it impossible to resist setting these absolute purity kinds of standards that make no sense. And not only that, but Congress sets deadlines for achieving the standards that doom the agency to failure before it starts" (Ember 1995, p. 23).

Considerable progress has been made, however, in improving the quality of the environment. For example, the portion of U.S. rivers and lakes that were designated as "clean" increased from 36% in 1972 to 62% in 1992. Emissions of major air pollutants have also been reduced. United States EPA data show that from 1970 to 1993, lead emission decreased by 98%, sulfur dioxide by 30%, carbon monoxide by 24%, and volatile organic compounds also by 24%. However, such gains have increasingly involved regulatory complexity, legal conflict, and puzzling cost-benefit trade-offs. In 1992, for instance, the U.S. Department of Agriculture paid $46 million to return 50,000 acres of farmland to their former wetland condition. In the same year, other regulatory enforcement restored about 3,600 acres of wetlands at a cost of $80 million (*The Economist* 1995).

The United States has only recently experimented with voluntary industry programs to improve environmental quality. As one example, in 1988 the EPA sponsored a voluntary program known as "33/50," which successfully achieved the goal of reducing the industrial release of 17 toxins by 33% in 1992 and by 50% in 1995. Also, recently, the U.S. Congress has shown some flexibility in certain strict regulatory standards. The Delaney Clause of the Food, Drugs, and Cosmetics Act of 1958 prohibited food from containing any amount of additives found to be carcinogenic. Many have contended that the zero requirement of the law has become obsolete and impossible to achieve since modern measurement techniques allow detection of chemicals in almost unbelievably minuscule amounts — parts per quadrillion. Farmers and pesticide suppliers lobbied for repeal of the clause, pointing out that continued adherence to the no-tolerance standard would lead to withdrawal of many pesticides and, in turn, to decreased crop production and higher food prices. The Delaney Clause was replaced in 1996 with a less restrictive requirement based on minimizing chances of developing cancer.

Implementation of environmental regulation in Germany is also more formal and rule-oriented than in the United Kingdom (for example, see Grant, Paterson, and Whitston 1988). Governments in Germany have enacted standards that are the strictest in Europe, often preceding directives at the European Community level. These standards are in many respects similar to the codified approach in the United States; that is, they have been applied uniformly with statutory deadlines. The strict recycling laws in Germany, which make recycling the responsibility of the manufacturer, are noteworthy. The "Töpfer Law," named after the environment minister Klaus Töpfer which took effect in 1991 has required waste collection to increase in stages until 80% of all packaging waste is collected by 1995. Of the waste that is collected, 90% of the glass and metal will have to be recycled and 80% of the plastic and paper (*The Economist* 1992). Many exporters point to recycling laws in Germany as obstacles to trade. However, in response to this and a newer packaging law that took effect in 1993, the chemical industry has been investing to develop new recycling processes.

Japan's approach for implementing environmental policy has been to combine demanding standards, in the American and German style, with a relatively decentralized administration, in the U.K. style. Firms in Japan can sometimes secure greater flexibility in meeting compliance deadlines than their counterparts in the United States or Germany. However, this flexibility is governed by requirements to meet strict standards, especially concerning air and water pollution.

Although rules and regulations in the four countries have been implemented in distinct national styles, some standards have been slowly converging. In particular, the United States, Japan, Germany, and the United Kingdom all have similar source-discharge standards and policies based on ambient levels of air and water pollutants. On a relative scale, Japan's ambient and source-discharge standards are more strict, and that of the United Kingdom the least strict of the four countries (Kopp, Portney, and Dewitt 1990). In response to domestic pressures and growing international appeals for harmonization of regulatory policies and procedures, particularly within the EC, environmental regulation in the United Kingdom is becoming more rule-based and centralized. However, regulatory policies have changed more incrementally in the United Kingdom and have been less demanding in recent decades than in the United States, Germany, or Japan. Enforcement of compliance standards has been more flexible in the United Kingdom and Japan, and more formal in the United States and Germany.

Government assistance to industry in complying with designated standards has been consistent with the national regulatory styles. United States policy has been to provide little direct industrial assistance for compliance. In contrast, some EC countries and Japan have opted to provide direct assistance for development of pollution-control technologies. German chemical firms have access to government-subsidized loans for pollution-control investments. In Japan, low-interest government-sponsored loans have accounted for about 35% of all industrial pollution control funds between 1975 and 1990. Germany provides grants to small- and medium-sized firms for covering one-half of all environmental consultant costs. The Japan Finance Corporation for Small Business assists smaller chemical firms with loans for environmental expenditures. Many other industrial assistance programs, including research and development grants and tax incentives, are also available (U.S. Congress 1994).

Land Contamination and Cleanup Liability

The United States has enacted a comprehensive set of measures for regulating the transport, disposal, and cleanup of hazardous waste. The Resource Conservation and Recovery Act of 1976 (RCRA) regulates the identification, transport, and disposal of hazardous wastes. The Comprehensive Environmental Response, Compensation and Liability Act of 1980 (CERCLA) was enacted to bring about cleanup of land sites that were previously contaminated. CERCLA established a revolving fund, commonly known as the Superfund, financed by a tax on chemical manufacturers and federal revenues, to be used by the Environmental Protection Agency in implementing toxic-site cleanup. (The Superfund program was amended with the Superfund Amendments and Reauthorization Act of 1986.) Penalties for noncompliance with hazardous-waste regulation have been severe in the United States.

Superfund's aim is to reduce health and safety risks by creating a pool of funds to be used for cleaning hazardous sites. Funds are raised by identifying and charging those responsible and are supplemented by a tax on polluting industrial sectors. The Superfund program has made progress. By 1995, 15 years after the passage of the original legislation, 324 hazardous-waste sites had been cleaned up at a government cost of about $15 billion, while at least 1,200 more sites are awaiting cleanup. But the cleanup has been an expensive undertaking marred by ongoing legal and political battles. Two main areas of conflict have characterized the program.

First, each site requires an assessment of the trade-offs between risks and expenses. For example, in some cases, containment of pollution may protect public health quite successfully at a far lower cost than complete cleanup. Recent proposals for Superfund reform would allow cleanup standards to be based on future uses of land, such as whether the future use of the site is a playground for children or an industrial factory. The Ciba-Geigy Corporation estimates that such a change would save the company 20% of its current annual Superfund-related costs.

But issues of legal liability surrounding Superfund have evoked the greatest contention within and among chemical companies, the insurance industry, government agencies, and environmental organizations. The law allows retroactive recovery of cleanup costs from current and previous owners and operators of contaminated land, as well as generators and transporters of hazardous materials to the waste site, even if the actions of these owners and companies were legal at the time they occurred (Church and Nakamura 1993). Another aspect of the Superfund liability system allows for joint and several liability, whereby a single party can be held liable for all cleanup costs caused by actions of several responsible parties. For example, a firm may be held liable for the entire cost of cleaning up a polluted site that it owns or operates, even though the contamination was caused by former owners of the land.

Superfund's liability provisions have arguably led to expensive and slow remediation of toxic land, in addition to enormous legal costs. Between 30% and 70% of expenditures on Superfund-mandated problems in the United States are for legal and transaction costs, with the remainder used for the intended treatment of hazardous waste sites (Rich 1985; Kriz 1995). The law has also been criticized for blocking or slowing the redevelopment prospects of older, and mostly urban, industrial sites, since potential developers are scared off by the prospect of legal liability. Assessing the impact of these liability provisions, Church and Nakamura (1993, p. 162) write:

> The liability system in Superfund profoundly influences almost every aspect of the program. Without question it is the program's most controversial element. Assigning liability lengthens the decision-making process and generates substantial transaction costs. The dictates of strict, joint and several liability serve as major criteria against which the EPA and others measure success in negotiations. The liability doctrine affects overall program organization, the EPA's conception of the problems it faces at specific sites, and the procedures the agency uses in dealing with PRPs [potentially responsible parties].

The combination of America's adversarial approach in formulating environmental regulation and its reliance on civil law for implementing regulatory policy has meant that lengthy and expensive court proceedings have become common in managing hazardous wastes. Various reforms have been proposed to reduce the degree of legal entanglement; for instance, rules to prorate liability (rather than attempting to make an actor who contributed only a portion of the pollution pay for the entire cleanup), to exempt marginal polluters, and to limit municipal liability. In addition, the insurance industry has proposed to pay $8.1 billion into a trust fund to cover cleanup costs of contaminated sites, on the condition that at least 85% of companies with insurance coverage agree not to sue their insurers. But ultimately, the liability-laden Superfund approach will remain attractive in times when government budgets are tight, because it promises a way to keep the costs of cleanup off the budget. Kopp, Portney, and Dewitt (1990, p. 92) note:

> It has been stated many times that the Superfund law is a bill collectors' statute. The law's provisions for strict, and several liability are designed to make it as easy as possible for the federal government to find a suitable deep pocket to pay for a site cleanup. These liability provisions are not written into Superfund to provide efficient incentives for generators, transporters, and disposers to take due care with respect to their activities (this is the role of RCRA); rather, the provisions are there to ensure that cleanup costs remain off budget.

The American Superfund law stands in sharp contrast to parallel laws in the United Kingdom, Germany, and Japan. No other country relies so extensively on a comprehensive liability system for hazardous-waste management as does the United States. In Germany, contamination of land has had a higher profile since reunification in 1991, since the region of the former East Germany contains numerous toxic-waste sites. German federal law, however, has left the determination of pollution accountability largely to the discretion of regulatory authorities. In most cases, the present owner or user of the contaminated land is held accountable, even though the pollution may have been caused by a former owner. As Wissel et al. (1993, p. 233) describe the German situation: "In the selection process, effectiveness comes before fairness, i.e., it does not matter who is the most at fault, but rather who is easily implicated and at the same time provides the best guarantee of effectively rectifying the damage from a financial point of view."

The doctrine of joint and several liability (by which anyone contributing to a polluted site is potentially responsible for 100% of the liability) is not applied in the cleaning up of hazardous-waste sites. Instead, in Germany the application of joint and several liability is limited to especially dangerous incidents; for example, cases involving several plants in water contamination or genetic-engineering-plant accidents. Moreover, legal practices such as suing banks that lend money to parties who contaminated sites are nonexistent in Germany. However, a U.S. court ruling has recently limited lender liability to cases in which the lender has been an active participant in managing the potentially responsible enterprise.

Like Germany, Japan and the United Kingdom do not have a system of legal liability for cleanup comparable with America's Superfund, and there is little indication that they will consider one (Kopp, Portney, and Dewitt 1990; Gresser, Fujikura, and Morishima 1981). (This subject is also addressed in the chapter by Campbell and Landau.)

Public Recourse and Redress

The pattern of administrative response and public involvement in the regulatory process differs across the four countries. The U.S. regulatory policy has allowed wider public access to information and been more open to public participation. The Freedom of Information Act has been in force in the United States since the 1930s. The Emergency

Planning and Community Right-to-Know Act (a part of the Superfund Amendments and Reauthorization Act of 1986) has further increased public access to information by requiring that plant operators disclose hazardous-substance data about their facilities. Both environmental and business-interest groups in the United States have had access to open hearings in a variety of regulatory proceedings, and they have often exercised the right to challenge regulatory-policy decisions. In a study of hazardous waste siting, Rabe (1994, p. 154) remarks: "Often the process becomes a magnet for stalling tactics, including litigation, with the main public role confined to delay of program implementation, whether it involves siting or site cleanup."

The regulatory process in Germany is less public than in the United States, but more than in the United Kingdom and Japan. Public pressure in Germany for a cleaner environment has become very strong in recent years. The number of Germans describing themselves as "environmentally aware" leapt from 35% in 1982 to 62% in 1989. In 1991, 57% of the German population claimed the preservation of the environment was more important to them than economic growth (*Marketing* 1991). A vivid example is the experience of Hoechst, the largest chemical company in the world. The Hoechst chemical plant complex in Germany is located near Frankfurt. Environmental capital expenditures account for 25% to 30% of Hoechst's total annual capital spending, while other chemical companies spend about 15%. But despite much improvement in energy use and waste reduction, Hoechst has experienced problems with negative public opinion. Frankfurt is a large metropolitan area, not especially hospitable to manufacturing and particularly not to chemicals production. Moreover, Hoechst has sometimes responded to incidents with limited disclosure of information, which has led to negative press and reduced public credibility. In contrast, operations in the Ludwigshafen and Leverkusen areas, where BASF and Bayer have plants, are smaller and are located in more manufacturing-oriented surroundings. Also, company policy seems to have been more in accord with public disclosure of information. More generally, German citizens' groups have had direct legislative representation in the German parliament through the Green party beginning in the late 1970s. Notice that German public involvement has tended to happen through legislative formulation rather than through the primary U.S. approach of judicial challenge or indirect legislative lobbying.

In Britain, environmental policy-making has been largely closed to environmental organizations, reflecting the private and cooperative relationship between industry and government. In addition, access to regulatory information has been extremely limited. Only recently, with the passage of the Environmental Information Regulations in 1992, has this situation been partly remedied.[2] While the problem of unavailable information has been improved in the United Kingdom, many difficulties still remain, as Symes and Phillips (1993, p. 120) discuss:

> This problem has been addressed to some extent by the public registers which the regulatory authorities now maintain under the integrated pollution control, waste, water and other environmental legislation. These contain details of applications for authorizations, together with copies of authorizations and details of enforcement action taken. However, companies are not obliged to submit inventories of the registers on a week-to-week basis. The public also do not have the same powers as regulatory authorities to obtain information from businesses concerning important matters such as the processes that are being operated on a site and the chemical composition of emissions to the environment. It

---

[2]These regulations implement the EC Directive on Freedom of Information Relating to the Environment.

is therefore often difficult for the public to find out which companies are breaking the law, since they do not generally have the capacity to monitor emissions and thus to collect the scientific evidence needed for a successful prosecution.

On the whole, however, it seems that the British public have been more satisfied with administrative response to health and environmental damage than their counterparts in the United States, Germany, and especially Japan.

Japan has made great strides in improving environmental quality for its citizens since its Basic Law for Environmental Pollution Control in 1967 and its amendments in 1970. Yet it has arguably experienced higher levels of public dissatisfaction with its environmental policy than have America, Germany, and the United Kingdom. One long-standing dispute between citizens' groups and the authorities has followed from an incident in the 1950s whereby thousands of residents of the Japanese fishing village of Minamata were poisoned after eating fish tainted by mercury sludge that was dumped into waterways by the chemical company Chisso (Gresser, Fujikura, and Morishima 1981). Settlement of the case has been very slow, with continuing, bitter court battles. In Japan, the courts have focused on problems of damages or injunctive relief in civil litigation. The review and reform of government's regulatory decision-making have not been primary objectives (in contrast to the U.S. pattern), although they have been important consequences of some damage-award decisions. Public dissatisfaction with the slow response of the regulatory system for health and environmental damage and the delayed recourse for the injured has been loudly expressed and at times expressed through violent demonstrations. Takamasa Nakano (1986, p. 275) wrote of the incident: "Administrative response to health damage caused by specific pollution episodes has been very slow, and there is a feeling that negotiations between government, prefectural governments and private companies take far too long to reach sensible conclusions." In 1995, the Japanese ruling coalition parties announced aid to approximately 5,000 Minamata victims who had been denied treatment under established compensation guidelines.

More broadly, public access to the Japanese regulatory process has been limited, as Gresser, Fujikura, and Morishima (1981, p. 232) have pointed out:

> The public's capacity to influence internal ministerial affairs is extremely limited... although citizens' groups have scored some recent successes in administrative litigation, decision making within the bureaucracy remains a closed process. Public participation is generally by administrative grace, not by right. The law does not require formal adversarial hearings on important rule making decisions, and agencies are not even legally required to maintain a record, although most agencies do keep records for internal purposes. Japan has not enacted a Freedom of Information Act, and a recent trend favors even greater governmental secrecy. Finally, the almost overwhelming legal presumption upholding the broad discretion of public officials preserves the integrity of the system from public intrusion.

Chemical producers have come to recognize the importance of public perception in policy decisions affecting them. A voluntary industry initiative called "Responsible Care," started in 1985 in Canada as a response to the combination of major incidents and changing public attitudes over the last decades, has spread to many industrial countries. Responsible Care embodies a public commitment to continuous improvement of environmental, health and safety performance, and to responding to public concerns about chemical products and operations. Two out of three major chemical producers operated environmental-management systems by 1996, up from only a few companies in 1993. Most member companies have already implemented the pollution-prevention, health-and-safety, and chemicals-transport codes of Responsible Care. Product stewardship and life-cycle assessment for products are lagging. Chemical companies, especially in the United States, have also become increasingly open about their environmental, health, and safety performance. In 1996, the Hamburg Environmental Institute, a nonprofit organization that monitors the environmental performance of large chemical companies, noted that U.S. companies have established greater public dialogue through implementation of the Community Awareness & Emergency Response program of Responsible Care than have their counterparts elsewhere. Chemical companies in Europe and Japan, it is perceived, have not yet openly communicated with the public (Freemantle 1996). Perhaps not coincidentally, public opinion of the European chemical industry has worsened.

NATIONAL LEVEL

Differences in the environmental-regulatory setting might conceivably confer economic advantages to some countries or states — say, those where permission to proceed with potentially hazardous projects is granted — at the expense of others, where permission is withheld. Differences in environmental policies have the potential to affect costs, trade flows, innovation, investment levels, industrial structure, resource allocation, and location of industry. A number of studies have examined the effect of environmental regulations at all of these levels. This section focuses on the environmental laws at the national level, on such issues as overall growth and trade. The next major section focuses on the effect of such laws at the industry level, on such issues as investment, innovation, and industry structure. The following major section then discusses how environmental regulation can affect location decisions of firms, both within a country and across national borders.

Economic Growth and Environmental Regulatory Costs

The pace and extent of environmental legislation dramatically increased in the 1970s, resulting in a shift of economic resources toward pollution control. For the chemical industry, average environmental spending rose from about 3% of total operational investment in the 1940s to more than 6% in the 1960s and reached about 20% by the early 1990s (Clark 1993, p. 59). For industry as a whole, the costs of pollution control and abatement are proportionally lower. For the four countries considered here, pollution-control expenditures in 1990 were about 1.1% of GDP in Japan, 1.5% of GDP in the United Kingdom, and about 1.6% of GDP in Germany and the United States.

Several studies have focused on the connection between a nation's environmental regulations and its economic growth.[3] Michael Hazilla and Raymond Kopp (1990) have compared the overall economic costs of complying with America's Clean Air and Clean Water Acts, taking into account changes in investment and labor supply by industry, with the original projections of compliance cost. They find that initial annual economic costs were smaller than the projected pollution-control expenditures, but because of

---

[3]Jaffe et al. (1995) provide an assessment of the evidence on the relationship between environmental regulation and competitiveness of the U.S. manufacturing sector. Their review includes a discussion of general equilibrium and international trade studies.

reductions in investment and employment, costs greatly surpassed those projections in the long run. The results of this study emphasize that industrial-compliance costs are not a good measure of total environmental costs, since if a firm simply stops investing or producing in a certain area, its compliance costs may be low, even though the loss of production is very real.

In a detailed study of the effects of regulation on the cost of capital and economic growth, Dale Jorgensen and Peter Wilcoxen (1992) simulated the long-term growth of the U.S. economy with and without various regulations, using a model that divides the economy into 35 commodity groups, including chemicals and allied products, and then examines production and consumption of these goods using the business, household, government, and "rest-of-the-world" sectors. The model allows consumers to substitute among goods as prices change and producers to substitute among inputs. The model makes various analytic simplifications: For example, consumers and producers have perfect foresight and there is a single stock of malleable capital, and so on. Using this model, Jorgensen and Wilcoxen find that the environmental legislation enacted before the Clean Air Act of 1990 reduced the annual growth rate of the U.S. economy by 0.19% over the period from 1973 to 1985 — for an accumulated reduction of about 2.6% in U.S. gross national product over that time. When they assessed the impact of the Clean Air Amendments of 1990, those researchers estimated that the amendments would reduce the U.S. gross national product by an additional 0.4% for the period from 1990 to 2005. In their estimate, by 2005, regulation affecting water, air, and solid waste will reduce capital stock by 4.3%, increase the cost of capital by 5.5%, and reduce real annual GNP by over 3 percentage points. Jorgensen and Wilcoxen found that pollution controls have caused the greatest reduction in long-term output for motor vehicles (a 15% reduction in long-term output), coal mining (7%), chemicals (5%), and then primary process industries such as petroleum refining, metals, and pulp and paper.

On the whole, the studies suggest that environmental regulation has a significant impact in the form of increased cost of capital and reduced investment. But two caveats are worth remembering here. First, the key issue in reconciling economic growth and the environment is what sort of environmental controls are enacted and how growth is achieved, as Arrow et. al. (1995) have pointed out. In concluding that economic incentives can encourage sensible use of the environment, they argue that attention should be directed toward the raw materials and resources — including environmental — and outputs and by-products — including pollution — used in achieving growth. A second important point is that wealthier countries tend to be more willing to enact laws and spend money on the environment. Gene Grossman and Alan Krueger (1995) find "no evidence that environmental quality deteriorates steadily with economic growth." Their study supports the hypothesis that while economic growth for very poor countries initially causes environmental deterioration, growth for middle-income and richer countries tends to correlate with improvement of the environment.

Regulation and Trade

Regulation may affect patterns of international trade. Environmental legislation, in essence, curtails the ability of industries to pollute freely. As a result, industries operating under strict regulation may have higher costs and lower exports. The average tariff for non-primary products imported to industrialized countries is below 5%. In comparison, U.S. pollution-abatement expenditures by the chemical industry were about 3.5% of chemical industry value added in 1991. Dow Chemical has estimated that the prices of its products reflect about a 3% increase due to environmental investments. Environmental laws thus are of a similar order of magnitude to tariffs and thus potentially important in affecting trade. Various models of trade that have been used to examine the effects of environmental legislation generally indicate a minimal effect on overall international patterns of trade (Baumol and Oates 1988; McGuire 1982; Pethig 1976). Comparative advantage on the aggregate national level is more dependent on factors such as cost of financing, available technology, energy and raw-material costs, transportation costs, labor, and social costs. In certain industrial sectors with especially high environmental-compliance costs, however, the effects on investment and exports can be larger.

Consider the range of available evidence. An OECD report (1985) concluded that pollution-control measures may have reduced total exports from the United States by 0.5 to 1.0 percentage points (see also D'Arge 1974). But using historical patterns of trade and foreign-investment flows and direct compliance costs across several industries and countries, other studies have found little to support the predicted trade and investment effects of environmental policy (Leonard 1988; Pearson 1985). Differences in direct compliance costs were judged to be insignificant compared with other factors such as labor and materials costs, proximity to markets, and industrial infrastructure. Similarly, Tobey (1990) tested the effect of the relative stringency of environmental policies on net exports using a set of resource endowments for 1975, direct pollution abatement costs, and indirect abatement costs embodied in intermediate-input-factor costs. Defining "pollution-intensive" commodities as those which require large environmental resources, he examined five pollution-intensive industries including chemicals, mining, paper, steel, and primary metals and found that environmental-policy stringency was not an important determinant of trade. The uncertain correlation between environmental policies and enforcement may have influenced the results. Jaffe et al. (1995) point out that "the results could theoretically be due to no more than the failure of the ordinal measure of environmental control costs, but Tobey's results are essentially consistent with those from other, previous analyses that employed direct cost measures."[4]

Kalt (1988) found an insignificant relationship between changes in net U.S. exports from 1967 to 1977 across 78 industrial sectors and changes in direct environmental-compliance costs. When the sample of industrial sectors was restricted to manufacturing industries, however, the effect of compliance costs on net exports became significant. Surprisingly, removing the chemical industry from the manufacturing sample increased the significance of the predicted negative effect even further. In the late 1960s to mid-1970s, the chemical industry continued to have a relatively strong net export performance, even with increasing environmental expenditures. However, it is possible that given the broad and varied product base and export performance of the chemical industry, a sector by sector breakdown would have yielded different results.

Grossman and Krueger (1993) used 1987 data on manufacturing industries and three indicators of economic activity to examine U.S.-Mexico trade. They found that pollution-abatement costs in the United States have apparently not affected imports from Mexico or activity in the *maquiladora* region, where Mexican laws regarding firm ownership and taxes are relaxed to attract qualified foreign investment along the U.S.-Mexico border. The results verified the importance of labor intensity but indicated that industry differences in environmental costs were small and statistically insignificant.

In evaluating the evidence on how environmental regulations affect trade, it is important to keep the limitations of this research in mind. As in all empirical studies,

---

[4]The "other, previous analyses" include Walter (1982), Pearson (1987), and Leonard (1988).

one difficulty involves distinguishing the effects of environmental regulations from effects of changes in other variables such as exchange rates. Another problem is the availability of reliable data on environmental expenditures, comparable across time, across industries, and across countries. Even when direct costs of compliance are available, the total will not include indirect administrative and legal requirements that will affect product and plant approval, litigation costs, and diversion of effort away from productive activities. There may also be a considerable gap between the environmental policy that is enacted and how (or whether) it is enforced. Empirical studies do not account for the social benefits arising from regulation, especially when, as with many health and environmental benefits, they may not be easily quantifiable. Moreover, most studies on environmental regulation and trade rely on data from the 1970s. Since the 1970s, environmental regulations have become progressively more strict and firms face greater global competitive pressure. Even if the effects of regulation were small in the 1970s, they may be more significant today.

There is some anecdotal evidence that environmental costs may be more burdensome in the 1990s. In 1992, the chief executive officer of Hoechst AG and president of the German chemical industry association (VCI), Wolfgang Hilger stressed that environmental-protection costs had risen to the point that it was hard for companies to compete internationally. In 1990, the German chemical industry spent nearly $4 billion, or about 4% of sales, on environmental-protection measures. Hilger estimated that by 1996 this figure would rise to $5.6 billion without new environmental legislation or taxes. In BASF AG, environmental operating costs rose from $260 million in 1981 to $666 million in 1991, while investments in environmental protection rose from $45 million to $242 million during that period. Similarly, environmental operating costs for Bayer AG increased from 4% of sales to 8% in that period (Ondrey 1992). German chemical companies have argued that it would be more cost-effective for Germany to pay to clean up Eastern Europe than to raise Germany's standards (*The Economist* 1993).

Whereas most assessments conclude that the initial economy-wide or national level effects of environmental regulation have been insignificant or nonexistent, evidence suggests that effects in certain sectors are more significant. Some smaller suppliers, as well as chemical-industry observers, charge that larger companies favor cheaper imports from suppliers in countries with lower environmental standards (Freemantle 1996). It is difficult to determine to what extent lower labor costs, lower overhead, or lower environmental costs are responsible for lower product costs. Nonetheless, the absence of environmental regulation can bestow considerable economic advantage. For example, David Robison (1988) found that from 1973 to 1982, the United States increasingly imported goods in industries that had high domestic environmental costs and less from those in which such costs were lower. In contrast, the net ratio did not change in that period for imports from Canada, which had environmental regulations that were similar to those in the United States.

A recent European example involves phosphate fertilizers. The mining and production of phosphates generate toxic pollutants, and the use of phosphate fertilizers potentially damages surface and ground water sources. In Western Europe, there has been growing pressure and regulation to halt discharge of contaminated phosphogypsum from phosphate fertilizer production. This has led some companies, such as BASF, to close units and choose alternate technologies when investing in new phosphate-fertilizer production capacity. However, only a small number of older units have been replaced with the cleaner nitrophosphate units in all of Western Europe. Meanwhile, Western Europe is importing increasing supplies of phosphate fertilizers and phosphoric acid used in the production of phosphate fertilizers, in large part from northern Africa. Production has rapidly expanded in this region where environmental regulation has not been strict (Heerings 1993).

INDUSTRY LEVEL

In all four countries discussed here, between 5% and 10% of the aggregate capital expenditures of the chemical industry were for environmental protection in the early 1980s. For Japanese chemical companies, that figure has remained at about 5%. For German chemical companies, the figure rose to 10% to 15% in the late 1980s. In the 1990s, American and British chemical companies have increased their environmental expenditures to a percentage of total capital spending comparable to that in Germany. In fact, by 1994, direct environmental spending by the U.S. chemical industry approached 20% of total capital spending. Figure 11-1 illustrates the role of chemicals in relation to some other industries in terms of value added, pollution-control expenditures, and toxic chemicals released. As a percentage of industry turnover, total capital and operating environmental expenses rose from 2.3 in 1985 to 4 in 1991 in Germany. For the same period, the percentage rose from 1.75 to about 2.2 in the United States. In the United Kingdom during the period for which data are available (1990–1992), total expenditures were about 3% of turnover, according to the European Chemical Industry Council. Environmental spending as a share of turnover is represented for selected companies in Table 11-2. The chemical industry's total compliance expenditures have been rising everywhere. For example, in 1991 a law defining corporate liability for pollution took effect in Germany. The law, which outlines civil damages for wrongful death, personal injury, or property damage caused by air, soil, or water pollution, has increased expenses for chemical companies in safety measures and insurance (Kirk 1994).

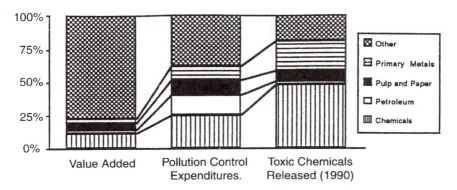

Figure 11-1. Industrial Share of Value Added, Pollution-Control Spending, and Release of Toxic Chemicals in the United States, 1991. *Sources:* U.S. Census Bureau, *Pollution Abatement Costs and Expenditures*, 1991 (Washington, DC: U.S. Government Printing Office, 1993); U.S. Census Bureau, *1991 Annual Survey of Manufacturers, Statistics for Industry Groups and Industries* (Washington, DC: U.S. Government Printing Office, 1992); U.S. Environmental Protection Agency (EPA), *1990 Toxics Release Inventory: Public Data Release* (Washington, DC: EPA Offices of Pollution Prevention and Toxics, 1992).

TABLE 11-2. Chemical-Industry Environmental Spending as a
Percentage of Turnover, for Selected Companies: 1992

BASF	7.1
Bayer	4.1
Dow Chemical	3.1
Du Pont	3.1
Elf Atochem	5.2
Hoechst	4.4
ICI	3.0
Rhône-Poulenc	2.2
Solvay	7.7

Source: EC (1994).

Sectoral Impacts

The impact of environmental regulation will differ for various sectors of the chemical industry, depending on many factors: the severity of environmental health-and-safety hazards associated with the chemical product or process; chemical product type (e.g., primary or intermediate); level of product differentiation in the sector; compliance and abatement cost levels; plant and product approval processes; size of the firm; level of competition; and geography. Here, I can offer only some examples of sectors within the chemical industry that, for one reason or another, seem especially susceptible to the rising costs of environmental regulation at the present time.

The primary chemical sectors, involving products such as commodity and agricultural chemicals, tend to compete principally on price differentiation; therefore, they may be particularly disadvantaged by cost increases due to regulation. The geography and density of a country can also affect the shape of environmental concerns. For example, environmental concerns regarding pesticides in food and ground-water contamination from fertilizers have inhibited the growth of the agricultural-chemical sector, in some instances with good reason. Recent amendments of legislation in the United States, for example, are proposed to prevent the export of pesticides that are either banned in the United States or not re-registered.

The story of Avtex Fibers, formerly the largest U.S. producer of rayon, provides a perspective on noncompliance and on the impact of mounting pollution-abatement costs. Avtex Fibers closed its only rayon plant at Front Royal, Virginia, in 1990, citing financial difficulties and environmental problems. Since 1975, it turned out, the plant had violated its operating permit by allowing an estimated 2,000 illegal toxic discharges into the Shenandoah River. Among other problems, PCB (polychlorinated biphenyl) leaks were discovered in 1986 and 1987, and the company failed to report the 1986 PCB leakage to state authorities. The state of Virginia sued Avtex Fibers for $20 million in fines. The company had already spent millions of dollars on the plant in recent years to meet state and federal mandates, and the suit could have required an additional $40 million in compliance expenditures. The closure terminated nearly 1,300 employees, leaving Avtex with a polyester fiber plant and a marketing division (Agoos 1988; U.S. Congress 1990).

An emerging issue of concern for many sectors of the chemical industry is the potential health effects of synthetic compounds. It is feared that hormone-disrupting chemicals—among them organochlorine pesticides such as DDT, dioxins, alkylphenol polyethoxylates, phtalate esters, and polychlorinated biphenyls (PCBs)—harm reproduction and sexual function in humans, as well as in wildlife. These compounds are utilized in a variety of products including crop-protection biocides, cleaning products, paints, and plasticizers.

Although there is ample evidence of links between chemicals and hormone disruption, the impact of synthetic hormones on reproduction is unclear. There is evidence that synthetic and naturally occurring chemicals cause reproductive and developmental damage in animals; a study in the United Kingdom by Zeneca for the German pharmaceutical company Schering found that ingredients in birth control pills may inhibit reproduction in fish. Nonmetabolized synthetic estrogens in birth control pills are passed into the environment and are less biodegradable than natural estrogens. In laboratory tests, dioxins damage reproductive function in animals; Vinclozolin, a fungicide, has been shown to harm sexual organs of rat offspring. A 1980 spill of the pesticide dicofol, which was contaminated with DDT, is suspected of causing underdevelopment in sexual organs of 60% of the male alligators in Lake Apopka, Florida.[5]

Many studies of hormone disrupters in animals have used large doses not found in the environment. However, the hypothesized health-damaging impacts of synthetic chemicals on the human reproduction system are at least plausible. Some of the reproductive-system changes observed in humans have been induced in laboratory animals with low doses. Other studies suggest that when some common agricultural and industrial chemical compounds are present in combination, they may produce intensified responses, raising the risk of cancer or reproductive defects by as much as three orders of magnitude.

On the other side, human exposure to natural hormone disrupters, which are abundant in nature, is many orders of magnitude greater than to synthetic ones. Naturally occurring hormone disrupters are common in plants such as parsley, wheat, rice, oats, hops, soybean, carrots, beans, potatoes, and many others. A variety of clove, for example, is thought to have induced widespread sheep sterility in Australia in the 1940s. It is clearly incorrect to look at increases in prostate cancer or breast cancer and immediately assume that they must be due to synthetic hormone disrupters in the environment. Some of the increase is surely caused by improved detection or, possibly, longer life. Changes in lifestyles, working habits, and stress, in addition to chemical exposure, may be contributing to the observed changes.

Further, as already noted in the discussion of the Superfund regulations in the United States, the enormous costs of litigation and waste-site cleanup have jeopardized the survival of many companies. Arguably, this is one case in which well-meaning measures have created a scenario where costs dwarf benefits. On the positive side, environmental regulations may enhance the competitive standing of sectors and companies in a position to gain efficiency and marketing and innovative advantages. These sectors tend to use greater technology and to produce outputs closer to public consumption than primary chemicals. For example, consider zeolite, a major product in Europe and Japan. Zeolite is used as a substitute for phosphates in detergents, and growth in the zeolite market has partly been spurred by bans on the use of phosphate builders in detergents. Similarly, DuPont committed itself to a complete phaseout of CFCs, preceding the 1987 Montreal Protocol agreement, which committed all countries

[5]See also *Chemistry and Industry* (1996), *The Economist* (1996), and Hileman (1996). Colborn, Dumanoski, and Myers (1996) investigate the issue of developmental impairment brought on by synthetic-hormone disrupters released into the environment.

to this step, and gained a competitive advantage by doing so (Barrett 1992). Not all CFC producers followed DuPont's example. The main European producers, such as Atochem and ICI, initially resisted the international agreement to limit CFC use. But chemical companies have often benefited by early investments in environmentally cleaner processes and products, as will be discussed further in the next subsection on innovation.

Effects on Innovation

Environmental measures can both enhance and hinder innovation. Companies that have made early investments in environmental-protection equipment, for instance catalytic technologies, have often found a market for their products. One of the fastest-growing services in industrialized countries has been that of environmental-protection technologies that are exported mostly by those countries which enacted strict environmental policies sooner. Cairncross (1993) notes that the United Kingdom had lower pollution-control investments, and its ratio of exports to imports in environmental technology declined from 8:1 to 1:1 in the 1980s. The United States provides another example. Cairncross notes that with little environmental legislation enacted between the mid-1970s and 1990, when the Clean Water Act of 1990 was introduced, the United States imported 70% of all of its air-pollution equipment by the early 1990s.

One set of leading examples involve air-pollution reduction. During the 1970s, Japan allocated an average of 14% of industrial investments to pollution control, mostly for cleaner air. This allowed Japan to make technological advances ahead of other countries and to create an export base in pollution-control products, for example, providing flue gas scrubbers to European countries. Similarly in Germany, with increasing air-quality standards, manufacturers have developed desulfurization equipment for sale in Japan and elsewhere.

As environmental regulations continue to tighten, the business opportunities for being ahead of the curve should grow. The market for environmental technology has been growing and is expected to reach $500 billion by the year 2000 (Fig. 11-2). New European Union environmental rules are especially instrumental in creating greater technology investments in Western Europe. A new EC directive on urban wastewater treatment could require a $225 billion investment by 2005 (Rubin et al. 1994, p. 42).

Occasionally, a stronger claim is made that environmental regulations indirectly induce innovation through higher overall research activity associated with improving environmental quality (Porter 1991). A recent study has examined the evidence on pollution-abatement compliance expenditures, research-and-development expenditures, and patent data for a sample of industries (Jaffe and Palmer 1994; Jaffe et al. 1995). Successful patent applications were used as a proxy for level of innovation. The authors found that in the period between 1976 and 1989, increases in compliance spending were positively correlated with spending for research and development, but they found no support for the hypothesis that increased spending produces greater innovation.

However, increased research on environmental priorities may occur at the expense of other value-creating research or endeavors, and strict environmental standards can also have a negative effect on other types of research-and-development investments. For example, European chemical-industry executives maintain that biotechnology research in Europe has been hampered by regulation. Prolonged new-plant and new-product approval processes are disincentives for innovation. Industry sectors that rely principally on innovation for growth can be hurt by delays and uncertainty in future

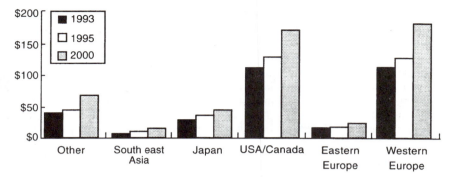

Figure 11-2. Global Environmental Markets (Billions of Dollars); 1994 Projections. *Source:* In Rubin et al. (1994), from Helmut Kaiser Consultancy, Tübingen, Germany.

regulation. Developing a new chemical product in Europe takes nearly two years; in Japan, the equivalent period is estimated to be closer to a year and a half (Schneider 1993). Regulation can also result in diversion of funds away from research and development activities to pollution prevention and cleanup.

Rather than developing new molecules, many biocide manufacturers in the United States are searching for new applications of existing product lines to avoid lengthy and expensive registration processes. Registration of new biocides can take more than five years and cost up to $5 million, while the cost for registering a new application for an existing biocide is about $300,000. The result has been a substantial drop in the introduction of active biocides.

Lengthy, complicated processes for plant and product approval can change company strategy and slow innovation. In the words of a chemical company chief executive, "If a marketing plan says a plant must go onstream in two years, and in Germany approval could take four years whereas in Belgium it would take two, this difference in timing plays a big role in making decisions." Another executive makes a similar point, "Processing applications for chemical plants takes considerably longer in Germany than it takes at any other European sites." It took seven years and DM 80 million for Hoechst to obtain approval for an artificial insulin product after starting plant operations. Building and preparing the plant for operation had taken about three years (Chynoweth 1991).

For the pharmaceutical industry, changes in health-care legislation can add considerable risk. As reported by the Bayer company (Bayer 1993, p. 36), "The heavy impact of new health-care legislation on the pharmaceutical market, particularly in Germany, has clouded the business environment for the research-based pharmaceutical company."[6] Bayer reports that recent changes in health-care legislation in Germany cost the company about DM 130 million in sales and cut its earnings in half for the fiscal year 1993. Thomas (1990) investigated the impact of U.S. Food and Drug Administration regulations on innovation and concluded that regulation had drastically reduced research productivity in smaller U.S. pharmaceutical companies. Larger firms had more moderate declines in research productivity and were able to offset losses with greater sales, due to reduced competition.

[6]See also Hoechst Annual Report 1996.

During the 1980s, chemical companies were able to take significant steps in pollution reduction and efficiency. Because industry had paid minimal attention to environmental problems in the previous decades, these improvements and innovations were aimed at obvious targets and have provided business opportunities for processes and products that could reduce pollution at relatively low cost. But as regulations become progressively stricter, such relatively easy solutions may become less likely. There are costs to preserving and promoting environmental quality. Walley and Whitehead (1994) offer a discussion that emphasizes why it is increasingly difficult to view environmental regulatory measures as profit-making business opportunities.

Effects on Industry Structure

The 1980s saw a wave of acquisitions and joint ventures in the chemical industry. There were nearly 13,000 acquisitions in the chemical industry between 1982 and 1991. In addition, new alliances have become more common. To be sure, the 1990s have seen some movement in the reverse direction, as companies have begun to divest and de-merge to concentrate on core areas. But nonetheless, there has been an overall general consolidation of the chemical industry.

This consolidation has a number of causes: economic conditions, new foreign markets, new pressures of global competition — and growing environmental demands. For example, small chemical firms that fear their potential legal liability have been driven into partnerships with larger chemical firms. Tough regulation and concerns about environmental liability have changed the rate and structure of mergers and acquisitions. For example, the paint and dye sector of the chemical industry, subjected to increasingly tougher environmental rules, has experienced a large number of mergers and acquisitions. The pesticide sector underwent similar changes through the 1980s. Development of a new pesticide can cost $50 million, nearly a tenfold increase in twenty years, and take an average of nine years. The biocide market provides yet another notable example. In the United States, the number of registered active ingredients in biocide applications dropped from 1,600 in 1976 to about 400 in 1993. Since the implementation of a re-registration requirement by the Environmental Protection Agency in 1987, about one-half of the biocide products on the market then are no longer commercially available in the United States. Companies estimate that it costs between $2 to $5 million and takes from three to five years to introduce a new biocide to market. As a consequence, companies are less likely to focus on market niches. More and more biocide producers have begun broadening their product lines and entering joint development and marketing agreements (Westervelt 1994; Fattah 1995).

Changes in the catalyst market emphasize the growing reliance of the chemical industry on innovative technology. Catalysis has become critical in reducing the environmental impact of chemical-industry operations; it is used in developing environmentally benign chemical processes. Catalyst manufacturers find that environmental, as well as economic, pressures have made partnerships more appealing, according to industry observers (Thayer 1994).

Environmental concerns can also stop some mergers from occurring, because of fear of potential liability and environmental responsibility. Consider the 1989 decision of the financial Sterling Group of Houston to back away from buying BP Chemical's nitrogen chemical operations in Ohio because of concerns about how liabilities of any future environmental concerns would be shared (Heller 1990, p. 14). The majority of such deals, however, do seem to be completed, albeit with higher costs and longer delays.

TABLE 11-3. Decline in Commodity-Chemicals Production Related to Environmental Concerns

| | | Percent Change (Peak Output Year to 1987) | |
	Peak Number of Producers	Output	Number of Producers
Acetaldehyde	9	−61%	−78%
Carbon disulfide	4	−60%	−25%
Chlorobenzene (mono)	8	−58%	−63%
Ethyl chloride	6	−76%	−17%
Hydrofluoric acid	9	−2%	−56%
Lead alkyls	4	−62%	−75%
Phosphorus	8	−42%	−50%
Sodium	3	−60%	−33%
Sodium phosphate	5	−48%	0%
Sodium sulfite	5	−51%	0%

Source: Lieberman (1990).

The alert reader may have noticed that many of these regulatory costs and pressures present a particular challenge to smaller chemical firms. Larger firms usually have greater contacts with government and labor unions, greater leverage within industry associations, and greater financial resources. Conversely, smaller firms have modest financial resources and limited access to market information, distribution channels, and services. Smaller chemical enterprises are often disadvantaged in acquiring advanced scientific information or are incapable of installing expensive sunk-cost pollution-control equipment. Cost of environmental liability and cleanup could easily equal or surpass the value of assets in smaller companies. It is not surprising that obtaining liability-insurance coverage has been cited as burdensome to smaller chemical firms in the United States. Bearing the costs of environmental compliance and abatement is proportionally more significant in the operations of these firms.

Research by Marvin Lieberman (1990) on several declining commodity chemicals has revealed that, faced with shrinking markets, smaller firms are more likely to exit the market, whereas larger firms are more likely to divest incrementally and reduce capacity. Table 11-3 lists chemical products in Lieberman's analysis whose decline is due wholly or partly to environmental reasons.

Regulations may also have an asymmetric influence between established and new chemical firms and serve as a barrier to entry, for a number of reasons (Dean and Brown 1995). The most direct economic rationale for the entry-deterrent effects of pollution-abatement standards is that they increase the capital required for entry. Compliance with environmental mandates is often capital-intensive and requires the installation of expensive equipment such as particulate scrubbers, waste-recovery and recycling systems, and wastewater treatment plants. This is especially relevant when costs of compliance decrease with firm size — in other words, when there are economies of scale. Such scale economies may exist in pollution-control capital expenditures (Brock and Evans 1985, 1986; Pashigian 1984). Some evidence on this point is presented in Table 11-4, which shows that pollution-control costs often represent a larger share of sales for small firms than for larger ones. Consequently, the optimal size of plants

TABLE 11-4. Chemical-Industry Pollution-Control Expenditures as a Percentage of Sales

	Pollution-Control Capital Expenditure as a Percentage of Sales		Pollution-Control Operating Costs as a Percentage of Sales	
	Company Sales		Company Sales	
	<$1 BN	>$1 BN	<$1 BN	>$1 BN
AVERAGE				
1993	2.2	1.6	2.3	2.3
1994	2.6	1.8	2.5	2.4
1995	2.9	1.6	2.7	2.2
MAXIMUM				
1993	15.0	5.0	10.0	5.0
1994	15.0	5.0	15.0	7.0
1995	20.0	4.0	15.0	5.0

Source: United States Chemical Manufacturers Association survey of member companies; Lenz et al. (1994).

tends to increase. The expected increase in minimum efficient scale may be exacerbated by the "technology-forcing" rules included in many environmental regulations, which require the use of particular pollution-abatement methods, regardless of the scale of the firm.

Scale economies are also important in the administrative aspects of environmental compliance, which is another reason that regulations may affect smaller organizations more severely than large ones. Small companies often lack the specialized administrative resources needed to handle compliance procedures and processes. Moreover, setting up systems to track and interpret relevant regulations and to deal with health, safety, and environmental agencies and the paperwork associated with the process impose fixed costs that increase the efficient scale of operations required for compliance. Firms that can pool their administrative resources and compliance costs over a larger production scale will likely gain cost advantages.

Examples of this record-keeping and compliance burden abound. Commenting in 1986 on record-keeping requirements in the United States, an executive with Dixie Chemical in Houston, Texas, observed, "It's obvious that without deep pockets you need an ongoing business of probably $7-15 million, depending on profit margin, to pay for inside and outside experts to deal with this area" (Gibson and Campbell 1986, p. 28). In a 1995 poll of over 2,000 small businesses, the U.S. Chamber of Commerce found that a majority of respondents (28%) find environmental protection and regulation to be the "most difficult" aspect of dealing with the Federal government. The same poll showed that 78% of those small companies believe that reducing federal government bureaucracy would be the most helpful area of reform (U.S. Chamber of Commerce 1996, p. 49).

Companies in the United Kingdom, especially fine and specialty chemical firms, face similar administrative challenges with the passage of the 1990 Environmental Protection Act and implementation of the Integrated Pollution Control (IPC) scheme. Estimates put the average processing cost of IPC applications at between £2,000 and

£15,000, with 100 to 500 personnel hours required for preparation of each application. Some fine-chemicals applications have reportedly taken up to 100 personnel days (Mullin et al. 1994, p. 21).

New legal-liability standards fall more heavily on small companies, as well. In the United States, rulings on the Superfund Act of 1980 have held banks liable when their customers pollute. Now, in some cases, banks and insurance companies are requiring environmental audits and proof that a potential customer has never polluted. The cost of site assessment falls on the prospective borrower. Moreover, "banks are cutting off small- and medium-sized businesses in industries that handle dangerous chemicals or produce contaminated waste" (Hector 1992, p. 107). Even in Germany, where no lender liability exists, banks have been affected, writes Brealy (1993, p. 232):

> Nevertheless, banks have in their own best interests become more cautious in the granting of credit; the responsibility of borrowers for expensive environmental protection measures affects their creditworthiness and is thus an integral part of every credit assessment. Soil and/or groundwater contamination also leads to a considerable reduction in the value of land, which in turn reduces its value as collateral security.

Recently, insurance organizations in the United States have increased their environmental-damage reserves, thereby recognizing the "immensity of the problem" of settling pollution claims, which are estimated to be anywhere between $30 and $100 billion (Scism 1995).

Some chemical firms have managed to encourage environmental regulations that encourage the use of their product. Zeolite, the phosphate substitute mentioned earlier, provides an example. In 1977, the German detergent manufacturer Henkel told the German government that it could reduce phosphates by 50% using zeolite. Soon after, the government required that the use of phosphates in detergents be reduced by 50% by 1984. Even before this regulation had been proposed, Henkel and DEGUSSA, its patent partner, had invested in and set up zeolite production facilities and planned capacity expansions. Zeolites completely replaced phosphates in detergents by 1989 and have become a major product. The market-entry barrier caused by phosphate regulation created a profit opportunity for Henkel and DEGUSSA (Barrett 1992).

A final effect of regulation on industry structure occurs through rules requiring approval for the construction of new chemical facilities. This process can be complex and lengthy, especially in the United States and Germany. Regulations requiring detailed impact assessments have led to considerable delays, and many new projects have been denied permits. In the United States, recent denials include a Dow Chemical facility planned for northern California, a Standard Oil of Ohio pipeline planned for the Southeast, and an oil refinery and marine terminal planned for Virginia. In a study of rules pertaining to industrial siting in the United States, Duerksen (1983) concluded that the requirements for obtaining plant permits have become dramatically more complex and cumbersome since the early 1970s. The problems include overlapping and contradictory permit reviews, changing laws and regulations, lengthy judicial reviews, and a general lack of understanding on the part of business planners, regulators, and environmental groups. Again, this can have a deterrent effect on the establishment of new operations, thereby providing an advantage to existing firms. This problem is exacerbated by what is called "new-source bias" meaning that the regulations place a heavier burden on new pollutant sources than on existing ones. Yet again, this can dissuade new firms from entering the chemical industry.

Thus, a range of initial evidence suggests that environmental regulations have a net deterrent effect on new firm entry into many pollution-intensive industries, including the chemical industry. Although a number of causes are plausible, the increases in capital expenditure required for entry and efficient operation are often cited as the major cause. According to Dean and Brown (1995):

> ...the ability of environmental regulations to create barriers to the entry of new firms implies that incumbents may be able to use environmental regulations strategically to enhance competitive advantage and profitability (Barrett 1991; Brock and Evans 1986; Porter 1980, 1990). Available evidence indicates that firms may have been able to influence environmental regulations to their favor (Barrett 1991), and some incumbents may be well advised to promote environmental legislation that gives them an advantage over new firms.

Since environmental regulations create various barriers to entry for new firms, there are proposals to offer some counterbalancing incentives to encourage small firms. In many industrialized and industrializing countries, small- and medium-sized firms constitute an important sector of the economy, and legislators have taken action to shield those firms from what is believed to be potentially disproportionate effects of regulations on smaller businesses. Entry into the chemical industry may be encouraged through mechanisms that include a tiered regulatory structure that favors small firms and through government actions such as regulatory exclusions of smaller firms, special subsidies, or reduced scrutiny and enforcement action. As described in the section on regulation, several programs have been initiated in Japan, for example, to address the problems of smaller firms in complying with environmental legislation. These programs support preferential tax treatments and access to special loans, as well as other forms of financial, legal, and technical assistance (Gresser, Fujikura, and Morishima 1981). In Germany, one of the major concerns put forth by the chemical industry in opposition to the EC's 6th Amendment was the impact of costly testing procedures on specialty chemicals that are often produced in small quantities by small firms (Grant, Paterson, and Whitston 1988).

To some extent, regulators already commonly differentiate between large and small firms and tend to put greater pressure on larger firms. After all, larger firms may be more likely to offer the greatest pollution reductions and, therefore, may attract higher scrutiny from regulators and environmental groups. Finto (1990) finds that limited budgets within the U.S. Environmental Protection Agency have resulted in larger firms being scrutinized more closely. Michael Greve (1989) analyzed notices of intent to sue under provisions of the Clean Water Act and concluded that environmental groups tended to target larger firms, who were not necessarily the worst polluters but were the most likely to accede to substantial settlement demands.

Steps to help small firms just might provide a less-expensive way to reduce pollution, as well. After all, a new firm can select its production and pollution-abatement technologies simultaneously, which can lead to greater efficiency relative to retrofitting and other methods that are applied to existing, earlier production technologies. In addition, the costs associated with replacement of older technology earlier than planned may inhibit existing plants from adopting more efficient integrated technologies (Kemp and Soete 1992). In fact, Evans (1986) found that the average pollution-abatement cost per employee in manufacturing industries was less for small firms than for large firms (although "small" and "new" are not identical, they often cover similar firms). However, it would seem difficult to reconcile this effect with the evidence that U.S. chemical industry figures for capital expenditures on environmental compliance are larger for small firms than large ones, as a age of sales. The question of encouraging new firms as a way of reducing pollution more efficiently is a topic that deserves further research.

Empirical evidence on the existence of a significant size-differentiation effect is inconclusive for the manufacturing industry as a whole. Although small manufacturing firms, on average, may have lower abatement costs per employee, environmental regulations seem to reduce both the number of plants in pollution-intensive manufacturing industries and the market shares of small plants (Brock and Evans 1986; Pashigian 1984).

MOVEMENT OF CAPITAL AND CHEMICALS PRODUCTION

Overall, political and economic factors are the most important determinants of investment decisions; in comparison, the savings from less strict environmental regulations are relatively small. However, there is evidence that environmental regulation may be contributing to the movement of investment within certain chemical sectors away from regions with highly stringent regulations. Patterns of divestment or reinvestment abroad are most visible when regulations are tightening and firms or sectors are hard-pressed by their competition.

Germany provides an example with strict state regulation, extensive plant and product approval procedures, high labor costs, and an expensive social system. According to Teltschik (1992), these factors have helped to damage the investment climate in Germany and, coupled with high energy costs and high taxes, have led to a greater incentive to move abroad. For example, it has been suggested that negative public opinion, unpredictable conditions, and the regulatory framework have encouraged chemical companies in Germany to invest in biotechnology abroad. The three major German firms have established biotechnology operations abroad, especially in the United States and Japan. Gottfried Plumpe of the Bayer Company has pointed out that public opinion is one of the key reasons for having relocated that company's biotechnology operations from Germany to the United States (Landau 1996, p. 406). A 1984 project by Hoechst to build a plant in Germany for producing human insulin was delayed for ten years because of public concern. In the 1980s, Hoechst and Bayer and others moved biotech operations to the United States and Japan (Miller and Hamilton 1995, p. 70). Recently, the prospect for this sector has been changing, as the German government position and public opinion have become more favorable to research and entrepreneurship in biotechnology. A 1990 German law governing biotechnology operations has been relaxed to reflect European Union rulings. Bayer, for example, is now reinvesting in pharmaceutical biotechnology products in Germany. But it remains true that the German government is pondering alterations in its environmental regulations, "partly because it can no longer afford a large regulatory and enforcement system, and partly because the high cost of regulation is part of what is driving German investment elsewhere" (Roberts 1995, p. 34).

In general, however, regulatory policies on air and water in these four countries have been converging. Much greater regulatory differences exist with respect to the newly industrializing nations. In recent decades, these differences have been sufficient to induce changes in the patterns of foreign investment in certain chemical sectors. Industry sectors with diminishing markets, whether due to environmental, health, and safety requirements or for economic reasons, seem to be more likely to relocate or invest abroad when faced with greater regulation. In the United States, these sectors include producers of highly toxic chemicals such as benzidine-based dyes and some pesticides.

A variety of studies substantiate this point. As mentioned earlier, in a study of U.S. trade patterns from 1973 to 1982, Robison (1988) found that imports from industries with high domestic (U.S.) environmental costs steadily increased, while imports from other industries with lower domestic-compliance costs decreased. In 1967, when public worry and media attention about the environment were beginning to induce greater political action, the United States had more pollution-intensive exports than imports. Kalt (1988) found the opposite true a decade later. In one of the few studies that cover the time range from the onset of heightened environmental regulatory activity in the mid-1960s to the late 1980s, Low and Yeats (1992) find that a similar trend holds for North America. They report that the North American share of pollution-intensive products in world trade fell from 20.5% in 1965 to 14.0% in 1988. For Southeast Asia, the respective shares rose from 3.4 to 8.4%. The share of all pollution-intensive trade originating in EC countries remained steady, at about 38%. The study describes a shifting pattern in which pollution-intensive activities are dispersed internationally, mostly in the direction of developing countries. Labor intensity was not found to be an important factor in the "locational pull" for developing countries toward pollution-intensive production.

Low and Yeats (1992) also show that the amount of pollution-intensive exports in world trade has fallen, and the share of such trade has been reduced in the total exports of industrialized countries. Over a twenty-year period, developing countries exported an increasing share of the total world exports from polluting industries. Table 11-5 illustrates this changing pattern for chemical-industry sectors using a measure of "revealed" comparative advantage, defined as share of that sector in its home country's total manufacturing exports relative to that sector's share in total world manufacturing exports.

In a study on the "migration" of toxic industrial pollution, Lucas, Wheeler, and Hettige (1992, p. 80) found evidence consistent with the movement of pollution-intensive industries to developing countries. They used data from a sample of 15,000 manufacturing plants in about 80 countries for the period 1960 to 1988.

> It is frequently asserted that stricter regulation of pollution-intensive production in the OECD countries has led to significant locational displacement, with the consequent acceleration of industrial pollution intensity in developing countries. All our results are consistent with this hypothesis. Both sets of estimates [pollution intensity time series estimates for developed and developing countries] suggest that the poorest countries have the highest toxic-intensity growth. The estimated toxic-intensity elasticity of income growth for a typical (midrange-distortion) LDC [lesser developed country] economy was apparently negligible in the 1960s, positive in the 1970s, and even higher in the 1980s. Of course, one cannot be certain of a causal connection between these decade patterns and the roughly concurrent shifts in OECD environmental policies. Yet the results are nonetheless suggestive of a potential contributory effect.

The closing warning of Lucas et al., that the correlation between stricter first-world environmental standards and greater third-world production of pollution does not prove causality, is worth emphasizing. Concerning pollution-intensive industries in general, Jaffe et al. (1995, p. 146) observe that trade-pattern changes may not be caused by increasingly strict regulatory requirements. Other economic and strategic factors, such as access to growing markets — including a willingness to suffer more pollution for

**TABLE 11-5. Share of Industrial Countries in all Exporters with "Revealed"
Comparative Advantage (%): 1966–68 and 1986–88**

Industry Sector	1966–68	1986–88	Change
Refined petroleum products	12	12	0
Organic chemicals	38	36	−2
Inorganic chemicals	50	28	−22
Other inorganic chemicals	32	34	2
Mineral tars & petoleum chemicals	28	21	−7
Manufactured fertilizers	35	22	−13
Insecticides, fungicides, etc.	50	72	22

Source: Low and Yeats (1992).

the sake of economic growth — also may be influential in the shift of pollution-intensive industries to the industrializing regions.

For the U.S. chemical industry in particular, overall foreign direct investment has grown at a slightly higher rate than that of manufacturing industries as a whole. Leonard (1988) documents that between 1973 and 1985, direct investment abroad by the chemical and mineral processing industries grew from 25.7% to 26.5% of direct investments abroad by all manufacturing industries combined. However, while the proportion of investments by all manufacturing industries in developing countries increased, the proportion of investment by the chemical companies fell; in other words, investment from chemical companies was not disproportionately headed to developing countries.

Anecdotal evidence abounds on the relationship between the relative stringency and enforcement of environmental regulation and chemical-plant location. Hilary French (1993, p. 32) recounts some of the available information on Mexico, Malaysia, and South Africa.[7]

On the books, Mexico's environmental laws are roughly comparable and in some cases stricter than U.S. ones. But enforcement has been lax; an official of the Mexican environment ministry estimated in 1991 that only 35 percent of the U.S.-owned factories along the border comply with Mexican toxic waste laws. Though maquiladoras are required to return to the United States any waste they generate, compliance with this mandate is believed to be the exception rather than the rule.

Investigators have revealed alarming conditions in the border area. At three-quarters of the maquiladoras sampled in 1991, the U.S. National Toxics Campaign found toxic discharges, including chemicals that cause cancer, birth defects, and brain damage, being emptied into open ditches running through settlements near the factories.

Though the maquiladoras have received more incriminating publicity than other such sites, they are far from unique. Environmentalists in Southeast Asia, for example, have sharply criticized Japanese industries for allegedly locating extremely harmful processes abroad because they can no longer pass environmental muster at home. A Malaysian subsidiary of the Mitsubishi Kasei Corporation was recently forced by court order to close after years of protests by local residents that the plant's dumping of radioactive thorium was to blame

[7]See also *Journal of Commerce* (1992), Lewis et al. (1991), *International Environment Reporter* (1992), and Motala (1992).

for unusually high leukemia rates in the region. Several multinational corporations operating in South Africa, including local subsidiaries of the German-based Bayer pharmaceuticals concern and the U.S.-based Duracell battery company, have been implicated by local environmentalists in toxic catastrophes that they believe caused severe health problems among workers.

Xing and Kolstad (1995) study the relationship between the level of foreign direct investments of the U.S. chemical industry and the relative stringency of environmental regulations. The emission of sulfur dioxide is used as a proxy for environmental quality and is shown to be highly correlated with the emission of other pollutants. Their results indicate that tough environmental regulations are associated with lower direct investment, and that regions with lax regulations tend to experience more capital investment from the U.S. chemical industry. Even though these results do not prove a causal relationship between capital outflow and environmental policy, they do confirm a trend.

It has been suggested that large corporations, aware of the social and political pressures of environmental responsibility, adhere to pollution-control standards in their foreign operations similar to those in their domestic operations. To the extent that this is true, it has the effect of reducing the importance of environmental regulatory differentials in investment decisions of the chemical industry, as well as other industries. However, there is some indirect evidence that executives are not fully confident of how their environmental practices abroad will measure up. In a survey of more than 400 senior executives, conducted by the firm of McKinsey & Company (1991), the majority of the respondents, 58%, indicated that potential liabilities were the major environmental barrier to higher investment in Central and Eastern Europe. Uncertainty about governmental environmental policy registered the largest number of votes as an obstacle to greater investment in developing countries.

Some plant-level analyses have been conducted in the United States to test the significance of state regulatory measures on plant-location decisions of firms. Bartik (1989) found a very small, negative effect of state environmental standards on the start of small businesses. In a study of plant-location decisions of multinational corporations during the period 1977 to 1988, Friedman, Gerlowski, and Silberman (1992) discovered that state-level environmental stringency in the United States was not a significantly negative influence. The plant-location decisions of Japanese multinational firms, however, appeared to be significantly and negatively affected by environmental regulation. Levinson (1992, as reported in Jaffe et al. 1995) reports that environmental stringency does not affect the location of U.S. manufacturing industries in aggregate, but he indicates that stringent state-level environmental regulations have a significant negative effect for the highly pollution-intensive industries including chemicals (and plastics) during the time period of analysis (1982–1987). Gray (1997) studies net "births" of manufacturing plants in the period 1963 to 1987 and finds a strong negative correlation between the number of new plants and strictness of state-level environmental regulation. The impact on highly polluting industries is only slightly greater than for all manufacturing industries.

These results are in agreement with Leonard's (1988) detailed study of the effects of environmental regulations on industrial investment activity. Leonard suggested that a few U.S. industries had moved production and investment abroad because of environmental requirements. Among them, he cited companies who elected to produce or purchase intermediate organic chemicals abroad for further processing in the United States. Producers of certain types of pesticides and dyes were also noted for overseas investments in the 1970s to 1980s.

Evidence on changes in the patterns of trade indicate that production of chemicals has shifted away from regions with strict environmental regulations. The complex and unresolved issue that remains is to what extent environmental regulations alone have influenced the movement of production capacity and changes in trade of pollution-intensive products, compared with other factors such as demand and proximity to markets and natural resources, quality of available infrastructure and workforce, labor costs, and political and financial stability. More detailed analyses of pollution-intensive commodities that are being increasingly produced in regions with less strict environmental standards may, in the future, allow a better assessment of the relationship between regulation and the chemical industry. It is also plausible that as industrialized-country environmental restrictions become more strict and costs increase, the importance of regulatory differences will be greater.

CONCLUDING REMARKS

Environmental issues have become a global concern for citizenry, industries, and governments. The chemical industry has done much in recent decades to reduce the industry's use of energy and output of waste. Chemical production in the United States, for instance, doubled from 1970 to 1993, while both emissions of major pollutants and energy consumption diminished by one-fourth. The panels of Fig. 11-3 show the pattern for Hoechst, the world's largest chemical company: steadily higher environmental operating costs, somewhat higher environmental capital costs, but steady reductions in air emissions and total waste.

Total waste quantity (million tonnes).   *Total air emissions (thousand tonnes).*

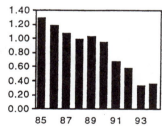
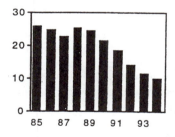

Environmental operating costs (DM millions).   *Environmental capital costs (DM millions).*

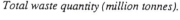
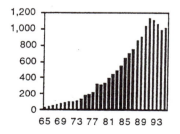
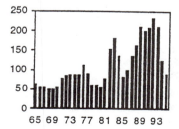

Figure 11-3. Sample Pollution Performance and Control Expenditures for Hoechst AG. *Source:* Hoechst AG.

The laws enacted since the late 1960s and early 1970s have made much progress in improving the quality of the environment and public health and safety in the United States, Germany, Japan, and the United Kingdom. However, it sometimes seems that environmental laws and progress have led to greater, rather than less, public apprehension and anxiety about the condition of the environment and human health. The progress that has occurred has often been overshadowed by incidents involving hazardous chemicals, unwieldy regulatory measures, misunderstood and exaggerated health risks, and frivolous litigation.

On the aggregate national level, environmental regulation has not been a significant factor in shaping patterns of trade. On the industry level, however, one observes differences. Chemical companies have faced far greater legal entanglement in the United States than in the United Kingdom, Japan, or Germany. German and U.S. companies have dedicated substantial resources to publicly communicate their positions and to influence policy on environmental, health, and safety issues. Their chief industry executives have had to devote a larger share of their managerial efforts to environmental-policy issues. Evidence over the last 25 years suggests that diverse regulatory measures can and do influence changes in industry structure, differences in cost of compliance; rate, and type of innovation; availability of opportunities, and evolution of the chemical industry. Differing geography, politics, histories, and levels of prosperity bring about varying tolerances for risk across countries.

As a general rule, it will often make sense to direct regulatory focus away from machinery and equipment requirements and toward environmental-quality standards that can be met in any of several flexible ways. With globalization of production, investment, and trade, it becomes increasingly more important that regulatory efforts are directed toward increasing efficiency and improving priority. After all, environmental laws are simply another way of allocating society's resources, and spending on environmental priorities should be compared with spending on other public priorities. For example, the 1994 U.S. Federal budget allocated $1.2 billion for heart and lung research, $0.9 billion for AIDS research, and $2.0 billion for cancer research. Spending for hazardous-waste cleanup in 1994, however, was $8.5 billion, although much of that money went to legal and transaction costs, and some of the remainder was spent cleaning up sites that posed no immediate danger to health. In many ways, current regulatory measures have been effective, but not efficient. There is potential for progress in the health, safety, and environment arena.

REFERENCES

Agoos, Alice (1988), "As Rayon Booms, Avtex, the Market Leader, Falls," *Chemical Week* (9 November): 6–7.

Arrow, Kenneth, Bert Bolin, Robert Costanza, Partha Dasgupta, Carl Folke, C. S. Holling, Bengt-Owe Jansson, Simon Levin, Karl-Göran Möler, Charles Perrings, and David Pimentel (1995), "Economic Growth, Carrying Capacity, and the Environment," *Science* (28 April) 268(5210): 520.

Badaracco, Joseph L., Jr. (1985), *Loading the Dice: A Five-Country Study of Vinyl Chloride Regulation*, Harvard Business School Press, Boston, MA.

Bartik, Timothy J. (1989), "Small Business Start-ups in the United States: Estimates of the Effects of Characteristics of States," *Southern Economics Journal* (April) 55(4): 1008–1014.

Barrett, Scott (1991), "Environmental Regulation for Competitive Advantage," *Business Strategy Review*, 2(1): 1–15.

Barrett, Scott (1992), "Strategy and the Environment," *Columbia Journal of World Business* (fall & winter): 202–208.

Baumol, William J. and Wallace Oates (1988), *The Theory of Environmental Policy*, Prentice-Hall, Englewood Cliffs, NJ.

Bayer Group (1993), *Annual Report*, p. 36.

Brealy, Mark, ed. (1993), *Environmental Liabilities and Regulation in Europe*, International Business Publishing Limited, The Hague, Netherlands.

Brock, William A. and David S. Evans (1985), "The Economics of Regulatory Tiering," *RAND Journal of Economics* (autumn) 16(3): 398–409.

Brock, William A. and D. S. Evans (1986), *The Economics of Small Businesses: Their Role and Regulation in the U.S. Economy*, Holmes & Meier, New York, NY.

Cairncross, Frances (1993), *Costing the Earth: The Challenge for Governments, The Opportunities for Business*, Harvard Business School Press, Boston, MA.

Carson, Rachel (1962), *Silent Spring*, Fawcet Crest, Greenwich, CT.

Chemistry and Industry (1996), "Hormone Mimics Pose Challenges," (20 May): 364–66.

Church, Thomas W. and Robert T. Nakamura (1993), *Cleaning Up the Mess: Implementation Stategies in Superfund*, The Brookings Institution, Washington, DC.

Chynoweth, Emma (1991), "BASF: Steaming Up the Green Path," *Chemical Week* (15 May): 49–50.

Clark, H. (1993), "The Chemical Industry's Image: It Doesn't Have to Be That Way, Does It?" presented in IIIe Forum Mondial de L'Industrie Chimique, Informations Chimie, April, pp. 59–62.

Colborn, Theo, Dianne Dumanoski, and John Peterson Myers (1996), *Our Stolen Future: Are We Threatening Our Fertility, Intelligence, and Survival?* Dutton, New York, NY.

D'Arge, Ralph (1974), "International Trade, Domestic Income, and Environmental Controls: Some Empirical Estimates," in *Managing the Environment: International Economic Cooperation for Pollution Control*, Allen Kneese (ed.), Praeger, New York, NY.

Dean, Thomas J. and Robert L. Brown (1995), "Pollution Regulation as a Barrier to New Firm Entry: Initial Evidence and Implications for Future Research," *Academy of Management Journal*, 38(1): 288–303.

Dominick, Raymond H. III (1992), *The Environmental Movement in Germany: Prophets and Pioneers, 1871–1971*, Indiana University Press, Bloomington and Indianapolis.

Duerksen, Christopher J. (1983), *Environmental Regulation of Industrial Plant Siting: How to Make It Work Better*, Conservation Foundation, Washington, DC.

EC (1994), *Panorama of EU Industry 94*, Office for Official Publications of the European Communities, Luxembourg, Germany.

Economist (1992), "Abolishing Litter: Environmental Protection in Europe," (22 August): 59–60.

Economist (1993), "The Money in Europe's Muck" (20 November): 81–82.

Economist (1995), "Can Republicans Fix It?" (11 March): 25–26.

Economist (1996), "Toxic Shock," (3 August): 67–69.

Ember, Lois R. (1995), "EPA Administrators Deem Agency's First 25 Years Bumpy but Successful," *Chemical & Engineering News* (30 October): 18–23.

Evans, David S. (1986), "The Differential Effect of Regulation Across Plant Size: Comment on Pashigian," *Journal of Law and Economics*, 24:187–200.

Fattah, Hassan (1995), "In with the Old: New Uses for Familiar Molecules," *Chemical Week* (17 May): 51.

Finto, Kevin J. (1990), "Regulation by Information through EPCRA," *Natural Resources and Enviornment* (winter) 4(3): 13–15, 46–48.

Freemantle, Michael (1996), "Environmental Performance Improves for Many Large Chemical Companies," *Chemical & Engineering News,* (20 May): 30–32.

French, Hilary F. (1993), *Costly Tradeoffs: Reconciling Trade and the Environment,* Worldwatch Institute, Washington, DC.

Friedman, Joseph, Daniel A. Gerlowski, and Jonathan Silberman (1992), "What Attracts Foreign Multinational Corporations? Evidence from Branch Plant Location in the United States," *Journal of Regional Science* (November) 32(4): 403–18.

Gibson, David W., and Jackie Campbell (1986), "Small Chemical Firms: A Balance Sheet," *Chemical Week* (10 December): 28–29.

Grant, Wyn, William Paterson, and Colin Whitston (1988), *Government and the Chemical Industry: A Comparative Study of Britain and West Germany,* Oxford University Press, Oxford, U.K.

Gray, Wayne B. (1997), "Manufacturing Plant Location: Does State Pollution Regulation Matter?" working paper no. 5880, National Bureau of Economic Research, January.

Gresser, Julian, Koichiro Fujikura, and Akio Morishima (1981), *Environmental Law in Japan,* MIT Press, Cambridge, MA.

Greve, Michael S. (1989), "Environmentalism and Bounty Hunting," *Public Interest* (fall) (97): 15–19.

Grossman, Gene M. and Alan B. Krueger (1993), "Environmental Impacts of a North American Free Trade Agreement," in *The U.S.-Mexico Free Trade Agreement,* P. Garber (ed.), MIT Press, Cambridge, MA.

Grossman, Gene M. and Alan B. Krueger (1995), "Economic Growth and the Environment," *Quarterly Journal of Economics* (May): 353–77.

Hazilla, Michael and Raymond J. Kopp (1990), "Social Cost of Environmental Quality Regulations: A General Equilibrium Analysis," *Journal of Political Economy* (August) 94(4): 853–73.

Hector, Gary (1992), "A New Reason You Can't Get a Loan," (21 September): 107–112.

Heerings, Hans (1993), "The Role of Environmental Policies in Influencing Patterns of Investments of Transnational Corporations: Case Study of the Phosphate Fertilizer Industry," in *Environmental Policies and Industrial Competitiveness,* OECD Publications Service, Paris, France.

Heller, Karen (1990), "Environment Tops the List of Merger Woes," *Chemical Week* (7 February).

Hileman, Bette (1996), "Environmental Hormone Disrupters Focus of Major Research Initiatives," *Chemical & Engineering News* (13 May): 28–30.

Hoechst AG (1996), *Annual Report.*

International Environment Reporter (1992), "Japan, Malaysia Embroiled in Dispute Over Alleged Pollution From Chemical Firm" (29 July).

Jaffe, Adam B. and Karen L. Palmer (1994), "Environmental Regulation and Innovation: A Panel Data Study," paper prepared for the Western Economic Association Meetings, June.

Jaffe, Adam B., Steven R. Peterson, Paul R. Portney, and Robert N. Stavins (1995), "Environmental Regulation and the Competitiveness of U.S. Manufacturing: What Does the Evidence Tell Us?" *Journal of Economic Literature* (March) 33: 132–63.

Journal of Commerce (1992), "Nafta May Bring 'Import' of Pollution to Mexico" (20 August).

Jorgensen, Dale W. and Peter J. Wilcoxen (1992), "Impact of Environmental Legislation on U.S. Economic Growth, Investment, and Capital Costs," in *U.S. Environmental Policy and Economic Growth: How Do We Fare?* Donna L. Brodsky (ed.), American Council for Capital Formation, Washington, DC.

Kalt, Joseph P. (1988), "The Impact of Domestic Regulatory Policies on International Competitiveness," in *International Competitiveness,* Michael Spence and Heather Hazard (eds.), Harper and Row, Ballinger, Cambridge, MA.

Kemp, Rene and L. Soete (1992), "The Greening of Technological Progress," *Future* (June) 24(5): 437–57.

Kirk, Don L. (1944), "German Law May Boost D&O Cover Demand," *Business Insurance* (24 January): 31–32.

Kopp, Raymond J., Paul R. Portney, and Diane E. Dewitt (1990), "International Comparisons of Environmental Regulation" in *Environmental Policy & the Cost of Capital*, American Council for Capital Formation, Washington, DC.

Kriz, Margaret (1995), "The Superfund Saga," *National Journal* (21 October): 2592–96.

Landau, Ralph (1996), "Strategy for Economic Growth: Lessons from the Chemical Industry," in *The Mosaic of Economic Growth*, Ralph Landau, Timothy Taylor, and Gavin Wright (eds.), Stanford University Press, Stanford, CA.

Lenz, Allen J., Kevin Swift, Keith A. Christman, and Kris Greenwald (1994), *The CMA Economic Survey: Outlook for 1994 and Beyond*, Chemical Manufacturers Association, Washington, DC.

Leonard, Jeffrey H. (1988), *Pollution and the Struggle for the World Product: Multinational Corporations, Environment, and International Comparative Advantage*, Cambridge University Press, Cambridge, U.K.

Levinson, Arik (1992), *Environmental Regulations and Manufacturers' Location Choices: Evidence from the Census of Manufacturers,* Columbia University Press, New York, NY.

Lewis, Sanford J., Marco Kaltofen, and Gregory Ormsby (1991), *Border Trouble: Rivers in Peril*, National Toxics Campaign Fund, Boston, MA.

Lieberman, Marvin B. (1990), "Exit from Declining Industries: 'Shakeout' or 'Stakeout'?" *RAND Journal of Economics*, 19: 538–54.

Low, Patrick and Alexander Yeats (1992), "Do 'Dirty' Industries Migrate?" in *International Trade and the Environment*, Patrick Low (ed.), The World Bank, Washington, DC.

Lucas, Robert E. B., David Wheeler, and Hemamala Hettige (1992), "Economic Development, Environmental Regulation and the International Migration of Toxic Industrial Pollution, 1960–1988," in *International Trade and the Environment*, Patrick Low (ed.), The World Bank, Washington, DC, pp. 67–86.

Marketing (1991) (28 March).

McGuire, Martin C. (1982), "Regulation, Factor Rewards, and International Trade," *Journal of Public Economics*, 17:335–54.

McKinsey & Company (1991), "The Corporate Response to the Environmental Challenge."

Meadows, Donella H., Dennis L. Meadows, Jorgen Randers, and William W. Behrens III (1972), *The Limits to Growth: A Report for the Club of Rome's Project on the Predicament of Mankind*, Universe Books, New York, NY.

Miller, Karen Lowry and Joan O'C. Hamilton (1995), "Biotech Blooms in Germany–Again," *Business Week* (13 January), pp. 70–71.

Motala, Mohammed (1992), "Bayer Poisons South Africa" and "Thor Chemicals Reopens Mercury 'Recycling' Facility in South Africa," in *Toxic Trade Update*, Greenpeace USA, Second Quarter.

Mullin, Rick, Michael Roberts, Allison Lucas, John Hallenborg, Harold Hyman, and Ray Pospisil (1994), "Manufacturers Head for Technology's High Ground," *Chemical Week* (2 February): 20–26.

Nakano, Takamasa (1986), "Environmental Policies in Japan," in Chris C. Park (ed.), *Environmental Policies: An International Review*, Croom Helm, London, U.K.

OECD (1985), *The Macro-Economic Impacts of Environmental Expenditures*, Organization for Economic Cooperation and Development, Paris, France.

Ondrey, Gerald (1992), "Germany Prepares for a Rebound," *Chemical Engineering* (December): 30–35.

Pashigian, B. Peter (1984), "The Effect of Environmental Regulation on Optimal Plant Size and Factor Shares," *Journal of Law and Economics* (April) 27: 1–28.

Pearson, Charles S. (1985), *Down to Business: Multinational Corporations, the Environment, and Development*, World Resources Institute, Washington, DC.

Pearson, Charles S., ed. (1987), *Multinational Corporations, Environment, and the Third World*, Duke University Press and World Resources Institute, Durham, NC.

Pethig, Rudiger (1976), "Pollution, Welfare and Environmental Policy in the Theory of Comparative Advantage," *Journal of Environmental Economics and Management*, 2: 160–69.

Porter, Michael E. (1980), *Competitive Strategy*, and *America's Green Strategy*, Free Press, New York, NY.

Porter, Michael E. 1990), *The Competitive Advantage of Nations*, Free Press, New York, NY.

Porter, Michael E. (1991), "America's Green Strategy," *Scientific American* (April): 168.

Rabe, G. Barry (1994), *Beyond NIMBY: Hazardous Waste Siting in Canada and the United States*, The Brookings Institution, Washington, DC.

Rich, Bradford W. (1985), "Environmental Litigation and the Insurance Dilemma," *Risk Management* (December) 32(12): 34–41.

Roberts, Michael (1995), "Bayer Pushes for Regulatory Reform," *Chemical Week* (2 August): 34.

Robison, H. David (1988), "Industrial Pollution Abatement: The Impact on Balance of Trade," *Canadian Journal of Economics* (February) 21(1).

Rubin, Debra K., Don Shapiro, Peter Reina, and Armin Schmid (1994), "Firms Gear Up to Think Globally, Link Locally," *ENR* (21 February): 42–44.

Schneider, Manfred (1993), "Challenges and Opportunities," *Chemistry & Industry* (6 December): 946–51.

Scism, Leslie (1995), *The Wall Street Journal* (13 July).

Symes, Tom and Victoria Phillips (1993), "England and Wales," in Mark Brealey (ed.), *Environmental Liabilities and Regulation in Europe*, International Business Publishing Limited, The Hague, Netherlands.

Thayer, Ann M. (1994), "Catalyst Industry Stresses Need for Partners as Key to Future Success," *Chemical & Engineering News* (11 July).

Thomas, Lacy Glenn (1990), "Regulation and Firm Size: FDA Impacts On Innovation," *RAND Journal of Economics* (winter) 21(4).

Teltschik, Walter (1992), *Geschichte der deutschen Grosschemie: Entwicklung und Einfluss in Staat und Gesellschaft*, VCH, Weinheim, New York, NY.

Tobey, James A. (1990), "The Effects of Domestic Environmental Policies on Patterns of World Trade: An Empirical Test," *Kyklos* 43(2): 191–209.

U.S. Chamber of Commerce, "Readers' Views on Uncle Sam," *Nation's Business* (January) 84(1): 49.

U.S. Congress, Office of Technology Assessment, Industry, Technology, and the Environment (1994), *Competitive Challenges and Business Opportunities*, OTA-ITE-856, U.S. Government Printing Office, Washington, DC, January.

U.S. Congress (1990), Senate, Environmental issues and Avtex Fibers, Inc., hearing before the Subcommittee on Toxic Substances, Environmental Oversight, Research and Development of the Committee on Environment and Public Works, S. Hrg. 101-837, U.S. Government Printing Office, Washington, DC, June.

Vogel, David (1986), *National Styles of Regulation: Environmental Policy in Great Britain and the United States*, Cornell University Press, Ithaca, NY.

Walley, Noah and Bradley Whitehead (1994), "It's Not Easy Being Green," *Harvard Business Review* (May-June): 46–52.

Walter, Ingo (1982), "Environmentally Induced Industrial Relocation to Developing Countries," in *Environment and Trade: The Relation of International Trade and Environmental Policy,* Seymour J. Rubin and Thomas R. Graham (eds.), Allanheld, Osmun, Totowa, NJ.

Westervelt, Robert (1994), "Biocide Market Pared Down by Environmental Regulations," *Chemical Week* (27 July): 30.

Wissel, Holger, Horst Schlemminger, Joachim Schktze, and Claus-Peter Martens (1993), "Germany," in Brealy, Mark (ed.), *Environmental Liabilities and Regulation in Europe*, International Business Publishing Limited, The Hague, Netherlands.

Xing, Yuqing and Charles D. Kolstad (1995), "Do Lax Environmental Regulations Attract Foreign Investment?" working paper in economics #6-95, University of California, Santa Barbara, May.

12 Evolution of Industry Structure in the Chemical Industry

ASHISH ARORA and ALFONSO GAMBARDELLA

INTRODUCTION

There is always a temptation to identify the success of individual firms with the success of their industry. An analysis of industry structure provides a healthy corrective to this temptation. To paraphrase Karl Marx, while firms may well shape their own destiny, they do not shape it in circumstances of their own choosing.

Industry structure has several dimensions: the number of firms in the market, the patterns of entry and exit, the breadth and diversity of product portfolios of firms, and the degree of competition. Our approach to understanding the evolution of industry structure combines economic and historical factors. At the firm level, the chemical industry offers a canonical example of firm-level economies of scale and scope (Chandler 1990), which offer advantages to size. At the industry level, chemicals provide an additional example of the advantages of bigness through industry-level economies of specialization (Young 1928; Stigler 1951). Typically, these industry-level benefits arise from an increase in specialization and division of labor.

A historical look at industry structure reveals reinforcing mechanisms that tend to preserve and accentuate initial differences across countries. But there are forces that make for convergence, as well. Technological innovations, the growth of international trade, and the growth of new markets are powerful forces that encourage greater convergence across regions.

We begin by describing the structure of the U.S., European, and Japanese chemical industries around the end of the 1980s and offer some explanations for the differences among countries. We then take up some dynamic factors of particular importance: antitrust law and related institutions, the effects of the two world wars, and the effects of certain technological advances on industry structure. We discuss the relationship between technological changes and corporate governance and their impact on the structure of the firm and industry. Finally, we offer some conclusions.

At different moments in time, Tsutomu Harada, Carolyn Chan, Namhoon Kwon, Rezal Reman, and Zhang Jianyu provided valuable assistance in the data collection and analysis. We thank Ralph Landau, Nathan Rosenberg, and Albert Richards for useful comments and suggestions.

379

INDUSTRY STRUCTURE: A SNAPSHOT

All-Around versus Focused Chemical Companies

The chemical industry is typically composed of two main types of firms. There are "all-around" companies, which are involved in different value chains, operate in several of those with large scales, and often occupy a number of stages of these value chains. There are also many "focused" companies, in areas such as electrochemistry, catalysts, surfactants, fertilizers, paints and surface coatings, engineered plastics, food, fragrances and other fine chemicals, herbicides and pesticides, and pharmaceuticals. The focused companies are typically found downstream, where product differentiation, brand name, and other product specific features are more important. The focused companies also tend to be smaller, although many of them are rather large in absolute terms.

To investigate the distribution of these different types of firms, we constructed a sample of leading U.S., West European, and Japanese chemical companies. These are the top five producers, ranked by production capacity, in each of the three regions in a sample of 123 chemical products (not including pharmaceuticals).[1] We used information from *Predicasts* (1991) and other company thesauruses to group the firms that were subsidiaries of other companies under the name of their parents and to obtain their nationalities. This produced a sample of 473 companies — 175 from the United States, 132 from Western Europe, and 166 from Japan.[2]

We divided these companies into three categories, which we labeled L2, L35, and L6. The L2 category includes companies that are leaders in no more than two products in their home region. The L35 companies are all the firms that are leaders in three to five products in their region. The L6 companies are leaders in six or more products in their home area. Thus, while the L2 firms are narrowly specialized and the L35 are typically focused companies, in the sense we defined earlier, the L6 category includes the largest chemical corporations such as Dow, DuPont, ICI, BASF, Bayer, and Hoechst.

Table 12-1, column 4 shows that the distribution of firm types is similar across the three regions. Europe and Japan, however, do have a relatively higher number of large firms than does the United States. Column 1 of Table 12-1 shows that the vast majority of the L2 companies are leaders only in their home region. However, while the all-around companies of the United States and Europe tend to be leaders in regions outside their own, in Japan 23 of the 25 L6 companies are leaders only in Japan.

Table 12-2 lists the major all-around European, U.S., and Japanese firms among the top 25 chemical companies worldwide, ranked by chemical sales in 1988 and the number of product markets (among the 123 in our sample) in which they are among the five leaders in their region. The large number of European companies in this group reflects the fragmentation of the European market, with each country featuring one or more national chemical producers. Moreover, although the three leading all-around

---

[1]The data are from *Chemical Economics Handbook* (*CEH*), SRI International, June 1991, and relate to the end of the 1980s. Our procedure can be best described as an informal, stratified random sample of all the products covered in *CEH*, subject to the constraint that complete and comparable firm-level market data be available for all three regions of interest. We selected our products to be as representative as possible of the chemical industry (excluding pharmaceuticals), and they cover all the major categories — basic inorganic, basic organic, basic petrochemical, thermoset plastics, synthetic fibers, surfactants, resins, surface coatings, paints and pigments, elastomers, and fertilizers.

[2]When we could not find information on some companies in the *CEH* list of leading producers in various commercial databases, we discarded these companies from our sample. This process is likely to mean that small firms (less likely to be covered by commercial databases) are underrepresented in our sample. In addition, we also excluded from our sample the handful of firms not from our three regions.

TABLE 12-1. Internationalization of Chemical Companies By Type*

Firm Types and Regions	Domestic Firms (1)	International Firms (2)	Global Firms (3)	Total (4)
L2 (Specialized				
U.S.	102	17	0	129
W. Europe	95	5	2	102
Japan	110	3	1	114
L35 (Focused)				
U.S.	19	11	1	31
W. Europe	7	4	1	12
Japan	26	1	0	27
L6 (All-around)				
U.S.	2	7	6	15
W. Europe	3	9	6	18
Japan	23	1	1	25

Domestic = Leader in no products outside home region.
International = Leader in at least one product in at least one other region.
Global = Leader in at least one product in all three regions.
L2 = Leading producers in two products at most in their home region.
L35 = Leading producers in three to five products in their home region.
L6 = Leading producers in more than 6 products in their home region.
*Based on our sample of 123 products from CEH (1991).
Source: Our calculations from CEH (1991.

Japanese producers are smaller than the European or U.S. firms, they operate in roughly the same number of products *in the chemical industry*. Put differently, Japanese chemical producers may be excessively diversified, when we restrict attention to the chemical industry.

While Table 12-3 looks at focused companies, Table 12-4 lists all the chemical companies among the top 50 in the world (ranked by chemical sales in 1989) that are typically associated with a major core business. These can be thought of as the largest focused chemical suppliers. Several themes emerge from these lists. Pharmaceuticals is the dominant industry among the largest specialized U.S. and European firms, while the large specialized Japanese manufacturers have more varied backgrounds, such as textiles and fertilizers. However, while many all-around European chemical companies also operate in pharmaceuticals — Bayer, Hoechst, and until recently, ICI — in the United States the chemical and pharmaceutical businesses tend to be the province of distinct firms. The "focused" Japanese companies are not much smaller than their European or U.S. counterparts. The number (and size) of the large specialized U.S. producers is similar to that of the European firms. Moreover, all of the large, focused European companies come from Germany, Switzerland, and the United Kingdom (Aftalion 1989, appendix).

Comparative Trends in Vertical Integration and Scale

To compare the extent of vertical integration of the major U.S., Japanese, and European chemical companies we focused on a sample of 13 chemical products in five main classes. These are: (1) polyethylene, polypropylene, and PVC (plastics); (2) acrylic fibers,

TABLE 12-2. Major European, U.S., and Japanese All-Around Chemical Companies

Companies	Rank by World Chem. Sales	World Chemical Sales*	Total World Sales*	No. of Products[†]
EUROPEAN COMPANIES				
Bayer (Ger.)	1	22,694	22,824	23
Hoechst (Ger.)	2	21,948	23,105	30
BASF (Ger.)	3	21,543	24,743	43
ICI (U.K.)	4	21,125	21,125	26
Royal Dutch-Shell (U.K.-N.L.)	8	11,848	79,643	26
Rhone-Poulenc (F.)	11	10,802	10,802	22
Elf Aquitaine (F.)	14	8,216	20,848	13
Akzo (N.L.)	16	7,846	8,283	12
Montedison (I.)	17	7,725	10,763	14
Solvay (B.)	19	6,836	6,836	3
U.S. COMPANIES				
Du Pont	5	19,608	32,917	20
Dow	6	16,659	16,682	28
Exxon	12	9,892	87,542	15
Union Carbide	13	8,324	8,324	20
Monsanto	18	7,453	8,293	12
Eastman Kodak	20	6,724	17,034	8
JAPANESE COMPANIES				
Mitsubishi Chemical	15	8,095	n/a	29
Sumitomo Chemical	21	6,532	7,178	24
Asahi Chemical	25	5,799	7,696	33
Showa Denko	33	5,076	n/a	n/a

*Millions of 1988 U.S. dollars.
[†]Number of products in which the company is among the top five leaders in its own region. Based on our sample of 123 products from *CEH* (1991).
Source: Aftalion (1989).

nylon 6, nylon 66, and polyester fibers (fibers); (3) styrene, polystyrene, and SBR elastomers (rubber products); (4) urea (fertilizers); and (5) epoxy resins and poly-urethanes (which represent more downstream, specialty categories). For each product we identified the vertical chain of the main chemical inputs that are used to manufacture them, and we looked at whether the manufacturers of the final good were also involved in these earlier steps.[3] We constructed an index of vertical integration by taking the ratio between the number of steps in which the producer was involved and the total number of steps,[4] looking at both the top five and the top three producers from a given

---

[3]For instance, for polyethylene, we looked at whether the producers were also manufacturing ethylene (step 1). For acrylic fibers we looked at whether they produced propylene (step 1), ammonia (step 2), or acrylonitrile (step 3).

[4]The index takes the value 0 if the producer is not involved in any of the previous steps ('buys" all the inputs), and it is equal to 1 if the producer is involved in all of the previous steps. Values in between denote intermediate levels of integration.

TABLE 12-3. Focused World Chemical Companies

Company	Rank by World Chemical Sales	World Chemical Sales*	Core Business
EUROPEAN COMPANIES			
Unilever (N.L.-U.K.)	7	12,338 (30,950)	Consumer products
Ciba-Geigy (Swi.)	9	11,018	Pharma
Sandox (Swi.)	22	5,960	Pharma
Henkel (Ger.)	26	5,782	Soap & detergents
Hoffman-La Roche (Swi.)	29	5,416	Pharma
Huls (Ger.)	37	4,645	Petrochem
Beecham (U.K.*)	41	4,532	Pharma
U.S. COMPANIES			
Proctor & Gamble	10	11,000 (21,400)	Consumer products
Merck & Co.	23	5,939	Pharma
Bristol-Myers ([‡])	28	5,500	Pharma
Abbott Labs	42	4,500	Pharma
American Home Prod.	43	4,500 (5,500)	Pharma & consumer products
Colgate-Palmolive	44	4,500	Soap & detergents
Johnson & Johnson	45	4,500 (9,000)	Pharma & consumer Products
Pfizer	46	4,485	Pharma
Eli Lilly	50	4,070	Pharma
JAPANESE COMPANIES			
Dainippon Ink & Chemical	30	5,412	Paints
Takeda	31	5,219	Pharma
Toray	32	5,208	Textile
Kao	47	4,335	Soap & detergents
Teijin	49	4,210	Textile

*Millions of 1988 U.S. dollars. In parenthesis are total sales, when different from chemical sales.
[†]Merged with Smithkline in 1989.
[‡]Merged with Squibb in 1989.
Source: Aftalion (1989).

region and across the products in the corresponding class. The leading Japanese firms are less vertically integrated than the U.S. and European firms in four out of five of our classes but more vertically integrated in fertilizers (urea). The U.S. and European producers have very similar degrees of vertical integration in three areas, but the U.S. producers show somewhat more vertical integration in plastics (area 1), and European producers show considerably more vertical integration in rubber products.

We also compared the scale of firms and plants in the main commodity resins (LDPE, HDPE, LLDPE, PP, PVC, and polystyrene) and in the main fibers (acrylic, nylon, polyester, polyolefin fibers, rayon, acetate fibers). As Table 12-4 shows, in practically all these products, the United States shows higher average capacities per

TABLE 12-4. Number of Producers, Number of Plants, and Capacities by Region, for LDPE, HDPE, LLDPE, Polypropylene (PP) PVC, Polystyrene (PS): Late 1980s

Product & Regions	Number Of Firms*	Total Capacity[†]	Avg. Cap. Per Firm	Avg. Cap. Per Plant
LDPE				
U.S.	10	3478	347.8	124.2
W.E.	19	5508	289.9	131.1
Jp.	10	1167	116.7	72.9
HDPE				
U.S.	11	3664	333.1	203.5
W.E.	18	2253	125.2	93.9
Jp.	10	703	70.3	70.3
POLYPROP				
U.S.	12	2949	245.7	173.5
W.E.	19	2855	150.3	109.8
Jp.	12	1315	109.6	101.2
PVC				
U.S.	13	4531	348.5	174.3
W.E.	22	5320	241.8	126.7
Jp.	17	2077	122.2	79.9
POLYSTYRENE				
U.S.	16	2630	164.4	79.7
W.E.	20	2670	133.5	80.9
Jp.	14	1229	87.8	64.7

*Data refer to the end of the 1980s.
[†]Thousands of metric tons.
Source: Our calculations from *CEH* (1991).

plant and per producer, and Europe tends to show higher average capacities than Japan.

Industry Structure in the Three Regions: A Summary

In each region there are a few large, diversified firms. These firms are market leaders in several product markets, and often they are market leaders in other regions as well. Many more firms are smaller, more focused, and market leaders in only a few product markets, typically in their domestic markets.

Within these similar structures, there are some differences. The large, all-around Japanese firms are smaller than their counterparts in the United States and Western Europe and less international in orientation, but they display the same degree of diversification. Moreover, the leading Japanese firms appear to be less vertically integrated. The picture is different when we look at the more specialized firms. Japan and the United States have a significant number of specialized producers that are of relatively large size. While in the United States these large, specialized firms are predominantly in pharmaceuticals (along with a few other specialized sectors such as rubber or consumer products), in Japan they are of more varied background. Europe tends to have fewer large specialized firms, and most of them come from Germany and the United Kingdom. Western Europe and Japan have a greater number of producers and a larger number of plants with smaller average production capacity. This appears especially true in Japan, where the manufacture of the sample of products that we examined is fragmented across several companies, each of them producing relatively small volumes.

In the last decade or so, developments that we discuss in this chapter have changed many of the firm boundaries, so that lists analogous to those in Tables 12-2 and 12-3 constructed for 1997 would look somewhat different, especially for Western Europe. They would include some new names, reflecting some large mergers and de-mergers, such as Montell (a Montedison-Shell joint venture in polyolefins), Novartis (the merged life-science businesses of Sandoz and Ciba Geigy), Zeneca (the life-sciences businesses of ICI), Borealis (the polyolefins business of Statoil and Neste Oy), and Marlene (the polyolefins businesses of BASF and Shell). Virtually all of these large companies were fashioned from existing businesses in the last couple of years; and if one were to construct such lists, the net result will be to make the European landscape appear even more similar to that of the United States.

INTER-COUNTRY DIFFERENCES: CAUSES AND CONSEQUENCES

Differences in Historical Origins

Firm strategies and the evolution of the chemical industry are conditioned by the historical origins of the industry and firms in the various countries and by the resulting patterns of industry structure. Although Europe and the United States display some interesting and subtle historical differences, the most prominent differences are with regard to Japan.

Many of the special features of today's Japanese chemical industry, which were documented in the previous section — lower economies of scale at the product level, excess diversification, excessive number of producers — appear to be a legacy of the conditions that characterized the industry's early development. The Japanese chemical industry developed later, when the world chemical industry was already dominated by Europe and the United States. Around the turn of the century, the Japanese chemical industry was fragmented into very specialized manufacturers in relatively low-technology downstream sectors: milling, liquor, glass, paper, pottery, matches. There were some concerted attempts by firms to develop domestic production capabilities, especially after the interruption of supplies of dyestuffs and soda by the First World War (as discussed in the chapter by Hikino et al.). The late industrialization had three important aspects: (1) Japanese firms developed a pattern of import and use technology developed elsewhere; (2) Japanese chemical firms had to face competition from European and U.S. firms, which had far greater technological capabilities; and (3) the Japanese economy as a whole developed behind trade barriers. These features left their mark on the structure of the Japanese industry.

For example, the combination of the generally backward state of the early Japanese chemical industry, the availability of imported technology, and the importance in Japan of groups of companies, commonly called *keiretsu*, acting together enabled firms in downstream sectors to play a more prominent role in the chemical industry than was common elsewhere.[5] To be sure, prominent firms from downstream sectors are not unknown in the chemical industries in other regions; for example, General Electric in plastics and Eastman Kodak in photography. But it is also common in the United States and Europe for general chemical firms to dominate particular sectors. For example, the major rayon producers in the United States were chemical firms such as DuPont. But in Japan, synthetic-fiber production, particularly that of rayon, was dominated by textile producers such as Toray, Teijin, and Kanebo. Other Japanese companies, for instance Kao (detergents), Lion (detergents) or Takeda (pharmaceuticals), integrated backward to acquire the in-house chemical capabilities that were increasingly necessary to run their businesses. Similarly, nitrogenous fertilizer companies in Japan integrated backward into chemicals (Molony 1990, pp. 35–36).

After World War II, many of Japan's downstream user firms further integrated into petrochemicals, largely for internal use. Thus, Showa Denko diversified from fertilizers into various polyolefins, and Daicel produced acetic acid and other derivatives to be used in its cellulose business; textile companies such as Toray, Teijin, and Kanebo made similar moves (Spitz 1988, pp. 375–84; Aftalion 1989, p. 205). Many petrochemical producers were themselves part of *keiretsu* that included firms operating in downstream sectors such as electronics and automobiles. Chemical companies belonging to a *keiretsu* sold a substantial fraction of their output to downstream companies in the same group, rather than in the open market. The influence exercised by the users is likely to have directed the chemical industry to the somewhat idiosyncratic requirements of special applications, and it also contributed to the high degree of product customization that exists in Japan. For instance, the number of grades of plastic resins is said to be an order of magnitude larger than in Europe or the United States, and Japanese automobile companies are credited (or blamed) for inducing such a high degree of product differentiation.

Japanese import protection also provided incentives for domestic firms to diversify into the production of a wide range of products. During the 1950s and the 1960s, Japan made an astonishingly rapid entry into petrochemicals, led by the *keiretsu* groupings of Mitsui, Mitsubishi, and Sumitomo, along with several other companies such as Asahi Chemical, Maruzen Oil, and Idemitsu. However, since each *keiretsu* attempted some measure of self-sufficiency in the basic chemical inputs, the *keiretsu* structure further exacerbated the tendency, already encouraged by import protection, toward production on scales that were too small to be internationally competitive.

Like Japan, the United States was also a "follower" in the chemical industry, in the sense that until well into the first part of the 20th century it relied on technology imports. The U.S. chemical industry also enjoyed some tariff protection during its formative years. The enormous size of the U.S. market, however, implied that chemical firms, like firms in other industries, could specialize without sacrificing size. In fact, a general characteristic of the U.S. industry was the application of technology to exploit the natural-resource base and produce on a large scale (as explored further in the chapter by Arora and Rosenberg). Thus, the U.S. chemical industry developed with large firms that were willing to invest in the commercial application of technology, that paid great attention to manufacturing costs (Wright 1990; Nelson and Wright 1992), and that took advantage of other suppliers in a way that created an extensive division of labor in the chemical industry.

[5]We are using the term *keiretsu* to signify both an industrial group and a business group. This is not quite accurate but is consistent with the usage of the term in the West. (See the chapter by Hikino et al. for more details.)

The structure of the U.S. chemical industry also reflects other aspects of the American growth experience. Unlike in Japan, chemical technology in the United States often arrived through subsidiaries of foreign firms, as well as through scientists and engineers who immigrated. Sectors such as fine chemicals, pharmaceuticals, and dyestuffs relied heavily on European technology, and the subsidiaries of German and Swiss dyestuffs and pharmaceutical producers played an important role. After World War I, many of these subsidiaries disassociated themselves from their parent companies and grew into full-fledged chemical companies such as Schering Plough, Merck, and Rohm and Haas. One result of this pattern, evident even today, is the fairly clear separation between chemical and pharmaceutical firms in the United States.

The origins of the first chemical industry in the world, in Britain, were largely in the inorganic sector. The production of inorganic chemicals such as soda ash, soda, and bleach grew because of the opportunities created by the growing demand for such products by the textile industry (as discussed in the chapter by Murmann and Landau). The production processes for inorganic materials are far less complicated than those for producing organic chemicals. With limited opportunities for the application of science and innovation to chemical production, British chemical firms (with some exceptions) de-emphasized investments in science and technology, as well as in complementary activities such as commercialization. Instead, British firms focused on the volume of output, sales, and distribution because profits and growth came from new markets (including overseas markets), cheaper sources for inputs, and control over pricing. Many of these strategies proved to be poorly suited to the rapid technological advances of the organic chemicals industry.

In Germany, dyestuffs represented the engine of the chemical industry until World War I. The German chemical industry grew through the systematic application of scientific discoveries to chemical manufacture. Advances in organic chemistry clarified how carbon atoms are linked to hydrogen and other atoms to form more complex molecules. These discoveries proved to be a general-purpose "method" for developing new products such as alizarin, azo dyes, and indigo. In addition, it was soon realized that the method provided by organic chemistry could be extended to other products such as pharmaceuticals, explosives, and photographic materials. Moreover, the dyestuffs companies undertook large investments in marketing to find the special and distinct grades of products that were needed by the users. Thus, the origin of the German industry in the dyestuffs "model" laid the basis for the creation of a sector that had considerable skills in applied chemistry and was skilled in applying chemistry to commerce. These competencies represented an important source of competitive advantage to German companies.

Since the new method of research and development had large fixed costs, German chemical companies sought new markets by expanding abroad soon after being established. This early globalization started with the end of the 19th century. Horizontal mergers also took place early in the history of the German industry. Extensive vertical integration up to the most basic materials for chemical production was deemed necessary to maintain proprietary control of the know-how about the production process that each dyestuffs manufacturer used. As the early producers started by integrating backward, there was little room left for independent suppliers. In turn, this meant that when pressures arose to achieve greater economies of production and to moderate competition, the solution took the form of horizontal integration, of which the creation of IG Farben in the 1920s is the best-known example.

Differences in the Size and Nature of Demand

In the 20th century, European chemical companies developed with a pattern of a few dominant firms, with other firms focusing on niche markets. In Germany before World War I, for example, the big three — Bayer, BASF, and Hoechst — were the large diversified firms in the industry. Although many other German companies were pulled along by the general developments of the industry, they remained as smaller niche players when compared with the big three (Chandler 1990, p. 475). This pattern was only reinforced by the formation of IG Farben as the dominant German chemical firm in 1926. The second-largest German chemical firm in that period, Deutsche Solvay Werke, was only one-tenth as large. IG Farben dominated a variety of organic chemicals and metals; explosives were produced by Dyanamit AG and soda and other inorganics by Deutsche Solvay. Within IG Farben itself, the lines of business of the companies that had formed the conglomerate remained fairly distinct. Indeed, when IG Farben was broken up after World War II into Bayer, BASF, and Hoechst, the overlap in the product portfolios of the new firms was minimal.

Similarly, in Britain, ICI acquired a dominant position as the domestic monopolist in a large number of products. Other companies such as Albright & Wilson, Laporte, and British Oxygen were forced into such narrower businesses, as phosphorus and metals, hydrogen peroxide, and industrial gases. ICI, however, was careful not to enter into more downstream markets, at least until the end of World War II, out of a reluctance to compete with its customers. This enabled companies in those areas, most notably Courtaulds in fibers and the Anglo-Dutch company Unilever in consumer products, to develop their own chemical operations. As in the United States, the British chemical industry also gave rise to a specialized pharmaceutical sector, which was dominated by such companies as Boots, Beecham, and Glaxo.

The size of the U.S. market allowed the formation and growth of many large chemical firms. However, an integrated domestic market implied that firms could not shelter themselves from competition in geographical niches, and this discouraged excessive diversification. At the same time, the large size of the U.S. market enabled firms to spread the cost of assets that were specific to some core activity over a large volume of output, which in turn encouraged specialization. When U.S. chemical firms diversified, they typically went into related products and businesses, so as to exploit the fixed assets of their core business. Thus, the U.S. chemical industry witnessed the growth of more specialized concerns, such as U.S. Rubber, B.F. Goodrich, Firestone, and Goodyear in rubber and Merck, Pfizer, and Eli Lilly in pharmaceuticals.

So far, the discussion of advantages of a larger domestic market have been associated with economies at the level of the firm; but important benefits to the U.S. chemical sector also arose from economies at the level of the industry. Especially with the rise of petrochemicals, the U.S. market provided opportunities for producing the same basic inputs for many downstream products. The resulting vertical specialization in production led to the growth of an upstream sector composed of a class of firms that were specialized in the design and engineering of large-scale plants. Eventually, this led to the development of a global market in process technologies, as will be discussed later in this chapter.

CARTELS, ANTITRUST, AND MARKET SIZE

Industry structure in any country reflects the antitrust regime. The chemical industry in Europe before World War II has been described as a "gentlemen's club." Most markets were dominated by a few large firms or cartels who would often negotiate agreements.

Until World War II, cartels remained an important feature of the chemical industry in areas such as nitrogen fertilizers, dyestuffs, and dynamite. These cartels were able to control the diffusion of technology and entry into the industry in Europe. The leading U.S. firms such as DuPont and Standard Oil could not, because of U.S. antitrust law, join these international agreements directly. However, they entered into technology-sharing relationships with firms such as IG Farben and ICI that also facilitated market-sharing.

Domestic mergers and cartels usually preceded the formation of international cartels. In Britain, consolidation in the LeBlanc soda industry gained impetus with the formation of the United Alkali Company in the 1880s. The process of mergers and consolidation continued in Britain until the 1920s, when it culminated with the formation of ICI. In Germany, mergers and horizontal arrangements started early as well. In addition to IG Farben, there were state-sponsored cartels in nitrogen fertilizers and potash. The U.S. industry, too, underwent consolidation during this time; for example, a number of firms were taken over by others in the recession immediately following World War I. The large size and geographical extent of the U.S. market, however, along with active U.S. antitrust enforcement, prevented anything approaching the European situation.

Vigorous antitrust enforcement after World War II helped induce U.S. chemical companies to license their technology more readily, thereby lowering entry barriers. In the aftermath of the war, the entry of U.S. chemical firms and technology helped reduce the concentration in European markets. While Europe never became completely open to full-fledged competition in chemicals and national producers were still sheltered in their domestic markets, World War II did mean the end of fully effective cartels and market sharing. However, Japan was able to maintain a greater degree of protection for domestic companies, so that while industrial structures in Europe and the United States converged, the Japanese industrial structure remained distinct.

DYNAMICS OF INDUSTRY STRUCTURE: WORLD WARS AND OTHER HISTORICAL EVENTS

World War I spurred nationalistic concerns, which led many countries to promote the development of domestic capabilities in areas previously served by international trade. The war also meant massive demand for chemical products and technologies. Futhermore, the war effort encouraged the dissemination of technological knowledge, as governments encouraged domestic chemical companies to pool their efforts for military purposes. At the international level, the release of German patents and technologies into the public domain during wartime created an additional intercountry diffusion of technology.

Because of this technological diffusion, World War I reduced the disparities among Germany, Britain, and the United States in dyestuffs and, more generally, in organic chemicals. In 1914, Germany had a virtual international monopoly in synthetic dyes, controlling 90% of world production. But during the war, the United States and Britain expropriated German industrial property, including extremely valuable patents on ammonia and dyestuffs. One important consequence of World War I was, therefore, to create greater *international* competition in sectors such as inorganics, explosives, dyestuffs, and even ammonia and nitrogenous fertilizers.

The Second World War had effects similar to those of World War I. In Germany, the Nazi government pushed IG Farben into intensive autarkic programs such as the production of synthetic rubber and synthetic gasoline (from coal). The U.S. government also launched a program for the research and production of synthetic rubber and created a massive demand for oil needed as aviation fuel.[6] These and other wartime programs (such as the Manhattan Project) explicitly promoted cooperation among domestic companies. For example, the rubber program involved coordination in information, research efforts, and downstream applications. In addition, at the end of the war, plants and other facilities managed under the Rubber Program were sold out to different companies, whether oil (for example, Exxon) or chemical producers (Dow, Monsanto, Union Carbide).

Other than the two world wars, the historical event that most shook the global chemical industry in the 20th century was the discovery of oil reserves in the Middle East, and the corresponding development of an international oil market in the 1950s. This market was important in a global switch from coal- to oil-based feedstocks.

The switch to oil began in the United States, which had abundant domestic oil and natural gas reserves. By 1950, one-half of the total U.S. production of organic chemicals was based on natural gas and oil; by 1960, the proportion was 88% (Chapman 1991, p. 82). The switch came later, but as rapidly, in Western Europe. In the United Kingdom, for instance, 9% of total organic chemical production in 1949 was based on oil and natural gas, and the proportion rose to 63% by 1962 (Chapman 1991, p. 82). In Germany, the first petrochemical plant was set up in the mid-1950s, and by 1973 German companies derived 90% of their chemical feedstocks from oil. In light of Germany's natural endowment of coal and the fact that German firms had made large, irretrievable (sunk) investments in coal-related technologies, the changeover of the German industry is especially remarkable (Stokes 1994).

The discovery of oil in the Middle East was significant not only because it meant that more oil was available but also because the additional reserves would be supplied through a global market, and therefore, control over oil was unlikely to be used as an economic or military weapon. The U.S. guarantee of an unhindered oil supply to Western Europe was also very important in this respect.

DYNAMICS OF INDUSTRY STRUCTURE: TECHNOLOGY

This section discusses two classes of technological innovations that have been crucial for understanding the evolution of industry structure in chemicals: polymer chemistry and chemical engineering.[7] In addition, it discusses another prominent (and related) technological development in the industry — the convergence between oil refining and chemical production during the 1950s and 1960s.

Polymer Chemistry

Polymer chemistry was the start of "materials by design." It was initiated by the theoretical work in the 1920s of Herman Staudinger and other German scientists, who postulated that natural and synthetic materials consist of long chains of molecules

---

[6]Morris (1994) provides a detailed account of the synthetic-rubber case. For aviation fuel, see, for instance, Spitz (1988) and Aftalion (1989).

[7]Another important class of technological innovations that is complementary to those in polymer chemistry and chemical engineering is in the area of catalysis. Indeed, many of the new products required advances in all three areas. For instance, a new polymer typically required both a new catalyst system and substantial research in process design. The chapter by Landau offers a further discussion.

linked by chemical bonds. These discoveries provided the industry with a method of developing new products by modifying the way in which molecules were connected. Producing commercially successful new materials was not simple. Long and systematic experimentation was still needed for commercial innovations. The new knowledge of molecular structure and physical properties, however, helped in making this search more productive.

Three characteristics of polymer chemistry are worth noting here. First, polymer science facilitated the development of many different applications of given substances, which created linkages among hitherto distinct and unconnected markets as well as knowledge-based economies of scope. Second, because applications had to be adapted for specific uses, the new science shifted the problem of innovation from "how" to produce different products to "what" to produce. This increased the relative importance of marketing and downstream links with users to find out how to tailor products for their needs (see for example, Hounshell 1995; Hounshell and Smith 1988). Finally, not only did knowledge about polymers diffuse at a rapid rate, but many companies worldwide had the in-house expertise to exploit it.

The first two features of this technological development were not qualitatively different from those observed during the rise of the dyestuffs industry at the end of the 19th century. Scientific discoveries in organic chemistry by Kekulé, Hoffman, and others had provided a basis for the systematic development of new organic dyes. The knowledge-based economies of scope, in turn, prompted expansion of the three leading German dye firms from dyes to technologically related areas such as pharmaceuticals and photographic materials. The "dyestuffs model" also required extensive investments in marketing to find out "what" users wanted and to train users in how to use the new products.

The early developments in organic chemistry and dyestuffs, however, were exploited mainly by the German companies, which had made the systematic investments and had the necessary "absorptive capacity" (Cohen and Levinthal 1989; Rosenberg 1990; Arora and Gambardella 1994a). By contrast, when polymer chemistry arose, there were many large, technologically advanced companies in the world. The earlier German success had led to a greater appreciation of the importance of scientific capabilities for commercial objectives. The events of World War I and the consolidations of the 1920s had created many large-sized firms — ICI in Britain, Montecatini in Italy, Solvay in Belgium, and Akzo in the Netherlands, not to mention IG Farben in Germany and several U.S. firms — which had the required size and scope to exploit polymer chemistry. Thus, unlike in the case of dyestuffs, the opportunities presented by polymers could be exploited by a substantial number of large firms worldwide with comparable commercial and technological capabilities. As Freeman (1982) points out, the presence of a large number of firms with comparable capabilities in polymers implied that even "small" information leaks allowed very rapid imitation.

Many chemical companies and some oil producers thus found themselves operating and competing in very similar markets. For instance, Union Carbide, Goodrich, General Electric, IG Farben, and ICI all performed research on PVC and produced the polymer. Similarly, Dow, IG Farben, and Monsanto were all involved in polystyrene from very early on. DuPont, ICI, Union Carbide, Monstanto, Kodak, and many others invested in various kinds of polyamides, acrylics, and polyesters (Spitz 1988; Aftalion 1989). The net result was an increase in competition in virtually every market segment. Competition also reinforced the search for product differentiation, which encouraged additional investments in R&D to develop new products or variants of existing ones.

Polymer chemistry is a good example of the economic potential of generalized knowledge bases (Arora and Gambardella 1994b). By establishing relationships between the properties of materials and their molecular structures, polymer chemistry provided a "method" for developing innovations in a systematic way. For instance, it was discovered that a polymer can be in either a "crystalline" state (with the molecules oriented in the same direction) or an "amorphous" state (with a random orientation of molecules). Crystallinity confers rigidity, strength, and thermal stability, whereas the percentage of the polymer that remains amorphous confers softness, elasticity, solubility, and absorptivity. Thus, with different degrees of crystallinity, the same polymer can have different properties. For instance, nylon can be used to produce soft shopping bags when it has less than 15% crystallinity, women's underwear with 20% to 30% crystallinity, sweaters with 15% to 35%, stockings with 60%–65%, tire cords with 75% to 90%, and fishing lines with more than 90% (Mark 1984). In addition, applying heat and pressure or controlling the density or melt indices enables a producer to make many polymers into any desired shape. The same basic material thus can be used as a fiber, sheet, or film or molded to form a component or product of a specific shape.

The striking feature of these examples is that the underlying technological basis not merely generated many opportunities but also it linked previously disconnected user markets. Ultimately, polymer chemistry supplied a common technological basis in three different areas: plastics, fibers, and rubber. Spitz (1988) discusses how different the activities involved in the development of synthetic fibers were from those for plastics. Fibers are used in ways that differ considerably from the ways in which plastics (and glass, metal, and wood) are used. Moreover, fibers required different processing equipment and completely different marketing strategies, both to sell the products and to get feedback from the users on how to further develop and improve the products, as described in Landau's chapter on the process of innovation.

The rise of polymer chemistry created an opportunity for the main chemical firms worldwide to exploit increasing returns in the use of their knowledge. However, economies of scope upstream, resulting from knowledge, usually tended to be much greater than those downstream, in production and marketing, because the new areas often were far away from the firm's core businesses. This meant that commercialization assets — product development, manufacturing, and marketing — became very important. Equally, such assets were expensive and lumpy. Thus, in spite of the significant economies of scope in science and technology, actual large-scale entry into the new areas took longer than did appreciation of the opportunity.

Chemical Engineering and the Specialized Engineering Firms

If polymer chemistry is the science of creating economical new chemical products, chemical engineering is the science of creating economical new chemical processes. The objective of chemical engineers was to design and optimize plants to lower manufacturing cost and improve quality.[8] The development of the concept of "unit operations" was critical to the development of chemical engineering. The idea was that all chemical processes could be conceived of in a unified framework as combinations of a small, well-defined number of operations, such as distillation, evaporation, drying, filtration, absorption, and extraction.

---

[8]On the origins and history of chemical engineering as a discipline, the role of MIT, and its relationships with the oil companies, see Landau and Rosenberg (1992), and Nathan Rosenberg's chapter in this volume.

By separating the task of process design from the details of the particular product being produced, chemical engineering freed process engineers to think about chemical processes in general, rather than about the particular chemical process for ethylene or ammonia. By providing a unified framework, chemical engineering allowed for the experience gained in one petrochemical process to be applied to others, further enhancing the benefits to specializing. As a result, chemical engineering made possible the rise of specialized firms that focused on engineering and process design services for chemical plants — the so-called specialized engineering firms (SEFs).

With a few notable exceptions, most SEFs did not develop radically new processes. Most major process innovations were typically produced by the large oil and chemical companies (Mansfield et al. 1977). However, the SEFs were effective at moving new processes down the learning curve. Equally important, SEFs served as independent vendors of process technology, thus facilitating a great deal of entry into the chemical industry in the decades after World War II. Initially, the entrants were firms from the developed countries themselves, but more recently, SEFs have helped firms from developing countries enter as well. By acting as independent licensers, SEFs also induced chemical firms to license their technology. In essence, SEFs helped create a market for technology, making process technology into a commodity that could be bought and sold.

The development of an independent engineering-design sector is an example *par excellence* of what we mean by economies of specialization at the level of the industry. A large market for basic petrochemicals, plus the relative independence of process design from products, implied that SEFs could design many more plants for a variety of closely related processes than any single chemical company could. The accumulated learning of SEFs was the basis of their comparative advantage in process design. In turn, the cost reduction due to cheaper and more innovative processes lowered prices; stimulated the substitution of chemical products for natural fibers, resins, metals; and caused the market to grow further. The presence of SEFs allowed entry by firms from other industries and countries, which further increased competition and lowered prices.

The first SEFs were formed very early in the 20th century. Their clients were large oil companies, which concentrated their energies on "searching for crude oil and establishing retail market facilities" (Landau and Brown 1965, p. 7). In the years before World War II, SEFs were mostly employed as suppliers of specialized equipment and the like, rather than for the design and engineering of entire production processes. In part, this was due to the traditional secrecy of chemical companies. Further, most chemical operations tended to be relatively low-volume "batch" production, embodying a great deal of company-specific know-how.

The situation changed markedly after World War II. The growing importance of petrochemicals and the increase in the scale of production raised the payoff to improvements in plant design. In turn, the growth in the size and complexity of plants, as well as the concomitant development of the discipline of chemical engineering, laid the foundations for division of labor in the chemical-processing sector. By the 1960s, SEFs had come to occupy an important place in the industry. In a pioneering study, Freeman (1968, p. 30) noted that for the period from 1960 to 1966, "[N]early three quarters of the major new plants were 'engineered,' procured and constructed by specialist plant contractors." Moreover, Freeman found that SEFs were an important source of process technologies: During 1960 to 1966, they accounted, as a group, for about 30% of all licenses of chemical processes.

TABLE 12-5. The Market for Engineering Services and Licensing in Chemicals—by Sector: 1980–1990

Sectors	Panel A % of Plants Engineered				Panel B % of Licenses	
	In-House	By SEF	By Other Firms*	Own Technology	By SEF	By Other Firms*
Air separation	32.4%	34.1	33.5	27.2%	33.7	39.0
Fertilizers	4.8	79.6	15.6	4.8	61.5	33.7
Food processing	5.0	74.8	20.3	20.4	38.8	40.8
Gas handling	5.0	78.0	17.1	4.9	62.3	32.8
Inorganic chemicals	14.1	66.9	18.9	24.4	29.2	46.4
Industrial gases	21.9	60.3	17.8	12.9	36.1	51.1
Minerals & metals	7.8	71.3	20.9	23.9	24.4	51.7
Miscellaneous	6.6	78.9	14.4	16.8	34.6	48.5
Organic chemicals	24.3	53.8	21.9	44.2	19.4	36.4
Oil refining	6.4	83.7	10.0	9.3	48.6	42.1
Petrochemicals	13.3	75.9	10.8	18.5	32.4	49.1
Pharmaceuticals	19.4	63.0	17.6	54.8	3.2	41.9
Plastics & rubber	23.8	63.1	13.2	41.2	6.1	52.8
Pulp & paper	4.0	79.0	17.0	3.8	46.2	50.0
Misc. specialities	31.0	52.1	16.9	61.5	2.9	35.6
Textile & fibers	7.4	72.2	20.3	17.9	52.9	29.2
Total	12.7%	71.6%	15.6%	21.5%	34.6%	43.9%

* = Typically chemical companies or other downstream manufacturers.
(A) = Total fraction of plants in which name of engineering company is reported.
(B) = Total fraction of plants in the database.
Source: Chemical Age Profile (1991).

TABLE 12-6. Market Shares of SEFs — Engineering Services: 1980–1990 (Shares of Total Number of Plants by Region)

Nationality of SEFs	Regions				
	U.S.A.	West Europe	Japan	Rest of the World	Share of Total World Market
U.S.A.	58.8	19.8	3.7	18.9	26.0
West Germany	1.9	18.5	4.6	12.7	11.7
U.K.	6.9	12.2	2.0	7.3	8.1
Italy	0.3	8.2	0.0	5.8	5.1
France	0.2	2.3	0.3	4.6	3.2
Japan	0.2	0.2	34.0	5.1	4.0

·*Source: Chemical Age Profile* (1991).

Freeman's (1968) findings are confirmed by more recent data. The first panel in Table 12-5 shows that for the period 1980 to 1990, almost three-fourths of the total number of plants in the world were engineered by SEFs.[9] Although the share of SEFs varies across sectors, in practically all sectors, it is above 50%. The rest of the plants were engineered in-house, by other companies, including the chemical-engineering divisions of chemical companies. The division of labor in licensing is less marked than in engineering, as shown in the second panel of Table 12-5; nonetheless, SEFs still account for about 35% of the licenses. The lower shares of SEFs reflect the difficulties of financing research into radically new processes, as well as the specialized, product-specific knowledge that such research may need. However, as we will discuss further below, the role of SEFs was perhaps more important than this figure suggests, because their very presence induced many downstream companies to license their processes. One warning must be made about both of these estimates: In a number of cases, the engineering or licensing firm was not reported in the database, and Table 12-5 gives the percentage of data in each sector for which a specific firm was reported.

SEFs started as an American phenomenon. Freeman (1968) estimated that U.S. firms accounted for more than 50% of the total value of engineering contracts worldwide during 1960 to 1966. Table 12-6 shows that more recently, U.S. SEFs accounted for 26% of the world market in engineering services, with German, British, Italian, French, and Japanese companies as important competitors, particularly in the third-world markets. While the U.S. SEFs have a sizable share of the market in Europe, the European firms have only a small market share in the United States. This is a legacy of the period following World War II, when American SEFs moved to Europe. Many American SEFs were able to establish local subsidiaries, and some of these subsidiaries even became full-fledged "national" companies (for example, Foster-Wheeler in Britain). Japanese SEFs account for the bulk of the engineering services in Japan, where even the U.S. SEFs have only a modest presence. Although American firms have been allowed to sell licenses in Japan, their entry in the market for engineering services has been restricted.

[9]The data are from a comprehensive data base of investments in the chemical sector in *Chemical Age Profile* (1991). The shares in Table 12-5 (and in Tables 12-6 and 12-7 are computed over the total number of plants in which the information about the engineer or the source of technology was available.

TABLE 12-7. Market Shares of SEFs — Licenses: 1980–1990 (Shares of Total Number of Plants by Region)

Nationality of SEFs			Regions		
	U.S.A.	West Europe	Japan	Rest of the World	Share of Total World Market
U.S.A.	18.0	10.3	6.5	16.9	15.1
West Germany	3.1	11.3	1.0	10.2	8.8
U.K.	1.2	3.0	2.7	2.4	2.4
Italy	0.1	1.4	0.0	2.2	1.6
France	0.1	0.6	0.0	0.9	0.7
Japan	0.1	0.1	1.5	1.1	0.7

Source: Chemical Age Profile (1991).

Table 12-7 reports the shares of SEFs from leading countries in licensing. The global market share of U.S. SEFs is about 15%. A comparison of Table 12-6 with Table 12-7, however, reveals that the world share of U.S. SEFs, relative to its competitors, is larger in licensing than in engineering, with Germany the only exception. The comparative advantage of U.S. SEFs in licensing is even more apparent if one compares their shares in Europe and Japan with respect to the corresponding shares of their competitors. In Europe, for instance, the share of U.S. SEFs in engineering is only about 1.6 times that of the United Kingdom, while it is 3.4 times that of the United Kingdom in licensing.[10]

SEFs have been a primary mechanism for the diffusion of chemical-process technology, whether as licenses or as embodied in engineering services (or both), first to Europe and then worldwide, to Asia, East Europe, Latin America, and the Middle East.[11] In addition, many SEFs also specialize in construction services and can provide "turnkey" plants to their clients. The implications for industrial structure have been profound. Spitz (1988, p. 313) notes that most major chemical products have between five and fifteen main producers. By contrast, in the pre-World War II era, it was unusual to have more than three manufacturers of a major product. In the United States, the sharp increase in the total number of producers of selected chemicals occurred during the 1950s. For example, in the U.S. market from 1947 to 1963, the number of producers of benzene rose from 11 to 36; of ammonia, from 12 to 47; of urea, from 1 to 19; of xylene, from 2 to 28 (Backman 1964). The fraction of this entry that can be attributed to the diffusion of technology through SEFs cannot be estimated precisely, because there are a number of intertwined factors. Demand was increasing rapidly at this time, but the minimum efficient scale of plants was increasing. Moreover, antitrust concerns at this time encouraged many chemical firms to license their technology, which also contributed to lowering entry barriers (Backman 1964, pp. 47–50).

Other studies provide evidence supporting the importance of SEFs as suppliers of technology and know-how to new entrants. In a study of 39 commodity chemicals in the United States in a period from the mid 1950s to the mid 1970s, Lieberman (1989)

---

[10]The reasons for the comparative advantage of U.S. firms in this sector are more fully explored in Arora and Gambardella (1997).

[11]The discussion of the next few paragraphs is based on Arora, Fosfuri, and Gambardella (1996).

found that after controlling for demand conditions, experience accumulated by incumbents did not deter new entry. Given the importance of learning by doing, this suggests that entrants had access to other sources of know-how, most likely from SEFs. In a related study of a subset of 24 chemicals, Lieberman (1987) found that high rates of patenting by nonproducers were also associated with faster rates of decline in prices. Once again, this evidence is consistent with an interpretation that patenting by nonproducers (especially SEFs) led to entry by new firms through licenses.

As SEFs became important sources of plant design, their importance as sources of process innovation also increased (Mansfield et al. 1977). SEFs have been particularly innovative in two areas: catalytic processes and engineering-design improvements.[12] SEFs have relied on licensing to appropriate rents from their innovations. This has encouraged entry by producers and also has had notable effects on how the chemical firms themselves have used their technologies.

In a marked departure from their pre-World War II strategy of holding on to their technology, after the war, several chemical and oil companies began to use licensing to profit from innovation.[13] As Spitz (1988, p. 318) put it: " [S]ome brand-new technologies, developed by operating (chemical) companies, were made available for license to any and all comers. A good example is the Hercules-Distillers phenol/acetone process, which was commercialized in 1953 and forever changed the way that phenol would be produced."

Technology licensing in the chemical-processing industry is widespread, as illustrated earlier by Table 12-7, and only a portion of it happens through SEFs. Some chemical companies that have been major licensers of their patented technologies include ICI in ammonia; Union Carbide in polyethylene, polypropylene, and air-separation technologies; Montecatini (including affiliates such as Himont) in polypropylene; and Mitsui also in polypropylene. Oil companies, especially Shell, Mobil, BP, and Amoco, have been active in licensing their technologies, as well. A typical large chemical company today — DuPont, Hoecht, ICI, Monsanto, or Union Carbide — has dozens of licensing agreements, many with firms located outside its home markets (see Arora 1997 for more details).

Convergence: Oil and Chemicals

The rise of petrochemicals in the post-World War II period laid the basis for the technological convergence of the oil-refining and chemical sectors. The blend was never complete: For example, oil prospecting and extracting have remained distinct activities, and mixing and formulating paints has remained distinct from oil refining. But large-scale chemical-processing technologies and large-scale oil refining became very close: The process of refining and producing gasoline and other oil fractions became a chemical process (the use of catalysis being a prime example); and the process of

[12]UOP and Scientific Design are two of the SEFs that have radical innovations to their credit. UOP has a number of catalytic refining and reforming processes that it has licensed widely. For a discussion of Scientific Design, see Landau's chapter in this volume. A number of other SEFs have contributed to advances in engineering design. For instance, Kellogg made significant contributions to developing high-pressure processes for ammonia in the 1930s, while Badger is associated with fluidized-bed catalytic processes (in collaboration with Sohio).

[13]Landau (1966, p. 4), writing two decades after the end of the war, noted that the "the partial breakdown of secrecy barriers in the chemical industry is increasing ... the trend toward more licensing of processes." In a recent interview, the CEO of Dow, Frank Popoff, confirmed that even Dow Chemicals would seriously consider licensing as a strategy of rent appropriation Popoff (1995).

producing large-scale chemicals involved insights and knowledge from refining (for example, the use of fluidized- and fixed-bed catalysis or heat exchange). Chemical techniques were applied to refining for choosing the mix of by-products, converting less desirable by-products into more desirable ones (cracking heavier fractions into lighter fractions, such as ethylene and propylene), and reforming — that is, converting aliphatic (straight-chain by-products) into aromatic (benzene-type) by-products. In addition, chemical engineering provided the basis for the design of both refineries and chemical plants for products such as polyethylene and nylon. Polymer chemistry showed how petroleum and gas-based feedstocks could be used to make very useful products. A precondition for this convergence was the discovery of major oil fields in the United States and elsewhere. Another factor of importance was the demand pressure for advances in oil refining caused by the impetus of war, the demand for aviation fuel, and the ubiquitous automobile in the United States.

The impact of this convergence in the oil and chemical sectors was asymmetric. Oil firms moved downstream into chemicals, but chemical firms were much less successful in moving upstream into oil refining. Oil companies succeeded for a couple of reasons. First, physical economies of scope between the refining and production of ethylene and polyethylene in integrated facilities arose from reduced costs of material handling and transaction costs. Second, knowledge-based economies of scope arose because the skills of chemical engineering offered extensive spillovers in designing and operating oil refineries and petrochemical plants. By the early 1960s, oil and gas companies had seized more than 90% of the market in certain chemical categories such as *ortho*-xylene, *para*-xylene, butadiene, xylene-mixed, and toluene. For a number of other chemical products, oil firms accounted for over one-half of the entire productive capacity (Backman 1964).

Chemical companies were less successful at becoming oil refiners. Most importantly, oil firms retained control of crude-oil supplies. In addition, oil companies had existing distribution channels and networks for by-products of oil-refining such as lubricants and heating oil, which were difficult for chemical producers to try to duplicate. Also, oil refineries are typically more expensive than chemical plants, and chemical firms are typically smaller than oil firms. Thus, it was much easier for oil firms to diversify into

TABLE 12-8. Shares of Plants Operated by Oil and Chemical Companies in Petrochemicals and in Natural Gas and Oil Refining: 1981–1990

Location of Plants	U.S.A.	Western Europe	Japan	Rest of the World
PETROCHEMICALS				
Total no. of plants	910	1073	480	2149
Oil companies	40.1%	30.8	25.4	19.7
Chemical companies	30.7%	34.4	49.8	46.7
NATURAL GAS & OIL REFINING				
Total no. of plants	2331	960	283	2055
Oil companies	51.7%	52.0	55.5	55.9
Chemical companies	5.0%	9.1	23.0	9.8

Source: Hydrocarbon Processing Unit (1991).

chemicals than for chemical firms to diversify into oil refining. Table 12-8 shows how oil companies have invaded petrochemicals, whereas chemical companies have made only small inroads into natural gas and oil refining. However, chemical companies have reacted to their inability to move upstream by moving further downstream into differentiated products.

THE CHEMICAL INDUSTRY STRUCTURE SINCE THE 1980s

Maturation, Competition, and Restructuring

The number of chemical firms in most markets has increased substantially in recent decades. During the 1950s and the 1960s, the industry could accommodate such increases, because demand for chemical products grew rapidly — typically two or three times more quickly than GDP. Even so, profitability in the chemical sector had already begun to decline in the early 1960s (Chapman 1991, pp. 234–36). The problem became severe when demand growth slowed in the 1970s and the 1980s. Firms in a number of chemical-product markets, especially basic intermediates, were confronted with excess capacity. Entry and competition from firms in developing countries, including Eastern Europe and the Middle East, aggravated the problem. Established international producers found their exports to those countries threatened and also faced a new source of potential imports in their own markets (Grant 1991).

For the established producers, the adjustment to declining demand and the rise of new sources of supply was slow and painful. Existing producers had sunk large-scale investments in capacity, especially in the basic intermediates. The adjustment problem was magnified because many chemical and petrochemical operations were highly integrated, both vertically and horizontally; thus, a reduction in capacity at one plant often left excess capacity at others or could even reduce the manufacturing efficiency of an entire production complex. To make matters worse, not all firms accurately foresaw the slow growth of demand, and many companies continued to invest in the 1970s as if they were expecting to return to the high growth rates of the 1960s. A comprehensive realignment of expectations occurred only during the 1981 to 1982 recession (Aftalion 1989; Chapman 1991; Lane 1993).

The pressures for capacity rationalization were not identical across the United States, Western Europe, and Japan. The problem of large installed capacities and rosy expectations about future demand growth were common to all three areas, but the threat posed by new producers was mainly a European and, to a lesser extent, a Japanese problem, because those areas began with greater export dependence (Grant 1991, p. 255). In addition, Europe faced the special problem that many of its large, diversified national companies were vulnerable to pressures from stakeholders and employees.

Capacity reduction and rationalization proceeded in different ways in the United States, Europe, and Japan. In the United States, the reductions were market driven and without explicit coordination. In Europe, both market forces and government intervention played a role, especially in Italy and France, where the state had large ownership stakes (Martinelli 1991). In Japan, MITI (Ministry of International Trade and Industry) played a major role by coordinating capacity rationalizations. Its Industrial Structure Council formed a committee, which suggested that the industry had to move away from petrochemicals to focus on high value-added products, particularly new materials

(Bower 1986, p. 202). Legislation was then enacted in 1983, which established targets for capacity reductions and for reductions in the number of producers by 1985, in petrochemicals as well as in other depressed industries (Chapman 1991).

The pressures for cost reduction were also reflected in R&D spending. In part because of derived demand from users in the electronics industry, the Japanese chemical firms have invested considerably in R&D during the past few years, and their R&D intensity today is comparable to that of the U.S. firms. By contrast, U.S. and European firms have been cutting down on R&D. The total R&D spending for the leading U.S. chemical firms between 1985 and 1995 increased at a little over 3% per year in nominal terms and even declined over 1993 to 1995 (*Chemical and Engineering News* 1996, pp. 53–54). The cuts have been deep in some leading companies such as DuPont, where R&D spending has fallen significantly in real terms, with cuts in basic research being particularly sharp.

Capacity reduction through plant shutdowns started in the early 1980s, and the bulk of it was finished before 1985. In the United States, restructuring via unilateral capacity reduction usually preceded changes in the ownership of plants (Lane 1993). This suggests that U.S. chemical companies responded via internal rationalization before attempting consolidation at the level of the industry. Capacity reduction and plant shutdowns took place in Europe, as well. Both in Britain and Germany, a number of plants were closed during 1981 to 1982, and the exit of many U.S. companies from the continent helped the process as well (Chapman 1991 p. 250).

In Europe, restructuring also involved interfirm agreements and an active role for interindusty associations.[14] In 1985, thirty-four large petrochemical firms formed the Association of Petrochemical Producers in Europe (APPE). APPE did not have any formal role in encouraging agreements; however, by collecting and providing data on capacities in different product markets, it diffused information, helping members take informed decision about capacity expansions and reductions. In fact, Europe's capacity rationalization in the chemical industry is said to have been far more successful than in other sectors, such as steel (Grant 1991, p. 265). In some cases, European firms tried to form a cartel, but with only limited success. The number of producers had increased considerably over time, which made enforcing agreements difficult. In addition, the pro-competitive attitude of the EC restrained such deals.[15] Thus, the problems of overcapacity were largely resolved by the second half of the 1980s.

Table 12-9 documents the effects of the ongoing restructuring process in selected petrochemical products for the United States and Europe. As Table 12-9 shows, the number of producers fell more sharply in Europe, while concentration of capacity, which was more substantial in the United States to begin with, also declined in the United States, whereas it rose slightly in Europe. This effect is consistent with the view that restructuring in the U.S. chemical industry took the form of reduction of capacity

---

[14]As one example of interfirm agreements, the restructuring of the European PVC market proceeded in this way. First, BP and ICI signed a deal in 1981, which led to the consolidation of their businesses. BP ceded its PVC operations to ICI, which pulled out of polyethylene by relinquishing its activities to BP. This concentrated PVC in ICI and polyethylene in BP. Then, in 1985, ICI formed a joint-venture, European Vinyls, with the Italian company Enichem, which merged the PVC businesses of the two firms. The new company became the major European PVC producer. Similarly in polypropylene, Statoil and Neste have merged their petrochemical operations to form Borealis (sales, $2.3 billion), Europe's largest, and the world's fifth-largest, polyelefin producer. Montecatini and Shell have merged their polypropylene businesses into a joint venture called Montell, with a very substantial market share.

[15]The most important intervention was the fine imposed by the European Court in 1986 against 15 leading polypropylene producers (among others, Montedison, ICI, Shell, and Hoechst), that were found guilty of market-sharing and price-fixing actions.

TABLE 12-9. Producers, Capacities, and Concentrations for Selected Petrochemicals in Western Europe and the U.S.: 1972–1990

	Ethylene	Polyethylene	Ethylene Oxide	Styrene
	Western Europe			
1973				
No. of producers	38	40	16	15
Share of top 3 producers in total capacity	24.3%	18.7%	41.5%	40.2%
1990				
No. of producers	29	26	10	12
Share of top 3 producers in total capacity	25.9%	27.2%	41.4%	47.6%
	United States			
1973				
No. of producers	25	21	13	12
Share of top 3 producers in total capacity	40.5%	40.8%	72.0%	54.1%
1990				
No. of producers	22	16	12	8
Share of top 3 producers total capacity	30.8%	37.9%	54.2%	52.1%

Source: Chapman (1991), tables 5.2 and 5.3.

by the largest producers. Moreover, the entry of specialized commodity manufacturers (discussed below) both reduced concentration and kept the number of producers from falling too far. In Europe, the figures reflect partly the exit of U.S. firms and partly the consolidations carried out through interfirm agreements. While we lack systematic evidence for Japan, the available evidence suggests that restructuring has not proceeded much beyond capacity reduction and some MITI-brokered swaps to promote rationalization (see the chapter by Hikino et al.).

Restructuring of Chemical Firms and the Industry

The restructuring of the chemical industry since the 1980s involves more than just capacity reduction. For example, firms are increasingly looking outside national boundaries, especially in moving production closer to the emerging markets in Asia and eastern Europe. Perhaps more interesting, chemical firms have reversed a long trend toward diversification and have narrowed their business portfolios. While some firms are focusing on commodity chemicals, others are focusing on the so-called specialty sectors: sectors marked by greater possibilities of product differentiation, less acute price competition, greater service content, higher margins, and often lower volumes. In other words, it appears that firms have to focus on competencies either in production and cost efficiency (commodities) or in product quality and customer satisfaction (specialties). The focusing also reflects the fact that there appear to be relatively few synergies between the life-science and chemical businesses.

To examine this restructuring systematically, we analyzed publicly reported data on acquisitions in the chemical sector between 1985 and 1993 and compiled a list of all acquisitions exceeding $1 million over that period. The results of our compilation are reported in Table 12-10. In the United States, we tallied 1,690 acquisitions over that time (12% of which were foreign), compared to 1,188 acquisitions in Europe (58% foreign) and 110 acquisitions in Japan (72% foreign).[16] While differences in coverage period and possible complicating factors do not allow us to draw firm conclusions about when restructuring starts in the different regions, the trends reported in Table 12-10 are consistent with the idea that restructuring began earlier in the United States.

Financial deals have been particularly important in creating new companies that focused on commodity chemicals. In the United States, many such companies were created in the 1980s from acquisitions of plants sold by larger producers that were moving downstream. For example, Huntsman was founded in 1983 by the acquisition of Shell's polystyrene business and three more plants, from Hoechst, in 1986; similarly Sterling Chemical and Cain Chemical were founded from the acquisition of petrochemical plants of larger chemical producers such as Monsanto, DuPont, and ICI. Vista Chemical was formed in 1984 through a leveraged buyout of plants from DuPont; Georgia Gulf was founded after a 1985 leveraged buyout of most of the assets of the chemical division of Georgia Pacific; Aristech emerged through a management buyout of the heavy-chemicals division of USX (formerly U.S. Steel) (Lane 1993).

Britain, too, had its "commodity specialists" in the form of the Hanson Group, which at one point attempted a takeover of ICI. In the rest of Europe, however, the formation of new companies played a much smaller role because of the different role of stock markets in corporate governance (see the chapters by Marco Da Rin and by Albert Richards for more details). Instead of appearing as buyouts of some form, whether leveraged or management, refocusing in Europe took place via reorganization and splits of existing chemical producers. The recent rounds of privatization of large state-owned companies in France, Italy, and Finland have been important in this respect. In France, for instance, Rhône-Poulenc developed an aggressive strategy of acquisitions and divestitures to enter into a number of new specialty business. Orkem, another French company, was split into Elf, which focused on petrochemicals and fertilizers; and Total, producing inks, adhesives, acrylics, glass, and paints. There have been some intra-group mergers in Japan, and some overseas acquisitions: For example, Dainippon Inks acquired companies in the pigments business in United States; Mitsubishi acquired Aristech.

While some firms focus on commodity chemicals — as do Huntsman, Borealis, and Quantum — many chemical firms are trying to move downstream into more differentiated product lines. This was achieved in part through acquisitions. For instance, Monsanto acquired G.D. Searle in 1985, which marked its entry into pharmaceuticals. Many large European chemical companies such as Rhône-Poulenc in France acquired specialty producers while withdrawing from commodity chemicals. This strategy is not new; such companies as Bayer have pursued it for for many years. Although both

---

[16]Our primary data cover deals in the chemical industry (defined as SIC 28) with value that were announced in the United States between 1984 and 1992 and in Europe between 1987 and 1992, but selected deals from earlier years are also included. Notice also that acquiring companies include nonchemical firms. However, chemical companies account for about two-thirds of our sample of acquirers. Our data source also provided up to 10 four-digit SIC codes of the acquiring companies, as well as the four-digit SIC code in which the acquisition can be classified. We matched these data with other publicly available data sources to obtain nationalities of both the acquiring and the acquired companies.

TABLE 12-10. Trends in Acquisitions in the Chemical Sector, by Acquirer[a]: 1985–1993

Year	United States[d]	Europe[b]	Japan[c]
1985	210	78	4
	(3%)	(29%)	(0%)
1986	138	32	6
	(5%)	(34%)	(0%)
1987	225	58	7
	(5%)	(34%)	(29%)
1988	206	79	12
	(14%)	(61%)	(92%)
1989	184	193	28
	(14%)	(69%)	(79%)
1990	210	269	18
	(15%)	(61%)	(100%)
1991	252	242	24
	(15%)	(55%)	(75%)
1992	262	235	11
	(23%)	(61%)	(73%)
1993[d]	3	2	0
	(0%)	(100%)	(0%)
Total	1690	1188	110
	(11.8%)	(57.6%)	(71.8%)

Notes:
[a]All deals announced worth more than $1 million, or that involve more than 5% of a firm, or of undisclosed value are covered. Deals that were announced but canceled are not included. Entries in each column refer to the nationality of the acquiring company. Fractions of overseas acquisitions are in parentheses.
[b]For Europe, systematic coverage begins in 1987.
[c]Coverage for Japan does not include possible acquisitions in Japan.
[d]Figures for 1993 include some acquisitions announced in 1994, as well. 1985 data are for Jan. 1984 to Dec. 1985 for the United States.

Source: Our calculations, from the *IDD Information Service Data Base.*

product differentiation and cost reduction can increase profits, the former has the advantage of relaxing price competition (Sutton 1991), which makes product differentiation a more attractive strategy. After World War II, rapid economic growth and economies of scale encouraged strategies aimed at large, homogeneous markets rather than at entry in several market niches. But once growth slowed down and economies of scale at the level of the plant became less important, companies tried to segment the market by creating market niches.

The Japanese adjustment to changing circumstances was much easier in this respect. As noted earlier, the important role played by users in Japan had resulted in an early emphasis on customization of product-to-user requirements and high-product quality, even at the expense of production efficiency. In the 1980s, these close links with users turned out to be an asset. For instance, the Japanese chemical industry forged links very early with its growing electronics sector and acquired a leading position in this field. (Although U.S. and European firms are also active in chemical applications to

TABLE 12-11. Specialty and Non-Specialty Acquisitions in the U.S. and Europe during the 1980s*

Nationality of Acquirer	All Target Regions		U.S. Target Companies		European Target Companies	
	(1)	(2)	(1)	(2)	(1)	(2)
U.S.	1576	118	1275	102	151	11
	93.0%	7.0	92.6	7.4	93.2	6.8
W. Europe	1180	56	265	28	785	26
	95.5	4.5	90.4	9.6	96.8	3.2
Japan	103	9	57	9	36	0
	92.0	8.0	86.4	13.6	100	0
Other	86	6	36	5	17	1
	93.5	6.5	87.8	12.2	94.4	5.6
Total	2945	189	1633	144	989	38
	94.0	6.0	91.9	8.1	96.3	3.7

*All acquisitions in the chemical sector (SIC 28) that occurred in the United States during 1984 to 1992 and in Europe during 1987 to 1992. (Financial and nonfinancial investors.)
Note: (1), (2), (3) = Nonspecialty targets, and totals, respectively.
Source: Our calculations from the *IDD Information Services Data Base.*

electronics, they have focused largely on the life-science business.) Of course, the market for electronics in Japan is far larger than the traditional chemical markets, a fact that affects the opportunities of the Japanese chemical manufacturers. Japanese chemical producers have also diversified, moving into the life sciences as well as into electronics.

We gathered evidence on the extent to which chemical producers used acquisitions to move into specialties. Table 12-11 shows the number of specialty and nonspecialty acquisitions, by nationality, of the acquiring company. (We were somewhat conservative in our definition of specialty sectors.) Most specialty targets are U.S. companies, and European and Japanese companies are more likely to acquire a U.S. specialty company than one from their own domestic market. However, because most specialty producers are in the United States, European and Japanese chemical companies moving into these fields could find suitable targets only overseas.[17] In carrying out this compilation, we also found that 22% of the acquiring companies had at least one line of business that could be identified as specialty. This percentage rises to 31.4% in the case of U.S. acquirers, but it drops to 8.5% in the case of non-U.S. specialty buyers. Thus, U.S. acquirers of specialty targets are more likely to have specialty businesses themselves. We also found that European specialty acquirers tended to be large and diversified firms, while U.S. acquirers represented, instead, a wider class of firms, which included large as well as smaller companies. Acquiring companies in a foreign market requires assets and capabilities that large firms are more likely to have, and hence European acquirers are larger on average than U.S. acquirers. In the United States, smaller firms could, instead, acquire specialty producers without having to incur the costs of international expansion.

[17]The low value of the U.S. dollar, especially in 1987 to 1988, made U.S. targets more attractive at this time. Indeed, our data indicate that acquisitions of U.S. companies by European firms grew significantly in that period. See also Lane (1993).

Restructuring: Increasing Focus

The separation underway between specialty and commodity chemicals is an instance of a prevailing wisdom which says that firms ought to concentrate on their "core competencies." To see whether firms are actually focusing, we used a sample of the largest 250 chemical firms in the world, as ranked by world chemical sales in 1988 (Aftalion 1989, appendix). Of these 153 are covered by our acquisition and mergers database. They are predominantly companies from the developed world, and particularly from Europe and the United States, since our data for mergers and acquisitions include few Japanese companies. The key here is to find a compact way of summarizing how focused a firm is. We used a version of the Herfindahl-Hirschman Index, which is more commonly applied in an antitrust context, for determining whether a market is concentrated, and is calculated in that case by taking the sum of the market shares of each firm. For example, if one firm controls 90% of an industry and a second firm has the other 10%, the HHI would be equal to $0.9^2 + 0.1^2 = 0.82$. If 50 firms have equal 2% market shares, then the HHI would be $50(0.02)^2$, or 0.02. In this way, the HHI ranges from nearly zero in the case of many small competitive firms to 1 in the case of a monopoly. This measure can also be applied within a firm, where the firm's lines of business are treated like the market shares in the preceding calculation. Thus, a firm in one line of business would have an HHI near 1, while a firm divided in many lines of business would have an HHI near zero. For our sample of firms, we found that after adjustment for the degree of diversification, a typical firm in the 1980s has acquired more businesses than it sold off. However, the large U.S. firms also have more acquisitions than sell-offs, but they are growing less rapidly through this route than are their European competitors.[18]

Are acquisitions being used to increase focus? We first examined the portfolio of acquisitions and sell-offs for our sample of large firms to see whether firms make most of their acquisitions in a few sectors, or whether these are spread across a range of sectors. We found that U.S. firms sell off on average a slightly wider variety of businesses than do European firms but do not differ significantly from the latter in the spread of their acquisition portfolios. Taken together, these results are consistent with the notion that large U.S. chemical firms are becoming more focused and that this occurs mainly through more diversified sell-offs.[19]

We also analyzed the degree of focus in the investment patterns of chemical firms: We used the same database on all investments in new plants in the chemical sector during 1980 to 1990 that we had used to investigate the importance of investments through SEFs earlier in this paper, from *Chemical Age Profile* (1991). This database includes investment patterns of 781 U.S. chemical firms, 648 West European firms, and 267 Japanese firms. As before, we analyzed separately data for the large firms from the United States, Japan, and Western Europe. Column A of Table 12-12 shows that, on average, the typical large Japanese firm invests in about 15 plants, while the corresponding figures for large European and U.S. firms are 51 and 41, respectively. Columns B

[18]Specifically, we ran three regressions with three different dependent variables: number of acquisitions, number of sell-offs, and number of acquisitions net of sell-offs. The internal HHI measure was our explanatory variable in the regression to control for the level of firm diversification. The other explanatory variables were a dummy for U.S. firms and a constant term. Details of the calculations are available on request from the authors.

[19]Specifically, we ran two regressions with two different dependent variables: number of acquisitions and number of sell-offs. The explanatory variables were a measure of the diversity of the firm, using the Herfindahl-Hirshman Index number as explained in the text, a dummy variable for U.S. firms, the size of the firm, and a constant term. Details of the calculations are available on request from the authors.

TABLE 12-12. Concentration in Investments in Chemical Plants, All Areas: 1980–90

Nationality of Investing Firm	Total Plants (A)	Average Number per Sector (B)	Average Number per Location (C)	Degree of Sector Concentration (D)	Degree of Geographical Concentration (E)
All U.S. (781)	7.08	1.77	1.81		
Large U.S. (56)	41.76	5.25	6.73	0.346939	0.623964
All W. Europe (648)	7.91	1.97	1.94		
Large W. Europe (44)	51.79	6.77	9.22	0.331699	0.556587
All Japanese (267)	4.69	1.86	1.31		
Large Japanese (34)	15.08	3.85	2.73	0.346939	0.962976

Notes: Concentration is measured as the Herfindal Index. Number of firms is given in parentheses.
Source: Our calculations from *Chemical Age Profile* (1991).

and C show that Japanese firms also tend to be involved in fewer sectors and to have fewer plants at a given location. However, column D shows that if one narrows the focus to large companies, there is no significant difference in the degree of focus. For the large chemical firms in these three areas, the average value of the Herfindahl-Hirshman Index for investments across different sectors is about 0.34.

The similarity in the degree of concentration in physical investment is consistent with the similarity in acquisition portfolios. Moreover, consistent with our earlier results, the United States, and especially the European firms, have very geographically diversified investments, compared with those of Japanese firms, as shown in column E.

A different but related question is whether an acquisition is in a business that is related to a firm's existing businesses. We looked at the different lines of business for each chemical company, as defined by the Standard Industrial Classification (SIC) codes of the U.S. Department of Commerce. We defined an acquisition to be in "related" businesses if the SIC code of the acquisition (at the four-digit level) was already one of the SIC codes of the acquirer. We restricted our sample to the 2,548 acquisitions made by non-financial investors. In Table 12-13, columns A and B show that two-thirds of acquisitions were in unrelated businesses. It appears, then, that acquisitions are primarily a means for achieving diversification. However, when U.S. chemical companies are acquiring European firms, as shown in the last two columns of the first row, about one-half of all acquisitions are in related businesses.

This last result suggests that U.S. firms are more likely to use acquisitions in Europe as a means of consolidating their existing lines of operations in the European market. In contrast, a larger fraction of U.S. companies acquired by European firms are in activities unrelated to the core business of their acquirer. This suggests that most of these operations aim at the acquiring capabilities, rather than reinforcing their commercial presence in the United States. When we restrict our attention to the large chemical firms only (not shown in Table 12-13), we find that U.S. firms are somewhat more likely than European firms to make related acquisitions, after controlling for firm size and the scope of existing lines of business. Our interpretation of these results is that some large U.S. firms are indeed focusing on acquisitions in the area of commodity chemicals, while others are diversifying into specialty chemicals. An important aspect of the consolidation process in commodity chemicals is the search for larger size. This trend is found

TABLE 12-13. Number of "Related" and "Unrelated" Acquisitions*, U.S. and European Markets in the 1980s[†]

Nationality of Acquirer	All Target Regions		U.S. Target Companies		European Target	
	(1)	(2)	(1)	(2)	(1)	(2)
U.S.	465	836	356	708	74	79
	35.7%	64.3	33.5	66.5	48.4	51.6
W. Europe	350	736	93	177	214	498
	32.2	67.8	34.4	65.6	30.1	69.9
Japan	29	70	14	47	12	21
	29.3	70.7	23.0	77.1	36.4	63.6
Other	17	45	6	23	1	12
	27.4	72.6	20.7	79.3	7.7	92.3
Total	861	1687	469	955	301	610
	33.8	66.2	32.9	67.1	33.0	67.0

Notes:
(1), (2) = Related and unrelated, respectively.
*Related = SIC code of acquisition is one of the SIC codes of the acquiring company; unrelated = not so.
[†]Acquisitions in chemical sector (SIC 28) that occurred in the United States during 1984 to 1992 and in Europe during 1987 to 1992. (Only nonfinancial investors.)
Source: Our calculations from the *IDD Information Services Data Base.*

not just in commodity chemicals such as polyolefins and PVC but also in specialty products. The merger of Akzo and Nobel has created the largest paint company, and the second-largest producer of catalysts, in the world. Similarly, ICI acquired Glidden and the Grow group to become a leader in the paints and coatings business. In agrochemicals, Hoechst and Schering merged their operations into Hoechst Schering AgEvo, while Bayer and Monsanto have a joint venture for R&D and commercialization in pesticides. In catalysts, Cytec (formerly American Cyanamid) has a joint venture with Shell-Criterion Catalyst (*Chemical Week* 1994, p. 27).

In sum, through a series of divestitures, acquisitions, mergers, and alliances, firms are attempting to increase both the absolute size and the market share of their remaining businesses. It is not clear whether the search is for volume or for market share. In other words, the trends could be driven by some efficiency motives, such as the desire to spread fixed costs of process research, sales, and management over larger volumes (Cohen and Klepper 1992). Alternatively, these trends could reflect an attempt to reduce the stringency of price competition from products that often have different brands but are perceived to be close substitutes.

The growing globalization of the industry also played a role in initiating the restructuring that we have described. The key elements of firm restructuring — increasing focus on core businesses and the search for size in individual markets — are complementary to a third element: global operations. Globalization is complementary to the search for size. It provides a way of increasing the size of a given business through geographical expansion. European chemical firms, especially German ones, moved into the United States in the 1980s and made a number of acquisitions, as well. The economic liberalization in a number of countries in Asia and Latin America, and in Eastern Europe, and their emergence as major markets, attracted investment from the leading chemical companies in Japan and the West.[20] Apart from market expansion, firms have also used their international investments as a part of global sourcing. India and China have recently emerged as the leading sources of fine chemicals because of lower labor costs and less stringent environmental regulations.

Taken as a whole, chemicals are truly a global industry. As noted earlier, West European firms were the first to move overseas. U.S. firms have tended to be a bit slower, mainly because they already enjoyed a large domestic market. Japanese firms have tended to lag behind both. Indeed, in our dataset on physical investments, the median Japanese large firm invested in two countries, and the maximum number of countries in which a Japanese firm invested was eight. The corresponding figures for the United States are 4.5 and 29. For West European firms, the corresponding figures are 8 and 26. In other words, large European firms are the most globalized (least geographically concentrated) in their investments, followed by U.S., and then by Japanese firms.

Industry Maturity and the Market for Corporate Control

The recent method of entry of new firms into the chemical industry, such as that used by Huntsman and Aristech in the United States during the 1980s, is noteworthy. Since the 1920s, entry has typically consisted of diversification by existing firms rather than entry by new ones. However, the new entrants were typically buying assets that had become unprofitable to incumbents and for which market prospects looked dim. How could they expect to do better than firms that had extensive experience in the business?

We do not have a complete answer. One likely explanation is related to the need for clearer separation between different types of products. It may well be that the managements of many of the existing chemical companies were paying inadequate attention to their commodity chemicals, perhaps because the rapid growth and diversification activities of the 1950s and 1960s (Chandler 1994) had made it more difficult to manage each of the many individual product lines. Financial investors like Huntsman, or management buyouts like Vista, had greater incentives to control the actual performance of their businesses, and hence they expected better financial results than did large diversified firms.[21] Gordon Cain (1997) also points out that these buyouts were highly leveraged, and the new managements (operating with high-powered incentives such as generous stock options) drastically cut R&D and other overheads.

In this respect, the restructuring in the chemical industry points to an important insight: Optimal management structure differs according to the type of product. Commodity chemicals need a different management style from that required by higher value-added chemicals. One way, therefore, to interpret the drive for focus is that the management structures of existing chemical companies were better suited to the

---

[20]For instance, Dow has made significant investments in Leuna in Eastern Germany. DuPont has made a serious push into Asia, especially in its fibers business (Tyvek, lycra, nylon), in fiber intermediates such as adipic acid, adiponitrile, and in titanium dioxide (*Chemical Week*, 1994, p. 24). The result is that 10% of its chemical sales are now accounted for by the Asia-Pacific region. Courtaulds, which initially made investments in marine coatings in East Asia and then expanded on that base, now derives 14% of total sales from that region.

[21]Scherer (1993) makes a similar argument. However, Verlag (1993) argues that by the early 1980s an "excess supply" of commodity-chemical businesses had developed, and such businesses were available very cheaply. The excess supply, in turn, was caused by the expectation of continued high energy prices and the desire of chemical producers to move downstream. Verlag's argument turns on the assumption that buyers of commodity businesses had more optimistic expectations of oil prices.

production of higher value-added, technology-intensive products. In the United States, with its liquid and broad-based equities markets, buyouts and similar financial strategies were the means of separating the commodity and specialized chemicals and restructuring the industry. Pure financial acquisitions through management or leveraged buyouts are a relatively small share of the overall number of acquisitions that have taken place in the chemical industry: In the last 10 years they have been less than 6% of all acquisitions by U.S. firms and less than 3% of acquisitions by European firms. The influence that these deals have exercised, however, is far greater than is suggested by their number. Two prominent cases come to mind, in which the effort at a financial takeover was unsuccessful, but the prospective target firm responded to the threat by selling off a variety of businesses and becoming more focused. In the United States, the abortive takeover by GAF of Union Carbide forced Union Carbide to sell off its consumer goods businesses (such as EverReady batteries and Prestone antifreeze), as well as to spin off its industrial-gases Linde division as Praxair and focus on its polyolefins business. In Britain, Hanson's failed attempt to take over ICI is credited with forcing ICI to restructure by dividing the company into Zeneca, focused on the life-science-based and R&D-intensive sectors such as pharmaceuticals and pesticides, and ICI, focused on the more traditional chemical business. More recently, ICI has tried to further define its focus away from commodity chemicals by acquiring Unilever's specialty-chemicals businesses in a multi-billion dollar deal.

Such financial engineering is not an entirely new phenomenon, although it has become more intense in recent years. For instance, a wave of consolidations and takeovers had characterized the chemical sector in the 1920s, albeit on a smaller scale. But apart from scale, the two periods differ in another respect. In the interwar period, a number of new technologies were emerging, and the "winning" companies were those that invested in R&D and technology — companies like DuPont, Union Carbide, and IG Farben itself. In times of rapid technological progress, even if mergers and acquisitions take place, they are driven mostly by the search for "real" synergies in production, R&D, or commercialization (Chandler 1994; Hall 1994). By contrast, in more "stable" industries, profits are more sensitive to cost efficiencies, and mergers are often aimed at reducing costs. In sum, while the chemical industry of the 1920s was a rising high-tech sector, in the 1980s technological opportunities were declining, most firms cut R&D, and production and cost efficiency became more important.

Will the more advanced U.S. financial markets then stimulate U.S. firms to become more efficient and cause industry structure to differ from that in Europe or Japan?[22] This appears unlikely, for two reasons. First, European chemical industries are undergoing a restructuring similar to that in the United States, even without the threat of hostile takeovers. Second, the world financial markets are becoming more integrated, so that large companies have to raise and manage capital on the world market. Even the large German chemical companies are making preparations to be listed on the New York exchange. While this does not guarantee that European management will be quite as responsive to their stock prices as their American counterparts, they will surely become more responsive.

It is clear that as a technology-intensive industry such as chemicals evolves, and segments of that industry reach maturity, the problems that management must solve change. A young industry must focus on innovation; a mature industry must pay greater attention to cost reduction and efficiency. Whether the stock market is the most efficient

---

[22]The different patterns of the 1920s and 1980s, together with specific firm examples, are discussed in more detail in the chapters by Chandler, Hikino, and Mowery and by Marco Da Rin in this volume.

way to bring about the appropriate restructuring is a separate matter (see the chapter by Richards). However, the experience of the United States has shown that the stock market, along with a market for corporate control, is at least capable of bringing about the necessary changes.

CONCLUSIONS

Perhaps the most important reason to look at the overall structure of an industry is to emphasize that the fortunes of the industry as a whole are different from those of individual firms. Firms may rise and fall as markets expand and shift. In particular, individual firms often must struggle with the inertia of their histories, with their established patterns, and their local institutions. Markets force change. They are an important mechanism for the diffusion of technologies and for allowing new competitors to enter. Entrants may bring new technologies or new organizational capabilities, effectively forcing existing firms to either evolve or succumb. The study of evolving industrial structure takes place where the pressures of markets run into the abilities and constraints of existing firms.

One obvious example in our context is the rise of petrochemicals. The diffusion of petrochemicals in Europe was clearly associated with the growth of overseas investments by individual American firms. This, however, is only part of the story. The rise of petrochemicals was accompanied by the formation of two very important international markets — the market for the technology (populated by the specialized engineering firms and by the oil and chemical companies themselves), and markets for the new feedstock (following the discovery of oil in the Mid-East). These markets produced both the "compulsion" and the "option" to switch. The compulsion stemmed from the fact that as American firms became international, bringing along new petrochemical technologies that used much cheaper inputs, they exerted competitive pressures on local firms. The option stemmed from the opening of the two markets. Through licenses and engineering services, American firms (whether engineering companies or downstream producers) made the technology widely available. Moreover, because of the relatively large number of firms that could offer these services, the technology was sold at close-to-competitive prices.

It is surely not new to point out the remarkable power of international markets. However, it is important to emphasize that the creation of such markets is not an obvious "deterministic" outcome of industrial dynamics. Various factors often seek to prevent the formation of new markets. An obvious impediment is protectionism and other impediments to trade. But in addition, markets for technology or technological knowledge disembodied from capital or final goods are less common than the markets for "material" goods. Thus, firms that generate new technology can take advantage of it only by also using the technology themselves. A market for technology is absent because technological information is often hard to codify and articulate in ways that can be effectively protected by patents and easily transferred (Arora and Gambardella 1994a). But this also means that a market for technological information can be spurred by greater opportunities for codification and generalization of knowledge, and by better protection for intellectual property.

The chemical industry is an interesting example of the effects produced by the rise of a full-fledged market for technology. The development of chemical engineering as a discipline was a fundamental precondition for this knowledge market, since chemical engineering meant that the knowledge about chemical processes became systematized and organized in ways that focused on the general-purpose properties of the processes rather than the special features of the products. In short, chemical engineering made knowledge about chemical plant technologies a "commodity" that could be bought and sold. This prompted the entry of the specialized engineering firms, whose main task was to produce the technology, and who could enter into this business without having to bear the (much higher) fixed costs of being also producers of the downstream products. Not only did these firms diffuse technology internationally, through licensing and engineering services, but their existence pressured other companies, and most notably the large chemical producers, to consider licensing as well. One important implication of the diffusion of knowledge was that the *strategic* importance of process technology has diminished over time in many sectors of the chemical industry. Other factors such as access to cheap capital and raw materials, and proximity to customers have instead become more important for commercial success.

The rise of SEFs highlights another important point. International trade in products and foreign direct investments by multinational enterprises are often regarded as important mechanisms for the diffusion of technology, and this is certainly so in the chemical industry. But another mechanism has not been appreciated to the same extent: Technology transfer can also occur through the intermediation of an upstream sector. In other words, opportunities for vertical specialization in one country can be transmitted by international markets to other countries or regions that did not have the market conditions for such specialization on their own. For instance, American SEFs arose because the large size of the U.S. market allowed vertical specialization; but this also meant that when an international market for these services opened up, European countries and Japan could import the new technologies and related services from SEFs and other companies and share in the benefits of specialization. A similar process is at work today, as SEFs from the industrialized nations move toward the developing countries. Ultimately, the presence of an upstream sector selling technology at competitive prices means higher demand for investment by the local downstream firms than if these firms had to rely on their home markets (Arora, Fosfuri, and Gambardella 1996).

Our analysis has also highlighted some key events that shape the configuration of industries when they enter more mature stages of their life cycle. Small innovative firms, often specialized in certain functions or subsectors, become less important as the industry matures, and larger firms that are better suited to manage in a more stable environment play a larger role. Even within firms, the types of managerial problems change. The story of the recent restructuring in the chemical industry suggests that when certain segments of industries mature and technological innovation is no longer the driving force, management styles and structures have to change, too. This means that institutions for corporate governance must find a way to encourage or require firms to separate the mature segments from the newer ones, because the two require different management structures.

This discussion does not extend to the pharmaceutical industry, which is also a part of SIC28. It is very R&D intensive, and far from the maturity of the chemical industry. In a sense, it is a large specialty sector.

REFERENCES

Aftalion, F. (1989), *History of the International Chemical Industry*, University of Pennsylvania Press, Philadelphia, PA.

Arora, A. (1997), "Patent, Licensing and Market Structure in the Chemical Industry," in *Research Policy*, forthcoming.

Arora, A., A. Fosfuri, and A. Gambardella (1996), "Division of Labor and the Transmission of Growth," working paper, Center for Economic Policy Research, Stanford, CA.

Arora, A. and A. Gambardella (1994a), "The Changing Technology of Technological Change:General and Abstract Knowledge and the Division of Innovative Labour," *Research Policy*, 23:523–32.

Arora, A. and A. Gambardella (1994b), "Evaluating Technological Information and Utilizing It," *Journal of Economic Behavior and Organization*, 24:91–114.

Arora, A. and A. Gambardella (1997), "Domestic Markets and International Competitiveness," *Strategic Management Journal*, forthcoming.

Backman, J. (1964), *Competition in the Chemical Industry*, Manufacturing Chemists' Association, Washington, DC.

Bower, J. L. (1986), *When Markets Quake:The Management Challenge of Restructuring Industry*, Harvard Business School Press, Boston, MA.

Cain, G. (1997), *Everybody Wins*, Chemical Heritage Foundation, Philadelphia, PA.

Chandler, A. (1990), *Scale and Scope*, Harvard University Press, Cambridge, MA.

Chandler, A. (1994), "The Competitive Performance of U.S. Industrial Enterprises since the Second World War," *Business History Review*, 68:1–72.

Chapman, K. (1991) *The International Petrochemical Industry*, Basil Blackwell, Oxford, U.K.

Chemical Age Profile (1991), Pergamon Financial Data Services, London, U.K.

Chemical and Engineering News (24 July 1991):53–54.

Chemical Economics Handbook (CEH) (1991), SRI International, Menlo Park, CA, June.

Chemical Week (10 October 1994).

Cohen, W. and S. Klepper (1992), "The Anatomy of Industry R&D Sensitivity Distributions," *American Economic Review*, 82:773, 799.

Cohen, W. D. and D. Levinthal (1989), "Innovation and Learning:The Two Faces of R&D," *Economic Journal* 99:569–96.

Freeman, C. (1968), "Chemical Process Plant:Innovation and the World Market," *National Institute Economic Review*, 45 (August):29–51.

Freeman, C. (1982), *The Economics of Industrial Innovation*, Francis Pinter, London, U.K.

Grant, W. (1991), "The Overcapacity Crisis in the West European Petrochemicals Industry," in A. Martinelli (ed.), *International Markets and Global Firms*, Sage Publications, London, U.K.

Hall, B. (1994), "Corporate Restructuring and Investment Horizons in the United States 1976–1987," *Business History Review*, 68:110–43.

Hounshell, D. (1995), "The Dialectics of New Materials:On the Relationship between Natural Fibers, New Knowledge, and Synthetic Fibers, 1840–1960," in Gerald L. Geison (ed.), *The Dialectics of Materials*, Princeton University Press, Princeton, NJ.

Hounshell, D. A. and J. K. Smith (1988), *Science and Strategy:DuPont R&D 1902–1980*, Cambridge University Press, Cambridge, U.K.

Hydrocarbon Processing Unit (1991), HPI Data Base, Gulf Publishing, Houston, Texas.

IDD Information Services Data Base, online service from Lexus-Nexus.

Landau, R. (1966), *The Chemical Plant. From Process Selection to Commercial Operation*, Reinhold, New York, NY.

Landau, R. and D. Brown (1965), "Making Research Pay," *AIChE-I. Chem. E. Symposium Series* No.7, London Institute of Chemical Engineers, London, U.K., pp. 35–43.

Landau, R. and N. Rosenberg (1992), "Successful Commercialization in the Chemical Process industries," in N. Rosenberg, R. Landau, and D. Mowery (eds.), *Technology and the Wealth of Nations*, Stanford University Press, Stanford, CA.

Lane, S. (1993), "Corporate Restructuring in the Chemicals Industry," in M. Blair (ed.), *The Deal Decade*, The Brookings Institution, Washington, DC.

Lieberman, M. (1987), "Patents, Learning by Doing, and Market Structure in the Chemical Processing Industries," *International Journal of Industrial Organization*, 5:257–76.

Lieberman, M. (1989), "The Learning Curve, Technological Barriers to Entry, and Competitive Survival in the Chemical Processing Industries," *Strategic Management Journal*, 10:431–47.

Mansfield, E. (1977), *The Production and Application of New Industrial Technology*, Norton, New York, NY.

Mark, H. (1984), "The Development of Plastics," *American Scientist*, 72:156–62.

Martinelli, A., ed. (1991), *International Markets and Global Firms*, Sage Publications, London, U.K.

Molony, B. (1990), *Technology and Investment:The Prewar Japanese Chemical Industry*, Harvard University Press, Cambridge, MA.

Morris, Peter J. T. (1994), "Synthetic Rubber:Autarky and War," in S. T. I. Mossman and P. J. T. Morris (eds.), *The Development of Plastics*, Royal Society of Chemistry, Cambridge, U.K.

Nelson, R. and G. Wright (1992), "The Rise and Fall of American Technological Leadership:The Postwar Era in Historical Perspective," *Journal of Economic Literature*, XXX:1931–64.

Popoff, Frank (1995), personal interview.

Predicasts 1991, Company Thesaurus, Predicasts, Cleveland, OH.

Rosenberg, N. (1990), "Why Do Firms Do Basic Research?" *Research Policy*, 19:165–74.

Scherer, M. (1993), comment to S. Lane, "Corporate Restructuring in the Chemical Industry," in M. Blair (ed.), *The Deal Decade*, The Brookings Institution, Washington, DC.

Spitz, P. H. (1988), *Petrochemicals:The Rise of an Industry*, Wiley, New York, NY.

Stigler, G. (1951), "The Division of Labor is Limited by the Extent of the Market," *Journal of Political Economy*, 59(3):190–201.

Stokes, R. G. (1994), *Opting for Oil: The Political Economy of Technological Change in the West German Chemical Industry, 1945–1961.* New York, Cambridge University Press.

Sutton, J. (1991), *Sunk Costs and Market Structure*, MIT Press, Cambridge, MA.

Verlag, P. K. (1993), comment to S. Lane, "Corporate Restructuring in the Chemical Industry," in M. Blair (ed.), *The Deal Decade*, The Brookings Institutions, Washington, DC.

Wright, G. (1990), "The Origins of American Industrial Success," *American Economic Review*, 80(4): 651–68.

Young, A. (1928), "Increasing Returns and Economic Progress," *Economic Journal*, 38(152):527–42.

13 The Evolution of Corporate Capabilities and Corporate Strategy and Structure within the World's Largest Chemical Firms: The Twentieth Century in Perspective

ALFRED D. CHANDLER, Jr., TAKASHI HIKINO, and
DAVID C. MOWERY

This chapter surveys the evolution of corporate strategy and structure in the global chemicals industry during the 20th century, focusing on the largest chemical firms in the United States (DuPont, Dow, Monsanto, Union Carbide, and Exxon Chemicals), Germany (Hoechst, BASF, and Bayer), Great Britain (Imperial Chemical Industries), and Japan (Mitsubishi Chemical, Mitsui Chemicals, Sumitomo Chemical, Showa Denko, and Ube Industries). These firms have been remarkably durable; most of them (or their constituent elements) predate this century. In fact, even the specific corporate capabilities that have underpinned these firms have tended to persist over time. For example, U.S. chemical firms have repeatedly tried to enter into the pharmaceuticals industry, with limited success, while two of the leading German chemicals producers have been major pharmaceuticals producers since before the turn of the century.

The development of these firm-specific technological capabilities will be a central theme of our discussion, which will allow us to explore many interesting dimensions of corporate capabilities. For example, firms have, at different times, acquired technological capabilities by internal research and development, mergers with other firms, purchases of licenses, and purchases of plants from specialized engineering firms (SEFs). Public policies have influenced corporate evolution, through channels as varied as governments' posture toward mergers or competition, and the degree of support for academic fields such as chemical engineering. The enormous size of the internal U.S. market, even early in the 20th century, helped make U.S. companies the leaders in large-scale production. World wars stimulated both innovation and production in the chemical industry and to impair the strengths of Germany's chemical industry. Differences have affected the industry: In the 19th century, Germany and the United Kingdom were drawn to coal as the main feedstock for the chemical industry, partly because of their extensive natural reserves; while the U.S. chemical industry led the shift to using petroleum as a feedstock in the 1930s, and 1940s, partly because of its comparatively abundant oil reserves. However, these national differences in resource endowments, which had been significant in the interwar period of restricted international trade flows, lost much of their influence in the 1950s and 1960s, with the revival of trade and investment flows.

Although technical capabilities have proven critical to maintaining corporate competitive advantage, they have always had to be reinforced by capabilities in other areas, including marketing, the acquisition of raw and semi-finished materials, internal accounting systems, and access to external sources of funds. Management skills across the different operating levels, from the plant to the functional department to the product division to the corporation as a whole, also were essential. The managerial function involved coordination, monitoring of operations at each level, and, most important of all, the planning and allocation of corporate resources — both human and physical — for the continued health and growth of the overall enterprise.

The sections that follow focus on corporate strategy among major firms in the United States, Germany, Britain, and Japan. Within each country, the discussion is broadly divided into four historical periods of unequal length: 1890 to 1914, 1914 to 1945, 1945 to 1970, and 1970 to 1990.

From 1890 to 1914, the chemical industries of the United States, Germany, and Great Britain differed substantially in their product mix, process technologies, and patterns of international trade and investment. German firms dominated the production of high-value dyestuffs that were based on coal-tar feedstocks and, by the end of the period, had begun to exploit high-pressure process technologies. In Great Britain, by contrast, the firms that had pioneered in dyestuffs had lost substantial global market share to firms from Germany and Switzerland, and the leading chemical firms were focused on bulk chemicals oriented toward agricultural and textile-industry markets, as well as on explosives. The U.S. chemical industry focused on production of agricultural bulk chemicals, inorganic chemicals based on electrolytic processes, and explosives in a domestic market protected by high tariffs. The Japanese chemicals industry began to emerge during this period mainly as a producer of fertilizers. The leading firms in the United States, Great Britain, and Germany all had begun the organized exploitation of science through industrial research laboratories.

Two world wars and the Great Depression caused a severe disruption in international trade and investment flows during the 1914–1945 period. However, government-led mobilization of the chemicals industry in wartime also led to significant technical advances and considerable diffusion of technology.

The great postwar boom of 1945–1970 sparked an enormous upsurge in investment in production capacity throughout the chemical industry. Such investment was further accelerated by the widespread adoption or expansion of petroleum-based processes for chemicals production and by the substitution of chemical-based materials for wood, glass, metals, and natural fibers. Competition among the large non-Japanese chemical firms became truly global in scope during this period. The growing integration of markets and broader availability of process technologies that characterized the postwar industry increased interfirm collaboration in new plant investment, technology licensing, and other tactics. Toward the end of the period, however, the substitution of chemical-based products and materials for other products had run its course, and the overall demand for commodity chemicals began to slow. More and more firms entered the chemical industry, and overcapacity became a serious drag on the profits of the long-established leaders. As markets matured, leading firms began to look to other markets and unrelated diversification to support continued growth, especially in the United States.

Finally, the post-1970 period was one of great turbulence in the chemical industry. In the 1970s, inflation, recession, higher oil prices, and continuing competitive pressure led to severe deterioration in the performance of many firms. The efforts of many firms to raise profits through diversification or vertical integration largely failed to restore the profit levels of the earlier postwar era. These disappointing results led to a restructuring of the chemical firms of Germany, Great Britain, and the United States. Although the methods for this restructuring differed in each country, reflecting differences in systems of corporate governance, the strategic goals of the restructuring were remarkably similar: The leading U.S., German, and British chemical firms sought to reduce their exposure to mature "commodity" chemical products and to enter (whether via acquisition or internal development) the production of higher-margin specialty chemicals and pharmaceuticals, while retaining older lines in which inherited company-specific capabilities gave them competitive strength.

THE UNITED STATES: DUPONT, DOW, UNION CARBIDE, MONSANTO, AND EXXON CHEMICAL

Four of the five American chemical companies whose experiences are reviewed here — DuPont, Dow, Union Carbide, and Monsanto — have, since the 1960s, been the nation's largest in terms of sales. The fifth, Exxon Chemical, has more recently been the oil company with the largest sales of petrochemicals. The four chemical companies were established within five years of each other at the turn of the century: Dow in 1897, Union Carbide in 1898 (although it took a 1917 merger to create the present-day enterprise), DuPont in its modern form in 1902, and Monsanto in 1903. Exxon Chemical was created in 1963 as a wholly owned subsidiary of the Standard Oil Company (New Jersey), the successor to one of the nation's first giant industrial enterprises, John D. Rockefeller's Standard Oil Trust, formed in 1881. The four chemical companies were established in different parts of the nation: Michigan, New York, Missouri, and Delaware. Each produced a different line of chemical products. These firms each developed specific corporate capabilities, and the capabilities developed in one period became the basis for the capabilities in the next period. During the four major time periods under discussion, however, these companies faced a common external environment, and each in its own way followed similar broad patterns of growth and adjustment.

Creating Initial Capabilities: The Period to World War I

From their beginnings until World War I, four of these five chemical firms created their initial technological, functional, and managerial capabilities (the predecessor of Exxon Chemical was founded in the 1930s).

Between 1902 and 1904, three duPont cousins created the base from which the learned capabilities of today's E. I. duPont de Nemours and Company evolved. By merging their enterprise with many other small family firms, they completely reorganized the American explosives industry. They concentrated 70% of the nation's production into a small number of large plants carefully located in terms of suppliers and markets, built a national distribution and marketing organization and a central purchasing department, and established a research and development group to focus on improving product and process. To assure its plants of a steady flow of supplies, the company built its own glycerin plants and purchased and operated nitrate beds in Chile. Its output grew with the American market for explosives in mining and railroad and urban construction.[1]

Dow evolved quite differently from DuPont. Dow began as an entrepreneurial start-up and grew through the development and continuous exploitation of its internal corporate capabilities rather than through acquisition of external resources. It focused more on process technology than on basic product research and concentrated on the production of feedstocks and commodity chemicals more than on end products. In 1897, after five years of perfecting an electrolytic process for extracting bromide and chlorine from brine, Herbert Dow formed the Dow Chemical Company in Midland, Michigan, to produce from these two ingredients bleaching power and caustic soda. In 1900 came the production of sulfur chloride, which in turn was the base for commercializing, in 1903, a synthetic chloroform. Out of that commercialization process came carbon tetrachloride, which became the base for insecticides and agricultural chemicals. The production of sodium chloride was used, in turn, to produce magnesium chloride and then calcium chloride (Haynes 1949, pp. 113–33; Trescott 1981, pp. 94–95; Whitehead 1968, pp. 97–102, 120–22, 191–92).

The initial learning base of Union Carbide and Carbon Corporation differed from those of both Dow and DuPont. The company was neither a start-up that built and expanded capabilities, like Dow, nor was it formed from horizontal merger that consolidated a number of established enterprises in a single industry, like DuPont. Instead, it was a merger, in 1917, of six U.S. leaders in different complementary electrochemical product lines, including electrolytically produced calcium carbide and acetylene in plants at Niagara Falls and Sault Ste Marie, Michigan (Union Carbide); carbon electrodes for lighting systems and EverReady batteries (National Carbon); bicycle and auto headlights and welding and cutting equipment (Prest-O-Lite); liquid oxygen used in the production of acetylene (Linde Air Products); and electrolytically produced metal alloys (Electro-Metallurgical Company).[2] Many of these products and processes were based on European, particularly German, technologies. On this base, the company quickly became the nation's leader in several broad product lines, including industrial gases, metal alloys, and such consumer products as batteries and antifreeze.

In its early years, Monsanto was an entrepreneurial start-up similar to Dow. Established in 1901 in St. Louis by John F. Queeney, it produced saccharine and then caffeine, vanilla, and other fine chemicals. Unlike Dow, however, Queeney relied on German processors for basic chemicals and its equipment, which meant the company would be heavily affected by World War I.

---

[1] Information on DuPont before World War II comes from Chandler and Salsbury (1971) and is reviewed in Chandler (1962, 1977, 1990). Dutton (1942) is also a useful source for this period. For the time since World War II, Hounshell and Smith (1988) and Taylor and Sudnick (1984) have valuable information.

[2] Strictly speaking, the birth of Union Carbide at the tail end of World War I would mean that it should be discussed in the next section, in the time period after World War I. But it is useful to compare and contrast the origins of all four chemical companies in this section.

Expanding Corporate Capabilities: From World War I to World War II

The outbreak of World War I caused shortages of chemicals, particularly in intermediate products, resulting from the removal of European producers, particularly the Germans, from world markets. These shortages, as well as the demands for explosives and other wartime materials, caused firms to expand in existing product lines and then to move into new product areas. Their development of new products benefited in several cases from the U.S. government's expropriation of German chemical firms' patents in 1918.

DuPont, for example, swiftly built new facilities for the production of propellants and military explosives after the outbreak of World War I. Production rose from 8.4 million pounds of propellants in 1914 to 455 million in April 1917, and the number of employees and salaried workers and the amount of capital expenditures grew proportionately.[3] Dow turned to the production of phenol, and other chemicals for military explosives. Although these plants were shut down at the war's end, they created a new learning base in organic chemistry that led to the development of chlorobenzyl phenol for biocides and agricultural chemicals. During the war, Dow also experimented in producing magnesium from brine, through an electrolytic production technology comparable to that used in the production of chlorine, one that came on stream in 1919. As late as 1920, however, Dow was still a relatively small company by world standards. Monsanto, cut off from its German suppliers during World War I, had to develop its own intermediates. It did so largely through the acquisition of a neighboring producer of heavy chemicals, including phenol, chloride, and caustic soda. It also made bulk aspirin based on Bayer's expropriated patents.

By war's end, the American chemical firms had begun to commercialize technologies that the Europeans had dominated, particularly those in organic chemistry and high-pressure process technology. At the same time, they expanded the potentials of their prewar product lines. During the interwar years, they added to their product portfolios through internal research and development and through acquisitions.

When DuPont's Board of Directors began planning for the postwar world in 1917, their initial strategic insight was that no existing industry could fully absorb their huge product-specific wartime capacity.[4] On the other hand, several new product lines could be based on the company's technological, functional, and managerial capabilities. These were industries whose production was based on nitrocellulose technology, including paints and varnishes, pyroxylin (celluloid), artificial leather (Fabricoid), and rayon. In addition, the company could enhance its capabilities in product lines whose development it had embarked on to meet wartime shortages, for instance, dyes and synthetic nitrates. To implement this plan, in 1921 the senior managers set up a new management structure that placed the operations of each of these industries into autonomous departments whose managers and staff were responsible for each department's production, marketing, research, and profit and loss.

During the 1920s, DuPont invested in both internal capabilities and acquisitions. In the product lines based on nitrocellulose technology, the paint and finishes department developed fast-drying Duco enamel to meet the demands of the swiftly growing automobile industry. In pigments, it made the move from lithophone to titanium dioxide. In the production of rayon and closely related cellophane and film, commercialization was advanced by joint ventures with French pioneers. Another joint venture permitted the company to use the Claude process for high-pressure production of synthetic nitrates and ammonia, a project that had begun during wartime shortages. By

---

[3]DuPont's World War I experience is summarized, with specific sources cited, in Chandler (1990, pp. 175–77).

[4]For more extended discussion of DuPont during the interwar years, see Chandler (1990) and Haynes (1949, pp. 133–37).

1929, DuPont accounted for 40% of the synthetic ammonia and 30% of the synthetic nitrates produced in the United States, but the Ammonia Department only became profitable in the early 1930s (Chandler 1990, pp. 184–85; Hounshell and Smith 1988, pp. 185–88). In dyes, DuPont had even greater difficulty in competing with a revived German industry than in ammonia. The acquisition of Grasselli Chemical to provide heavy chemicals used in the production of its end products also enhanced DuPont's strength in pigments, particularly titanium dioxide. In 1930, the acquisition of Roessler & Haaslacher, the U.S. subsidiary of a pre-World War I German leader in inorganic chemicals, DEGUSSA, gave DuPont an entry into electrochemicals and a base for the production of herbicides and fungicides.

During the 1930s, new products emerged from each of these technological learning bases, and innovations also began to come from DuPont's central research and development department. For example, research and commercialization of polymer-based products brought the first nylon and neoprene products onstream in 1939, through an impressive combination of internal research and large subsequent investments in development and commercialization.

Dow continued to focus on internal expansion through the 1920s. By 1930, it was the nation's leading producer of calcium chloride, magnesium chloride, and related products and had acquired a near monopoly in the production of magnesium. During the later 1930s, process development in magnesium products led to the commercial introduction of a new silicone which led to the formation of a joint venture with Corning Glass in 1942 to produce silicones and silicon-related products.

During the 1930s, Dow also began to move in the direction of petrochemicals and polymers. In contrast to DuPont, where new products came more from scientific research, at Dow innovations such came much more from process (chemical) engineering. Dow's initial entry into petrochemicals began with the development of chemicals for increasing the yields of gas and oil fields, leading to the formation in 1932 of a subsidiary, Dowell, to market these products. Of longer term importance was the formation, in 1933, of a 50/50 joint venture with the Ethyl Corporation, producer of the recently developed "no knock" gasoline, to extract bromide from sea water in a plant on the North Carolina coast. That venture, in turn, became the learning platform for the construction of a much larger works at Freeport, Texas, to produce magnesium and then petrochemicals, which came on stream in January 1941.[5] Such process development encouraged the production of polymers, which could be produced most efficiently from oil and natural gas. As early as 1932, experimentation to improve benzene led to the commercial introduction of a monomer, liquid styrene. Then in 1937, experiments "in controlling the styrene production process with precise accuracy" led to the commercialization of the versatile end product styron, of which Styrofoam is only one commercial product. In the meantime, investigations in chloride chemistry brought forth vinylidene chloride as another versatile end product, saran, that could be extruded into pipes and tubing, injected into molded parts, made into sheets of varying thickness (as in the household Saran Wrap), or woven into fabric. In the same way, work on ethyl cellulose created Etocal, a plastic with foil and coating uses.

From its start in 1917, Union Carbide had been a company that brought together several learning bases within a single enterprise. Although their activities supplemented one another, each of the firm's product lines concentrated independently on improving existing products and developing new ones.[6] Union Carbide took the step, in 1920, of

---

[5]For greater discussion of Dow in the 1930s, see Whitehead (1968, pp. 144–46, 193–95, 227), Haynes (1949, pp. 120–23), and Spitz (1988, pp. 89–96).

[6]For a summary of Union Carbide's history to the 1930s, see Chandler (1990, pp. 103, 173–75, 180–81) and Haynes (1949, pp. 429–38).

forming a new division called the Carbide and Carbon Chemical Company, which initially pioneered in producing propane from natural gas. In the late 1920s, by setting up plants next to Standard of Indiana's largest refinery at Whiting and another at Texas City, Texas, that division became the most focused pioneer in petrochemicals. In the 1930s, the division led the way in the development of such basic chemicals as butadiene and ethylene (from grain alcohol) and polymers including polyvinyl chlorine (PVC), vinyl chloride monomers, ethyl vinyl chloride, and polystyrene. By 1936, Union Carbide had begun production of Vinylite for flooring, phonograph records and fabric coatings, and then in 1939 a synthetic fabric, Vinyon. That same year it purchased the Bakelite Company which produced Formica and other molded formaldehyde-based plastic products.

Monsanto differed from Dow, and even DuPont, in that during the 1920s and 1930s its growth rested more on acquisitions and vertical integration than on internal investment. For example 1929 saw the acquisition of three companies: one in Akron, Ohio, producing rubber chemicals; one in Newark, Ohio, making intermediates for food and perfume products; and the third in Woburn, Massachusetts, whose products paralleled its own and provided an eastern production base. During the 1930s, Monsanto continued to grow through acquisitions, entering plastics through the purchase of Fiberloid in 1938 and two smaller units that produced cellulose acetates and nitrates, vinyls, and polystyrene. Another acquisition included firms producing phosphates and fertilizers.

Boom Years and Petrochemicals: World War II to the 1970s

The exploitation of polymer petrochemicals, which was just getting underway in the 1930s, took off with the wartime crash programs that used petroleum-based processes to produce huge volumes of synthetic rubber and high-octane gasoline. Petrochemical technology transformed the chemical industry. The early postwar years brought an unprecedented flow of new products and processes, including a cornucopia of synthetic fibers, plastics, and other materials that opened up huge new markets. During the postwar boom, annual sales of chemicals rose 2.5 times as fast as GNP; but by the 1960s, output capacity had increased sharply, new product development had leveled off, and firms entered from other industries, particularly oil. Chemical firms began to search for new markets, often diversifying into product lines in which their existing capabilities gave them little competitive strength.

DuPont was not directly involved with the massive government-funded synthetic-rubber and high-octane-gasoline programs that drove the petro/polymer revolution, so that World War II itself had less of an impact on DuPont than it had on the other companies described here. Its major government program was the Manhattan Project, but building the nuclear bomb did not develop large-scale chemical capabilities.[7]

After the war, DuPont continued to focus more on the commercialization of new products rather than on new and improved processes. In fact, it dropped its prewar policy of using acquisitions to move into new product lines and decided instead to rely on its own basic research, as well as product and process development. This new strategy reflected, in part, the strong antitrust thrust by the government in the immediate post-war years.

In the 1950s, the engine of DuPont's growth became synthetic fibers, an area in which the firm enjoyed powerful first-mover advantages. In the 1950s, the fiber

---

[7]For DuPont's wartime developments and postwar plans, see Hounshell and Smith (1988, ch. 6) and Haynes (1949, pp. 134–37).

department used its nylon capabilities to bring forth Dacron and Orlon, which gave DuPont first-mover advantages in the new wash-and-wear clothing market (Hounshell and Smith 1988, pp. 394–407, 407–20). In 1960, the company produced 2.5 billion pounds of nylon and 1.9 billion pounds each of Dacron and Orlon. These new products led DuPont to phase out its own rayon products, but this was intentional: A policy of replacing (cannibalizing) its old products with new, higher-value ones was a keystone of DuPont's postwar competitive strategy.

In plastics, DuPont developed a much broader line of polymer-based products. These included high-performance plastics that replaced metals in machinery, motor vehicles, and appliances. Among the most successful were Deralin, termed "synthetic stone"; Teflon, initially used in the insulation of the wiring of jet engines, space missiles, and computer cables, and later for cookware and clothing; and Mylar, a polyester film (discovered in the development of Dacron) whose resistance, insulating properties, and strength made it a base for photographic film, magnetic tapes, capacitor dialectics, and packaging products (Hounshell and Smith 1988, pp. 482–91). Along with other new polymer films, Mylar replaced cellophane in DuPont's product line.

In addition to exploiting the potentials of polymer chemistry, DuPont expanded its lines based on organizational capabilities created before 1940 — in chemicals, including ammonia and methyl alcohol; in paints, finishes, and nonporous fabrics; as well as in biological chemicals, insecticides, and herbicides. Its innovative weed-control products permitted it to capture 20% of the U.S. herbicide market by 1960 (Hounshell and Smith 1988, pp. 451–64). By the early 1990s, DuPont had become the second-largest and most profitable producer of herbicides in the world.

Nevertheless, by the late 1950s the swift growth of the polymer markets was leveling off, competition had intensified, and the number of products commercialized began to decline. DuPont responded in 1959 with two strategic moves. One was to concentrate more on overseas markets; by 1963, foreign sales accounted for 18% of total sales, as compared to 5% in 1939. The second was to search for new product lines for markets not yet reached by operating departments (Hounshell and Smith 1988, p. 533; Taylor and Sudnick 1984, pp. 187–94). A "New Ventures Program" was created. But its projects — including development of an office copier, scientific instruments, and mass-produced home-construction materials (roofing, sidings, trims, windows, doors, and interior furnishings) — were widely recognized by 1969 to have been unsuccessful. Thus, DuPont was one of the first U.S. industrial firms to become aware of the ways in which a firm's strategic options were limited by the potential of its existing organizational capabilities. (Hounsell and Smith 1988, ch. 22; Burgelman and Sayles 1986, ch. 1, 8, 9; Fast 1979).

During the 1960s, DuPont's operating divisions continued to develop a stream of new, profitable products. In fabrics, Lycra spandex fiber, an extremely strong and elastic fiber, enjoyed a return on investment of over 30% by 1972. It was followed by Kevlar fabric, which had four times the strength of steel, and then by new paper-like fabrics including Tyvek and Typar. In plastics, Zytel replaced Mylar. In photographic products, a new coloring process produced Cromaline, which became the market leader in its niche. Other departments enjoyed comparable success (Hounshell and Smith 1988, pp. 424–39, 480, 488, 499, 548, 582). Until the mid-1960s, DuPont's profit margin was twice that of other chemical companies.

In contrast to DuPont, which was less affected by World War II, Dow used the capabilities developed in the 1930s to become the leader in the massive wartime crash programs for high-octane gasoline and synthetic rubber. It produced unprecedented amounts of magnesium for the huge expansion of aircraft output. Its giant plants at Freeport and Velasco on the Gulf Coast produced from brine, oil, and natural gas massive amounts of styrene, polystyrene, ethylene, ethylene dibromide, ethylene glycol, vinylidene chloride, and magnesium chloride. Dow also built and operated a synthetic-rubber plant at Sarnia, Canada (Whitehead 1968, chs. 12–14; Haynes 1949, pp. 122–24). As the result, Dow's assets, which had been $21.5 millon in 1930, rose to $271 million in 1948, making it the country's fourth-largest chemical company behind DuPont, Union Carbide, and Allied Chemical (Chandler 1990, pp. 653, 660).

As Bower (1986, p. 105) has pointed out, Dow's postwar strategy of expansion was well defined and successful. The company's goal was "to be the world's largest and most profitable producer of commodity chemicals and plastics based on low-cost leadership and aggressive marketing in the businesses in which it competed. Low cost, in turn, was achieved by vertical integration, technical excellence, near or maximum scale, and leveraged financing, all on a worldwide basis." Thus, in the 1950s Dow enlarged the Freeport and Velasco plants and in 1958, built a still-larger plant near Baton Rouge, Louisiana, using the new "single train technology," which had much lower per-unit costs. To assure a stable flow of oil and natural gas into these works, Dow integrated backward, purchasing 265,000 square miles of oil and gas holdings. This commitment to vertical integration, so critical in these scale-dependent process technologies, has remained a Dow hallmark into the 1990s. (Whitehead 1968, pp. 230–37; Spitz 1988, pp. 398–99; Moskowitz et al. 1980, pp. 601–603).

Dow's overseas growth was carried out by direct investment and by joint ventures with foreign firms, but rarely by acquisition. Particularly innovative at the time was the use of debt in financing the swift expansion.[8] The joint ventures of the 1950s included the formation of Asahi-Dow in Japan in 1954, a comparable venture with Britain's Distillers (Dow did purchase 50% of Distillers in 1968), ventures with Germany's BASF in 1958 and with France's Pechiney in 1959, and also a venture between France's Schlumberger and Dowell. At the same time, Dow made large direct investments in plants and in distribution, marketing, and development facilities in the Netherlands, Greece, Italy, Brazil, South Korea, and Yugoslavia. To provide a European financial clearing house for its transactions, Dow obtained a 40% interest in a Dutch bank. In 1965, it formed the Dow Banking Corporation in Zurich, Switzerland, where it had established its overseas headquarters (Whitehead 1968, pp. 248–51; Spitz 1988, pp. 356–57; Moskowitz et al. 1980, p. 602).

During the 1960s and into the 1970s, Dow continued to concentrate on its basic strategy of growth, so that its efforts to expand its portfolio of end products were minimal. However, it did make an initial move into pharmaceuticals with the acquisition of Allied Laboratories in December 1960, and subsequently acquired two small, but international, companies that produced vaccines.

Union Carbide did not start producing substantial amounts of basic petrochemicals and polymers until World War II; but by the immediate post-World War II years, Union Carbide was the nation's largest producer of petrochemicals, and a particular leader in the production of butadiene, styrene, ethylene, and then polyethylene. It was also a major player in the production of industrial gases and alloys, as well as of finished plastics and plastic consumer products (Spitz 1988, pp. 148, 251–53, 308; Haynes 1949, p. 436).

[8]In Ralph Landau's interview with Frank D. Popoff, Chairman, (Landau 1995b), Popoff (p. 1) recalled that "our financing has been a shameless address to debt financing for as long back as we can remember." For Popoff's comments on the commitment to vertical integration, see pp. 4–5.

Possibly because of this diversity of products, Union Carbide lacked focus and began to fall behind DuPont and Dow during the years of the most dramatic growth of polymer/petrochemical products. As Union Carbide enlarged its activities in engineering plastics, pesticides, and agricultural chemicals, it did commercialize the new acrylic fiber Dynel, the packaging film Glad Wrap, and the car polish Simonize. But the firm soon faced difficulties on all fronts. It was unable to compete with DuPont in fibers and high-performance plastics. It was losing market share in basic chemicals and polymers to Dow's large-scale investments. Competition was increasing rapidly in its alloys and industrial-gases businesses (Bower 1986, pp. 101–102).

In response, Union Carbide embarked on a search for new ventures in the 1960s, but it did so in a very unsystematic manner. It purchased a Canadian mattress bedding company in 1964, a pharmaceutical enterprise in 1965, and a maker of lasers in 1966. It expanded its Nuclear Division, created to operate the U.S. government's facilities at Oak Ridge, by investing in uranium-mining and -processing equipment. Few of these ventures either grew or could nurture the organizational capabilities of Carbide's several operating divisions. Most of these businesses were sold off in 1969. Then came attempts to expand into unrelated consumer businesses, with investments in deodorants, motor-oil additives, and disposable diapers. These, too, soon met the same fate (Moskowitz et al. 1980).

Monsanto also participated in the World War II synthetic-rubber programs, although to a much smaller extent than Dow, based on its rubber chemical operations in Akron and other acquisitions it had made in the 1930s. By the end of the war, it had become a major provider of styrene and polystyrene (Forrestal 1977; Haynes 1949, pp. 282–87; Moskowitz et al. 1980; Spitz 1988, pp. 109, 251–56). Monsanto had also become the nation's sixth-largest industrial chemical company — behind the three just reviewed, along with Allied Chemical and American Cyanamid (Chandler 1990, pp. 652–53).

In the early boom years of the polymer/petrochemical revolution, Monsanto, like Dow, concentrated on basic polymers. It enlarged its production of polystyrene, invested in polyvinylchloride plants, and began large-scale production of polypropylene. To assure itself of supplies, it integrated backward into oil and gas, through the acquisition of the Lion Oil Company. It expanded abroad, albeit less aggressively than Dow, by entering into joint ventures with firms in France, Belgium, and Spain (Spitz 1988, pp. 245, 256, 355–56, 400; *Moody's Industrial Manual*, various years).

Monsanto moved more slowly into higher value-added polymer products, beginning with the formation of Chemstrand in 1949, a 50/50 joint venture with American Viscose (a leading rayon producer) to make acrylic fibers. The joint enterprise was unsuccessful until 1951, when DuPont, under antitrust pressure, not only licensed its techniques to produce nylon but had its engineers assist in the design, building, and start-up of a large plant, so that it could assure the government that it had a strong nylon competitor. In 1961, Monsanto acquired its partner's half of this activity. At the same time it enlarged its agricultural business by moving into herbicides and phosphates. It made its first move into consumer products by commercializing a low-sudsing detergent, All. A commercial failure, All was sold off to Lever Brothers in 1957 (Spitz 1988, pp. 289–90; Taylor and Sudnick 1984, pp. 179–80; Moskowitz et al. 1980, pp. 610–14).

As competition intensified in the 1960s, Monsanto, like DuPont, set up a New Ventures Division. That division made exploratory investments in many areas — graphic systems, protein foods, educational toys, engineered composite systems, and electronic chemicals — but only an investment in the production of instruments used in chemical operations proved successful. Monsanto then expanded into this area by purchasing the Fisher Governor Company, expanding its facilities, and turning it into Fisher International, a worldwide producer of valves, regulators, and other process controllers (Forrestal 1977, pp. 187–91; Moskowitz et al. 1980). But the singularity of this success by the late 1960s had taught top management at Monsanto the same lesson learned at DuPont: To be profitable, new product lines must rest firmly on the learned capabilities of the operating divisions.

By the mid-1980s, Exxon Chemical was the third-largest chemical producer in the country, and six of the fifteen largest U.S. producers of chemicals in terms of sales were oil companies (with Atlantic Richfield ranked fourth, Shell eighth, Amoco ninth, Mobil thirteenth, and Chevron fifteenth). However, from the vantage point of the mid-1940s, chemicals did not look like a promising business to most of the petroleum industry. As Haynes (1949, p. 211) observed:

> "Production of chemicals by the petroleum industry appeared to be economically and technically sound, but most petroleum executives could not see what appeared to them to be a tiny market for a multitude of chemicals produced by a complexity of operations and sold to a long and diversified list of customers, tasks for which they neither had the technical nor the sales staffs."

Nonetheless, four U.S. oil companies had begun to move into petrochemicals before 1940. Shell and Standard of California (Chevron) produced by-products and fertilizers supplied by their California refineries for local markets, while Phillips Petroleum and Jersey Standard began systematic research into gasoline-production technologies during the 1920s. Jersey Standard was the name commonly used to refer to the Standard Oil Company (New Jersey), which in 1973 would change its name to Exxon.

As early as 1927, a Jersey research group worked with Germany's IG Farben to develop a synthetic gasoline, which allowed the company to develop capabilities based on German technology to produce butadiene and toluene in volume. By 1937, Jersey Standard had begun to produce high-octane gasoline by polymerization of natural gas, using a technology commercialized by Phillips Petroleum Company. In 1939, it began large-scale production of ethylene. These capabilities ensured the company of having a major role in the government's crash synthetic rubber and high-octane gasoline during World War II and a strong position in providing basic and intermediate chemicals for the swiftly growing, postwar polymer-based industry. Indeed, by 1950 Jersey Standard was already the nation's twelfth-largest chemical company in terms of sales.[9]

As early as 1943, senior managers at Jersey Standard urged the company to define an explicit strategy of diversification into petrochemicals (Wall 1988, ch. 5). A set of 1950 recommendations, for instance, stressed that Jersey's 15.8% return on investment in petrochemicals was already much greater than that in gasoline and other petroleum products. Indeed, it was greater than the average return on 14 chemical companies studied. In basic olefins and aromatics, Jersey had "the longest experience, the best research and production talent" in the industry. These talents, it was argued, could easily be transferred to large-scale production of basic polymers and even end products (Wall 1988, pp. 175–84, also pp. 225–27).

Top managers agreed, but they did little to implement the proposal. Meanwhile, Dow and other chemical companies were rapidly expanding their output of commodity

---

[9]The information on Jersey Standard and the other petrochemical companies mentioned up into the 1990s comes from Larson and Porter (1959, pp. 559–61, 593–98), Gibb and Knowlton (1956, pp. 544–46), Haynes (1949, pp. 400–403), and Chandler (1960, pp. 268–71).

and intermediate petrochemicals, and other oil companies were pulling ahead of Jersey. In 1960, Jersey's chemicals accounted for 4% of gross revenues as compared to 15.4% for Phillips, 8.2% for Standard of California, and 6.5% for Shell (Wall 1988, p. 200). At the same time, the massive expansion of oil-refining capacity worldwide had driven down revenues and income for Jersey's primary products, making the production of petroleum chemicals more attractive.

In 1960, at the height of the polymer/petrochemical boom, a senior manager who had been involved in petrochemicals through the 1940s and 1950s, Monroe "Jack" Rathbone, became CEO and Chairman of the Executive Committee. Under his leadership, operating managers were urged to expand production in basic polymers, particularly polypropylene, and also to seize opportunities in the markets for end products. Jersey Standard was soon making investments in all parts of the world for new plant construction and for acquisitions of producers of fertilizers, containers and other plastic goods, fibers, packaging film, and laminated products. For example, agricultural-chemical plants were built in Colombia, Aruba, Costa Rica, El Salvador, Argentina, Spain, Greece, Lebanon, Holland, and Canada. In Europe and the United States, factories came on stream to produce nylon and polypropylene fibers, plastics, and film and often the basic and intermediate chemicals and polymers needed to supply them. By 1965, 30% of the company's chemical revenues and almost 50% of its capital expenditures for chemicals were associated with the new product lines (Wall 1988, pp. 192–97, 215).

In their enthusiasm for growth through diversification into chemicals the company's operating managers, in the words of Exxon's historian Bennett Wall (1988, p. 207), often failed to follow their normal procedures for a "thorough analysis of markets, technologies, and labor supply." In 1965, it became apparent that the move into petrochemicals was seriously flawed: Capital expenditures were soaring and profits were far short of forecasts. In response, Jersey folded essentially all of the company's petrochemical business into its subsidiary, Esso Chemical, put new senior management in place, and defined a new strategy of investing primarily in existing businesses where good returns had been demonstrated. However, profits continued to drop, and in 1968, Esso Chemical reported losses (Wall 1988, pp. 215–29). Top management then instituted a program of divesting those end products for which Esso Chemical had not developed the capabilities to compete effectively with mainline chemical companies for instance, fertilizers, plastics, and fibers. Moreover, management set about consolidating personnel and facilities in the core areas of basic chemicals and intermediates (Wall 1988, pp. 229–38), to take advantage of economies of scale that seemed ever-increasing.

The failed move of Esso Chemical into end products therefore reveals the consequences of a mismatch between a company's strategy and its existing organizational capabilities. Indeed, almost all the petroleum companies that moved into petrochemicals made an attempt to move into higher value-added end products at some point, and it was almost always a strategic failure. Although the processes of production of petroleum products were similar to those of producing feedstocks and commodity polymers, the development, production, and marketing of higher value-added end products apparently calls for different functional, technological, and market skills.

Corporate Restructuring in the Industry: From the 1970s to the 1990s

The 1970s were a difficult time for chemical companies. The huge jump in oil prices in 1973 and again in 1979 intensified the pressures of overcapacity. The rate of new-product development continued to decline. The postwar worldwide boom came to an emphatic end. All five U.S. chemical companies began, in the late 1960s and early 1970s, to restructure and realign their product portfolios. By the 1990s, such restructuring appeared to have been completed.

DuPont senior managers had learned the limits of their corporate capabilities for new product development by 1970, thanks to the failure of their New Ventures Program. Clearly, the product-specific technical and managerial capabilities developed in operating departments were a more reliable source for future process improvement and the commercializing of new applications of existing technologies than were basic research or attempts by corporate headquarters at product diversification. As a result, the company reduced its expenditures for basic research: Its overall R&D expenditures, as a percent of sales, dropped from 7.1% in 1970 to 3.6% in 1980 (Hounshell and Smith 1988, pp. 405, 582–83, 587). The soaring petroleum prices of the 1970s also encouraged DuPont to acquire a stable source of raw materials, which led to the acquisition of Conoco in 1981. As the oil crisis lessened, Conoco proved a profitable enterprise whose revenues helped to even out the earning fluctuations in the two very different industries: chemicals and petroleum (Taylor and Sudnick 1984, pp. 205–206).[10]

In the early 1980s, DuPont began reshaping its lines of businesses by moving out of basic chemicals and polymers and focusing on higher value-added products and products in which it still enjoyed first-mover advantages. For example, since Conoco's petrochemical output was in polyvinyl chloride, vinyl chloride monomers, and related products that were not intermediates used by DuPont, the company spun off these lines in a leveraged buyout to the Vista Chemical Company, headed by the former manager of Conoco's petrochemical division, Gordon Cain. It also divested itself of a variety of commodity polymers. DuPont also sold off its less profitable end-products: its plastic packaging to Amelen; its urethane rubber business to Uniroyal; its consumer paint to Clorox; some of its pigments to Switzerland's Ciba-Geigy; and some of its industrial finishes to Whittaker.[11]

These divestitures provided funding for a wave of acquisitions that reinforced lines in which DuPont had been expanding its product-specific capabilities, at least since the 1950s. It purchased Exxon's carbon-fiber business in 1984, Exxon's film-composites activities in 1985, and Hercules' olefin-fiber carpets in 1989. Later, in 1992, it traded its acrylic assets to ICI for the latter's nylon business (plus cash). DuPont acquired Shell's Agricultural Chemicals Division in 1986 and Ford's North American Automotive Paint Division in 1986 (*Chemical Week*, May 9 1990, March 4 1992).

One expansion that has worked less well is DuPont's major move into pharmaceuticals and related products. By the early 1970s, biology and biochemistry had already become the largest single focus of DuPont's central R&D facility. The 1980s saw major investments in facilities to expand these activities, together with a move into the production of analytical and surgical equipment through acquisition of two smaller companies and medical divisions of two major companies (Hounshell and Smith 1988, pp. 589–90; *Chemical Week*, May 9 1990; *Business Week*, Oct. 28 1989)[12].

The capabilities needed in research, production, and especially marketing of pharmaceuticals and related lines differed from those of DuPont's chemical lines, and profits have been slow in coming. To enhance these capabilities, DuPont formed a joint venture

---

[10]See also *Chemical Week*, February 20, 1980 and September 2, 1981, and Ralph Landau's interview with Edward Jefferson (Landau 1995a, pp. 11–13).

[11]These transactions are reported in the leading business journals and can be followed by examining the DuPont entry in *Predicast* for these years.

[12]Moreover, Edward Jefferson notes that it proved more difficult for DuPont to recruit outstanding biochemists than first-rate chemists and chemical engineers (Landau 1955, pp. 17–18).

with the pharmaceutical leader Merck in 1989, but whether the capabilities required to become and remain profitable in pharmaceuticals can be developed at DuPont is still open to question.

In the early 1990s, DuPont is an oil and chemical enterprise, although it manages its activities in these two different industries as autonomous units. In 1994, of DuPont's sales of just under $40 billion and operating income of $4.4 billion, 43% of sales but only 24% of income came from Conoco's petroleum business.

In chemicals products, DuPont has built on its long-established lines. For example, DuPont was working on automotive finishes in the 1920s, and they remain a large contributor to profits in the 1990s. In chemicals, the leading revenue producer in 1994 was titanium dioxide, a product in which DuPont still held one-half of the North American market and one-fifth of the world market after 70 years of production. DuPont's capabilities in fibers, which have developed from nylon and polyester to Lycra and Kevlar, are still a substantial money-maker. In polymers, DuPont's sales in the 1990s still come primarily in familiar areas, including elastomers (such as neoprene), nylon and other resins; fluoropolymers (including Teflon) and specialty ethylene polymers; and performance film, including Mylar and Cromalin (Hounshell and Smith 1988, pp. 549–50). Other products such as agricultural chemicals, with roots in the 1920s, are still an integral part of the company in the 1990s. DuPont's experience shows that long-established, learned organizational capabilities are a strong base for maintaining competitive advantage.

Dow Chemical had long pursued a strategy of building capabilities in basic petrochemicals and commodity plastics; by 1979, 85% of Dow's businesses remained in these areas (Whitehead 1968, pp. 254–55; *Chemical Week*, Sept. 9 1988). In 1978, Dow's senior managers agreed that the market for basic petrochemicals had become overcrowded and that the future lay in moving to higher value-added products. They set a goal of having 50% of Dow's revenues coming from these new areas by 1987. The plan was to sell some of the lower value-added polymer and basic-chemical lines and use the money to acquire in new areas (MIT Commission 1989, p. 78; Bower 1986, pp. 106–108).

Dow's first major move was into pharmaceuticals, where corporate strategists believed that 20 years of managing Allied Laboratories had developed, the planners believed, a modicum of capabilities. In 1981, Dow bought the Merrell Drug Division of Richardson-Vicks, a pioneer in the commercializing of nonsedative antihistamines. In the same year, it formed a joint venture with Otsuka Pharmaceutical of Japan to market jointly in Japan and the United States. In 1986 came the acquisition of an 84% interest in Funai Pharmaceuticals, another Japanese firm. As the Merrell Dow division expanded worldwide, it soon built an impressive reputation for marketing capabilities.[13]

Divestitures at Dow began in 1982. Oil and gas properties were sold that year to Apache Petroleum and to Dome Petroleum of Canada. Next came the shucking off of its overseas joint ventures: its 50% holdings of Asahi-Dow Chemical in 1982 and its share of comparable ventures in Korea, Saudi Arabia, and Yugoslavia. In 1984, it sold its 50% share of Dowell to Schlumberger. (It had sold its holdings in the joint venture with Pechiney in 1969 and with BASF in 1978.) In 1986, Dow divested itself of its foreign banking and financial services.

---

[13]See Bozdogan (1989, p. 79. The transactions described in this and the next two paragraphs are summarized in Moskowitz et al. (1990, p. 24); *Hoover's Handbook (1993), The International Directory of Multinationals, Moody's Manuals*, and in greater detail in review of the articles listed for Dow in *Predicast*. The most useful articles are *Chemical Week*, September 9, 1988; November 23, 1988, April 6, 1989, and *Business Week*, August 7, 1989. Also valuable is Lane (1992).

Then, in 1985 Dow began a series of acquisitions funded in part by these divestitures. It acquired the Tersize Division of Morton Thiokol, a manufacturer and marketer of home cleaners and household products, which added to Dow's existing line of Saran Wrap and other household packaging films, cleaners, and detergents. It acquired the Italian firm Dompak for a presence in the European household market. It acquired the polymer businesses of the Upjohn Company and Film Tec: The first provided a product that was essential to obtaining a worldwide presence in polyurethane foams; the second gave the company access to a new technology in spiral-bound membranes for water and fluid purification and separation. Next, in 1986 it purchased for $50 million Bromide Products, a leader in super-hard ceramics technology, and Haeger and Kaesner, a German manufacturer of specialty chemicals.

By the end of 1986, the strategy defined at the 1978 meeting had been fully implemented. By 1989, chemicals and commodity plastics accounted for 48% of Dow revenues, and 52% came from higher value-added products.[14] Dow continued to implement its value-added strategy both by acquisitions and by joint ventures. The acquisition of Marion Labs in 1989 and the formation of Marion Merrell-Dow made Dow the nation's ninth-largest pharmaceutical producer. In the same year, came a joint venture with Eli Lilly, Dow Elanco, was formed to produce agricultural chemicals based on crop genetics and biochemistry. (Dow bought out Lilly in 1997). The acquisition of Essex Chemicals, a premier supplier of adhesives and sealants, gave Dow its first entry into the automobile market, while joint ventures with Exxon Chemical, BOC of Great Britain, and Sumitomo expanded its activities in specialty chemicals, particularly overseas.

By the 1990s, it was clear that Dow had largely succeeded in walking a delicate line. In nearly all its acquisitions, Dow's existing capabilities provided essential technological bases for further product development.At the same time, Dow maintained its strength in the lines of petrochemicals and polymers, which it had decided to retain. In commodity plastics, Dow shut down old plants but built new, far-more-efficient petrochemical plants to replace them. In the early 1990s, Dow was still the nation's largest producer of chlorine and caustic soda and one of the largest in the production of styrene and other commodity polymers. Indeed, Dow's sales and income in the 1990s rest, even more than DuPont's, on capabilities developed in the 1920s and 1930s and in the immediate post-World War II polymer revolution.[15]

Dow's long-term strategy has been based on building its corporate capabilities over a full line of products from basic feedstocks, chemicals, and polymers to complex end products. Dow remains a company that excels in design and manufacturing, and it has always had strong chemical-engineering skills.[16] Clearly, the one major product line that has not succeeded for Dow is pharmaceuticals, where the scientific base, the procedures of product development, and the ways of production and marketing differed widely from those for chemicals. Thus, in 1995 Dow sold 71% of Marion Merrell Dow for $7.1 billion to the German giant Hoechst, which had been developing its capabilities in pharmaceuticals for over a century. From 1920, when Dow was a third-tier company (Plumpe 1990), it has become the largest American chemical company, even larger than DuPont without Conoco.

---

[14]See *Hoover's Handbook of American Business*, 1991.

[15]See *Hoover's Handbooks* (1991 to 1995); *Moody's Industrial Manuals;* Bozdogan (1989, p. 79); *Chemical Week* (July 29, 1989, and September 18, 1991).

[16]See Landau, (1995b, pp. 4–5, 28–29, 34).

In the mid-1970s, Union Carbide began to slim down and refocus its heterogeneous product portfolio in a more systematic manner. Unable to compete with DuPont, it withdrew from fibers and packaging film. Unable to compete with Dow, it exited from a number of basic commodities. The company sold off its PVC operations in 1977 to Tenneco, its vinyl unit to American Hoechst in the same year, its styrene in 1976 and its polystyrene in 1977. At the same time, it withdrew from Europe by selling its petrochemical assets to British Petroleum Chemicals (Bower 1986, pp. 102–104).

By 1979, Union Carbide and Carbon was using funds from these divestitures to concentrate in products for which it had strong learned capabilities. These included polyethylene (both HDPE and the older LDPE), ethylene oxide end products, and its long-established business in industrial gases, calcium carbide, and alloys, as well as some of its older consumer products: batteries, flashlights, headlights, and the like. Moreover, in the late 1970s Carbide's long-term development work in polyethylene brought on stream a revolutionary new process, Unipol, to produce linear, low-density polyethylene (LLDPE), which sharply reduced the cost and greatly improved the quality of that product. Using the Unipol technology, Union Carbide (working with Shell Chemical) commercialized a comparable process for linear polypropylene (LPP) that promised high income through licensing. In 1982, Union Carbide sold off a portion of its ferro alloy and metals (tungsten, vanadium, and uranium) businesses for $240 million (Spitz 1988, pp. 507–11; *Chemical Week*, Nov. 25 1987). But as the licensing income began to flow and Carbide began to reshape its product portfolio, disaster struck. In December 1984, a gas leak in its pesticide plant in Bhopal, India, killed and injured thousands. The stock price plummeted, and Samuel Heyman, fresh from a successful takeover of General Aniline and Film (GAF, the former U.S. subsidiary of IG Farben) began to stalk the company. In fall 1985, he made a $4.3 billion tender offer at $68 a share, later raised to $72. (It had been selling at below $30 before the raids began.) Carbide's management responded in January 1986 by offering to buy back 55% of its shares at $85. The buyback netted GAF $268 million, doubling its net worth.[17]

Financing of the buyback forced a fire sale of many of Carbide's most valuable assets. It sold its engineering (high-performance) polymers and composite division, its carbon business (to Amoco), and its packaging-film business (to Envirodyne). During the spring, the EverReady battery operations (the nation's largest producer) were sold to Ralston Purina, making the purchaser the world's largest producer of batteries. Carbide's nonpetrochemical home and automotive products, including Prestone antifreeze (also the nation's largest) went to an investment group headed by First Boston and its worldwide chromium interests to General Mining Union. In the fall, it sold off its worldwide agricultural-chemical business to the French chemical company, Rhône-Poulenc. It disposed of its headquarters building. In addition, Carbide wrote off $650 million worth of assets, shutting down most of its older commodity works, concentrating production in its new Unipol LLDPE and LPP plants. Even with these sales, the company's increase in debt was so large that, for a brief time, it enjoyed the dubious distinction of being the largest issuer of junk bonds of any major U.S. corporation.[18] The cycle of divestments has continued in the 1990s.

[17]The course of events can be followed in the Wall Street Journal, especially in 1985 Aug. 15, 2:3, Sept. 3–5:1, Dec. 9, 3:3, Jan. 10, 3:1 and almost daily until mid-1986.

[18]The chronology of the transactions and the amount involved in the fire sale can be followed in *Predicast's* lists of articles on Union Carbide, supplemented by those in the index of the *Wall Street Journal.* Union Carbide's $2.5 billion recapitalization was so significant that it changed the practice of investment-banking houses in the financing of mergers and acquisitions. See the introductory chapter of Eccles and Crane (1988).

Union Carbide has yet to recover from Heyman's raid. In the early 1990s, its debt ratio was higher than those of any of its competitors. Its revenues were one-half what they had been before 1985. By 1994, sales and income came entirely from basic chemicals and petrochemicals. Of these, only 15% came from goods listed as specialty polymers and products. Union Carbide had lost the capabilities and the funds to implement competitive strategies comparable to those of DuPont, Dow, or Monsanto. Whatever the outcome of Heyman's raid, the fact that it happened at all ensured the dispersal of Union Carbide's strength. After all, if Carbide had permitted Heyman to acquire the company, a comparable sell-off would have been necessary to bring down Heyman's more than $5 billion debt of more than $5 billion in high interest junk bonds (*Hoover's Handbook of American Business*, 1991 to 1996).[19]

In the 1980s, Monsanto rose from the fifth- to the third-largest U.S. chemical company, and benefited from an especially successful realignment of its product lines during the 1970s and 1980s. In the 1970s, the company concentrated on building a balanced portfolio in the manner of Dow. Like Dow, it maintained its basic and commodity products and expanded its oil-drilling activities by moving into the North Sea oil fields. It enhanced capabilities in basic chemicals and polymers, including acrylonitrile, butadiene, styrene, and nylon 6-6 and developed new products in engineering plastics with a focus on thermoplastics used in appliances and automobile production. In fibers, it sold off its European nylon business and focused development on its acrylic-based line. With the purchase, in 1977, of Rohm & Haas' acrylic-fiber business Monsanto became the country's leading producer of carpet fibers, including a new product, Astro Turf.

But Monsanto's major move was in agricultural chemicals. Its herbicides, Round-Up and Lasso, quickly made the company the world's largest producer of herbicides. The company aggressively pursued R&D efforts in plant biology, hiring prominent academics, forming its own biotechnology R&D subsidiary, and acquiring stakes in biotechnology start-ups (such as Biogen). These moves made Monsanto a pioneer in genetically engineered agricultural chemicals (Moskowitz et al. 1980; *Chemical & Engineering News*, Feb. 23 1981; *Journal of Commerce*, October 29, 1981).

Only in 1984, with the appointment of Richard J. Mahoney as CEO, however, did Monsanto begin to carry out a systematic policy of reshaping its product portfolio, by concentrating much more narrowly on its proven learned capabilities and divesting itself of the rest of its lines. Mahoney began by selling off basic chemical plants and polyester latex operations (to Morton Thiokol), the phenolic-resins business (to Borden), the plasticizer activities (to Witco), and then the North Sea oil holdings and gas interests (to Amerada Hess).[20] The divestitures went on and on. Monsanto Oil Company was sold to Broken Hill Properties of Australia, several specialized polymer units to the Dutch State Mine (DSM), much of its polystyrene business, including three plants, to Polymers (a Canadian firm); a single acrylonitrile plant went private in a leveraged buyout; its original vanilla business went to Rhône Poulenc, its nonwoven fiber business to James River, and its paper chemicals to AKZO, the leading Dutch chemical producer. Divestitures of higher value-added units whose profits were waning followed, including the sale of Astro Turf to Balsamsports Lattenbau; its electronic materials business to another German firm, Huels; its global analgesics (bulk aspirin and related products) business to Rhône-Poulenc; its polyethylene nonfood bottle

[19]A series of comparable transactions in the early 1980s reduced Allied Chemical, in 1950 the third-largest U.S. chemical company, to producing plastic "engineered materials" as the chemical division of Allied Signal (Moskowitz et al. 1990, pp. 571–74).

[20]These transactions and those listed in the next two paragraphs can be followed in *Predicast* for these years.

business to Innopack; and its hot-plastic-bottle technology to Johnson Controls. The only major acquisition in this time was the purchase of the pharmaceutical firm G. D. Searle. However, beyond producing prescription drugs, Searle also made aspartame-based sweeteners, including Nutrasweet, which then took saccharin's place in Monsanto's portfolio.

By the end of 1988, Monsanto's product portfolio had been reshaped. It was a world leader in acrylic fibers, rubber chemicals, and other thermal plastics. In agricultural chemicals, it was using its profits from Round-Up and Lasso to expand biotechnology research. By 1994, just under 70% of its operating revenues came from high-value agricultural, nutrient, and health-care products. During the 1990s, it continued to focus on its strengths, especially its growing capabilities in biochemistry and biotechnology, and on commercializing agricultural and nonhuman biogenetic products. At the same time, it concentrated on increasing its foreign sales, which grew from less than 10% of total sales in 1989 to 49% in 1994 (*Chemical Week*, September 28, 1988, August 21, 1989, December 6, 1989; January 24, 1990; September 2, 1991; January 27, 1992; *Hoover's Handbook of American Business*, 1991 to 1996). These strategic moves culminated in December 1996, when the company announced that it would spin off its chemical businesses to its stockholders. Monsanto's current product lines — agricultural products, nutrients and pharmaceuticals — are based in good part on the capabilities it has developed, particularly in biogenetic technology, since the late 1970s (Davis 1996).

By the mid-1970s, the former Jersey Chemical had reined back its 1960s policy of helter-skelter diversification. Instead, the product portfolio of what was now Exxon Chemical again resembled the chemical operations of the firm in 1960. Its primary business was the production of ethylene, basic polymers, plasticizers (particularly for the vinyl industry), elastomers, and synthetic rubbers. Profits had returned; in 1974, Exxon Chemical was one of the five most profitable chemical companies worldwide (Bower 1986, pp. 112–17; Wall 1988, p. 237). By the 1980s, Exxon Chemical had sold off most of its more specialized petrochemicals to DuPont and other chemical companies. It concentrated its R&D on intermediates and polymers, often seeking to share the cost of product development with joint ventures. By focusing on these basic product lines, Exxon grew from the fourteenth-largest U.S. chemical company in 1960 to the third-largest in revenues in 1986. Since 1989, Exxon Chemicals has contributed an average of 10% of Exxon's revenues and 20% of its profits.[21] It is widely believed that Exxon has the best integration of refineries with chemical plants among all oil companies. The story of Exxon Chemicals indicates that the capabilities in producing and marketing gasoline, heating oil, and other products differed from those required in high-end chemical lines

These histories of U.S. chemical firms' corporate strategies strongly support the argument that successful corporate strategy must be based on accumulated, firm-specific capabilities: functional, technological, and managerial. When companies attempt to stretch beyond their capabilities — as in the attempts of Exxon (and other petroleum companies) to integrate forward into higher-value end products during the 1960s, or the attempts of DuPont and Dow to move into pharmaceuticals in the 1980s — they often fall short. By the mid-1990s, the broad division of labor in the U.S. chemical industry had been defined on the basis of such long-established organizational capabilities. The petroleum companies concentrated on producing feedstocks, commodity petrochemicals, and polymers. The long-established drug companies continued to dominate

[21] Articles listed in *Predicast* for the more recent years and in *Hoover's Handbooks* since 1990.

pharmaceutical markets. The old-line chemical companies focused increasingly on the capabilities they had developed over the previous century, producing chemical products in which their capabilities gave them the greatest competitive strength, with companies such as Monsanto shifting to biotechnology to improve their position in older agricultural and nutrients markets.

In Europe, the pattern differed. There, the head start in organic chemicals that gave the German and Swiss firms first-mover advantages in pharmaceuticals and, in dyes gave them an advantage they still hold, while the slower start in petrochemicals called for different requirements and capabilities that led to different paths of growth.

GERMANY: BAYER, HOECHST, AND BASF

The German Dyestuffs Industry: 1890–1914

Three firms have dominated the German chemicals industry since the 19th century: Bayer, Hoechst, and BASF. All three were founded as dyestuffs companies in the 1860s and established their headquarters and primary production facilities along or near the Rhine, with its cheap water transport, to facilitate access to nearby coal fields and other inputs characterized by relatively low value-to-weight ratios. A more detailed treatment of the 19th-century history of the German chemicals industry is provided in the chapter by Murmann and Landau.

By the late 1870s and 1880s, all three firms had developed a new approach to competition that relied on intrafirm industrial-research laboratories (see Beer 1959). The impetus for this strategy came from several directions. The passage of an 1877 patent law had strengthened patent protection for new dyes. Government support encouraged the education of university-trained chemists. Indeed, the research laboratories of the three leading German chemical firms retained close links to leading university-based research faculties within Germany (Liebenau 1992, p. 61). For example, the synthesis of indigo that was to prove critical to BASF was based on patents licensed from Adolf Baeyer of the University of Munich (Beer 1959). These early R&D laboratories also worked closely with dyestuff customers in improving their products and developing new applications for dyestuffs (Beer 1959, p. 93). Finally, buoyant market demand for new dyes ensured that the successful innovators would reap a return on their research-and-development efforts.

By the mid-1880s, however, profit margins were declining in some major classes of dyestuffs, partly because a convention governing the pricing of red dyes broke up in 1885, leading to a price decline of one-half, while the price of a coal tar, a key input, was rising (Liebenau 1992, p. 59). Along with continuing their search for new dyes, these firms used their technical expertise to diversify into pharmaceuticals products. Hoechst had begun R&D in pharmaceuticals development within 20 years of the firm's foundation and by 1900 marketed anti-malaria drugs and a diphtheria serum. In addition to developing aspirin in 1888, Bayer commercialized an array of sedative drugs and in 1902 began to commercialize photographic film (Chandler 1990, p. 478). BASF was less active in this product area: To this day, the firm derives a much larger share of its sales from chemicals, as opposed to health-care products, than do either Bayer or Hoechst. Instead, BASF focused intensely on synthesizing an indigo dye and, after the turn of the century, on the high-pressure technologies that resulted in the synthesis of ammonia in 1913.

All three firms grew rapidly. By 1904, BASF employed more than 7,500 people (Beer 1959, pp. 134–35). By the start of World War I, German firms dominated global production of dyestuffs, accounting for nearly 90% of global dyestuffs production in 1914 (Beer 1959). German firms were also the world technology leaders in dyestuffs: "By the decade after the turn of the century, Germans had taken out 1,754 [U.S.] patents in the 10 largest classes of chemical patents, as opposed to only 212 British and around 1,550 acquired by Americans" (Liebenau 1992, p. 65).

As in the case of the U.S. firms in chemicals and other manufacturing industries that grew to great size during this period, the three leading German chemical firms developed centralized management structures that were dominated by professional (nonowner) managers. The production and R&D operations of these firms, however, remained relatively centralized, thanks to the large minimum efficient scale of production, strong economies of scope in dye and pharmaceutical production, and the fact that their products had a relatively high value-to-weight ratio, so that transportation was relatively cheap. German chemical firms built plants abroad for the final processing of end products, for packaging and branding pharmaceuticals and films, and for repacking bulk dyes (Schroter 1993, p. 370). However, the German chemical firms relied primarily on exports rather than on direct foreign investment to gain access to non-European markets for their dyestuffs and pharmaceuticals products (Chandler 1990). Their foreign operations in non-European markets were largely confined to marketing and distribution.

Concern over the costs of maintaining their ambitious R&D programs, as well as a desire to maintain profit margins, led the German chemical firms to form two "mini-cartels" in 1904 and 1907. (Carl Duisberg, the director of Bayer's R&D activities, sought the participation of all three leading chemical firms in a single agreement, but this proved impossible.) The *Dreibund*, enlisting the participation of BASF, Bayer, and a smaller firm called AGFA, called for profit-pooling and coordination in patenting, technology development, and sales practices. The *Dreiverband*, formed in 1907, joined Hoechst with two smaller dyestuffs firms, Cassella and Kalle, in an exchange of stock. The effects of these cartels on pricing, market growth, and technology development have not been well documented. Beer (1959, p. 133) suggests, "After their formation they [the two cartels] did not raise the price of dyes. On the contrary, prices dropped during the entire decade preceding the war, though the curve of their fall was gradual and showed none of the erratic pattern and below-cost dips that had been so often characteristic before 1905." Hayes (1987) makes a complementary argument that the reduction in interfirm competition in pricing and production development in the dyestuffs area of their businesses enabled each of the three leading firms to direct their R&D efforts to other products: Bayer developed hard rubber tires; Hoechst entered the electrothermic production of carbide and began a major progrem in acetylene-based chemistry; and BASF expanded its costly development program in ammonia synthesis. In addition, these mini-cartels provided a foundation for the formation of the IG Farben combine after World War I.

Further Cartelization and the Creation of IG Farben: 1914–1945

The outbreak of World War I severely threatened the dominance of world dyestuffs markets by the leading German chemical firms. Other industrial economies, especially Great Britain, France, and the United States, moved rapidly (albeit with mixed results) to create domestic-production capabilities in these critical products and related ones (such as synthetic ammonia), which were essential to modern warfare. To maintain their dyestuffs business in the face of excess capacity and to share in the costs and profits of the high-pressure technologies developed by BASF for ammonia synthesis, the *Dreibund* and the *Dreiverband* agreed in 1916 to combine their activities in a confederation that provided for profit-sharing, cost-sharing for the large BASF Leuna ammonia works, and some reduction in production capacity.

This alliance was a loose one; any member could withdraw at will (Chandler 1990; Hayes 1987). Nevertheless, formation of the so-called Little IG improved the profitability of the German chemical firms and facilitated their shift from dyestuffs to other products, particularly synthetic ammonia. Hayes (1987) argues that the formation of this alliance ended competition among its members in dyestuffs output and pricing within Germany and sustained their profitability in the face of sharp declines in dyestuffs exports after the end of World War I. Even in the face of the extraordinary disruption of World War I, the growth trajectory of these firms continued, although the composition of their output began to shift. According to Hayes (1987), overall sales of the alliance grew by roughly 50% in real terms, while the share of dyestuffs in total revenues shrank and revenues from synthetic ammonia grew.

Despite the apparent success of the Little IG in supporting the German industry's adjustment to postwar chemicals markets, senior managers at both BASF and Bayer — respectively Carl Bosch, one of the codevelopers of the synthetic-ammonia process, and Duisberg, who had played a key role in developing Bayer's early dyestuffs products nearly 50 years earlier — pressed for a stronger management structure. IG Farben, formed in 1925, represented a merger of the major participants in the 1916 confederation. Bosch became chair of the "Working Committee," which oversaw and implemented operating decisions. Indeed, Hayes (1987) argues that BASF's need for ever-larger budgets for its project in synthetic fuels program was a central motive for the IG Farben merger.

Behind the facade of centralized management at I.G. Farben, the merged firm retained many of the federal characteristics of the Little IG. For example, three of the major "divisions" of IG Farben were conveniently located at Leverkusen (headquarters of Bayer), Ludwigshafen (BASF), and Frankfurt (Hoechst), allowing each of the "big three" member firms to maintain the location and the product focus of their established R&D and production operations. This "federalist" management structure contrasted with that of DuPont, as Chandler points out (1990, p. 578): "Unlike DuPont, they [IG Farben] had no corporate or general office where general executives, largely free from operating duties, concentrated on monitoring performance; on allocating resources on the basis of that monitoring and of their own understanding of economic, technological, market, and political considerations; and then on defining and implementing long-term strategies."

IG Farben did little to rationalize production, probably because it was substantially self-financing and faced limited domestic and international competition (Chandler 1990, pp. 569–73). Instead, Farben addressed the problem of international competition by negotiating international cartel agreements. Within Europe, this meant agreements covering market-sharing and pricing of dyestuffs and synthetic ammonia. Outside of Europe, especially in the United States, it often meant trading access to technology for access to particular markets (Smith 1992, p. 143) and support for I.G. Farben research in a depressed economy. For example, a technology-sharing agreement covering hydrogenation technology for synthetic-fuel production enlisted the participation of ICI, Shell, and Standard Oil of New Jersey, all of whom saw this technology as a viable substitute for the petroleum resources then believed to be at the point of exhaustion. Standard Oil of New Jersey and DuPont agreed to a technology-sharing agreement with IG Farben that covered work in synthetic rubber, based on the high-pressure hydrogenation technologies originally developed at BASF and extended by IG Farben. All of these international agreements utilized technology-sharing as a vehicle for dividing international markets and maintaining stability in pricing and capacity utilization. In this era,when government barriers to market access were rising, "private" interfirm agreements were used to maintain some stability in the strategic planning of giant firms like IG Farben.

After about 1929, IG Farben increased its focus on domestic markets. This domestic orientation was supported by the increasingly bellicose German government, with the Nazi takeover of power in 1933, which saw self-sufficiency in fuels as an indispensable component of war mobilization. In 1933, for example, the German government agreed to guarantee domestic prices for the synthetic fuel that was beginning to flow in commercial quantities from its hydrogenation plants in Leuna and elsewhere. Another reason for this greater emphasis on domestic markets was the rise in tariff barriers during the Great Depression, which made reliance on foreign markets increasingly difficult.

Although IG Farben controlled some of the most advanced technology and R&D complexes of any chemical firm in this century, it produced surprisingly few major fundamental research advances. Much of the technology that formed the basis for synthetic fuel and rubber, for example, drew on advances made one or two decades earlier. In contrast to DuPont, IG Farben never established a corporate central-research laboratory, and by the end of the 1930s, the IG Farben product portfolio relied more heavily on older products than did that of DuPont. Chandler (1990, p. 580) points out: "By World War II more than half of DuPont's sales volume came from lines that the company had not been producing commercially twenty years earlier—some of which had not been invented twenty years earlier.... By contrast, IG Farben's sales in the late 1930s came primarily from product lines developed well before World War I—chemicals, dyes and film—and from such massive projects as the production of nitrogen (i.e., ammonia) from the air just before that war, aluminum and magnesium during the war, synthetic gasoline after 1925, and synthetic rubber in the 1930s." In fact, IG Farben gradually reduced its R&D intensity during the 1930s: "After 1930, it never again devoted more than 7 percent of its sales turnover to research. Indeed, throughout the Nazi period, the trend was for research expenditures as a percent of sales to decrease..." (Stokes 1988, pp. 25–26).

Shifting Feedstocks in the Early Postwar Era: 1945–1970

At the end of World War II, each of the three major firms whose merger had formed IG Farben came under the control of a different Western government. The British military government controlled the headquarters and main production facility of Bayer; the U.S. military occupation zone included Hoechst; and the French occupied the area surrounding the enormous BASF production complex. The Russian occupation zone included the synthetic-fuel plants at Leuna and other production facilities, some of which were dismantled and shipped to the Soviet Union. The structure of the immediate postwar occupation thus weakened ties among these three German chemical firms, but importantly, it did not result in the dismantling or major restructuring of the major chemical-production complexes that each firm had constructed over the preceding 80 years along the Rhine. Even more important was the preservation within each of these entities of the learned organizational capabilities developed over those eight decades.[22]

IG Farben was formally broken up in 1950, thus ratifying the separation that had already occurred with the end of the war. After 1949, all three firms experienced rapid growth in total output and exports. Typical rates of turnover were 40% to 50% for these three companies from 1949 to 1951. Further, by 1951, exports accounted for roughly one-third of total sales for each of the three firms, a share that exceeded that of IG Farben during the 1930s (Stokes 1988, pp. 160–61). The three newly independent firms not only retained most of their major production facilities (some of which were heavily damaged); they also retained the same senior management. The chairman of the supervisory board for each of these three firms had occupied high positions within IG Farben, and all members of the first supervisory board for each firm after 1950 had served as senior managers in the same organization when it was part of IG Farben (Stokes 1988, p. 183).

One of the most significant technical challenges faced by all three firms in the early postwar period was that of shifting their feedstocks from coal to petroleum. This task posed major challenges and uncertainties. First, petroleum-based feedstocks were of limited use for many of the products that had contributed a large share of corporate profits for all three firms, particularly pharmaceuticals and dyestuffs. Indeed, even in the U.S. chemicals industry, coal feedstocks accounted for approximately 75% of all organic-chemicals production as late as the mid-1950s (Stokes 1994, p. 4). Second, the characteristics of petroleum-based chemicals-manufacturing processes differed significantly from those using coal-tar feedstocks. According to Stokes (1994, p. 247), these differences had significant implications for process and product-technology development and for management. The scale of petroleum-based chemicals-manufacturing processes generally was much larger than that of coal-based plants, and among other things, this increase in scale required new approaches to plant layout and materials handling. In addition, chemists charged with developing new processes had to adopt a new approach, one that emphasized the rapid development of process "solutions" that worked, regardless of their theoretical elegance or even their yield.

Finally, reliance on petroleum-based feedstocks would force all three chemical firms to depend on foreign sources for a critical input to their manufacturing processes, in sharp contrast to the situation with coal. The shift of these German firms to petroleum thus rested in part on the political settlement that developed in the shadow of the Cold War, one that included guarantees by the United States and other recent adversaries of assured access to supplies of petroleum from the Middle East, as well as substantial Marshall Plan aid for the construction of oil refineries in Germany (Chapman 1991; see also Stokes 1994, p. 98; Stokes 1988, p. 198).

These formidable uncertainties of demand, technology, and politics led both BASF and Bayer to maintain technical programs in chemicals production that relied on coal, rather than petroleum, feedstocks. But these German firms had at least one significant advantage in shifting from coal to petroleum feedstocks — the development of a competitive global market in the supply of the necessary technology and expertise (Stokes 1994, p. 174). A number of petrochemicals and chemical firms, mainly from the United States and Great Britain, had developed considerable expertise in

---

[22]During the first years of its military occupation of Germany, the U.S. military did attempt to break up the Hoechst production complex into self-contained, individual production facilities (Stokes 1988), but this "radical deconcentration" policy was short-lived.

petroleum-based chemicals production (often as a result of wartime programs), and they sought to use this expertise to enter the West European chemical industry.

BASF and Bayer formed alliances with major petroleum firms to acquire petrochemicals technologies and to develop the necessary large plants. BASF acted first, forming a joint venture with Royal Dutch Shell, which culminated in a 1952 agreement to create the Rheinische Olefinwerke. Since early in the century, BASF had been a leader in high-pressure processes for the manufacture of ammonia, synthetic fuels, and (since the 1930s) polyethylene. These technological capabilities were closely related to the large-scale, continuous-process technologies needed for the exploitation of petroleum-based feedstocks. Thus, BASF's central motive for pursuing a collaboration with a petroleum firm was not the acquisition of technology, but an assurance of feedstock supplies, especially for polyethylene production (Stokes 1994, p. 137). Although BASF's joint venture with Shell represented a substantial corporate commitment to petrochemicals, the firm continued to rely heavily on coal as a feedstock for other products through the 1950s. In its polyethylene production, BASF strove to maintain flexibility well into the 1960s, to shift back to feedstocks derived from coal (Stokes 1994, pp. 150–52). BASF also acquired Wintershall, a German oil firm with several domestic refineries, in 1969, showing its long-standing concern with vertical integration of its manufacturing operations and feedstock supplies.

Bayer was slower to enter petrochemicals, partly because of the firm's historic focus on dyes, pharmaceuticals, photographic film, insecticides (developed in the 1920s), and other higher-margin chemical products that benefited from Bayer's marketing expertise and global reputation. These products relied on small-batch production technologies for which coal-based feedstocks retained some advantages in the early 1950s (Stokes 1994, p. 155). Nevertheless, Bayer did launch internal programs during the late 1940s and early 1950s to master petrochemicals technologies and to acquire licenses for these technologies from outside sources. By the mid-1950s, Bayer built a "cracker" for the production of chemicals feedstocks from refined petroleum products. However, the firm had decided that a partner was necessary to share the high costs of the facility and to guarantee access to the needed petroleum inputs. Lengthy negotiations with Standard Oil of New Jersey failed to produce an agreement. Bayer then reached an agreement with British Petroleum, which was considering the construction of a large refinery near Bayer's Dormagen facility, to form a joint petrochemicals-production venture, Erdolchemie, in 1958.

In contrast to the alliances formed by others, Hoechst entered the petrochemicals industry largely through internal development of the necessary technology and chose to obtain its feedstocks through market contracts rather than joint ventures. This path was shaped by the firm's technological weaknesses in the early 1950s, which placed it in an unfavorable position for negotiating a joint venture with foreign petrochemical firms. This strategy, however, also allowed Hoechst to benefit from the growth of "independent" sources of chemicals technologies during the postwar era. For example, the firm was among the early licensees of the Ziegler high-density polyethylene technology, developed by Karl Ziegler, director of the Max Planck Institute at Mülheim. (This alternative polyethylene technology was especially important, since both BASF and ICI refused to license their polyethylene processes to Hoechst.) Working with Ziegler and researchers from a smaller U.S. company, Hercules, Hoechst commercialized high-density polyethylene and then a still more versatile polymer, polypropylene, an achievement that assured Hoechst a strong position in polymer chemicals (Stokes 1994, p. 184; see also the chapter in this volume by Ralph Landau). Thus, Hoechst, which entered the 1950s in a relatively weak technological position, ended the decade with profitable new capabilities in petrochemical production that did not involve significant revenue-sharing with another firm. Like BASF, Hoechst retained flexibility to shift among coal, gasoline, or crude oil as a feedstock well into the 1960s.

In addition to licensing foreign firms' process technologies in petrochemicals, German firms also licensed polymer-product technologies, especially from ICI of Great Britain. Two of ICI's most important innovations, polyester fiber and polyethylene, were licensed to Hoechst and BASF, respectively. Major German firms also formed joint ventures during this period to support expansion of their foreign manufacturing activities, especially in the United States: Hoechst with Celanese, Hercules, and Stauffer; Bayer with Monsanto; and BASF with Dow. BASF also acquired a U.S. subsidiary, Wyandotte. All of these joint ventures resulted in the German firm acquiring the jointly owned production capacity or (as in the case of the Hoechst-Celanese venture) acquiring the parent U.S. firm. These actions appear to reflect the German firms' growing conviction that a wholly owned presence in the large, dynamic U.S. market was of critical importance to their global competitiveness.

By 1960, 40% of the volume of German organic chemicals production was petroleum-based, a reasonably high share given where these firms had started, but still only one-half the share of U.S. organics production that utilized oil (Spitz 1988, p. 360). By the end of the 1960s, the three successors to IG Farben had developed contrasting product portfolios which reflected core capabilities that could be traced back at least 50 years. BASF retained its focus on process expertise, specializing in vertical integration in the manufacture of basic chemicals and feedstocks, including polyethylene technology licensed from ICI. Bayer was concentrated in pharmaceuticals, agricultural chemicals, synthetic rubber and polyurethane, and health-care products. Hoechst was less vertically integrated, remained a leading pharmaceuticals firm, and had become a major producer of synthetic fibers, based in large part on its license agreement with ICI for polyester fiber and a low-pressure polyethylene process based on Ziegler's invention.

Maturing Markets and Oil Shocks: 1970–1990

The two decades following 1970 were difficult for chemical firms everywhere, including the three major German firms. By the late 1960s, growth of demand had slowed as the possibilities for low-cost substitution of plastics and synthetics for other materials were becoming exhausted. However, capacity continued to expand and profit margins came under severe pressure, especially in bulk chemicals and feedstocks. The oil shocks of 1973 and 1979 made matters worse. The share prices of the three major German chemical firms underperformed the German stock market during the 1975–1993 period.

All three firms responded to the pressures of the 1970s by attempting to reduce their production of commodity petrochemicals such as polyethylene, ethylene, and polypropylene. Capacity cutbacks were most difficult for BASF because of its high levels of vertical integration, which included significant oil refining operations.[23] As part of their

---

[23]According to Gernot Winter, managing director of polyolefins and PVC at BASF (quoted in Bower 1986, p. 170): "The exit barriers for an oil company [from petrochemicals] are lower than for a chemical producer. In Shell, for example, the chemical business is a small part of the whole. With us, if you cut off petrochemicals, you lose a major part of the company."

efforts to reduce costs, all three firms shrank their workforces between 12% and 17% from 1990 to 1993. These firms have also expanded their foreign operations as they seek to lower their operating costs. In addition, all three forms are diversifying away from low-margin bulk petrochemicals and into specialty chemicals, although BASF still adheres to its basic strategy of integrated chemicals production.

All three German firms have undertaken substantial restructuring in recent years. Even dyestuffs, long highly profitable product lines for German chemical firms, has lost much of its attractiveness because of increased competition, and the distance between the major German dyestuffs production operations and the leading market, the growing textile industries of Asia. Bayer and Hoechst in 1995 announced plans to combine their dyestuffs operations in a joint venture (DyStar), which will be headquartered in Singapore. The German firms have also continued their push into ownership in the U.S. market. Hoechst acquired full control of the fibers producer Celanese in 1989 and of Dow's giant pharmaceutical subsidiary, Marion Merrell Dow, in 1995. In 1994, Bayer enlarged the initial U.S. base it had obtained through the acquisition of Cutter in 1974 and Miles in 1978 with the acquisition of Winthrop Sterling Drug's over-the-counter business from Eastman Kodak, which permitted it to use the trade name Bayer aspirin, a right it had lost when its U.S. business was expropriated in World War I.

To a remarkable degree, these three firms have continued to specialize in areas that they have dominated for nearly a century, drawing on technical and other capabilities that first started developing in the late 19th and early 20th century. For example, by the 1890s, Hoechst and Bayer were leaders in phamaceuticals, and after the turn of the century, they diversified into production of photographic film and other products. In 1991, Bayer's most profitable lines were pharmaceuticals and imaging technology; for Hoechst, the most profitable lines were health care and polymer fibers and films. As of 1995, Bayer and Hoechst both derived more than 25% of corporate sales and a much greater percent of profits (Bayer 55% and Hoechst 44%) from health-care products. In contrast, BASF concentrated on the production of chemical intermediates for the dye makers and other chemical producers in the late 19th century. Just before the outbreak of World War I, it invented and commercialized a revolutionary process, the high-pressure synthesis of ammonia from the air. In 1991, only 20% of BASF's profits came from consumer products, which included paints, coatings, batteries, and carpet fibers, and only one small pharmaceutical unit. However, 14% of profits came from oil and gas, 21% from plastics and fibers, and the remaining 44% from products it had been producing before World War I. Of the three major firms, BASF has remained the least diversified geographically, with only 35% of corporate sales generated outside of western Europe today; the most heavily dependent on basic chemicals, which now accounts for 65% of total sales (Bernstein 1996); and the most highly vertically integrated. The restructuring of these firms in the 1990s has proceeded to a remarkable degree along lines laid down before World War II — or even before the start of the 20th century (Hoover's Handbook of World Business, 1993).

GREAT BRITAIN: IMPERIAL CHEMICAL INDUSTRIES

The dominant British chemical firm for much of this century has been Imperial Chemical Industries. The strategy, performance, and structure of ICI reflect influences dating back more than 100 years.

The Predecessors and Formation of ICI: 1890–1914

The rise of Germany's "Big Three" in the late 19th century severely weakened the British dyestuffs industry, the logical source of commercial developments based on organic chemistry. At the beginning of the century, the two largest British chemical companies were both based on imported technology. Nobel Industries, formed in 1887 by Alfred Nobel and a group of Scottish industrialists, exploited Nobel's explosives technologies. Brunner Mond, the British licensee of the Solvay patents for alkali production, as well as soda ash and bleaching powder, dominated its industry after the establishment of its production facility at Winnington in 1882. The third major British chemical firm at this time was United Alkali Corporation, formed in 1891 in a confederation of other British alkali producers, which were using the older smaller-scale LeBlanc process. Both Brunner Mond and UAC were dominated by members of the owning families well into the following century. In contrast, because of its foreign ownership and its membership in the multinational cartel, Nobel Explosives was more under the control of professional, nonowner managers.

Formation and Early Operations of Imperial Chemical Industries: 1914–1945

The creation of Imperial Chemical Industries in 1925 involved Nobel Industries, Brunner-Mond, UAC, and a fourth firm: British Dyestuffs Ltd. This last firm was created by the 1919 merger of British Dyes, a government-owned enterprise established during World War I to replace dyestuffs that could no longer come from Germany, with Levinstein, one of the few surviving private producers of dyes in Britain. Thus, British Dyestuffs was an uneasy confederation between two firms with deeply antagonistic managements. During World War I, British dyestuffs production had made a modest comeback. Domestic production of dyestuffs accounted for roughly 20% of British consumption in 1914, although this was concentrated in the least sophisticated products. By 1919, British dyes accounted for nearly 40% of British dyestuffs consumption (Haber 1971, p. 188). Postwar British governments' continual wartime import restrictions to protect British Dyestuffs, in the face of strong opposition from British textile firms, the major customers for dyestuffs. There was some talk that British Dyestuffs would organize a national R&D program in dyestuffs research, but nothing came of it (Reader 1970, p. 275).

Following the end of World War I, British Dyestuffs proved incapable of developing the technology, management, and products needed to compete with German firms. Given its technical weaknesses, British Dyestuffs saw no alternative except to negotiate with German firms for access to German chemicals technologies, in exchange for access to the British market. However, a draft agreement between British Dyestuffs (in which the government retained a 17% ownership stake) and the "Little IG" was rejected by the British government in late 1923.

The competitive outlook for Britain's chemicals industry was further clouded by the formation of IG Farben in 1925. The formation of ICI in 1926 was a direct response to the creation of IG Farben. With strong encouragement from government advisers, Brunner Mond and Nobel Industries merged and acquired United Alkali and British Dyestuffs. Both policymakers and senior ICI management (dominated by personnel from Brunner Mond and Nobels) hoped to build a competitive British chemicals industry out of their own and British Dyestuffs' capabilities. But from the start, the senior managers of the merged companies had difficulty integrating their own, very different, set of learned capabilities and creating new ones in organic chemicals. Indeed, the defining difference between ICI and the German and American chemical companies flowed from ICI's origins in a merger of large enterprises that for the previous 40 years had been developing different technical and functional capabilities: Nobel Industries as a producer of explosives and propellants and Brunner-Mond as a producer of soda ash for fertilizers and bleaching powder. This mixed heritage created barriers to the merger's ability to commercialize new products and processes.

Until the 1926 merger, Nobel Industries had remained primarily a producer of explosives and propellants. Unlike DuPont after World War I, Nobel Industries did not diversify into other nitrocellulose products such as paints, pigments, finishes, rayon, and plastic products. Conversely, by 1926, Brunner Mond had become even more committed to the production of basic ingredients for fertilizer than it had been in 1914. Immediately after the war, with the encouragement of the British government, it had begun to build a plant to produce synthetic ammonia using the Haber-Bosch process initially developed by BASF (much as DuPont and Allied Chemical did in the United States). In April 1920, Brunner Mond and Nobel agreed to subscribe £2 million to begin plant construction at Billingham. Shortly afterward, Nobel Industries withdrew. As William Reader (1970, p. 357), ICI's historian, points out: "Brunner Mond was thus left sole owners of an enterprise which looked like it was becoming much the biggest thing they had ever taken on. It was certainly much the biggest project in the British chemical industry of the day... Once Brunner Mond had set out along this road, there could be no turning back, and by bringing Billingham with them into ICI they dictated where a large part of a new concern expansion must inevitably lie, for with so much capital, by that time, committed there, it was impossible not to commit more."

Although the senior managers of ICI remained committed to explosives and the production of synthetic nitrates and ammonia, they did attempt to create a learning base for the production of dyestuffs and other chemicals, including polymers. Their increased attention to systematic research and development led to the invention of new dyes and polymer chemicals, including polyethylene and perspex (polymethylmethacrylate). But few managers at ICI had well developed skills in commercializing such products, which required production, marketing, management, and distribution capabilities that differed considerably from those of importance to ICI's heavy-chemicals divisions.

During the 1930s, ICI maintained its commitment to the synthetic-ammonia plant at Billingham, even as costs rose and the anticipated market failed to appear (Reader 1977, p. 232). Because of its financial losses and the growing concern of the British military for gasoline and other fuels, the Billingham project was shifted to the production of synthetic fuels. Even then, only the British government's commitment in 1934 to guarantee a price floor for the fuels produced at Billingham—a decision that followed a similar commitment by the German government to IG Farben in 1933— enabled ICI to recoup a portion of its investment.

It was settled policy in ICI to avoid competition with customers or suppliers, which effectively meant avoiding competition with virtually every manufacturing company of any importance in the United Kingdom. For example, ICI kept clear for many years of chemical products associated with industrial alcohol, supplied by the Distillers' Company, Ltd., which were of considerable importance to the fields of organic activity opening up in the late 1920s and early 1930s. Similarly, to avoid giving offense to Courtaulds, important customers for caustic soda, ICI kept well clear of rayon and cellophane (Reader 1979, p. 174).

ICI's ambivalent attitude toward competition was replicated in its international markets, where the firm joined technology-sharing agreements with DuPont, IG Farben, and other chemical firms that would limit entry into its markets in the British Empire but also restricted ICI's opportunities to expand in the United States and Germany. In some cases, these international agreements also limited ICI's access to new technologies that might otherwise have accelerated the firm's diversification away from its dependence on heavy chemicals (Reader 1979, p. 174).

In the late 1930s, ICI's sales remained dominated by the products of predecessor companies. For example, in 1938, 24% of sales came from metals and 11% from explosives, the legacy of Nobel. Another 30% of sales came from alkalis and fertilizers, the legacy of Brunner Mond. Another 10% of sales were from dyes, the legacy of British Dyestuffs. Thus, less than one-quarter of sales were in other areas such as paint, synthetic rubber, combustion materials, salt, and so on (Plumpe 1990). These older product lines tended to have relatively low margins; indeed, sales per employee were always 10% lower at ICI than at IG Farben or DuPont over the years from 1929 to 1943 — and the gap was typically closer to 35% lower (Plumpe, 1990).

ICI's central management — like that of IG Farben — was smaller and less powerful than that of such firms as DuPont, and the firm's internal structure largely reflected the product lines and responsibilities of the leading firms in the 1926 merger rather than any technological or market-based relationships and interdependencies. The firm's R&D operations were decentralized and under the control of its product divisions; ICI established a corporate R&D laboratory only in 1949. This inherited structure often created barriers to commercialization. For example, ICI made some remarkable technical advances in plastics, but the commercial development and marketing of new products was controlled by the parent Alkali Division, with deep roots in heavy industrial chemicals and relatively few marketing or technical capabilities that were relevant to plastics. ICI's plastics products remained under the control of the Alkali Division until 1958. There were many similar stories of unexploited technical advances. These problems persisted for decades, and eventually forced a large- scale restructuring of ICI.

The Early Postwar Era for ICI: 1945–1970

Although ICI remained profitable through the mid-1970s, its postwar performance left much to be desired. Like the German firms, ICI faced a global chemicals industry that in the postwar era was characterized by much stronger competition. The cartel agreements of the interwar period were dissolved by wartime or antitrust action. Postwar liberalization of the global trading system created opportunities for entry by competitors. Competition within former Commonwealth markets also was intensified by the development of new domestic chemical firms that could exploit the availability of process technologies from petrochemical firms or specialized engineering firms. ICI's postwar performance was further hampered by its dependence on the British domestic market, which grew slowly during the postwar years.

ICI's senior management was slow to respond to these challenges. The firm's heavy-chemicals divisions slowed exploitation of internally generated technical advances in polymer chemistry (such as polyester synthetic fibers, polyethylene) and pharmaceuticals (for instance, "beta-blocker" drugs). The heavy-chemicals product divisions were slow to diversify into U.S. and West European markets. Senior corporate managers relied heavily on financial indicators and targets, with little substantive scrutiny of the strategic directions of the product divisions and virtually no effort to provide a broader strategic vision (Pettigrew 1985, p. 198). Moreover, these managers focused on the firm's mature product lines and allocated insufficient investment funding to support new products. In the case of polyethylene and polyester fibers, for example, ICI may have unwisely licensed technologies with enormous commercial potential to foreign firms, because it had not developed the financial and managerial capabilities for the volume manufacturing, marketing, and distribution in West European markets before the patents expired (see Landau, this volume).

During this time, ICI shifted its primary feedstock from coal to petroleum, as did the rest of the chemical industry. Like BASF, ICI had developed significant internal expertise in managing high-pressure, large-scale chemical processes for ethylene and propylene and thus was able to make the transition to petroleum feedstocks without a petrochemical partner firm. A large petrochemical complex was established at Wilton, close to the Billingham site. By the mid-1960s, ICI had entered into a joint venture with Burmah Oil to explore for petroleum deposits in the North Sea, and by the 1970s, like many U.S. chemicals producers, ICI was engaged in extensive petroleum production operations in the United States and elsewhere.

Restructuring: 1970 to the 1990s

The competitive pressures within the chemical industry and the oil price shocks of the 1970s created difficulties for the entire chemical industry; however, the problems were often worse at ICI. First, despite some efforts to diversify into foreign markets, ICI was still very much a British company trapped within a slowly growing domestic market. As Pettigrew (1985, p. 461) points out, as late as 1978, more than 50% of ICI's sale and 75% of its profits originated in the United Kingdom. Second, ICI had remained dependent on sales of heavy-chemicals products — petrochemicals, general chemicals, agriculture and plastics — whose markets were growing especially slowly. By the early 1980s, ICI had begun to accumulate significant losses, and it was clearly weaker than many large international chemical companies, particularly those of the United States. For example, ICI averaged a 12.8% return on assets from 1970 to 1978, while DuPont averaged 17.5%, Dow 18.7%, and Monsanto 15.7% (Pettigrew 1985, p. 80, reporting findings from Vivian, Gray & Co., 1980). ICI nevertheless did outperform the reported profits and returns on assets of its major German competitors. But another important measure of performance, sales per employee, reveals ICI in last place among major producers in 1978, trailing DuPont, Monsanto, Bayer, and Hoechst by only 10% to 20%, but lagging BASF and Dow by 40% to 50%. These dismal results reflect ICI's reliance on low-profit products in slow-growing markets.

Beginning in the 1980s, ICI finally underwent an extensive restructuring that sought to expand its activities in the more profitable specialty-chemicals and pharmaceuticals areas, align its product divisions more closely with new markets, and diversify more aggressively into foreign markets. In 1983, ICI and its major domestic competitor in petrochemicals, British Petroleum, agreed on a "swap" of capacity that effectively removed ICI from the manufacture of polyethylene (discovered by ICI) and expanded its activities in polyvinyl chloride; the effects on the product portfolios of ICI and BP were dramatic. During 1982 to 1987, ICI shrank total employment by 50,000 and expanded into the U.S. market by acquiring Glidden, a major paints producer, and Stauffer Chemicals, followed by the sale of all of Stauffer's product lines other than agricultural chemicals. These changes, however, brought new difficulties. "The push into specialties and the purchases in the United States only created new management problems, while the booming markets of the mid-1980s blunted the immediate pressure to withdraw from commodities. ICI still carried the burden of a geographical and product diversity that was exceptional among the world's leading chemical companies. Moreover, the international competitiveness of parts of the portfolio remained weak" (Owen and Harrison 1995, p. 135).

There was one bright spot: ICI's pharmaceutical business. Its first pharmaceutical products came, as such products had in Germany, out of research in dyes. But the initial-large scale commercialization of drugs came only with the demands of World War II. ICI pioneered in penicillin and was particularly successful with its antimalarial drug, Paludrine (Reader 1979, pp. 286, 460–61; Breckon 1972, pp. 39–41). In 1955, the pharmaceutical unit became a full-fledged integrated division. In 1957, the division set up new, extensive research facilities and then built its own production facilities and global-marketing organization. During the 1960s, that division became an international enterprise with manufacturing and processing plants in the United States, Europe, Latin America, Asia, and Africa.

ICI's pharmaceutical business also prospered because of the postwar therapeutic revolution, with its plethora of new prescription drugs, the large British home market provided by the National Health Service, and the fact that its two major competitors — Beecham and Glaxco — had not created their prescription-drug production and marketing organizations before ICI did. Through the 1970s, ICI's pharmaceuticals remained the most profitable of ICI's operating units, even though its share of corporate sales fell off from the heights of the late 1960s. During the corporate reconstruction of the 1980s, the pharmaceuticals division of ICI adopted a strategy of concentrating on the capabilities that it had developed since the 1950s, "areas where proven expertise exists." This was particularly true in cardiovascular agents and cancer treatments, which by 1988 accounted for 74% of ICI's pharmaceutical sales (Taggart 1993, pp. 270–71). In the second half of the 1980s, the pharmaceutical division's trading profits averaged just under 30% of sales, far greater than those of its other product lines. In pharmaceuticals, ICI did what it had failed to do in polymers/petrochemicals — it created the functional production and marketing capabilities to commercialize the products of its research investments.

Indeed, the success of ICI's pharmaceuticals unit led to a dramatic restructuring of the company in 1992. In 1991, the British conglomerate Hanson bought 2.8% of the stock of ICI. To discourage any threat of a takeover, ICI spun off its pharmaceutical and smaller biomedical units into a separate corporation, Zeneca. ICI retained the chemical business, which was still dominated by heavy chemicals (Hoover's Handbook of World Business 1993, 1996). The decision to split ICI into separate pharmaceutical and heavy-chemical firms in many respects illustrates the ability of equity markets to instigate far-reaching changes. It also highlights the continuing curse of inheriting two original, somewhat incompatible, bases and the value of creating a new base to exploit higher value-added products.

Since the separation of Zeneca from ICI, the chemicals firm has continued to diversify away from the less profitable heavy-chemicals product lines, while seeking opportunities for expansion into North America and Asia. Even before the split, ICI had swapped its European nylon production for DuPont's acrylics business. Since then, it has divested the polypropylene production complex in northeast England that was acquired in the British Petroleum swap and has sold other mature product lines, in an effort to focus more intensively on paints, polyester intermediates, titanium dioxide, and alternatives to CFCs. Significantly, the first CEO of the "new" ICI, Charles Miller

Smith, has a background in consumer products rather than heavy chemicals. Smith has made various statements endorsing the primacy of enhancing shareholder value and meeting tough targets for return on assets; however, it seems likely that he will also preside over cutbacks in the firm's basic research activities (Layman 1996). In 1997, Miller Smith moved and acquired the specialty-chemicals business of Unilever, from whom he had come, for $8 billion. He announced plans to sell off heavy-chemicals operations. Ultimately, the new ICI would be very different.

JAPAN: MITSUBISHI CHEMICAL, SUMITOMO CHEMICAL, MITSUI CHEMICALS, SHOWA DENKO, AND UBE INDUSTRIES

The basic evolutionary pattern of the strategy and structure of the largest chemical companies in Japan differs significantly from the pattern in the United States, Germany, and Great Britain. This difference reflects the fact that Japan's modern chemical industry of the nation developed after those three advanced economies had firmly established their own chemical industries. But the contrast has more to do with the basic capabilities and strategies that Japanese enterprises had to adopt and nurture in order to survive and grow in the international markets dominated by the enterprises discussed earlier.

The five enterprises reviewed in this section — Mitsubishi Chemical, Sumitomo Chemical, Mitsui Chemicals, Showa Denko, and Ube Industries — are the largest chemical firms in Japan that have integrated from basic ethylene cracking to various derivatives and intermediate products. When the term "all-around chemical enterprises" (*sogo kagaku gaisha*) is used in Japan, it customarily refers to these five companies. Historically, the five enterprises (and their predecessors) were the major forces in the development of Japan's modern chemical industry.[24] The strategic and structural problems of these firms for the last quarter of the century also exemplify the fundamental weaknesses that the entire Japanese chemical industry has had to confront in order to achieve international competitiveness in technological progress and product development.

Two major factors have distinguished the growth strategy of these five Japanese firms, from those of their competitors elsewhere. First, the critical capabilities for the commercial success of Japan's chemical enterprises, as in many other industries, rest on quick learning and incremental improvement of imported technology. While the giant diversified chemical firms of Germany, Great Britain, and, particularly after World War II, the United States, developed mainly as developers and large-scale commercializers of new products and processes, Japanese chemical firms imported commercialized technologies. Second, the leading Japanese firms differ from their Western counterparts in terms of their group affiliation. All five Japanese firms are parts of large diversified corporate or business groups (*kigyo shudan*). (For more details, see the chapter by Hikino, Harada, Tokuhisa, and Yoshida) For example, Mitsubishi Chemical belongs to the Mitsubishi Group, which forces the company (even after its merger with Mitsubishi Petrochemical in 1994) to share chemical markets with other important chemical-related enterprises within the Mitsubishi Group such as Mitsubishi Gas Chemical, Mitsubishi Plastics, Asahi Glass, and Mitsubishi Rayon.

---

[24]The chairperson of the petrochemical industry's association has always been chosen from among these five companies.

Diverse Origins and Developments: To 1945

Japan's chemical industry before World War II contained two basic types of large enterprises. Firms within old diversified *zaibatsu* groups had a long but rather desultory history of involvement in chemicals — for instance, Sumitomo Chemical, Mitsui Chemical, and Mitsubishi Chemical; and enterprises controlled by emerging entrepreneurs — such as Showa Denko and Ube Industries. The caution of the old *zaibatsu* resulted mainly from the conservatism shared by the owning families and senior managers, who emphasized the maintenance of family fortune and income. During the pre-1940 period, a small group of aggressive entrepreneurs seized this opportunity by importing the latest technology available and became leaders of Japan's chemical industry (Udagawa 1984; Morikawa 1990). In fact, the new major industrial groups such as Nissan, Nichitsu, Nisso, and Mori were organized around large electrochemical firms.

Among the old *zaibatsu* groups, Sumitomo's involvement in chemicals started earliest. The Sumitomo family had long operated a huge copper-smelting facility at the Besshi-Niihama mine on the island of Shikoku. The smelting process involved heavy emissions of sulfur dioxide, and in the early 20th century, research was conducted to find a way to reduce this health hazard. In 1913, the Sumitomo Fertilizer Factory was established to introduce a new technology that could transform sulfur dioxide into sulfuric acid and calcium superphosphate, which were to be utilized as fertilizers (Iijima 1981; Shimotani 1992). The chemical operation at Besshi-Niihama was reorganized 1925 as an independent company, Sumitomo Fertilizer Manufacturing. The company used public stock offerings to raise funds for entering ammonia production, which it did in 1928. In 1934, it added a new product line of nitric acid and changed its name to Sumitomo Chemical Company. Sumitomo Chemical became one of the biggest and most comprehensive chemical producers when, as a part of wartime industrial reorganization, it acquired Japan Dyestuff Manufacturing in 1944. The company's product line then consisted of inorganic, organic and agricultural products, and pharmaceuticals.

Mitsui's involvement in the chemical industry originated immediately before World War I when the Miike mine coking plant delivered the first domestically produced synthetic dyestuffs. Thanks to a shortage of imported dyes during the war, Mitsui's manufacturing operation expanded in spite of the poor quality of its products. The dyestuffs division became financially successful around 1926, and it became legally independent in 1941 as Mitsui Chemical Industries. Mitsui was the biggest dyestuffs manufacturer in Japan throughout the pre-1940 period.

In contrast to the case for Sumitomo and Mitsui, Mitsubishi's extensive involvement in chemicals came only in 1934, when Mitsubishi bought the Makiyama coking factory in northern Kyushu and reorganized and modernized it to become the main dye-making plant of the newly formed Nippon Tar Industries. In 1936, when the firm diversified into such coal chemicals as coke and related products, fertilizer, and ammonia and its derivatives, it changed its name to Nippon Chemical Industries (Yamaguchi and Nonaka 1991). Nippon Chemical Industries grew rapidly during the 1930s and 1940s, when the company started explosives manufacturing and followed Japan's military expansion by launching the production of cokes, magnesium, and agricultural chemicals in northern China. In 1942, the company took over Shinko Rayon, and in 1944, it absorbed Asahi Glass, both Mitsubishi-affiliated companies. It then became Mitsubishi Chemical Industries, a diversified manufacturer of coal-based chemicals, agricultural chemicals, glass, and rayon.

Showa Denko represents a set of aggressive emerging enterprises that were known as the "new zaibatsu." The company differs from the other four in that its technological background was electrochemical rather than coal-chemical. The company was formed in 1939 with the merger of two enterprises of the Mori group: Japan Electrical Industries and Showa Fertilizer. The first had originated in 1908 (as Sobo Marine Products) to manufacture iodine and potassium chloride from kelp. In 1926, the company was reorganized as Japan Iodine and produced a wide range of electrochemicals such as calcium carbide, electrodes, abrasives, and ferro-alloys. It became an original producer of aluminum in Japan in 1934 and soon changed its name to Japan Electrical Industries. In the meanwhile, Showa Fertilizer was founded in 1929 within the Mori group, for manufacturing fertilizers such as ammonium sulfate, which the company pioneered in Japan (Udagawa 1984; Molony 1990).

Ube Industries belongs to yet another group of large firms that developed in local areas and so often were called "local *zaibatsu*." Ube Industries was a war-related merger of four local companies of the Ube region in western Japan. The oldest and largest was Okinokyma Coal Mine; the others were Ube Cement, Shinkawa Iron Works, and Ube Nitrogen Industries. The development of Ube Industries was heavily shaped by the firm's experience in manufacturing coal-mining machinery from Shinkawa and by the technological skills of Ube Nitrogen, which in 1933 was producing ammonia and sulfuric acid by using coal and in 1936 succeeded in producing nitric acid and in distilling gasoline from the low-temperature carbonization of coal (Morikawa 1988).

When World War II largely stopped the inflow of new technical information and had a leveling effect among the firms discussed here, by allowing the old *zaibatsu* chemical firms to catch up with the technological leaders. However, the war also had a destructive impact on the newer chemical groups. Since the late 1920s, many of the groups and enterprises, seeking a cheaper source of energy, had aggressively transferred their operations to Korea and northern China. As the war ended, the groups not only lost a substantial part of their production facilities but also became a political target, because of their colonial aggression. Moreover, the organizational design created by these new groups did not enable them to weather the turmoil that resulted from the war. As groups, therefore, they simply collapsed, although many individual constituent companies, such as Asahi Chemical, Hitachi, and Nissan Automobile, survived.

The Emergence and Development of Petrochemicals: 1945–1970

The five Japanese companies discussed were instrumental in propelling the nation's chemical industry into the age of petrochemicals, a change that transformed the basic structure of Japan's chemical industry. Despite their diverse origins, the five firms all undertook massive investments in upstream, basic petrochemicals. Through the 1960s, domestic demand for upstream basic chemicals grew quickly, as downstream users such as producers of synthetic fibers, plastics, synthetic rubber, automobiles, and electronics developed rapidly. Because the industrial policy of the government did not allow unlimited entry, the result was years of high profitability for petrochemical producers.

Following these five firms, many enterprises attempted to enter into basic petrochemicals. Most could not make it, mainly because of financial constraints. They then had to find their own market niches in downstream fine and specialty chemicals. A few actually succeeded and, ironically, became prominent chemical producers after the age of basic petrochemicals came to an abrupt end in the early 1970s and the five petrochemical enterprises started struggling financially.

During the period of rapid growth in petrochemicals, Japan's five companies followed a common strategy. All were somewhat hesitant to leave behind their trusted coal technology for petrochemicals. Once the decision was made, however, they made continuous efforts to locate and import the latest product and process innovations from the United States and Europe (Mason 1992, pp. 209–18). They invested heavily in their own incremental-process improvement capabilities and built ever-larger facilities for ethylene production to exploit scale economies. To finance this strategy, they drew upon their substantial profits and also on low-cost loans from the government and the so-called main banks of groups (Kawade and Bono 1970; Watanabe 1972; Hamasato, 1994). However, within this common strategy, each firm took a somewhat different path.

Mitsui played a pioneering role in Japan's commercial entry into petrochemicals when in April 1958, Mitsui Petrochemicals' Iwakuni ethylene plant started its naphtha-cracking facility constructed with Stone & Webster technology. Mitsui's entry into ethylene production had originally been conceived in the context of coal-chemical technology, which had been Mitsui's strong background. Financial commitment and technical difficulties associated with a naphtha-cracking facility were initially thought to be too heavy a burden for Mitsui to accept to get into petrochemicals (Iijima 1981).

When Toyo Rayon, another member of the Mitsui Group and a major user of basic petrochemical products, expressed concern about the future of coal-based chemicals, the group reluctantly reached an agreement on the establishment of Mitsui Petrochemicals. Because there still was a strong inclination toward coal chemicals within Mitsui Chemical, and because the initial cost of an economically viable naphtha-cracking facility was too high for the company, a separate firm was organized by the investment of seven Mitsui members: Mitsui Chemical, Toyo Koatsu, Mitsui Mining, Toyo Rayon, Miike Synthetic, Mitsui Metal Mining and Mitsui Bank. Koa Oil, the Caltex-associated oil-refining company, also had a minority stake. The negotiating process among the constituent Mitsui companies that led to the formation of Mitsui Petrochemicals caused those often-antagonistic parties to reunite and reintegrate into the group as a unified business entity.

After World War II, Mitsubishi Chemical Industries was divided into three parts under a law intended to reduce the economic power of the old *zaibatsu* groups. The basic chemical operation became Nippon Chemical Industries, while Shinko Rayon took over rayon manufacturing; glass-making was separated as Asahi Glass. When the American occupation was formally terminated in 1952, however, Nippon Chemical Industries changed its name back to Mitsubishi Chemical Industries, and Shinko Rayon became Mitsubishi Rayon. Although the three enterprises kept their independent identities, they engaged in strategic coordination within the Mitsubishi group.

Mitsubishi's involvement in petrochemicals began in 1954, when Mitsubishi Chemical and Shell proposed a joint venture with the narrow focus of producing isopropyl alcohol, acetone, and derivatives. The limited product line partly came from Shell's uncertainty — shared by other foreign oil companies operating in Japan — about both the future growth of Japan's petrochemical markets and the technical capabilities of domestic chemical firms (Kudo 1990). However, MITI wanted to nurture more comprehensive and integrated ethylene-based petrochemical combine, so it declined the Mitsubishi-Shell proposal and asked Mitsubishi to come up with an alternative plan (Kudo, 1990).

By 1955, Mitsui and Sumitomo had already initiated petrochemical programs, and when a huge piece of coastal land in Yokkaichi near Nagoya became available, Mitsubishi Chemical drew up a concrete plan to launch a large-scale petrochemical facility. Shell remained dubious about the feasibility of Mitsubishi's idea and withdrew. Since Mitsubishi Chemical was not large enough to finance the plan alone (Mitsubishi Yuka 1988), Mitsubishi Petrochemical was formed with the equity held by core Mitsubishi enterprises such as Mitsubishi Chemical, Mitsubishi Rayon, Asahi Glass, Mitsubishi Bank, Mitsubishi Metal Mining, and Mitsubishi Trading. Shell eventually offered technical assistance on ethylene and styrene monomer in exchange for a 15% equity stake in the new company (Mitsubishi Yuka 1988; Kudo 1990).

Sumitomo Chemical also went through name changes after the war. The company was renamed as Nisshin Chemical Industries in 1946, but then revived its old name in 1952. The company positioned itself as the only chemical company within the Sumitomo Group, which was unusual among group-affiliated chemical firms, and Sumitomo Chemical, along with Sumitomo Bank and Sumitomo Metal, became the core of the group (Hadley 1970; Kudo 1990). During World War II, Sumitomo had a strong tie to the engineering department of Kyoto University, the site of research on synthetic rubber and high-density polyethylene. Sumitomo Chemical bought high-density polyethylene know-how from ICI, and that move was followed by the introduction of the Stone & Webster naphtha-cracking technology (Iijima 1981; Shimotani 1992). Sumitomo Chemical alone, rather than as part of a group, invested in the new technology on a relatively small scale. This was partly because of Sumitomo's tradition of "one company for one trade" and because the group lacked any rayon or synthetic-fiber manufacturers that could provide a guaranteed outlet for ethylene and derivatives, as Mitsubishi and Mitsui did.

Showa Denko established Showa Petrochemical in 1957, when the parent company formalized an extensive technical licensing agreement with Phillip Petroleum concerning entry into petrochemical products in Japanese markets. Showa Petrochemical was a joint venture of 17 companies loosely associated with Fuji Bank and targeted the commercialization of high-density polyethylene. MITI, fearing an oversupply of ethylene in Japanese markets, did not allow Showa Petrochemical to establish its own ethylene-production facility. When Nippon Petrochemical, a subsidiary of Nippon Oil, expanded a petrochemical complex in Kawasaki to facilitate ethylene production in 1959, therefore, Showa Petrochemical participated in the complex by building a plant for high-density polyethylene production. The Kawasaki complex was termed a "new corporate group," because, in contrast to the previous cases of Mitsubishi, Mitsui, and Sumitomo, it contained many companies of diverse backgrounds and different group affiliations.

Showa Petrochemical ultimately constructed its own petrochemical complex in Oita, in 1969. In addition to Phillips Petroleum, Stone & Webster, Ethylene Plastics, and Eastman Chemical supplied necessary technology. Because the financial burdens of the complex were too heavy for the Japanese participants, Phillips Petroleum agreed to participate. The Ministry of Finance and MITI were not happy with the large-equity stake of the American company, because the ministries were seriously concerned about the possibility of domination of Japanese industry by global enterprises. Only when Phillips Petroleum substantially lowered the extent of capital participation did the ministries permit the construction of the Oita complex (Showa Denko 1981).

As early as 1960, Ube Industries launched a 15-year plan for diversifying from coal-mining and cement into chemicals. Given the rapid growth of Japan's domestic demand for chemicals in the 1960s, the timing worked out very well. Furthermore, the company's affiliation with the emerging Sanwa Bank group helped as well, because the group was then particularly weak in the upstream part of the chemical industry, while Sekisui Chemical in plastics and Teijin in textiles guaranteed a stable chemical demand within the group (Iijima 1981). Ube Industries entered petrochemical production in 1964, when the company participated in the development of the Chiba petrochemical complex, where Maruzen Petrochemical established an ethylene-production facility. In contrast to the other four major chemical companies, however, Ube Industries thereafter relied on outside suppliers for ethylene. In 1967, Ube Industries built a new chemical plant in Sakai, near Osaka, that produced ammonia, urea, and caprolactam, and it added a new facility for polypropylene in the following year, relying at various points on technical know-how from Dart, ICI, Texaco, and Kellogg. By 1970, Ube Industries had closed its last coal-mining facility (the Sanyo anthracite mine), and by 1972, petrochemical products, particularly caprolactam and fertilizers, accounted for 40% of the company's sales revenues, while cement also accounted for 40%.

The rapid development of basic petrochemical production had a critical converging effect on the strategy of these five Japanese companies: All the enterprises imported advanced foreign technology and fully exploited it for operational efficiency and improvements. At least among these five firms, the basic capabilities and strategies became generic. The investment competition in Japan to build larger and newer petrochemical facilities did not last forever. As early as 1970 or 1971, Japan's market could not absorb the vastly expanded supply of ethylene, as MITI had feared. Japan's basic chemicals industry entered a period of overcapacity, slow growth, and poor financial performance (Watanabe and Saeki 1984).

Continuing Struggles: From 1970 to the Present

Since the early 1970s, Japan's major chemical firms have struggled to restructure and rationalize their petrochemical facilities, to expand into overseas markets, and to diversify into fine and specialty chemicals and pharmaceuticals. None of the five enterprises, however, has found these strategies easy to implement (Watanabe and Saeki 1984; Tokuhisa 1995). After all, new product development and international manufacturing historically have not been strong capabilities of Japanese firms, so many of these restructuring efforts were carried out through joint ventures (Ward 1992). It also turned out to be institutionally and organizationally difficult for these large enterprises to restructure their commodity petrochemicals operations.

Mitsubishi Petrochemical, for example, was the largest petrochemical manufacturer in Japan and was hit hard in the 1970s. The company moved into electrochemicals, fine chemicals, engineering chemicals, and other specialty items, including biochemicals, pharmaceuticals, and agricultural chemicals (Mitsubishi Yuka 1988; Yamaguchi and Nonaka 1991). But at the same time, a member of the same group, Mitsubishi Kasei, was also producing petrochemicals, while placing particular emphasis on pharmaceuticals, high-performance materials, and electronic products such as optical disks (*Chemical Week* November 27, 1991; Ward 1992). Since the early 1970s, the need for reorganization and consolidation of the chemical businesses has been apparent. It took more than 20 years, however, to consolidate Mitsubishi Chemical and Mitsubishi Petrochemical. When the two firms merged in October 1994, they formed the eighth-largest chemical producer in the world. The delay of this merger symbolizes the difficulty of reorganizing Japan's petrochemical industry within the country's business and institutional contexts.

In fact, until 1997, the Mitsui group failed in combining its two petrochemical interests. Mitsui Petrochemical had adopted a strategy of expanding specialty product lines such as containers, packaging materials, and industrial supplies. Mitsui Toa tsu, which was the first company among the major petrochemical producers to recognize the seriousness of the overcapacity problem, since the 1970s, has undertaken a strategy of exiting from commodity petrochemicals and expanding its fine-chemicals and specialty-chemicals business lines, as well as expanding overseas sales (Ward 1992). Despite urging by the group and the government, the senior management of the two companies refused to merge, to consolidate and rationalize their operations.

Sumitomo Chemical was relatively slow in responding to the deteriorating business environment of the 1970s and was surpassed by Mitsubishi Kasei's rise as the country's largest chemical producer. But in 1977, aluminum manufacturing was spun off, and the basic fertilizer business was curtailed. In 1983, the Ehime Plant scaled down its ethylene and ethylene derivatives to concentrate production at the newer and more efficient Chiba Plant. Simultaneously, Sumitomo Chemical expanded its business lines in fine chemicals, particularly agricultural chemicals and pharmaceuticals, and new materials and fibers. In the early 1980s, for instance, Sumitomo Chemical established a joint venture with Hercules for the production of carbon fiber. In 1983, Sumitomo Chemical spun off pharmaceuticals manufacturing to Inabata & Co., with which the company later established a joint marketing enterprise, Sumitomo Pharmaceuticals. Sumitomo Chemical has also worked since the mid-1970s to develop overseas businesses. It is a major player in a new petrochemical complex, PCS, in Singapore. In 1987, Sumitomo Chemical established Valant USA in a joint venture with Chevron Chemical for marketing its products in U.S. markets. In the same year, the enterprise entered into an agreement with Rohm & Haas to manufacture new methacrylates. Sumitomo Chemical further strengthened its European marketing networks by founding subsidiaries in the United Kingdom and Netherlands in 1988 and in France in 1990.

Showa Denko's restructuring has been particularly difficult, since the two strategic areas the company had invested in through the early 1970s — petrochemicals and aluminum and ferro-alloy refining — were among the most depressed.[25] In a volatile environment of excess supply of basic petrochemicals, the inflation of the early 1970s, and the first oil shock of 1973, Showa Denko continued to invest in the expansion of its Oita complex. Phillips Petroleum, partner of Showa Petrochemical, became ambivalent about further investments in ethylene facilities and eventually withdrew from the joint venture altogether in 1974. Showa Denko then reabsorbed Showa Petrochemicals in 1979, as a part of a rationalization program (Showa Denko 1981). Showa Denko sought to enter such diverse fields as housing construction and sales, pollution control, new ceramics and materials, agricultural products, and biotechnology areas. The results, however, were mixed at best. Pollution control remained a popular but unprofitable business, and the building-materials business had to be spun off in 1995. Compared to the other four enterprises, Showa Denko has also been slow in its internationalization

---

[25]Showa Denko spun off its aluminum operation in 1976 as Showa Light Metal, which was eventually closed in 1986.

drive in chemicals. Showa Denko has only two small overseas manufacturing enterprises in plastic molding: one in Taiwan and another in Singapore (Ward 1992).

Ube Industries suffered from low profitability in the two major industries, petrochemicals and cement, in which the company had invested heavily up to the 1970s. The firm has attempted to diversify and globalize through joint ventures with enterprises whose technology is attractive to the company. For instance, in 1984 Ube Industries and Marubeni, a general trading company, entered into a joint venture with Wormser of the United States to manufacture fluidized-bed combusters. In 1989, Ube Industries and Kemira, a specialized chemical company from Finland, formed Kemira-Ube to produce hydrogen peroxide in Japan. In the following year, Ube Industries established two joint ventures with technology-oriented chemical firms: UBE-EMS with Ems-Chemi of Switzerland and Ube Rexene with Rexene of the United States. Ube Industries has also established joint ventures with two major Japanese pharmaceuticals-manufacturing companies, Takeda Pharmaceutical and Sankyo Pharmaceutical, to employ their marketing networks for pharmaceutical products developed by Ube Industries (Ward 1992).

The lengthy struggles of the five large Japanese chemical companies since the early 1970s are the major outcome of their lack of core technological capabilities. Lacking distinctive product-development capabilities, the five enterprises adopted similar strategies of locating and importing promising foreign technologies for commercialization. This common strategy, which had worked brilliantly in the 1950s and 1960s, did not bring any differentiating or defensible product or process innovations that ensured economic viability for the companies. The Japanese enterprises thus were forced to enhance their own product-development capabilities after 1970. All of the five firms extensively invested in fine and specialty chemicals and diversified into pharmaceuticals and related fields. These strategic shifts have so far turned out to be unrewarding, partially because of the slow progress of overall global chemical technology since the 1970s (*Chemical Week*, November 27, 1991. The contemporary Japanese chemical industry includes many struggling enterprises with similar capabilities whose reorganization has been difficult, because of the strategic constraints within a group structure and, constraints imposed by government policy.

CONCLUSION: CORPORATE CAPABILITIES AND CORPORATE STRATEGIES

The corporate capabilities of today's leaders in the United States, Germany, Britain, and Japan are the result of a long evolutionary process. After creating their initial learning bases, these companies became and remained national leaders by developing strategies that permitted them to build, enhance, and reshape their capabilities as the industry's technologies and markets changed and as global wars, economic booms and depressions, and other macroeconomic developments reshaped their national political environments. The history of the evolution of their product-specific capabilities (their "genetic code") provides an insight on the overall evolution of one of the most dynamic of today's science-based industries.

All of these firms were long-lived. The first to appear were the Germans, who played an essential role in making their nation Europe's industrial giant during the 1880s and 1890s. In those decades, they pioneered in the production of synthetic dyes, pharmaceuticals and their intermediates, and other coal-based organic chemicals, and they did so for increasingly global markets. Somewhat later one firm, BASF, developed the new high-pressure processes for producing ammonia, nitrates, and other inorganic chemicals. The American and British firms began to compete with the Germans in global markets during the 1920s on the basis of capabilities initially created before World War I. The Americans pioneered in the commercialization of inorganic chemicals produced through electrolytic processes and those based on nitrocellulose technology, and continued to maintain leadership in these product lines. In Britain, neither of the two leading firms developed capabilities in such broad fields of chemistry before World War I. After the war, Brunner Mond attempted, with little success, to follow the Germans into the new high pressure technology for the production of ammonium nitrate. Nobel Industries failed to use its nitrocellulose learning base to develop new product lines in the manner of DuPont. As the Japanese firms had to rely wholly on imported technology to move into the modern industrial world, their leading firms during the years before World War II sought to adapt these long-established technologies to the nation's industrial and military needs.

World War II transformed the world's chemical industry by creating a sudden, massive demand for products based on polymers and for petrochemical technologies for producing organic chemicals that had only just begun to be commercialized in the 1930s. Their nation's wartime requirements and abundant domestic oil supplies gave the American companies the lead in exploiting the potentials of these new technologies. In the 1950s, German firms drew on long-established capabilities in research, production, distribution, and marketing of organic chemicals to move more quickly into the new product lines than did Britain's ICI. The latter's inherited capabilities provided a much more limited base for commercializing polymer/petrochemicals. The company's R&D technological capabilities were first-rate, but its executives had far less experience in building the essential production and marketing facilities and personnel necessary to compete with the Germans and Americans in global markets. The Japanese firms, once again starting almost from scratch, concentrated on building their production capabilities in feedstocks and basic polymer commodities and again did so through imported technology, acquired often through joint ventures with Western firms.

By the late 1960s, the potential of the new technologies had been fully realized. The number of new products commercialized declined, as production capacity soared and competition intensified on an increasingly global basis. As price and profits fell off, the leading Western firms began to move into related or even unrelated product lines, largely through acquisition, even though their existing capabilities gave them little competitive advantage. The oil shocks, particularly that of the late 1970s, created crisis conditions in the early 1980s. During the 1980s, all of these leaders began to reshape their product portfolios by concentrating on products in which their existing capabilities gave them a strong competitive advantage. They maintained the lines in markets in which they were still strong and moved into higher value-added products. They sold off the less profitable lines and expanded, through internal investment, acquisitions, and joint ventures, those in which they had established an effective learning base. Only Union Carbide failed to make the move, and it was forced to sell off the lines that had made it the leader in relatively high value-added products. For the Japanese firms, which during the 1950s and 1960s had concentrated on enhancing the capabilities of their basic product lines, the challenge was greater than for their Western competitors. The Japanese firms had not yet developed to a sufficient extent skills needed to come up with higher-value end products. As they had done in the past, they resorted to foreign sources as the means to acquire and develop such capabilities.

This pattern of adjusting strategies to corporate capabilities was much the same for the smaller competitors of the corporate leaders that have been described here. They were integrated enterprises that competed globally, as well as locally. These chemical companies were older and, on the average, larger and more numerous than the leaders in other major global industries, including such high-tech industries as computers, consumer and industrial electronics, aerospace, and other medium-tech industries, such as motor vehicles and nonelectric machinery.

The detailed examination of the largest international chemical enterprises makes two basic points. First, all are large: the smallest has over a billion dollars in sales. Second, they are all long-established enterprises. The technological innovations that transformed the industry, and with it the global economy, have been commercialized almost wholly by large existing firms. Of course, commercialization could not take place without smaller firms acting as researchers, designers, suppliers, distributors, and providers of services. But the effectiveness, utility, and even existence of such networks of ancillary companies depend on the viability of the central nexus.

The lesson learned by the chemical producers during the past three decades has been that success depends on maintaining product-specific capabilities, not just technological, but — as important — those in production, marketing, and management, and doing so in a carefully selected small number of closely related product lines.

REFERENCES

Beer, I.I. (1959), *The Emergence of the German Dye Industry*, University of Illinois, Urbana, IL.

Bernstein, Sanford C., Inc., (1996), *BASF Group*, Bernstein Research.

Bower, Joseph L. (1986), *When Markets Quake: The Managerial Challenge of Restructuring Industry*, Harvard Business School Press, Boston, MA.

Bozdogan, Kirkor, (1989), "The Transformation of the U.S. Chemicals Industry," Working paper MIT Commission on Industrial Productivity.

Breckon, William (1972), *The Drug Makers*, Eyre Meuthen, London, U.K.

Burgelman, Robert A. and Leonard R. Sayles (1986), *Inside Corporate Innovation: Strategy, Structure and Managerial Skills*, Free Press, New York, NY.

Chandler, Alfred D., Jr. (1960), "Development, Diversification, and Decentralization," In Freeman, Ralph E. (ed.), *Postwar Economic Trends in the United States*, Harper & Brothers, New York, NY, pp. 268–71.

Chandler, A.D., Jr. (1962), *Strategy and Structure*, MIT Press, Cambridge, MA.

Chandler, Alfred D., Jr. and Stephen Salsbury (1971), *Pierre S. duPont and the Making of the Modern Corporation*, Harper & Row, New York, NY.

Chandler, A.D., Jr. (1977), *The Visible Hand*, Belknap Press of Harvard University Press, Cambridge, MA.

Chandler, A.D., Jr. (1990), *Scale and Scope*, Belknap Press of Harvard University Press, Cambridge, MA.

Chapman, K. (1991), *The International Petrochemical Industry*, Blackwell, Oxford, U.K.

Davis, Nigel (1996), "Spin-Off Is Inevitable Result of Monsanto's Long-Term Strategy," *Chemical Insight*, no. 596 (late December): 1–4.

Dutton, William S. (1942), *DuPont – 140 Years*, Scribner, New York, NY.

Eccles, Robert G. and Dwight B. Crane (1988), *Doing Deals: Investment Bankers at Work*, Harvard Business School Press, Boston, MA.

Fast, Norman (1979), *The Rise and Fall of New Venture Divisions*, UMI Research Press, Ann Arbor, MI.

Forrestal, Dan J. (1977), *Faith, Hope and $5,000*, Simon & Schuster, New York, NY.

Gibb, George S. and Evelyn H. Knowlton (1956), *The Resurgent Years, 1911–1927*, Harper & Brothers, New York.

Haber, L. F. (1971), *The Chemical Industry. 1900–1930*, Clarendon Press, Oxford, U.K.

Hadley, Eleanor, M. (1970), *Antitrust in Japan*, Princeton University Press, Princeton, NJ.

Hamasato (1994), Hisao, *Ronshu Nippon no Kagaku Kogyo*, Nippon Hyoronsha, Tokyo, Japan.

Hayes, P. (1987), *Industry & I.G. Farben in the Nazi Era*, Cambridge University Press, Cambridge, MA.

Haynes, Williams (1949), *American Chemical Industries, Vol. V, Decade of New Products*, and *Vol. VI, The Chemical Companies*, Van Nostrand, New York, NY.

Hounshell, David A. and John K. Smith, Jr. (1988), *Science and Corporate Strategy: DuPont R&D, 1902–1980*, Cambridge University Press, Cambridge, U.K.

Hoover's Handbook of American Business Reference Press, Austin, TX (1991–1996).

Hoover's Handbook of World Business, Reference Press, Austin, TX (1993, 1996).

Iijima, Takashi (1981), *Nippon no Kagaku Gijyutsu: Kigyoshi ni miru sono Kozo*, Kogyo Chosakai, Tokyo, Japan.

Kawade, Tsunetada and Mitsuisa Bono (1970), *Sekiyu Kagaku Kogyo*, new edition, Toyo Keizai Shimposha, Tokyo, Japan.

Kudo, Akira (1990), "Sekiyu Kagaku," in Shin'ichi Yonekawa, Koichi Shimokawa, and Hiroaki Yamazaki (eds.), *Sengo Nippon Keieishi*, vol. II, Toyo Keizai, Tokyo, Japan, pp. 279–336.

Landau, Ralph (1995a) interview with Edward Jefferson, March 22.

Landau, Ralph (1995b) interview with Frank P. Popoff, September 28.

Lane, Sarah, (1992), "Corporate Restructuring in the Chemical Industry," in Margaret Blair, (ed), *Takeovers, LBOs and Changing Corporate Forms*, The Brookings Institution, Washington, DC.

Larson, Henrietta M. and Kenneth W. Porter, (1959), *History of the Humble Oil & Refining Company: A Study in Industrial Growth*, Harper & Brothers, New York, NY.

Layman, P. L. (1996), "Customer Awareness Becomes Byword at Rapidly Internationalizing ICI," *Chemical & Engineering News*, no. 613: p. 14.

Liebenau, J. (1992), "The Management of High Technology: The Use of Information in the German Chemical Industry, 1890–1930," in T. Hara and A. Kudo (eds.), *International Cartels in the New Industries*, University of Tokyo Press, Tokyo, Japan.

Mason, Mark (1992), *American Multinationals and Japan: The Political Economy of Japanese Capital Controls, 1899–1980*, Harvard University Press, Cambridge, MA.

MIT Commission on Industrial Productivity (1989), "The Transformation of the U.S. Chemical Industry", working paper.

Mitsubishi Yuka Kabushiki Kaisha (1988), *Mitsubishi Yuka Sanjyunen Shi*, Mitsubishi Yuka, Tokyo, Japan.

Molony, Barbara (1990), *Technology and Investment: The Prewar Japanese Chemical Industry*, Harvard University Press, Cambridge, MA.

Moody's Industrial Manual, various years.

Morikawa, Hidemasa (1988), *Chiho Zaibatsu*, Toyo Keizai Shuppansha, Tokyo, Japan.

Morikawa, Hidemasa (1990), *Zaibatsu: The Rise and Fall of Family Enterprises*, Tokyo Daigaku Shuppankai, Tokyo, Japan.

Moskowitz, Milton, Robert Levering, and Michael Katz (eds.), (1980), *Everybody's Business: An Almanac*, Harper and Row, San Francisco, CA.

Moskowitz, Milton, Robert Levering, and Michael Katz, (eds.), (1990), *Everybody's Business: A Field Guide to the 400 Leading Companies in America*, Doubleday, New York, NY.

Owen, G. and T. Harrison (1995), "Why ICI Chose to Demerge," *Harvard Business Review*, (March/April):133–42.

Pettigrew, A. (1985), *The Awakening Giant*, Blackwell, Oxford U.K.

Plumpe, Gottfried (1990), *Die I.G. Farbenindustrie AG: Wirtschaft, Technik und Politik 1904-1945*, Duncker & Humblot, Berlin.

Reader, W. J., *Imperial Chemical Industries: A History*, 2 vols. (Oxford, 1970, 1975).

Reader, W. J., "Imperial Chemical Industries and the State, 1926-1945" in B.E. Supple, ed., *Essays in British Business History* (Oxford, 1977).

Reader, W. J., "The Chemical Industry," in W.E. Alford and Buxton, *British Industry Between the Wars* (Scolar Press, 1979).

Schroter, H. G. (1993), "The German Question, the Unification of Europe, and the European Market Strategies of Germany's Chemical and Electrical Industries, 1900–1992," *Business History Review*, 67:369–405.

Shimotani, Masahiro (1992), *Nippon Kagaku Kogyoshi Ron*, Ochanomizu Shobo, Tokyo, Japan.

Showa Denko Kabushiki Kaisha (1981), *Showa Denko Sekiyu Kagaku Hattenshi: Showa Yuka no Setsuritsu kara Kappei made*, Showa Denko, Tokyo, Japan.

Smith, J. K., Jr. (1992), "National Goals, Industry Structure, and Corporate Strategies: Chemical Cartels Between the Wars," in T. Hara and A. Kudo (eds.), *International Cartels in the New Industries*, University of Tokyo Press, Tokyo, Japan.

Spitz, Peter H. (1988), *Petrochemicals: The Rise of an Industry*, Wiley & Sons, New York, NY.

Stokes, R. G. (1988), *Divide and Prosper: The Heirs of I. G. Farben under Allied Authority, 1945-1951*, University of California Press, Berlekey, CA.

Stokes, R. G. (1994), *Opting for Oil: The Political Economy of Technological Change in The West German Chemical Industry, 1945–1961*, Cambridge University Press, New York, NY.

Taggart, J. (1993), *The World Pharmaceutical Industry*, London, Rutledge, U.K.

Taylor, Graham D. and Patricia E. Sudnick (1984), *DuPont and the International Chemical Industry*, Twayne Publishers, Boston, MA.

Tokuhisa, Yoshio (1995), *Kagaku Sangyo ni Miraiwa Aruka*, Nippon Keizai Shinbunsha, Tokyo, Japan.

Trescott, Martha M. (1981), *The Rise of the American Electrochemicals Industry: Studies in American Technological Environment*, Greenwood Press, Westport, CT.

Udagawa, Masaru (1984), *Shinko Zaibatsu*, Nippon Keizai, Tokyo, Japan.

Wall, Bennett H. (1988), *Growth in a Changing Environment: A History of Standard Oil Company (New Jersey) Exxon Corporation, 1950–1975*, McGraw-Hill New York, NY.

Ward, Mike (1992), *Japanese Chemicals: Past, Present and Future*, Economic Intelligence Unit, London, U.K.

Watanabe, Tokuji (1972), *Sekiyu Kagaku Kogyo*, 2nd edition, Iwanami Shoten, Tokyo, Japan.

Watanabe, Tokuji and Yasuharu Saeki (1984), *Tenki ni tatsu Sekiyu Kagaku Kogyo*, Iwanami Shoten, Tokyo, Japan.

Whitehead, Don (1968), *The History of the Dow Chemical Company*, McGraw-Hill, New York, NY.

Yamaguchi, Takashi and Ikue Nonaka (1991), *Asahi Kasei and Mitsubishi Kasei: Sentan Gijutsu ni kakeru Kagaku*, Otsuki Shoten, Tokyo, Japan.

PART V
The Modern Chemical Industry and Corporate Governance

14 Connecting Performance and Competitiveness with Finance: A Study of the Chemical Industry

ALBERT D. RICHARDS

If market competition is to encourage corporate efficiency, it is essential that the allocation of capital be weighted toward those firms with the best performance. But this raises immediate problems, as there are widely differing opinions around the world as to just what "best" is (or even should be!). This chapter explores several fundamental issues regarding corporate performance and capital markets.

Sensible corporate comparisons require consistent methodologies for measuring both performance and value. These methodologies are not obvious, especially when comparing companies that are active in different product lines and countries. To address this issue, we lay out a framework which can be used to circumvent these problems. Utilizing this framework in the analysis of a large amount of data from various leading firms in the chemical industry indicates that capital markets actually do a reasonably good job of evaluating not just the past performance of a company but also its likely future performance. This result, while perhaps disappointing from an investment-opportunity point of view (getting rich is never easy), does give us considerable additional confidence in our performance and valuation measures. Another conclusion from our analysis is much more provocative: Markets do a poor job of disciplining managers through the pricing mechanism alone. This chapter will explore how markets are perceived to discipline companies and then explain why they do not succeed especially well. In the absence of effective market discipline, other influences must be used to discipline managers (i.e., active shareholders), and we describe some mechanisms that might accomplish this goal.

The proper analysis of these issues requires understanding the relationships among finance, accounting, corporate performance, and capital market valuation. To develop this understanding, I have given a rather technical primer on how to evaluate companies. The reader unfamiliar with modern finance will find this somewhat challenging but at the same time all the more rewarding, as he or she will then be in the position to more fully grasp how capital markets function.

I will begin my exposition by explaining why many conventional performance measures are wrong. This will allow me to show that Internal Rate of Return (IRR) is

I would like to thank Peter Blair, Leslie Chow, Lewis Gasorek, and Judi Wind for their excellent assistance in preparing this manuscript.

461

the right performance measure[1] and to explain the not-insignificant difficulties in calculating this measure from the published accounting statements of an ongoing corporation. This discussion will be complemented by a brief discussion on how inflation and tax policies can further complicate the calculation of IRR, as well as a company's ability to earn an appropriate IRR. I next define a company's Economic Replacement Cost (ERC) and show how valuation is best described as a multiple of ERC. Having developed the proper performance measurement and valuation tools, I will then apply these tools to the chemical industry, showing how major chemical companies in Europe and the United States have performed in recent years and how markets have reflected this performance through the firms' market values. Some readers may indeed find these results surprising! I will then develop a central thesis to this chapter, namely that *markets alone do not have the power to force companies to deliver high economic returns.* In the final section of this chapter I will develop the implications of this seldom-appreciated fact for management and public policy. I should also state at the outset that while this chapter focuses on the chemical industry, the basic message is applicable to companies in every industry around the world.

DEFINING CORPORATE COMPETITIVENESS

Is Size or Growth a Proper Measure of Corporate Performance?

There is a natural tendency to assess performance by size of corporate revenue. Thus, let's begin the assessment of corporate performance by looking at the size and growth rates of a broad set of chemical companies, as shown in Figure 14-1. Notice that growth rates have been very similar for several of these companies — particularly the three German firms and DuPont (in the latter case, excluding the effect of the Conoco acquisition). Dow has also kept pace until the past few years, although it has always been a slightly smaller company. Others flourished in the past only to apparently flounder more recently — at least in turnover terms — e.g., Union Carbide and Monsanto. Thus, if one judges by size alone, it would be natural to conclude that DuPont, Bayer, BASF, and Hoechst have "outperformed" their (now) smaller competitors.

But this judgment immediately raises tough questions. Union Carbide diminished in size because many of the firm's businesses were sold off. However, if these businesses are now thriving — particularly if they are doing better than they would have done had they remained part of Union Carbide — then it is difficult to say categorically that Union Carbide has performed poorly. In general terms, the problem with size as a measure of corporate performance is that it is often the result of things most people would not associate with performance. Just because two companies merge, for example, does not suddenly make them more successful, nor are they necessarily unsuccessful because they have spun off or sold off certain operations, although unhappy shareholders are more likely to force a de-merger on an unsuccessful company. Growth rates of companies are correlated with size, and many observers argue that growth matters more than size. However, actually discerning growth rates from the previous charts is not particularly easy; therefore, we have calculated compound growth rates for our universe of major chemical companies for several different time periods, as illustrated in Table 14-1.

[1]Technically this measure should be Perpetual Equivalent Return (or Net Present Value), but IRR is much easier to use and perfectly adequate for the present analysis.

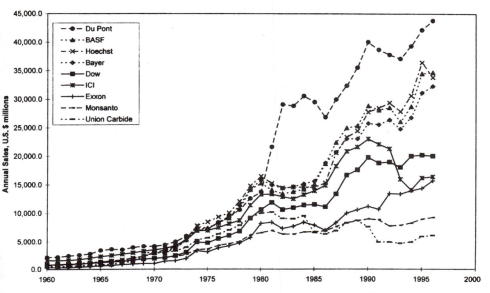

Figure 14-1. Chemical Company Sales, 1960–96 (U.S. $ million).

As shown, individual companies have grown at dramatically different rates over the past few decades. For the group as a whole, the boom years were obviously the 1960s and 1970s (the latter decade being aided by the rise in oil prices that pushed up chemical prices as well). The 1980s were more difficult, with the collapse in oil prices in 1986 resulting in significant price declines in chemicals; while the most recent decade has seen low growth rates.

There are also significant differences in growth rates depending on how we look at the data. For example, DuPont owns Conoco, which is an oil company. Including the acquisition of Conoco increases DuPont's average growth rate meaningfully, but the temptation for many industry analysts is to look only at DuPont without Conoco, since Conoco is not a true "chemical" company. To take another example, ICI spun off Zeneca, its pharmaceuticals division, in 1993. In this case, the temptation for many analysts in measuring the performance of ICI is to continue to include Zeneca's sales; after all, many other chemical companies still retain their pharmaceutical arms. But if it makes sense to adjust for these prominent corporate actions, what about the numerous smaller acquisitions and divestitures made by these companies in recent decades? Furthermore, should it matter whether these acquisitions were funded by internal cash flow or by raising capital on debt and equity markets?

The answer to these questions lies in the driving factor behind corporate growth: *Companies grow because they invest or acquire, not because they have been successful.* Corporate growth is therefore an indication of the level of investment — particularly if acquisitions are counted as investment — not a measurement of performance per se. If performance is indeed independent of growth, it naturally follows that value is independent of growth, as well. Indeed, one thing that I have learned, often the hard way, is that value is not a function of the growth rate, either in absolute or per-share

TABLE 14-1. Compound Growth Rates for Nine Major Chemical Companies (Sales in U.S.$)

Company	1960–69	1970–79	1980–89	1990–96	1960–96
Dow	9.2%	17.5%	4.9%	0.9%	10.3%
DuPont	8.0%	12.4%	6.8%	1.7%	9.7%
DuPont w/o Conoco	8.0%	12.4%	4.3%	3.0%	7.4%
BASF	14.5%	17.7%	7.4%	3.5%	12.6%
ICI	8.9%	12.6%	5.9%	−7.3%	7.9%
ICI w/Zeneca	8.9%	12.6%	5.9%	1.3%	8.5%
Bayer	11.7%	17.7%	5.5%	3.9%	11.0%
Hoechst	13.8%	17.9%	5.1%	4.0%	11.4%
Exxon	15.3%	19.0%	2.3%	5.9%	11.7%
Union Carbide*		13.3%	−3.3%	−1.0%	1.7%
Monsanto[†]		13.5%	3.0%	0.6%	5.2%
Average[‡]	11.6%	15.7%	4.2%	1.3%	8.9%

*From 1969 only.
[†]From 1971 only.
[‡]Excludes "Du Pont w/o.

terms of sales, earnings, or book value. Since this statement runs counter to the intuition of 99% of investors, it is probably best explained further. Consider two companies sitting on equal piles of cash: One company pays out the money to shareholders, while the other either buys another company or invests directly in new plant and equipment. Clearly, the second company will grow more. The value of the second company, however, will increase only if the retained cash is invested in projects with a positive net present value. Indeed, if a company engages in investments that do not keep pace with the opportunity cost of capital, i.e., projects with a negative net present value (an all-too-common occurrence), a higher growth rate actually decreases value.

The key concept here is that growth is a result, not a cause. A company's growth rate is determined by two things: return on investment and payout ratio. Furthermore, *it's a company's return on investment relative to its cost of capital that determines value.* Thus, if a company increases its growth rate merely by increasing its cash-retention rate, value increases only if the retained cash is invested in positive NPV projects — an all-too-rare occurrence in many segments of the chemical industry.

Is Share Price an Indicator of Performance?

The current craze in the Anglo-Saxon world is to assess corporate managers based on the price performance of the company's shares. This is clearly serious stuff, as multimillion dollar option packages often hang in the balance between the bulls and the bears. So how have the various chemical companies performed? Figure 14-2 shows share price progressions (in U.S. $) over the past few decades for each firm in our designated group of larger chemical companies. For each company, the share price is indexed to the average share price of that company over the 1960 to 1979 time period. This chart shows stock price performance rather than total return — which would include dividends — but the dividend yields of the various firms have been fairly similar, and thus a total-return chart would look much the same.

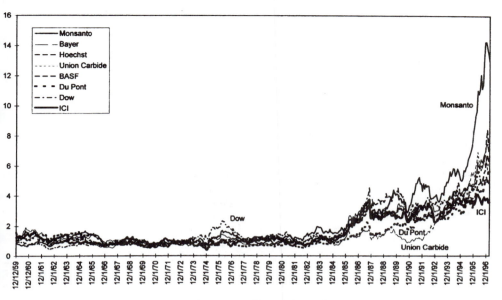

Figure 14-2. Chemical Company Share-Price Performance.

This chart may appear rather difficult to interpret, as the lines cluster tightly together, but their clustering is actually the main message of the figure. While share prices have varied somewhat, the variations are for the most part not nearly as great as many people expect. To put it in perspective, a mere 1% difference in performance per year would result in a 43% difference in performance over the 36 years of the table. Thus, despite the considerable variations in the growth rate of corporate sales shown earlier, the share-price performance of most of these companies has been nearly identical, for practical purposes. Furthermore, the main standout—Monsanto's terrific performance over the past decade—has been delivered by the company with the second-worst sales growth! Even Union Carbide, with all of its problems, has kept up with the pack quite well. While the ranking of share-price performance is reflected in the order in which the companies are listed in the legend (Monsanto first and ICI last), one should not read too much into this, since the order would have been completely different if the end point had been just one year earlier. The basic message therefore remains the same: Aside from Monsanto (ICI has recently recovered), no company has particularly distinguished itself in long-term share-price terms.

Do the remarkable similarities in changes in share price thus mean that managerial performance has also been roughly equivalent over this time period? The answer, perhaps unsurprisingly, is a resounding "no." Share prices are great at telling when corporate managers have outperformed or underperformed expectations, but they yield little information about what those expectations were in the first place. In other words, a terrifically performing company that is expected to remain terrific will have the same share-price performance as a terribly performing company that is expected to remain terrible—as long as both companies live up to their respective expectations. It's only after something unexpected occurs, such as the terrific company weakening just a little bit or the terrible company showing some signs of hope, that share prices exhibit underperformance or outperformance. Thus, the similar performance levels of the chemical companies shows only that the market is fairly efficient at putting expectations into the determination of asset values.

It's Internal Rate of Return That Counts

If corporate performance is not properly measured by either the level or growth of sales, or by share price performance, what is the appropriate measure? A hint was given earlier in the discussion of growth: A company that invests wisely will create returns above and beyond what would normally be available in the capital markets (for that level of risk) and will see its value grow. The rate of return on a company's own internal investments is called the *internal rate of return*. Internal rates of return are also a useful indicator of the relative performance of two companies, as long as these firms have equivalent asset lifetimes.[2] While the logic of comparing a firm's internal rate of return to the cost of capital is straightforward enough, the practicalities of actually applying this technique to existing corporations is considerably more difficult. The real problem, however, is not the theory, but the measurement: Traditional accounting is woefully inadequate as a basis for calculating or reporting a company's internal investment returns.

Accounting's Conflict with Internal Rates of Return

Accounting is merely a method of reporting history, and no single method of reporting will serve all possible purposes. For example, there are considerable differences between German and American accounting standards. Even after studying these two systems for many years, however, I do not know which standard is actually better; they are just very different. What I do know is that a blind comparison of German accounting figures on earnings, book value, and so on with American figures is often a very dangerous proposition. The inadequacies of traditional accounting are perhaps best illustrated by looking at the example of a "single investment company," or a firm that is formed solely to operate one single plant or facility, after which the company is liquidated. The most important thing to remember about such a project is that it has but one internal rate of return (IRR). This return is defined as the discount rate that must be applied to the future cash flows so that their discounted value is equal to the required investment. (When the internal rate of return of a project is equal to the cost of capital, the project has a net present value of zero.) If the future cash flows are known, we can calculate the IRR at the beginning of the project. If we allow volatility and uncertainty in cash flows, on the other hand, the project will still have only one IRR, we just won't be able to calculate it until project completion. The "single value" characteristic of internal rate of return that arises in this example is very useful in that we can compare this known value to the various accounting-generated return figures in order to assess their accuracy and stability.

The result of just such a comparison is shown in Figure 14-3. A complete understanding of all of these data is not necessary, because the main point of this exercise is to highlight that there are numerous different ways of reporting return and none of them work particularly well. The principal assumptions made in constructing Figure 14-3 were: (1) The plant has an asset lifetime of 10 years; (2) the working capital

[2]For companies different asset lifetimes, one must use perpetual equivalent return or a net present value-type analysis such as Stern, Stewart's Economic Value Added (EVA).

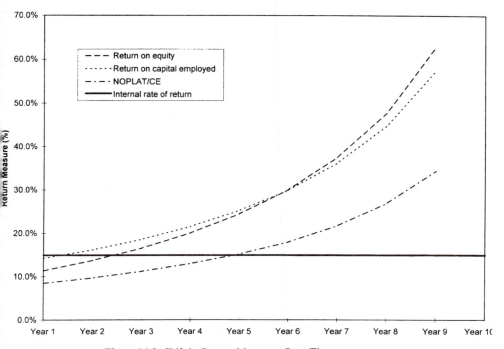

Figure 14-3. Shift in Return Measures Over Time.

requirement is 20% of the initial fixed investment; (3) depreciation is straight-line over the life of the asset; (4) the investment generates a constant real operating cash flow of 25 per year; (5) inflation is constant at 3%; and (6) the investment is funded 50% with debt and 50% with equity, with this debt-to-equity ratio held constant throughout the lifetime of the company.

Return measures included in Figure 14-3 are return on equity (ROE, equal to net income for the year divided by the average book value of shareholders' equity), return on capital employed (ROCE, equal to operating income divided by the book value of capital employed, the latter being the book value of assets plus net working capital), and net operating profit less adjusted taxes return on capital employed (NOPLAT-ROCE—essentially an attempt to make ROCE an after-tax number). Note that ROCE is sometimes called *return on net assets* or RONA.

Several problems in using most of the traditional accounting measures of performance are clearly evident from Figure 14-3. The first is that these measures are not constant over the life of the asset; indeed they generally increase with time. This often encourages managers whose performance measurement is based on these ratios to let their facilities age instead of investing for the future. The ratios rise because fixed assets (the denominator in most return calculations) are frequently depreciated at rates much more rapid than the erosion in earnings and cash flows generated by those assets (the numerator in the same calculations). In addition, fixed assets are usually recorded in the balance sheet at historic nominal values, whereas earnings and cash flows generally increase with inflation.

The second problem of traditional accounting is that many of the accounting-based measures overstate the return earned by the company, particularly near the end of the project's useful life. Again, this is principally caused by rapid depreciation policies and inflation (note that we did not use accelerated depreciation in the previous figure). This problem is also not particularly serious, as long as it is recognized. The danger, of course, is that uninformed managers will judge themselves or their business by comparing these (overstated) returns with more accurate benchmarks such as bond yields or cost-of-capital calculations. As a result, companies may continue a program of reinvestment, even though it is in a series of negative net-present-value projects, simply because their measurement technique is wrong.

The third problem is the killer: Quite simply, accounting ratios vary widely depending on the accounting principles used. If accounting standards were indeed "standard" and consistently applied, one could circumvent both the first and second errors by creating a simple set of correlations between the observed figure (such as return on equity) and the actual achieved internal rate of return. However, this is not the case. The reality of corporate financial statements is that large variations in accounting ratios often result from differences in both the interpretation and application of accounting standards, as well as from regional or national differences in the standards themselves. Return on equity is probably the most abused of the accounting ratios, as it is easily the most prone to manipulation. Particularly in the United States, it seems as if every new chairman also comes with an extraordinarily large write-off, usually a unilateral decision that the book value of assets is too high (i.e., the previous CEO was overstating profits). As a result, the reported return on equity jumps quite nicely in the year after the write-off, but the true returns being earned in the business more often than not remain unchanged. Return on capital employed is more difficult to manipulate, but it nonetheless can also be affected by both depreciation policy and extraordinary events.

The above example also highlights a conflict between the desired accounting goal of depreciation and the analyst's goal of using the resulting return-on-equity and return-on-capital figures to estimate IRR. The purpose of depreciating an asset is first to apportion the appropriate cost to the relevant time period (the income-statement impact) and second to reflect the fact that the net present value of the remaining future cash flows declines with time as the asset ages (the balance-sheet impact). In other words, accounting was designed to calculate income and remaining asset values, not return on investment. As a result, even if the cash flows and earnings from an asset are exactly equal in each and every year, a "proper" depreciation schedule will result in a steady decline in the balance-sheet value of that asset so as to reflect the decrease in remaining asset life. Under this scenario, any return measure based on reported book values will increase with time, even though the project IRR is a single figure.

For individual companies, there may be a way around at least part of this problem. When we move from a single investment company to a fully functioning organization, we begin dealing with a number of different assets, each with a different asset age. As a result, the observed accounting ratio for the entire company is actually a weighted average of the ratios generated by many different investments made in different years, with the weights representing the amount of investment in that year. If the asset-age profile (i.e., the distribution of assets with respect to asset age) is reasonably uniform with time, we could calculate a correlation between the observed ratio and the actual return figure to get a reasonably stable and accurate measure of true return. While this might work for an *individual* company, comparisons *between* companies would still not be possible, unless the correlation between the observed ratio and the actual return was recalculated for every firm to adjust for differences in the application of accounting standards, as well as differences in the standards themselves.

What to Measure?—The COCF/CCI Ratio

If most traditional accounting ratios are flawed, principally as a result of variations in the application of accounting standards, the problem is to find real-world variables that will be reported identically on essentially any accounting statement, regardless of the particular accounting system being used, and then to specify how to use these factors to estimate an internal rate of return. The key here is accounting-system independence. Neither managerial performance nor a corporation's value should be a function of the way the numbers are written down. As IRR is return on investment, we need to look for two terms, one that indicates profit and another that indicates asset base.

An immediate problem arises here. Most accounting statements are issued on an annual (or even a quarterly) basis. However, from an analytical viewpoint, the simple question, What was Bayer's internal rate of return last year? is actually impossible to answer, because returns on long investments are not earned in a single year. The true internal rate of return for Bayer is not only a function of what happened last year; it is also a function of earnings from those same assets in previous years, as well as of any future earnings that will occur in years to come. To overcome this problem, we instead ask the semi-theoretical question: What would Bayer's internal rate of return be if each and every year was just like last year? The difference is subtle but essential, as we can only calculate an IRR after we first assume constant business conditions throughout the full life of the assets.

This approach also introduces significant volatility in the return calculations for the chemical industry, much more so than actually occurs in real life, because annual cash flows in this industry vary considerably from year to year. No sane person would assume peak (or trough) cash flows for 20 straight years, yet this is the assumption we must make when calculating the IRR for a peak (or trough) year. As a result, it is prudent to establish true performance either by looking at relative performances of several companies in a given year or by looking at time-averaged returns for an individual company (in order to smooth out the peaks and troughs) compared to that company's cost of capital.

With this assumption in mind, the best solution we have found for practical calculations of internal rates of return is to use consolidated operating cash flow (COCF) as the measure of return and the firm's current cost investment (CCI) as the measure of asset base. While dividing these two figures does *not* give a proxy for IRR itself, the ratio does satisfy our need for accounting independence, and thus we can calculate a simple correlation (presented later) between COCF/CCI and IRR for a given asset lifetime. To give credit where credit is due, we should note that this approach appears quite similar (if not equivalent) to that used by HOLT Associates, a consulting company. Unfortunately, we are not aware of any papers by Holt that describe their technique. Additional thoughts on measures both COCF and CCI are given in Appendices 1 and 2, respectively.

Recent press coverage (and aggressive marketing by the consultants) has increasingly focused managerial attention on the overall subject of corporate-performance measurement, with the most popular methodology—at least based on air time—clearly being the EVA analysis of Stern, Stewart. There are also many other competing methodologies, and thus a word about why we have chosen our particular approach is probably warranted. In addition, we should also explain why our technique is so much more closely aligned to that of HOLT Associates, as opposed to that of Stern, Stewart.

Our view is that there is actually one, and only one, method of valuing companies, just as the Navier-Stokes equations are the one and only method of modeling fluid flow. As with the Navier-Stokes equations, however, it is solving the equations themselves that is so problematic, not the understanding the theory, per se. Under a very restrictive set of assumptions, the solution to the Navier-Stokes equations reaches the simple parabolic velocity profile seen for viscous fluid in a pipe, while under a similarly restrictive set of assumptions, the valuation equations reduce to the traditional dividend-discount model. When the complexity of the problem increases, chemical engineers usually resort to breaking the problem down into a finite-element analysis. Investment analysts do essentially the same thing, by creating a discounted cash-flow model with finite (usually annual) elements.

Applying the valuation equations to practical situations is also fraught with measurement error. The theory of valuation is rather simple: a corporation's value is determined by a company's return on capital relative to its cost of capital. In most situations, one can estimate the cost of capital with at least a reasonable degree of accuracy, but measuring or estimating the return of capital is exceedingly difficult. Some serious assumptions regarding return on capital—implicit or explicit—are required for *any* proper valuation technique, including EVA. While the "full application" of EVA may indeed work reasonably well, nearly all of the practical applications of this technique that we have seen have been based on book value, a basis that often creates considerable error, as we have already highlighted. Thus, we strongly prefer our technique (or that of HOLT) for the return-on-capital portion of the problem.

Furthermore, once we have estimated the return on capital using the techniques described, we could in theory calculate our own "adjusted" (and hopefully more accurate) EVA by multiplying the excess (or deficit) return by the capital employed over the course of the period. (Actually, it would be more technically correct to use the CCI, although this comes with some implicit assumptions.) We don't do this because we have not found the resulting figure of much use in the overall analysis: Being absolute, it makes a comparative analysis of two companies of different size quite difficult (just comparing the returns themselves is much easier). We could also have approached the valuation part of our analysis by summing the net present value of future (estimated) EVAs, but again we have found our own analysis much easier to apply. One should also note that the "sum the future EVA" analysis is essentially equivalent to the familiar NPV of FGOs approach, described later.

For readers interested in more detail, additional discussions concerning the measurement of COCF and CCI are given in Appendices 1 and 2, respectively, of this chapter. Note that COCF is essentially equal to the more familiar EBITDA (earnings before interest, taxes, depreciation, and amortization), although the former excludes income from nonconsolidated operations. Minor adjustments to German accounts also are required to correct for differences in pension costs.

Model Results and Return Sensitivities—Converting COCF/CCI into IRR

We should stress again that a company's estimated internal rate of return is not identical to the practical calculation of dividing the consolidated operating cash flow (COCF) by current cost investment (CCI). Moreover, any measurement of the internal rate of return will be sensitive to asset lifetime, as well as to the profile of cash flow over that lifetime, inflation, tax rates, and depreciation. Let us briefly discuss each of these issues.

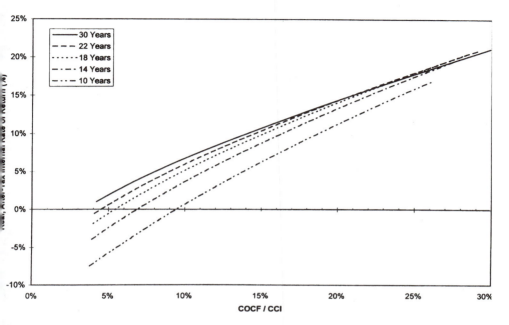

Figure 14-4. Internal Rate of Return vs. Consolidated Operating Cash Flow/Current-Cost Investment.

Asset Lifetime

When estimating IRR, the most important variable to take into consideration, in addition to the COCF/CCI ratio, is asset lifetime. In Figure 14-4, we show the correlation between IRR (real, after-tax) and the ratio of consolidated operating cash flow to current cost investment (COCF/CCI).[3] Each line represents a different asset lifetime. In our view, this is perhaps the most important chart in the entire chapter, as it gives managers a quick means of getting a reasonable estimate of their true performance from a relatively simple and readily available ratio.

As expected, an increase in either the COCF/CCI ratio or the overall asset lifetime results in an increase in the achieved IRR over most of the range under review. The latter sensitivity is the highest at low asset ages, with relatively small variations observable once the asset age passes about 20 years because additional cash flows in distant years have little impact once they have been discounted back to the present.[4] For the most part, the average lifetime of assets in the chemical industry is between 18 and 22 years, and thus this portion of the chart provides a handy reference tool for the quick estimation of IRR using the readily obtained variables of consolidated operating cash flow and current cost investment.

[3]The y-axis represents an after-tax return, even though COCF is a pre-tax figure. To do this, we have embedded an assumed tax rate into the correlation. The sensitivity of IRR to the actual tax rate and depreciation policy is discussed later.

[4]The fact that these lines cross at relatively high COCF/CCI ratios may seem somewhat confusing. This effect arises because depreciation in the early years declines as asset lifetime increases, thereby causing a slight decline in near-term cash flows due to a slightly higher tax charge. At high discount rates, this (really tiny) change in near-term cash flows actually affects the net present value more than adding four more years of cash flow to the end of the project.

In addition to asset lifetime, there is also the issue of cash-flow profile over that lifetime. In the earlier example of a single investment company, we assumed a constant operating cash flow over time. But in most real-life situations, cash flows decline with increasing asset age, thereby implying a declining COCF/CCI ratio, as well. This creates problems when we are looking at single investments. Such problems are somewhat overcome, however, when we look at ongoing corporations, because in that case the observed COCF/CCI ratio will be the weighted average of many assets of different ages. Since the age profiles of assets are usually rather similar between companies, variations in the observed ratio will be low, as well. Although the results are not shown here, we have also performed this same analysis assuming an exponential decline in cash-flow profitability as the assets approach the end of their useful operating lives. Such an assumption tends to result in a 0.5 percentage points increase in the estimated IRR at a given COCF/CCI ratio (note also that if we assume this for one company, we should assume the same for all other companies, with the net effect being no shift in our estimates of relative performance).

Inflation

The inflation rate can have a significant impact on industrial returns, with higher inflation depressing achieved IRRs, all other things equal. This "inflation disadvantage" results because depreciation is recorded in nominal, rather than current-dollar terms. Thus, even if operating cash flow is perfectly indexed to inflation (i.e., operating cash flow is constant in real terms), a higher inflation rate will result in an increase in *real* pre-tax earnings, since the nominal value of depreciation remains unchanged. This, in turn, results in higher real taxes being paid and a correspondingly lower after-tax return. Notice that the issue is not the impact of inflation on cash flow; real depreciation is still reduced and real taxes are still raised, even if operating cash flow is perfectly indexed to inflation. In other words, inflation does not change the total "social" value of the company; it just gives more of that value to the government and less to the owners — a result identical to that of higher corporate taxes. A rough estimate, based on the levels of taxation and depreciation prevailing in the industrial world, is that *a rise in inflation from 3% to 8% would reduce the achieved real after-tax IRR by over a full percentage point, just from the loss of the tax shield alone.*

The implications of this observation should not be underestimated. Companies operating in high-inflation countries — particularly capital-intensive companies with long asset lifetimes, such as those in the chemical industry — are at a significant disadvantage relative to otherwise equal competitors operating in lower-inflation countries, other things equal. While it is only part of the story, one cannot help but wonder how much of an effect Britain's inflation had on the relative underperformance of ICI's chemical division over the past few decades. It is also interesting to note that the country with the best inflation record in recent times — Germany — also has a preponderance of capital-intensive industries with long asset lifetimes.

Depreciation Schedule

In calculating an after-tax internal rate of return, the choice of depreciation schedule can also make a noticeable difference. With a reasonable set of assumptions for the chemical industry (e.g., an asset lifetime of 22 years, etc.), the switch from straight-line to accelerated depreciation can raise the achieved internal rate of return by over 0.5

percentage points. As a result, companies in countries that allow accelerated depreciation for income-tax purposes are at a competitive advantage relative to companies in other countries that are bound to slower depreciation schedules, other things equal. This observation should not be used in isolation, however, as depreciation policy is almost always inextricably linked to other elements of the tax system a company faces, principally the tax rate itself.

Taxes are a source of complexity and confusion when it comes to the performance assessment and valuation of corporations. Clearly, the assumed tax rate will have a significant effect on the after-tax rate of return. There are two principal ways of approaching the tax issue: Either calculate a pre-tax IRR, or choose a plausible tax rate and calculate an after-tax IRR. Pre-tax returns are probably the more relevant figure for managers, since they focus on the business factors that are most under managerial control. The principal argument against using a pre-tax rate-of-return figure is that most managers tend to compare their performance with other returns that are observable in society, and these are usually after-tax returns. Since pre-tax numbers are obviously larger, there is a tendency toward complacency if they are used as the internal benchmark.

Our creation of an estimated after-tax IRR may seem rather problematic because of the wide differences in tax rates across many national boundaries. However, countries with high tax rates tend to have rapid depreciation schedules and vice versa. As a result, if one looks at taxes paid as a percentage of cash flow, there is relatively little variation among chemical companies around the world. Making a basic assumption of a 35% tax rate and using a sum-of-the-year's-digits depreciation approach results in a balance-sheet structure very much like those in the United States. Increasing both the depreciation rate and the tax rate creates a balance-sheet structure more Germanic in appearance, but the estimated IRR is little changed.

Shifting tax rates and depreciation schedules can, however, shift the relative competitive advantage of entire industries. For example, imagine two neighboring countries that are identical in every way. If one country suddenly shifts to a more rapid depreciation schedule and a higher tax rate in such as way as to be neutral in overall tax revenue, this region will have a comparative advantage (i.e., higher internal rates of return relative to its neighbor) in its capital-intensive, long-asset-lifetime businesses and a comparative disadvantage (i.e., lower internal rates of return) in its less-capital-intensive service industries. One can't help but think of the United States on the one hand and Germany and Japan on the other.

Additonal Types of Investment

Fixed assets and working capital are not the only types of investment. Research and development spending (R&D) and advertising should also be included in this category, although accountants treat them only as current expenses. In particular, R&D has a large impact on the chemical industry. Adding R&D to the rate-of-return calculations involves several steps. First, one must estimate an asset lifetime for R&D; my preference is to assume that the life of R&D is equal to that of the fixed assets, which at least has the virtue of simplifying the other calculations. The next step is to add R&D expenses back to cash flow (as if they had been investment expenditures) to get an adjusted COCF. Then, for the calculation of CCI, the level of the fixed asset portion of current cost investment (the other portion being working capital) must be raised by the average ratio of R&D spending to capital expenditure. Once this calculation is completed, the ratio of COCF to CCI can again be calculated, and after-tax IRRs can again be estimated from the relationships previously shown in Figure 14-4.

IRR Conclusions

Clearly, a number of assumptions are required to convert observed accounting variables into IRR estimates. Thus, a reasonable reader might initially view this whole methodology with considerable caution. The argument against this caution is not that these methods are without flaw or question, but instead that the alternatives are considerably worse. In particular, the assumptions inherent in using any return measure based on traditional accounting ratios are much more onerous and much less logical than those described here. At least, the assumptions of this approach are an attempt to calculate a true rate of return: something that traditional accounting makes no pretense of doing. Two other insights provide comfort about this method. First, modifications to the model's basic assumptions (such as allowing for decreasing profitability as assets approach the end of their useful life) result only in relatively minor shifts in the estimated IRRs. Second, even if certain assumptions are questionable, as long as common assumptions are used across companies that compete within a single industry, the methods can be useful for assessing the relative position of each company. Of course, the relative positions are the key to assessing competitiveness and to an efficient allocation of capital.

CHEMICAL COMPANY PERFORMANCE

We now have a methodology for estimating internal rate of return, using both the observed profitability level, expressed in terms of cash flow divided by current cost investment, and the set of financial characteristics relevant to the chemical industry. These are discussed in detail in Appendix 3. The next step is to examine the companies themselves to ascertain which firms are delivering the best performance. We do this in two stages, first by looking at the highly volatile single-point returns and second by averaging these returns over time to get a more realistic picture of true performance.

A "single-point" IRR is calculated using only one year's worth of data and, most importantly, with the implicit assumption that each and every year is just like the year being examined. This naturally results in unrealistic highs and lows in the reported returns, since peaks and troughs in business cycles never last long relative to the asset lifetimes of most industries. Figure 14-5 shows the single-point IRRs for a number of chemical companies, based on data from 1996, as well as the five-year average for the period from 1992 to 1996. As already described, these data are based on first calculating the COCF/CCI ratio and then seeing what the ratio implies about IRR. The chart reveals a much wider range in performance than most people would initially expect. For the most part, the 1996 data are relatively close to the five-year averages, but there have been some notable "improvers" (particularly Monsanto, but also DuPont, Hoechst, BASF, and PPG), as well as some meaningful "decliners" (Great Lakes and Lawter). Of course, the business cycles of the various segments of the chemical industry are not always synchronized, and thus the five-year returns might be more meaningful data points than the 1996 observations. But as we will see, for the most part the relative performance positions remain reasonably stable.

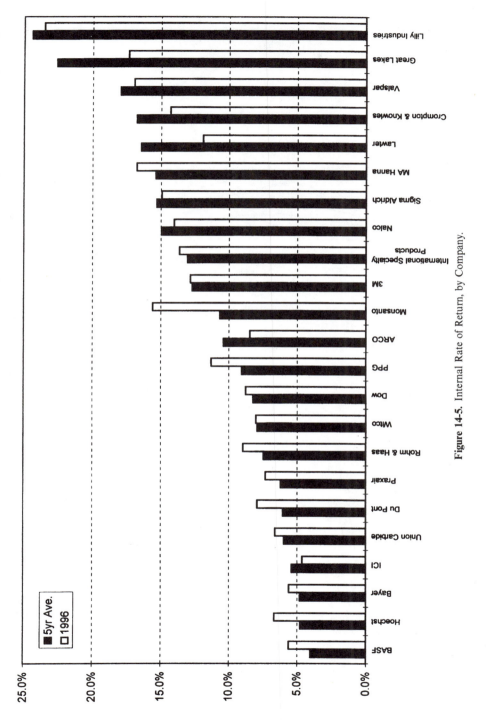

Figure 14-5. Internal Rate of Return, by Company.

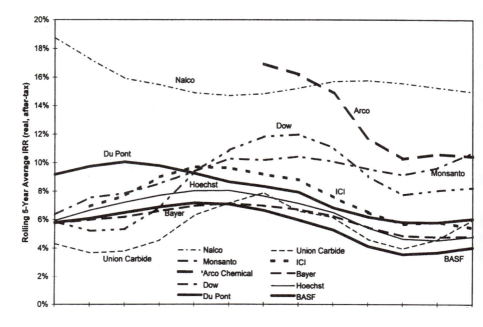

Figure 14-6. Rolling Five-Year Average Internal Rate of Return.

Shown in Figure 14-6 are rolling five-year-average internal rates of return (based on COCF/CCI calculations) for a number of global chemical companies; thus, the data for 1985 represent the average IRR for the years 1981 to 1985 and so on. Several observations can be made from Figure 14-6. First, even using five-year rolling averages, there are still quite discernible trends from several of the companies. Dow clearly stands out as the most cyclical company, with both the peak in 1988 and the trough in 1991 quite defined. The company's share price outperformance in the late 1980s and its recent underperformance (back to trend) make more sense in view of these figures. Union Carbide was clearly the underperformer of the early 1980s; perhaps it was right to shake the company up! The steadily declining trend at DuPont is certainly worrisome, and serious questions may soon be asked of management if the recent improvements (reflected in Figure 14-5) are not maintained as a serious trend. Monsanto, in contrast, has split apart from the pack on the upside in terms of returns, particularly in recent years, so that perhaps the firm's previously highlighted tremendous share-price perform-ance may indeed be deserved. The Germans, in general, have much steadier, if somewhat mundane, return profiles, yet they remain below the cost of capital on a rolling five-year basis. Hoechst, however, seems to be pulling away from the other two companies in 1995 and 1996, although this change is not yet fully reflected in the rolling five-year average. This trench is perhaps a positive result of the company's recent aggressive restructuring efforts. Finally, the initial results from ICI (ex-Zeneca) do not paint a happy picture, indicating that much restructuring remains to be done. ICI is, however, working on this, and the recently acquired Unilever businesses have much better returns (although this was reflected in the purchase price).

Another interesting point about Figure 14-6 is that the average of all of the data points, weighted for corporate size (Nalco is quite small), is not that far away from 7%. This time- and size-weighted average can be used as a rough estimate of the average full-cycle return in the bulk chemical industry (specialty companies earn higher returns, at least for now). This return level makes sense, since in highly competitive, reasonably mature industries, returns should be close to the cost of capital, which is also about 7%, in real terms, for these companies.

In my discussions with corporate managers in the chemical industry over the past several years, I have found that their targeted returns — sometimes called "the hurdle rates" — are fairly similar. For the most part, corporate boards appear to be shooting for real rates of return of around 15%, with 10% acceptable for core businesses, while rates of 20% are expected for more speculative ventures. From the data presented here, it's apparent that these hurdle rates are rarely, if ever, met.

CHEMICAL INDUSTRY PERFORMANCE

The previous section analyzed the performance of individual chemical companies. Additional insights can be gained by looking at how these individual results aggregate into a view of the industry as a whole. Figure 14-7 summarizes the results of the present analysis. In analyzing Figure 14-7, we can use a shortcut described in Appendix 2 to this chapter: Instead of trying to calculate CCI for each company in each year, we instead rely on the ratio of COCF to sales, based on a reasonably stable proportion between CCI and sales for many of the larger, more diversified firms. This measure of profitability is called the "cash-flow margin." Figure 14.7 shows the average annual cash-flow margin for four of the largest competitors in the chemical industry (BASF, Dow, DuPont, and ICI) since 1960. In addition, the horizontal line at 16%

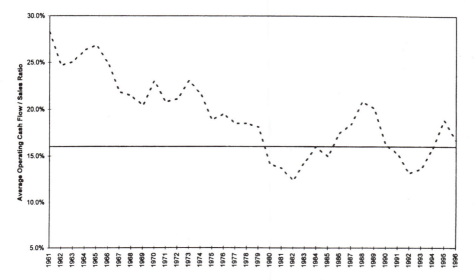

Figure 14-7. Average OCF Margin for Dow, DuPont, ICI, and BASF from 1960 to 1995.

represents a rough estimate of the cash-flow margin required for the average chemical company to earn its cost of capital.

Several important features of this plot all center around one theme: *Over the past several decades, a large portion of the chemical industry has made a transition from growth to maturity.* This transition has four principal features, which this section will cover in turn: (1) a downshift in averaged investment returns, (2) returns which cycle around the cost of capital, (3) industry returns that are set by global markets, and (4) an increase in the peak-to-trough level of cyclicality.

The Return Downshift

The 1960s and the 1970s were clearly a pleasant time for chemical-company managers. Cycles were relatively mild. Furthermore, the average cash-flow margins in all of these years imply that nearly any investment achieved a return in excess of the cost of capital. There was relatively little distinction between success or failure in these years: The differential was instead only the level of success. Things changed quite dramatically, however, in the early 1980s, and the industry climate has been harsher ever since. Not only did the chemical industry enter into a particularly severe cyclical downturn in the early 1980s but chemical-company returns (at least in the developed world) fell below the cost of capital.

Cycling Around the Cost of Capital

Looking ahead, we do not believe that the time-averaged returns for significant segments of the chemical industry will ever again be persistently above cost-of-capital levels, at least not without the creation of an effective oligopoly or monopoly. However, average returns should not consistently fall below the cost of capital, either, since free-market forces would tend to push rates of return for a mature industry toward the cost of capital, which is a reasonably constant figure in real terms. As a result, large portions of the chemical industry have become part of a zero-sum game, with excess returns for an individual company only possible if they are counterbalanced by a competitor's sub-par performance.

The push toward average returns at the level of cost of capital does not, however, mean a stabilization of returns at this level. Chemicals remain inherently cyclical, partly due to the "lumpiness" of many investments and the time lag between the investment decision and the date the plant actually comes on stream. In general, markets make ample cash available to those industries which are earning returns in excess of the cost of capital (as the chemical industry experienced in 1989), both through the large internal cash flows that are implied by high returns and improved access to external funding. The usual (inevitable?) result is overinvestment and a decline in average rates of return. At the other end of the cycle, returns below the cost of capital attract calls for lower investment or even the closure of existing capacity. The usual result is a tightening of the supply/demand balance and an increase in industry IRRs. Chemical engineers might say that the chemical industry fluctuates as if it had an inadequate control loop on a valve, with the result a correct amount of average flow over time, but (unwanted) short-term oscillations.

An Increase to Peak-to-Trough Cyclicality

Another final point to be made from Figure 14-7 is that the volatility of cyclical variations appears to have increased in recent years. This observation comes as no surprise to chemical company executives. While we do not have hard evidence of the reasons for this increased cyclicity, we do have a plausible theory. For many chemicals, there has been an increasing uniformity in production technology and, hence, a much greater "flatness" of the cost curve. Ethylene, the bellwether of the petrochemical industry, is perhaps the best example of this trend. The most advanced ethylene-plant technology is in the hands of the Specialized Engineering Firms (SEFs). As a result, anyone (even an investment bank) can commission a cost-efficient ethylene facility, and just about all facilities have the same level of overall costs, unless unusually low-cost feedstocks are available at a particular location.

Therefore, for a large part of the business cycle, the marginal cost of production is quite close to the average cost of most producers, and all firms exist on extremely thin margins. Thin margins result in no additions to capacity, and thus utilization rates inevitably rise with increases in demand. Only when supply becomes extremely constrained do we see a rapid escalation in marginal costs, as nearly all producers are constrained at once. Then everyone starts to plan new facilities, but because of the two-to-three-year time lag required for ethylene-plant construction, utilization rates increase further in the short-term, leading to supernormal profitability. This profitability is usually short-lived, however, and the ride back down the cost curve is just as rapid. Solving this flat-cost curve problem is one of the major challenges for the industry. Some of the more innovative suggestions include treating ethylene plants as utilities or establishing a futures market in ethylene to improve the market signals of future prices.

Finally, in global industries such as chemicals, it is the average *global* return that is pushed toward the cost of capital. Thus, it is not inconceivable that one region of the world (for instance, Europe or Japan) remains mired in a sub-par return environment for some time while other regions (such as non-Japanese Asia) enjoy a prolonged period of returns in excess of the cost of capital. This appears to be just the situation in the mid-1990s, and it doesn't appear likely to change rapidly. One characteristic of industries with long asset lifetimes is that they are very difficult to move geographically: Once the plant is there, it will continue to operate as long as cash margins remain positive, even though IRRs may be quite low. This is one reason European plants with low returns are unlikely to shut down quickly. The mobility problem is further enhanced by the "large site" effect, in that incremental investment is almost always cheapest at an existing, integrated facility as opposed to a greenfield operation. Thus, a series of logical, incremental investment decisions can actually result in the illogical conclusion of a large facility addition in an existing high-cost region, as opposed to a new, integrated facility in a much lower-cost area. A final factor contributing to the difficulties of shutting down plants in Europe are the high regulatory costs imposed on European factories that attempt to downsize.

LINKING PERFORMANCE WITH VALUATION

The general philosophy of valuing corporations is simple: Value is a function of company performance, and the better the performance the higher the value. However, the practical application of this philosophy is much more difficult, not least because of the problems discussed in this chapter surrounding the identification and measurement of performance. As a result, investors have a tendency to do what I call "correlating observables," rather than focusing on fundamentals. The chemical industry has its own version of this process. Invariably, there are a number of things going on in a chemical system which the chemists and the chemical engineers do not really understand. Nonetheless, they are able to get the process working — often quite well — by finding correlations between observed response and a changing environment, even if they do not understand the basic mechanism behind that response. Similarly, stock-market analysts have developed a wide range of correlations, most of which work most of the time. But one needs to look somewhat deeper into valuation theory — in particular, past price earnings and cash-flow multiples — to gain real insight into corporate performance.

A Value Paradox

Much of my thinking on corporate performance, competitiveness, and valuation has been driven by an apparent valuation paradox common not only in the chemical industry but in many other industries as well. Let me illustrate the paradox by looking at Dow Chemical and BASF. Dow and BASF are actually very similar in terms of overall size, products, technology, and geographical spread. They also have virtually identical capital intensivities and asset lifetimes. Furthermore, these two companies have exhibited remarkably similar "performance" when measured by growth in company size (as shown earlier, in Figure 14-1), share-price performance (Figure 14-2), and growth in real book value per share. However, in spite of these similarities, the capital markets have awarded these companies hugely different corporate valuations. When I first started following these companies many years ago, Dow had a firm value (equity market value plus net debt) that was roughly 120% of the company's sales. BASF, in contrast, had a firm value that was only around 40% of sales. (It should be emphasized that this was market value, not book value, and hence this discrepancy was not the result of different accounting policies.) Furthermore, the ratio of firm value to operating cash flow was 7 for Dow and only 3 for BASF. To my inexperienced eye, BASF seemed like an easy investment choice. However, when I visited the fund-management community, I was informed that this had always been the case, and as investors they would only be interested when I could tell them either why these differences had persisted so long or when they would change. The apparent value gap was, however, too great to ignore. Indeed, if BASF was awarded a multiple of sales equal to that of Dow, its value would have increased by an astounding 40 billion Deutsche Mark, or roughly 1.3% of Germany's annual GDP! Furthermore, this situation was not just restricted to BASF. Indeed, the entire German (if not European) chemical sector seemed to be facing a similar differential in market valuation. Continuing with the earlier comparison, if all three of the largest German chemical companies underwent a similar revaluation, the increase in market value would reach nearly 4% of annual German GDP.

So here is the paradox: Either BASF (and the rest of the European chemical sector) was the best investment since nylon, or many conventional measures of performance — such as growth in company size and share price performance — are wrong (or at least misleading). Indeed, the solution to the paradox is that the performance measures are flawed, not the market pricing mechanism. Further, in spite of the *apparent* valuation differentials, a more rigorous analysis shows that the market values of these companies are actually quite fair and that the markets are indeed reasonably efficient, even across international boundaries. Obviously, no investment banker would say that markets are totally efficient, because then all of our clients would head to index funds and we'd be unemployed, but the strong implication from this analysis is that markets do a very good job of assigning the appropriate value to the vast majority of securities.

Debt Valuation

The "science" of fixed-income valuation is quite advanced, particularly relative to the art of equity valuation. The nice thing about debt valuation is that analysts actually know what to argue about: the appropriate discount rate and its term structure. Once these have been agreed on, the value of the bond is easily calculated. Equities are considerably different. Even after two analysts have reached agreement on a set of future income-statement, balance-sheet, and cash-flow projections, they often still argue about valuation methodology.

Because of the simplicity of debt valuation, a quick review can be a very useful way to begin a discussion of the more complex but related issue of equity valuation. To illustrate some simple valuation relationships, consider two similar bonds, both with 25-year maturities. One bond will be "normal" in that principal is repaid at the end of the period, while the other will be the same except that the principal is not repaid. The value of either bond is calculated by discounting the expected future cash flows back to the present, using the discount rate. For simplicity, assume a constant discount rate with time (a flat-yield curve), although this discount rate will obviously be higher with an increasing default risk.

For the normal bond, when the "coupon" payment during each time period is equal to the discount rate, then the bond's value is equal to the par value of the instrument. A bond with a coupon above the market discount rate is worth a premium to par, while a low coupon bond is worth less than par. The valuation of the interest-only bond is similar to that of the normal bond, but it will always be less, with a value difference equal to the discounted value of the normal bond's final principal payment. If the time period is quite short, or the discount rate is quite low, then the discounted value of the final principle payment will loom relatively large and there will be a considerable difference between the two bonds. But if the time period is long or the discount rate high, the final period payment will matter less.[5]

Corporate Valuation

Single industrial investments are similar to interest-only bonds. They generally give an annual cash flow, although it's usually in real, instead of nominal, terms, and they are usually nearly worthless at the end of their lifespan; that is, scrap value is usually close to zero. A corporation, however, is not just a single industrial investment; it is a series of investments across time. In fact, in a corporation a substantial portion of the cash flow is internally reinvested in additional assets. Thus, the value of a corporation is not only the net present value of the remaining cash flows from the existing businesses but it also includes the net present value of any future investments. This latter component of valuation is often referred to as the "net present value of future growth opportunities," or the "NPV of FGOs," and as we will demonstrate later, this component can often be well over one-half the value of the entire corporation.

[5]Actually, the theoretically correct measure for the return on a bond is the "perpetual equivalent return (PER)," which is calculated by finding the perpetual series of equal cash flows that are required to achieve a net present value equal to that of the bond. This works equally well for single projects and other investments of any length. With ongoing companies, however, using internal rate of return also has some advantages. Unlike PER, IRR can, for example, be calculated separately from the discount rate (or cost of capital), which is required to calculate the perpetual equivalent return. Also, IRR is a reasonably well accepted and understood descriptor of a project, while perpetual equivalent return may have many traditional managers looking somewhat cross-eyed. As a result, we have chosen to stick primarily to IRR throughout this chapter, with the penalty being that some of our correlations are applicable only to a subset of companies with similar investment durations.

To calculate both the net present value of the existing assets and the net present value of the future growth opportunities, the investment analyst must define four "valuation drivers":

1. The existing asset base, including size and age profile
2. The current and expected future levels of investment returns
3. The amount of future investment (reinvestment rate)
4. The relevant discount rate or cost of capital

Once these four variables have been fixed, all other balance-sheet, income-statement, and cash-flow-statement items can be calculated. Note further that this set does *not* include the growth rate of the company's sales. While we could add this variable (substituting it for either current and future returns or, more likely, the level of future investment), we instead prefer to view growth as the *result* of returns and investment decisions, not as the *cause* of these decisions. Indeed, assumptions about future growth trips up many analysts, as they sometimes assume future growth rates that are mathematically inconsistent with the assumed levels of future investment. Obviously, any valuation methodology that is not internally consistent will yield unreliable results.

The Existing Asset Base and the Concept of Economic Replacement Cost

The starting point in a valuation calculation is the existing asset base. With debt valuation, the existing base is a relatively simple concept: It is almost always taken to be par value. With fixed assets or other investments, however, the concept becomes considerably more complicated. What we are looking for is something we have chosen to call "Economic Replacement Cost," or ERC.

For brand new assets, economic replacement cost is simply the required investment. Note, however, that this is *not* the same as the open-market purchase price of that asset after the investment is made; in an efficient market, the purchase price of the asset should, instead, be the net present value of the expected future cash flows, and this will equal the required investment only when the expected IRR is equal to the cost of capital. As a result, looking at public merger and acquisition (M&A) transactions is not a correct methodology for estimating ERC, although it is a good reality check for calculated valuations.

Partially used (as opposed to new) assets create additional problems, and hence we need to expand our definition of ERC. Obviously, the economic replacement cost of an asset halfway through its lifespan is not the same as when it was first built. Again, public transactions do not yield the desired figure, as they again are principally an indication of the net present value of the remaining cash flows and hence the actual value of the asset, as opposed to any notion of "par."

A clue to the methodology used to calculate the economic replacement cost for a partially used asset is given by the fact that the net present value of the cash flows from a new project will be equal to the initial investment when the project in question is earning the cost of capital. Pursuing this line of thought, our definition of economic replacement cost for any asset is the net present value of the remaining cash flows that would be created *if* the asset was indeed earning a return equal to its cost of capital. To put it another way, if a company is earning a return exactly equal to the economic cost of its capital, then the value of that company will be exactly equal to the ERC. However, companies seek a situation in which their internal rate of return will exceed their cost of capital, in which case the value of the company will exceed its ERC.

Several points should be noted about ERC. First, it is obviously a theoretical value, as opposed to a necessarily realizable value. Second, ERC is *not* a function of the returns management actually earns on this investment, although the value itself will be a function of managerial ability. Note further that economic replacement cost is meaningfully different from current cost investment (CCI). Whereas CCI is the investment that would be required to replace each and every corporate asset with a completely *new* asset, ERC corrects this investment for asset age under the assumption that the corporation is earning a return equal to the cost of capital.

Because it is a function of future cash flows, ERC will be a function of an asset's assumed cash-flow profile. This said, we have found that under most reasonable assumptions of future cash flows and depreciation schedules, the ERC of a fixed asset is generally an approximately linear function of asset age.[6] In practical terms, the calculation of ERC involves starting with gross fixed assets; adjusting them for inflation to get current cost investment (CCI); and then correcting this figure for asset age, either by looking at the actual levels of historic investment over time or by making some reasonable assumptions about past investment-growth rates. As a rough rule of thumb for companies with asset lifetimes and growth rates similar to that of the chemical industry, CCI is around 130% of gross fixed assets, and the fixed-asset portion of the ERC is around two-thirds of CCI.

Future Investment and Payout Ratios

Most corporations finance their investment requirements from internal cash flow. As such, the level of future investment — and hence the future growth rate — is over time determined by a combination of the firm's internal rate of return and the payout ratio that management decides to adopt (external capital-raising can be thought of as a negative payout ratio). Note that this is counter to the way in which most financial models are constructed; i.e., most models start with an assumed growth rate and back up to calculate the required investment. However, one of the most common mistakes in financial analysis is a disconnect between growth and investment: All too often they are forecast independently of each other, essentially resulting in some rather odd inherent assumptions about the marginal productivity of investment. Our view remains, however, that growth is the *result* of investment, and hence we have a strong preference for forecasting investment returns and payout ratios instead of growth rates. This approach also has the added benefit that payout ratios tend to be relatively stable over time, whereas growth rates are much more difficult to forecast, as they can undergo significant short-term swings due to shifts in capacity utilization.

Cost of Capital

A detailed discussion of cost of capital is well beyond the scope of this chapter. For simplicity, we make the (some may say rash) assumption that most large, diversified chemical companies have similar costs of capital in real terms. The intuitive argument for this assertion is that in this world of global capital flows, it would be unreasonable to believe that investors would have substantially different expected real returns for an investment in either Bayer, Dow, or ICI. If they did, shares would be sold in one and bought in the other until prices and expected returns were realigned.

We therefore take a simplistic approach to estimating the cost of capital for typical, large chemical companies. Equity investments in the developed world have, over the past several decades, delivered a real rate of return of around 8% annually, although

---

[6]Note that the ERC of working capital is always equal to the level of working capital, as working capital does not depreciate and is released at the end of the project. Beware, however, of LIFO accounting.

actual achieved returns have varied quite substantially both from decade to decade and from country to country. It is therefore plausible to assume that this is a realistic measure of the expected future return. Corporate debt, in contrast, has delivered an annual real rate of return of around 4% over this same time period, again with fairly substantial variations from decade to decade and region to region. As a rough estimate, most corporations have been financed with approximately one-quarter debt and three-quarters equity. Thus, a reasonable "first-cut" approximation for the weighted average cost of capital for the average chemical corporation is 7% in real terms. For the sake of simplicity, we will use this figure in all of our calculations, although I'm sure there will be numerous finance professionals cringing somewhat at the thought.

Terminal Values
Terminal values are often a significant source of controversy in valuation calculations. This is particularly true if the time frame under consideration is relatively short — say, less than five years. In this case, the net present value of the assumed terminal value can frequently be 80% or more of the total value calculation. Furthermore, if a simple valuation multiple is being used to calculate terminal value (such as value to sales or value to cash flow), the entire valuation can then hinge on this somewhat arbitrary number.

One way around the terminal-value problem is to just assume perpetuity. However, this assumption is highly dangerous and obviously incorrect for high-return companies, since the model tends to predict that their unrestricted growth will, in a few decades, make them larger than the entire global economy. A second approach, discussed in more detail in the next section, is to assume that corporate returns will eventually trend back to the average cost of capital. At this point, the terminal value becomes equal to the firm's economic replacement cost, and the levels of future investment/divestment do not matter, as any action takes the form of a zero net present value project.

Valuation-Model Results and Sensitivities

Now let's look at the sensitivity of corporate values in response to shifts in each of the valuation drivers. Not surprisingly, a corporation's value increases with an increasing internal rate of return. This relationship is shown in Figure 14-8 for a constant cost of capital (at a real rate of 7%) and a constant payout ratio of 50%. The horizontal axis shows the internal rate of return, while the vertical axis shows the value of the company divided by the economic replacement cost, which normalizes for the size of the company's assets. The different lines show the difference between assuming constant perpetual returns or assuming returns that fade to the cost of capital over a defined period of time.

Value versus IRR
Note from Figure 14-8 that when the company achieves an internal rate of return of 7%, which is exactly the cost of capital, then all the lines show the ratio of corporate value to ERC will be exactly 1. Low returns result in value lower than ERC; high returns earn a premium. Furthermore, at low internal rates of return, value is relatively insensitive to shifts in the returns being earned by the company (the slope of the line is low); but as IRR increases, the sensitivity of value to the assumed IRR becomes quite pronounced.[7] This is due to the compounding nature of the reinvestment that is taking place.

[7]The use of this chart is limited to investments of the duration specified in the model, principally defined by an asset lifetime of 22 years in this example. In other words, while correct for most chemical companies, the shown correlation is not applicable to all industries, as industries with different investment durations (generally caused by different asset lifetimes) will have slightly different valuation functions.

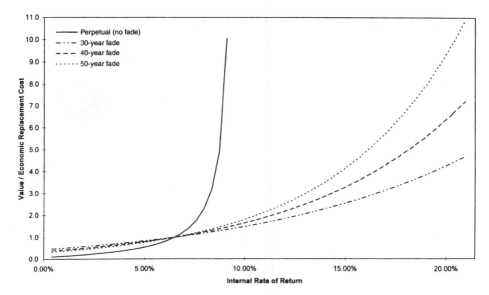

Figure 14-8. Value/Economic Replacement Cost vs. Internal Rate of Return.

The perpetual return line of Figure 14-8, seems to show that valuation explodes at a reasonably achievable IRR (less than 10% real). Are these companies really this valuable, or is something wrong with the model? Not surprisingly, the answer is the latter. In reality, high returns inevitably invite additional competition and overinvestment, and hence in valuing high-return companies it is sensible to include an anticipated decline in future profitability. (The reverse is true for low-return companies.) Three things must be specified when modeling shifts in future returns: the end-point, the time frame, and the path.

The logical end point for this return adjustment is the cost of capital itself, which should be the natural result of market forces. Moreover, once returns have reached the cost of capital, the terminal value becomes equal to economic replacement cost and also independent of future reinvestment rates. The time frame for future return variations (i.e., number of years before cost of capital returns are reached) is perhaps the most important variable in the valuation of high-return companies. Indeed, his abilities as a judge of this time frame are what distinguish Warren Buffett as one of the world's most successful investors. Using the chemical industry as an example, it took at least 20 years (and probably longer) for industry internal rates of return to trend toward the cost of capital (recall Figure 14-7). Finally, there is the path that the return adjustment takes. For the chemical industry, the path of long-term return erosion appears to have been reasonably linear.

If we start with the assumption of perpetuity and then restate the value-versus-IRR curve to allow returns to fade in a linear manner toward the cost of capital returns, the curve itself "pivots" about a central point. The result is that the low-return companies increase in value, because their returns are expected to increase in the future, while the high-return companies decrease in value, for the opposite reason. Not surprisingly, the largest variations in value are observed in the highest-return companies, and the exponential tail so evident in the perpetuity line of Figure 14-8 is substantially reduced.

From a practical point of view, the resulting curve is much more logical in shape and, as we will see shortly, it fits the empirical data much better as well.

Value versus Economic Replacement Cost

Value increases linearly with economic replacement cost. This fact is reflected in our presentation as the y-axis of the previous figure, which shows value as a percentage of economic replacement cost. This fact also highlights the importance of proper estimation of ERC, not the least element of which is the inclusion of some form of capitalized research-and-development cost.

Note again the slightly self-correcting mechanism in this valuation methodology. Assume for a moment that we overestimate the asset base (both CCI and ERC). The overestimation of CCI will result in an underestimated internal rate of return. This, in turn, will result in an underestimated valuation multiple, which will then be applied to our inflated ERC. Whether the resulting valuation ends up higher or lower than the correct value will depend on whether the slope of the Value/ERC versus IRR line is above or below 1.0.

Value versus Reinvestment Rate (Payout Ratio)

Dividend policy nearly always matters. Indeed, the only time it doesn't matter is when the company in question is earning the cost of capital. In that case, management can either reinvest at this expected return or give the money back to shareholders, who will reinvest it at an equivalent expected return (assuming an equal level or risk and no taxes). A shift in the payout ratio will, however, affect companies differently, depending on whether the IRR is below or above the cost of capital. For the poor-return company, an increasing payout ratio actually results in a higher valuation; because *a higher payout ratio means a lower number of ill-advised investments.* Indeed, this fact is an integral part of the "value enhancement" achieved by many leveraged buyouts of poorly performing businesses. Of course, the opposite is true for high-return companies. In that case, increasing payout ratio actually decreases value, because good investments are now being curtailed.

This relationship should not, however, encourage all managers of reasonably performing companies to rush out and curtail the dividend payment. Indeed, the act of paying out cash as a dividend and then returning to shareholders to ask for a capital increase to fund investments has the same effect on value as a trimming of the payout ratio, but the process of returning to the capital market may be a better form of corporate governance (if distasteful to many managers), in that it encourages outside-investor scrutiny of a company's investment plans.

Value versus Cost of Capital

A decrease in the assumed cost of capital—that is, the discount rate—results in an increase in corporate value (and vice versa). There are, however, significant differences between equities and bonds in how value reacts to changes in the discount rate. Because bonds effectively promise a defined set of *nominal* future cash payments, the reason for the change in nominal discount rate, be it a change in real expected returns or a change in inflation expectations, is irrelevant. With equities, however, the promise is closer to a defined set of real operating cash flows (defined, that is, under a set of constant business conditions). As a result, the change in a company's value with respect to a change in inflation is relatively low, because the real cash flows are usually not significantly affected. The valuation effects that do occur are principally the result of an increase in the effective tax rate, which in turn results from the loss of the depreciation tax shield (depreciation being reported in historic dollars) and the taxation of noncash inflation gains on inventory, as described earlier. However, the change in value of equities in response to a change in the real discount rate can, instead, be extremely high.

PLACING CHEMICAL COMPANIES IN THE VALUE GRID

We can now examine the extent to which empirical data match the theoretical expectations. Figure 14-9 shows the ratio of IRR to corporate value for a number of major European and U.S. chemical companies. The horizontal axis graphs the IRRs for 1996, and the vertical axis shows the ratio of value to ERC. The two lines on the diagram represent three different assumptions about the time period over which returns fade to the cost of capital rates.

For the most part, the empirical observations tend to fit the theoretical predictions quite well. Of course, in the real world, one should not expect each company to sit perfectly on the theoretical valuation lines, due to both the backward-looking nature of our IRR estimates and the assumptions and approximations built into these calculations. For example, if a firm has just announced or has just started to carry out a major restructuring program designed to enhance internal investment returns (Witco and Praxair, for example), then the likely market response would be for the value to increase. Yet, the historic IRR would not change. In this situation, the data point would shift upward on Figure 14-9 to a position above the theoretical "fair value" curve; yet the company itself would not necessarily be overvalued. In effect, the vertical "gap" between the empirical point and the theoretical line would be the value the market has ascribed to management's restructuring program. To put it another way, the horizontal gap between the empirical point and the theoretical line would be the extent to which management's restructuring program would have to improve internal rates of return in order to justify the current valuation.

In contrast, a firm should sit below the theoretical valuation curves if its returns are expected to decline quite rapidly in the near future. Great Lakes neatly fits into this example, as much of its current profitability is the result of earnings from tetraethyl-lead, the antiknock compound in leaded gasoline. As these returns are not expected to last (they have already slipped considerably in recent years, as evidenced in Figure 14-6), this fact must also be reflected in the company's valuation.

Valuation shifts over time can also be explained quite well by Figure 14-9. Monsanto's recent stock-market outperformance has, for example, clearly been driven by a substantial increase in its achieved IRRs. Note, however, that this methodology does not forecast future share-price movements (i.e., it is not the holy grail of stock-market investing). Indeed, one of the messages from this analysis is that *markets are actually rather efficient at reflecting the available information in current valuations.* Thus, to actually have profited from Monsanto's terrific performance, the investment analyst would have had to have correctly *forecast* the improvements in returns. Understanding future IRRs better than the market (generally through a detailed analysis of companies, products, and management) and not black-box modeling is therefore the key to investment success.

In general, European chemical companies tend to be at the lower end of the range in terms of average IRR, and their low corporate valuations appear to be a fair reflection of these low returns. This offers some insight into the original, apparent paradox in the valuations of Dow and BASF. The historically low firm value-to-sales ratios seen for the European chemical companies seem entirely justified, given that sales

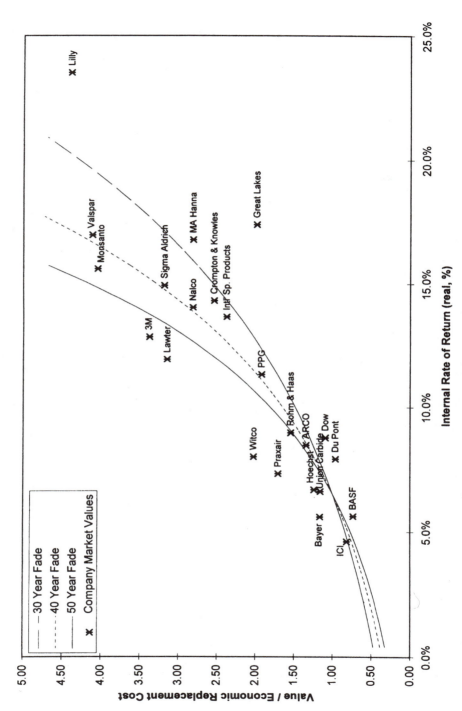

Figure 14-9. Value/ERC as a Function of Internal Rate of Return.

is a reasonable proxy for ERC in the chemical industry and that the European companies deserve a value discount to ERC, as a result of their historically low investment returns. As yet, however, we have not explained the discrepancies in cash-flow multiples. To do this requires a somewhat greater analysis of the more traditional valuation ratios.

Insights into Traditional Valuation Ratios

The basic assertion behind valuation is that any asset — be it a bond or a corporation — is worth the net present value of its expected future cash flows. Why, then, do analysts traditionally use earnings and cash-flow multiples rather than doing more rigorous, discounted cash-flow models? The answer is that these multiples are essentially short-cuts to discounted cash-flow valuations. Rather than doing a rigorous, discounted cash flow utilizing our principal valuation drivers, an analyst who takes a particular multiple to be "fair value" is making some implicit assumptions about the levels of future returns and reinvestment rates (or at least about the relationship between these variables).

Earnings Multiples

By far the most common valuation tool is the simple price-to-earnings (P/E) ratio. This multiple is crude, but it does excel in simplicity, as both a company's share price and its earnings per share are easily observed. The use of P/E as a valuation tool also has numerous disadvantages, not the least of which being that the "fair" level of this ratio is a function of just about everything imaginable, including the corporation's IRR. In Figure 14-10, we show the theoretically fair levels of P/E and firm value to operating

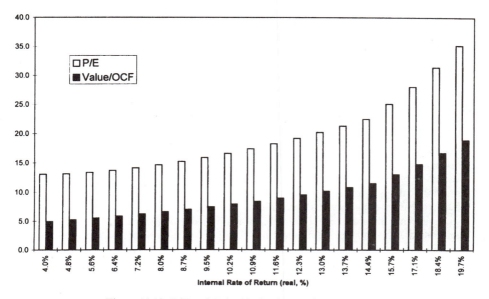

Figure 14-10. P/E and Value/Cash-Flow Ratios vs. IRR.

cash flow (FV/COCF, where firm value is the market value of equity plus the company's net financial debt) as a function of the company's internal rate of return. Once again, we have used an assumption set that is fairly representative of the chemical industry, including such items as an asset lifetime of 22 years and an IRR that fades to the cost of capital over a period of 30 years.

As can be seen from Figure 14-10, P/E increases with an increasing IRR. The simple logic is that earnings reinvested in high-return projects are worth more than earnings reinvested in low-return projects. Unfortunately, analysts often erroneously extended this relationship to conclude that P/E is also positively related to earnings growth. At the extreme, this latter relationship is also assumed to be linear, and thus an attempt is made to make relative valuation decisions based on the ratio of P/E to expected growth (often called the PEG ratio). Reality, however, is considerably more complex.

For example, decreasing the payout ratio will, in all cases, increase a company's growth rate. But remember that growth is not the same as value. When a company's internal rate of return is less than the cost of capital, this additional growth will be at the expense of value, and hence the P/E will tend to fall. When the company is earning above the cost of capital, the opposite will be true. When a company is earning the cost of capital, its P/E will be independent of its payout ratio. Furthermore, the more a company is outperforming its cost of capital, the more the P/E will expand for the same shift in growth rate. Thus, even at positive IRRs, the PEG ratio remains a strong function of the company's internal rate of return.

When looking for empirical evidence of these relationships in companies with varying capital structures, it is important to make sure that the payout ratio is very broadly defined — namely, as dividends and interest as a percentage of operating cash flow, and not just as a simple percentage of earnings. Empirically, one finds large variations in this broad payout ratio, particularly for the larger, more developed companies. As these companies also tend to have rather large variations in achieved IRR, it is not surprising that the P/E-versus-growth relationship frequently yields erroneous valuation results. There is, however, a subset of companies (growth stocks) that tend to have zero payout ratios. These firms will therefore have growth rates equal to the internal rate of return, and thus the positive relationship between P/E and growth rate will be more likely to hold.

Cash-Flow Multiples

Some years ago, traditional valuation analysis received a significant boost with the "discovery" of cash accounting. While earnings could vary quite widely because of the different ways in which accounting standards were applied (leading, in some cases, to outright manipulation), cash was cash. Thus cash-flow multiples became a principal valuation tool. Unfortunately, the principal assumption inherent in most cash-flow multiple-valuation methodologies is that two companies in the same business should have the same cash-flow multiple. This is, however, frequently not the case. As with earnings, cash flow that is reinvested in a good business is worth more than cash flow reinvested in a bad business, and thus the "fair" cash-flow multiple can vary quite widely, even within the same industry. Hence, to believe that all companies with low cash-flow multiples are "undervalued" and will rebound, based on crude industry averages, can often be a significant mistake. The historic case of Dow and BASF shows that very different cash-flow multiples can persist for decades among similar firms in a common industry.

Existing Assets versus Future Growth Opportunities

As a final bit of insight into equity valuation, it is useful to divide a company's value into two pieces: one deriving from the company's existing assets, and another deriving from the firm's future growth opportunities. The value of the existing asset base is the net present value of the remaining cash flows from those assets. The net present value of future growth opportunities (NPV of FGOs) is the additional value that the company is expected to create by making future investments at returns in excess of the cost of capital. It is important to realize that the NPV of FGOs can also be negative if the firm is investing at returns below the cost of capital. The relative size of these two valuation components is shown as a function of IRR in Figure 14-11. A higher IRR will raise both the value of the existing asset base and the net present value of future growth opportunities. However, this effect creates far larger variations in the NPV of FGOs. For example, while a low IRR will hold down the value of existing assets, that value will always remain positive. However, a low IRR for future growth opportunities means that the firm will be subtracting value in the future, and so the NPV of FGOs can readily become a negative number. Conversely, while a higher IRR does raise the value of existing assets, it has an even greater effect on the NPV of FGOs. After all, reinvestment with a high IRR will compound that rate over time, leading eventually to even higher growth. Remember, these high future returns would be discounted back to the present only at the lower cost-of-capital interest rate; so when the IRR is high, the NPV of FGOs will far outstrip the value of current assets.

Many investment funds (particularly in the United States) will alternatively describe their investment philosophy as either "growth" or "value." Unfortunately (or perhaps fortunately for the fund managers), there is no easy distinction between these two fund-management styles. The above methodology does suggest, however, a possible

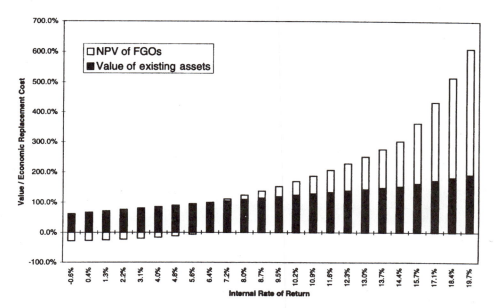

Figure 14-11. Value of Existing Assets and NPV of FGOs vs. IRR.

differentiation in that a "value" stock is one whose market value is dominated by the value of the existing assets while a "growth" stock is one whose NPV of FGOs meaningfully exceeds the existing asset values. Under this definition, valuation of growth stocks would be particularly volatile, as the sensitivity to shifts in future-return expectations would naturally be quite high.

THE MYTH OF MARKET POWER

One of the common reactions to Figure 14-9 is to notice that the chemical industry has a number of larger companies in the so-called "problem corner"—that portion of the chart that represents returns below the cost of capital and valuations below replacement cost (no prizes for readers who note that Microsoft, Intel, General Electric, and Kellogg aren't here). Furthermore, many of these companies are commonly thought to be the most successful in the industry, and several of these firms are now among the largest companies in the world. The question that naturally follows this observation is how these companies managed to stay in the problem corner of low returns and low valuation—and indeed, in some cases, to grow and apparently prosper—for so long. Surely markets must discipline companies with poor returns, so that they slowly give way to their more successful competitors?

These questions have led me to the conviction that capital markets by themselves have very little, if any, power. To many, this view violates one of the key claims of market economics, which is that a well-run firm is rewarded with a high share price and hence a low cost of capital, while a poorly run firm is given a low share price and hence a high cost of capital. This "cost of capital advantage" is then supposed to give the better firm a competitive edge, thereby forcing the poorly run firm out of business.

Unfortunately, this supposed cost of capital advantage is a myth for the majority of the marketplace. Capital markets do only one thing: they indicate value (i.e., price). Markets give the price of BASF as X and the price of Dow as Y. But price alone carries very little power. As long as owners fail to exert pressure on management, either because they are unwilling or unable to exert this pressure, managers can ignore their low returns and correspondingly low share price and be at very little, if any, competitive disadvantage. Furthermore, since it is owner action—or the threat of owner action—that is required, it naturally follows that economies which limit ownership power are indeed missing an essential ingredient of capitalism.

This argument also extends to a corporation's access to capital. In general, an industry's internal cash flow significantly outweighs the annual amount of debt and equity issues, even in the hottest of markets. In simple terms, markets can't deny managers access to capital, because it is something the managers already have. If management is willing to accept low margins, they can still invest, and hence grow, just as much as the competition simply by retaining a higher percentage of internally generated cash flow.

The myth of market power can be easily demonstrated through the simple example in Table 14-2. Imagine two companies, A and B, of equal size in terms of sales and assets and with identical investment histories. Assume A is nicely profitable, with an operating cash-flow-to-sales ratio of 15%, while B is somewhat less efficient (perhaps because it is overstaffed) and generates a corresponding ratio of only 11%. Given that these two companies are in the same industry (thereby having the same asset lifetimes), it naturally follows that A's IRR is comfortably ahead of that of B—although knowledge of the exact IRR level is not needed for this analysis. Identical investment histories will

TABLE 14-2. The Myth of Market Power

	Company A	Company B
Sales	200	200
EBITDA	30	22
EBITDA margin	15%	11%
Depreciation	(10)	(10)
EBIT (=pre-tax)	20	12
Tax (50%)	(10)	(6)
Net income	10	6
After-tax cash flow	20	16
Dividend	(8)	(4)
Cash for reinvestment	12	12
Maintenance capital expenditure	(6)	(6)
Capital expenditure for growth	6	6
Real growth rate = 6/200 =	3%	3%
Value = D/(r − g)	200	100
Value/sales	1.0x	0.5x
Value/EBITDA	6.7x	4.5x
P/E	20.0x	16.7x

result in identical depreciation charges. Furthermore, for simplicity, we have assumed that both of these companies are debt free. Working down through the income statement, we find that A's net income of 10 is 67% higher than B's net income of 6. Adding depreciation back to net income results in after-tax cash flows of 20 and 16 for A and B, respectively.

Now let's assume that A pays a dividend of 8, thereby leaving 12 for investment. Furthermore, assume that half of this investment goes to replace worn-out equipment, while the other half goes toward growth. For complete simplicity, assume that $1 in sales is generated for every $1 invested, and thus the real growth rate is 3% (6/200). If we assume the pattern continues into perpetuity, the value of Company A simply becomes the sum of a series of dividends, which will be growing at a real rate of 3% per year. Valuation in this case is very clearly defined by the dividend discount model ($V = D/[r − g]$), and if we use a real discount rate of 7%, the end result is that Company A has a value of 200. This value is 1.0 times sales, 6.7 times operating cash flow, and the resulting price-to-earnings ratio is 20 — all perfectly reasonable multiples.

Now, let's turn to Company B. The management of Company B looks at Company A and says, "Gee, they're going to invest 6% of sales this year. So we'll invest 6% of sales this year." Warren Buffett explained the logic behind this process for setting corporate strategy in this way: "Lemmings as a group have a terrible reputation, but no individual lemming has ever been singled out for bad press." Martin Liebowitz, formerly of Salomon Brothers and now Vice Chairman and Chief Investment Officer with Teachers Insurance Annuity Association-College Retirement Equity Fund (TIAA-CREF), added the corollary, "There's no prize for being the fastest lemming."

So how does Company B match Company A's investment level? It just pays a dividend of 4. Remember, growth is entirely determined by how much is invested, not by how well that investment is subsequently managed. If two chemical companies each build a new 600,000-metric-ton-per-year ethylene cracker, incremental sales will be the same, even if one firm staffs its plant with 50 people and the other firm staffs its plant with 500 people. So sales at Company B will also grow at 3%.

For company B, we once again assume perpetuity and use the dividend-discount model; but because of the lower dividends, this time we arrive at a value of just 100. Note also the difference in valuation multiples, in spite of the fact that A and B are similar in many ways. Company B is worth only 0.5 times sales, 4.5 times operating cash flow, and 16.7 times earnings.

The key point here, however, is that the valuation multiples are completely fair and understandable for both A and B. These are the values at which investors are indifferent between investing in Company A or Company B. As a result, the share-price perform- ance of these two companies from this starting level will be exactly the same (equal expected market returns), in spite of the fact that the expected *internal* investment returns for Company A significantly outweigh those of Company B. Even poor management and low rates of return have a fair price.

One message from this analysis is that share-price performance is *not* an absolute indicator of managerial performance. Instead, it is a measure of management perform- ance *relative to expectations*. If one company is expected to have a poor internal rate of return and another is expected to have a high internal rate of return, future share price performance will be exactly the same, as long as both companies live up to expectations and the shares are correctly priced at the outset. There are provocative implications here for the issue of management incentives. Pity Company A's management. They have to do extremely well just to be a market performer, because the market already expects them to do well. Company B, on the other hand, can just improve from "terrible" to "rather bad," and managers with share options will make a fortune. The obvious conclusion is that managerial reward should be based more on internal investment returns than on share-price performance. However, since it is easier to make money in some industries than in others, individual company IRRs should probably be benchmarked against those for the industry as a whole to get a more accurate picture of managerial success.

So how does Company B keep up with Company A, despite being less efficient and having a lower internal rate of return? The answer is quite simply by restricting the payout ratio. Notice that Company A is paying out 27% of its cash flow (8/30), while Company B's payout is only 18% (4/22). This difference may not seem like much, but it's still enough to compensate for Company B's low returns. This is how BASF, in spite of its low historic returns, has kept up with Dow. It is not a matter of better management or long-term planning; it is a simple matter of higher cash retention. The average payout ratio — that is, dividends and interest as a percentage of operating cash flow (EBITDA) — was 35% for Dow from 1980 to 1992, and only 11% for BASF over that same period.

The example of Table 14-2 raises other hard questions. Company A is clearly more profitable than Company B, as illustrated by higher margins and better internal-investment returns. Yet these two companies will not only grow at the same rate but they will also have identical share-price performances! Further- more, Company A won't put Company B out of business; it won't even be able to take incremental market share. So where is the competitive advantage in higher returns? The unfortunate answer to this question is that without active shareholders, it does not exist. As long as shareholders do nothing to restrict Company B, this situation can continue indefinitely. The obvious conclusion is that markets do not work by themselves; instead they require active shareholders to implement proper corporate governance.

While the example in Table 14-2 dealt only with internally generated cash flow, this argument extends itself to external capital raising, as well. For example, Company B could do a rights issue tomorrow and it would be successful, as long as the price was low enough — even if investors knew that the proceeds would be used to invest in projects that failed to earn the cost of capital. The resulting "fair" value for the company after the new share issue would now be less than the previous value of 100, since the expected poor investment would have dragged down the value; in effect, existing shareholders would suffer because management raised capital externally. For new investors, however, the expected share-price performance once again would be exactly equal to that of Company A.

Indeed, the only time that market pricing alone appears to exert some pressure on management is when a very poor-return company attempts a large rights issue. At this point, the negative net present value of the future investments may be so low that it drags the fair value of the shares down below the rights-issue price, thereby causing the issue to fail. Even in this situation, however, it's usually active (and irate) shareholders who become the catalyst for management change.

WHAT DETERMINES CHEMICAL COMPANY IRRs?

The analysis of corporate valuation to this point suggests investigation of a deeper question. Can insights based within this framework help to explain why some chemical companies are so much more profitable than others? I began my attempt to answer this question (at least partially) with the vague idea that profitability is somehow linked to productivity. In addition, I have found it useful to think separately about the productivity of labor and of capital, although in the end I have found it virtually impossible to completely separate the two.

Putting the Importance of Labor Costs in Context

In general, the chemical industry is not thought of as a particularly labor-intensive business, but this perception can be misleading. Labor costs — that is, wages, salaries, and benefits — are 24% of sales for the chemical industry (a rough estimate, based on a global sample of about 20 firms). For comparison, labor costs are 36% of sales in electronics, 26% of sales in pulp and paper, 21% of sales in autos, and 19% of sales in food manufacturing. The fact that labor costs measured this way are higher in the chemical industry than in the automobile industry is generally surprising to most observers. Indeed, Bayer has a wages, salaries, and benefits bill that reaches a tremendous 35% of sales!

Labor costs also stand out for the chemical industry because many other costs are rather difficult for management to control, at least in a way that significantly differentiates any one company from the competition. Raw-material cost positions do vary somewhat meaningfully from company to company (e.g., Gulf Coast ethylene and Green River soda ash), but costs saved on raw materials often seem to be counterbalanced by increased shipping costs. Energy costs were the buzz in the early 1980s, but aggressive investment in efficiency improvements (often sold to all companies by the same SEF — see the chapter by Arora and Gambardella) have narrowed many of the previous gaps.

One way of potentially differentiating chemical products is by adding in varying service levels, but my experience has been that the additional revenue achieved by the improved service levels is typically balanced by the additional costs of such a service, so that the impact on overall returns is quite low. Nalco is perhaps an exception here, as they made a technological advance in the delivery of a service (water-treatment chemicals) that their competitors were slow to adapt.

Product and processing technology is perhaps the one area, aside from labor, that offers managers meaningful possibilities for performance differentiation. Indeed, there are numerous examples of significant comparative advantage being gained from the discovery and development of novel products and production processes; these include the Unipol polyethylene technology of Union Carbide, the Spheripol polypropylene technology of the former Himont (now Montell), and the propylene oxide/styrene monomer breakthrough discovered and commercialized by one editor of this volume, Ralph Landau.

Unfortunately, many of the older technological breakthroughs are now well known and well copied throughout the industry (and often for sale to all by the SEFs), and so the remaining comparative advantage imbedded in these developments is now quite low. Further, while I believe that both new products and new processing technology still hold significant opportunities for future value creation, managers need to realize that a significant portion of their industry has matured (recall Figure 14-7) and, thus, the commodity aspect of the business has increased substantially. This in itself creates a management dilemma, for it is culturally difficult to have a people-oriented, high-labor-cost, research-and-development focus on the development side when the rest of the organization is best operated as a lean, mean production machine. Perhaps not surprisingly, we have recently seen many firms separate out these two activities through asset sales and de-mergers.

A purely cynical view of the chemical industry thus acknowledges only two types of chemical products: commodity and soon-to-be-commodity. This view is particularly relevant to the big, diversified firms, since a number of smaller companies do have specialty products in defensible market niches. The larger companies, however, have a problem in that they can all point to some exciting new product in their portfolio that has huge growth rates and astronomical returns, and yet such products are frequently too small to have much of an impact on the overall organization. In addition, the showcase product usually changes every year, as overinvestment and competition drive down profit margins almost as soon as one or two other large producers enter the market.

Measuring Labor Productivity with Ratios

There are many different ways of approaching the problem of measuring labor productivity. The two most common are simple ratios: sales per employee and value-added per employee. Neither is perfect, but both are worth some explanation.

The simple ratio of sales per employee can be an extremely useful indicator of a company's labor productivity. Not surprisingly, there are considerable variations in the observed levels of labor productivity across the rather broad spectrum of chemical companies. The problem is to determine which variations are due to differences in product mix and which are due to differences in actual efficiency. After all, it makes

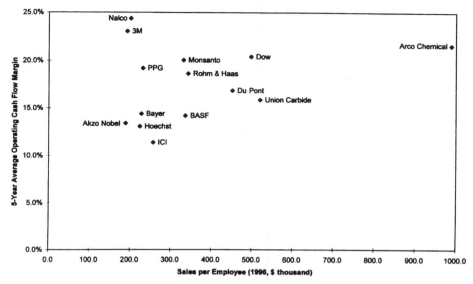

Figure 14-12. Five-Year Average OCF Margins vs. Sales per Employee.

sense that a traditional "blow it out the door" petrochemical company should have a much higher sales-per-employee ratio than a more specialty-oriented firm. However, we have found that interesting insights can still be gained by looking at profitability as a function of sales per employee, as we have done in Figure 14-12.

Figure 14-12 shows sales/employee on the horizontal axis, as a measure of productivity, and cash-flow margin (that is, the five-year average operating cash flow divided by sales) on the vertical axis, as a measure of return on investment. Note that we have chosen only a subset of our total chemical universe — namely, those companies with a gross fixed-assets-to-sales ratio of between 80% and 120% (see Figure 14-14), because operating cash-flow margin will only be an indicator of relative return under conditions of constant capital intensity. The trend is reasonably clear: For the larger, more diversified companies a higher sales/employee ratio appears to be closely associated with higher returns. The most notable exceptions are Nalco and 3M, which achieve their excellent profitability levels at low sales/employee ratios as a result of strong technological positions (R&D investment).

Some companies (usually those that do poorly in this chart) claim that their lower sales-per-employee figures are due to higher levels of vertical integration; that is, they have lower sales per employee because a greater share of their chain of production happens in-house. It is surely true that an increase in vertical integration will result in a lower sales-per-employee ratio, all other things being constant. However, greater vertical integration should also result in an increase in cash-flow margin, as more assets require a return behind each unit of final sales. As a result, if higher vertical integration were really the main factor causing differences across companies in the sales-per-employee ratio, the relationship between employee productivity and profitability, shown in Figure 14-12, would be *downward* sloping — exactly the opposite of the observed trend.

TABLE 14-3. The Dangers of Value Added per Employee

	Commodity	Speciality
Sales	100	100
Non-labor costs	*(60)*	*(30)*
Value added	40	70
Labor costs	*(20)*	*(50)*
EBITDA	20	20
Employees	100	250
Sales per employee	1.00	0.40
Value added per employee	0.40	0.28

An alternative measure for determining employee productivity is to look at value added per employee as an improved measure of labor productivity. Even this ratio, however, does not fully compensate for differences in product mix. This fact is illustrated in Table 14-3, which shows hypothetical partial-income statements for both a commodity company and a specialty company. The example is constructed to reflect common differences in the cost structures of commodity and specialty-chemical companies. Commodity companies tend to spend a great deal on raw materials and less on labor, while specialty companies do the opposite. However, in contrast to these differences in cost structure, the capital intensities of these two types of firms can be remarkably similar; for example, BASF and Bayer have almost identical fixed investments per unit of sales. Thus, given the assumptions of equal capital intensity and asset lifetimes, the equal cash-flow margins shown in Table 14-3 imply identical levels of internal rate of return. But despite the equal IRRs, the ratios of sales per employee and value added per employee remain much higher for a commodity company than for a specialty company. The implication is that neither sales/employee nor value added/employee are fully adequate ways of looking at labor productivity.

Calculating Overall Productivity

Given the difficulties with sales/employee and value added/employee, how should we measure labor productivity for an individual company? A basic problem in any reasonably easy labor-productivity calculation is that it almost inevitably requires the assumption of equal capital productivity, which may or may not be the case. To illustrate this point, imagine two chemical companies, each consisting of only one large styrene plant. Assume further that the starting point of our analysis is equivalent sales, staffing levels, cash flows, IRRs, and so on. Now consider what happens if one firm loses its largest customer, resulting in a meaningful loss of sales and income. The sales per employee and value added per employee would drop, along with other simple measures of labor productivity. But in reality, the problem is just that the plant isn't running at capacity any longer; it's a capital-productivity problem, not a labor-productivity problem. Unfortunately, the only way in which this problem can be addressed is to do a detailed benchmarking study between the two facilities — obviously great news for the consultants.

There does, however, appear to be a meaningful way of looking at the overall combined capital and labor productivity of an individual company: Just use the internal rate of return! When I reached this conclusion (as many others have doubtless done before me), I was suddenly reminded of a consulting study that compared differences in U.K. and German GDP per capita. After an exhaustive analysis of some 360 pages, this report concluded that the vast majority of the difference in U.K. and German GDP per capita was explained by differences in productivity. But what is productivity? For a country as a whole, it is essentially GDP per capita. It was, however, comforting to know that in 360 pages of analysis, a consultant could produce a tautology.

Does Value Added Matter?

Management themes tend to move in faddish ways. In the chemical industry in the mid-1980s, "value added" was a buzzword in management circles, and everyone was clamoring to get into higher value-added products. But remember that a corporation's value added is only the difference in value between what is sold and the nonlabor inputs that are purchased. Mathematically, value added is operating cash flow plus wages, salaries, and benefits. If value added per employee rests on dubious ground as a measure of employee productivity, how much does value added itself matter?

One way to look at this dilemma is to examine whether or not companies with higher value added ratios tend to deliver higher internal rates of return. Thus, Figure 14-13 plots the value added and operating cash-flow margins (that is, value added and operating cash flow as a percentage of sales, respectively) for the three German majors and Dow Chemical. For those familiar with these companies, this chart holds no surprises. Bayer makes specialty products with a higher value-added component than those of either Hoechst or, particularly, BASF, which specializes mainly in lower-value-added commodity chemicals. Bayer's particular strength in the early 1980s was in agrochemicals, an area that suffered in the middle of the decade, as represented by the

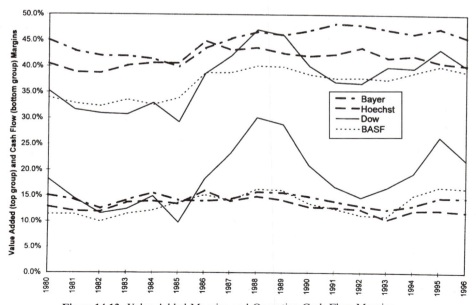

Figure 14-13. Value-Added Margins and Operating Cash-Flow Margins.

dip in Bayer's data in the top half of Figure 14-13. Fortunately for Bayer, the agrochemical profits were subsequently replaced with a successful pharmaceutical operation, thereby reinstating the firm's preeminence as the highest value-added producer of the major German chemicals. In value-added terms, Dow stands in the middle of the group, near Hoechst but with more cyclicity.

But is Bayer's position more advantageous? Does this higher value-added position enable the company to earn higher returns? The answer, it seems, is no. Note the operating cash-flow margins at the bottom of Figure 14-13. (Remember, the cash-flow margin is a reasonable indicator of internal rate of return, because of the similarities in asset lifetime and capital intensivity of these four firms.) The results are striking: There just isn't much difference among the German "Big Three," in spite of the significant differences in product lines and value-added margins. Dow shines above the others (in spite of its cyclicity), mainly because of its previously described level of apparent efficiency; note Dow's position in Figure 14-12. Thus, it seems that value-added isn't that important in determining chemical company IRRs. To put it another way, it's not *what* you make, but *how* you make it that counts.

While this conclusion holds true for the large chemical majors, we would also observe that a number of highly profitable small chemical companies seem to be focused firms with clearly defensible market niches and high value-added margins. Most of these market niches seem to be defined by technological positions, in either product or process innovations. However, a serious question still remains about cause and effect. Indeed, we believe the most plausible interpretation is that strong profits are the result of judicious investments in technology — that is, good IRRs on R&D — and that high ratios of value added to sales and value added per employee are the observable results, not the cause.

This final point has some important implications for chemical company management. In particular, increasing the value added of your products simply by adding labor — for example, by increasing the level of service — is not likely to add significant value to the corporation, since much of the increased selling price will be adsorbed by an increase in labor costs, leaving the overall IRR relatively unchanged. In contrast, a well-executed research and development program can result in a significant increase in IRR (and value added along with it) in a much more sustainable way. In other words, you must pursue the technology investment itself, rather than targeting the observable result of higher value-added margins. Note that it is difficult, however, to assess the "appropriate" level of research-and-development investment simply by looking at the smaller, highly profitable chemical companies. These are, after all, the self-selected successful ones; many others tried to develop through R&D and failed. As a result, the appropriate level of R&D spending is probably higher than many managers think.

IMPLICATIONS FOR MANAGEMENT AND CORPORATE GOVERNANCE

Modern capitalism is evolving into what I call a "circle of governance." The first stage of this model is the traditional management of employees by corporate managers, including determination of staffing levels, pay scales, and so on. The second stage is the employees' management of their own wealth. This can be direct, which only works well in a corporate governance sense when the individual is quite wealthy, or indirect, through intermediaries such as pension and mutual-fund managers and life insurance companies. The key, however, is that these intermediaries need to be actively "managed." Examples of this management are easy to find; witness the vast deposits and withdrawals that occur in response to the investment performance of the various fund managers. The final stage is the "management of the corporate managers" by the owners or their representatives. This management must be *active*, since the markets themselves are principally involved in pricing and not corporate discipline. When the circle of governance is working well, it creates an upward spiral of increasing social wealth. However, there are impediments to the efficient functioning of this system at each of the three stages.

The Need for Efficient Labor Markets

One key ingredient for the first stage in the circle of governance is efficient labor markets. Managers must be able to adjust their staffing levels to current needs, and labor-market pricing must broadly reflect supply and demand. Many countries in Europe are presently locked in a situation that combines high unemployment with poor industrial returns. Europe reached this position by having labor markets that were too restrictive; companies were unable to downsize or to reallocate labor as needed, and labor costs were prevented from adjusting to new market realities by rules and regulations. The result is that foreign competitors have made greater productivity gains, and Europe's IRRs have declined in a globally competitive marketplace. The current deadlock is particularly unpleasant, as poor returns cause the region to be an unattractive location for investment — yet investment is just what is needed to enable Europe to emerge from the current doldrums of high unemployment. Unfortunately, the higher returns required to attract additional investment may have to come from further reductions in employment. Thus, as the situation currently stands, Europe's path from 12% to 7% unemployment may actually have to pass through 18% unemployment, a fearful prospect indeed.

The ironic result is that well-intentioned attempts to protect the wealth and income of employees individually (or within certain companies or industries) actually damage the overall wealth of society, as restrictive labor laws and regulations create a region that slowly but steadily becomes increasingly unproductive relative to the rest of the world. Clearly, less social disruption occurs when industries maintain or increase employment. However, in many segments of the economy, including many basic chemicals, productivity growth is outstripping volume growth, so that fewer employees are needed. Society as a whole will be better off if these employees are more efficiently deployed in other areas of production. This redeployment should be done in a compassionate way, but companies that avoid the issue are only ensuring the competitive strength of their rivals, whether across the country or around the world. Further, if the process of change is bottled up for long periods of time rather than being allowed to happen gradually, the eventual sudden correction may cause significant angst in society as a whole.

Restrictive labor markets also create difficulty in that change becomes possible only when corporate performance becomes absolutely horrendous. For most firms, this point generally coincides with a period of poor macroeconomic conditions, which from a social view, is the worst time for companies to lay off employees. Indeed, corporate downsizing ideally should occur during a period of general economic expansion, as this is the time at which displaced workers would most likely find suitable reemployment. For example, America's corporate restructuring of the 1980s took place largely during a period of economic expansion. As a result, the majority of those who were laid off in the 1980s found new jobs reasonably quickly. In the early 1990s, Europe was faced with the dismal prospect of restructuring during a period of economic stagnation or even contraction, and the effect on society was much more severe (and still ongoing).

There is an understandable tendency for the general public to focus on those who experience employment dislocation, and social policies in response to such changes are appropriate. But the overall wealth of society as a whole is best served by social policies that encourage and facilitate labor-market flexibility rather than attempting to freeze the economy in place.

Individuals Must Manage Their Own Wealth

It may seem obvious that individuals should manage their wealth, but it happens all too infrequently. For example, numerous people have a sizable proportion of their net worth in government-managed retirement funds, many of which are in many respects "unfunded" (Social Security in the United States being an example). In Germany, pension funds remain on the balance sheet of the employer instead of being placed under external management. Not surprisingly, these funds inevitably get reinvested right back into the business, regardless of the net present value of those investments compared with the alternatives.

Historically, the individual investor has had only a tiny (and usually uninformed) voice with which to influence corporate management. However, the recent development of the pension and mutual-fund industries has made the management of wealth much more active. These funds are highly sophisticated and carry much greater clout. The net result is that the regions of the world with well-developed independent pension and mutual fund systems generally have much more efficient overall circles of governance, and countries in which the influence of owner over manager is highly restricted generally have an overdose of poorly performing companies. At this point, it is worth pointing out the dichotomy in Germany, where many of the large, publicly traded firms have rather dismal long-term returns, and the economy instead is carried by the strength and robustness of the famous group of small- and mid-sized firms known as the *Mittelstand*. The Mittelstand, of course, provides is a perfect link between the owner and manager, as they are usually the same individual.

Managers Must be Managed

Finally, managers themselves must be managed, which means being held accountable for poor long-term results. Unfortunately, in all too many regions of the world, managers can do whatever they please, with little outside oversight or accountability, despite the presence of large corporate boards (frequently comprising the previous managers, who are loath to admit a mistake).

All corporate managers face a direct conflict of interest: On one hand, they have a fiduciary responsibility to shareholders as managers of their wealth; on the other, managers have a personal incentive to increase the size of the company or division they manage, and this requires investment, even if it is in projects with a negative net present value. (All too often, managerial salary and prestige are based on assets under management and total cash flow rather than return on internal investment.) This conflict of interest is enhanced because most managers believe their own abilities at reinvestment are much better than the financial market's ability to reallocate resources toward more efficient uses.

When corporate managers are criticized for a dismal current performance, many of them (particularly in Europe) often respond by saying, "We're investing for the long term." Similarly, the United States often receives considerable criticism for the alleged short-termism of Wall Street. But the empirical evidence for this short-termism is actually rather scarce. The stock market can indeed shift considerably in response to earnings announcements, but more often than not these shifts result from changes in long-term expectations and not in response to the profit announcement per se. One only need look at the huge sums that have been raised for long-term investment in such industries as cellular communications and biotechnology to realize that analysts and investors do have the ability to value and fund the rather distant future. In general, the empirical evidence unambiguously indicates that markets are indeed quite good at perceiving returns and setting values and prices accordingly.

While capital markets don't explicitly discipline managers (this requires active shareholders), they do pass judgment on who is actually able to invest efficiently in long-term projects. This judgment can be "found" by comparing a firm's total market value (equity plus net debt) to its economic replacement cost. If this ratio is less than one, the basic message is that the market does not believe that future internal rates of return will equal the cost of capital. It's here that many (particularly large) companies seem to lose out, much to the chagrin of some very vocal and very influential big-company chairpersons. Most people, particularly those who take great pride at sitting atop the corporate ladder, do not like to admit that they have a boss.

Perhaps the strongest response to those managers who claim to be misunderstood long-term investors is to remember that, as Alice told us from Wonderland, sooner or later, jam tomorrow must mean jam today. Where, then, are the benefits of the long-term investments these companies were supposedly making 5, 10, and 20 years ago? Usually, they don't exist. Companies that persist with this line of reasoning are usually those with the poorest long-term performances and the greatest need for a corporate shake-up.

Our final point is that the overall wealth of society is indeed owned by society itself, and thus employees are the ultimate beneficiaries of increased IRRs. While a manager may not personally agree with the current distribution of this wealth in his or her own society, any process of wealth redistribution is much more efficiently done at a governmental level, through staggered tax rates, adjustment assistance, and so on, than at the company level through overstaffing and poor returns on capital.[8] Governments around the world increasingly are realizing the importance of strong IRRs and the benefits of the circle of governance in achieving these returns; thus, they are taking steps to increase the efficiency of this fundamental part of capitalism. This move includes increasing labor flexibility, expanding the presence of independent pension and mutual funds, and removing the protective barriers surrounding so many entrenched corporate managers. While, in the short run, such steps can feel more painful than peaceful, in the long run, these measures will result in a more efficient industry and a wealthier society.

---

[8]There is a widespread consensus amongst economists that having the firm maximize the return on investments — essentially taking all positive NPV projects — is the proper strategy for maximizing social benefits as well as private benefits. This is rooted in the basic statement of economic efficiency. To paraphrase, for example, from *Principles of Corporate Finance*, by Richard A. Breley and Stewart C. Myers (McGraw-Hill, Fifth Edition, New York, June 1996, p. 24) the best way to achieve efficiency "is to seize all investment opportunities that have a positive net present value." Of course, it is also recognized that there are a variety of reasons why this is difficult to achieve including the different motivations, information and incentives of the individuals making real investment decisions (Steven Ross, private communication).

APPENDIX 1: MEASURING CASH FLOW—AVOIDING CROSS-BORDER PROBLEMS

Calculating consolidated operating cash flow from reported financial statements is relatively straightforward. Usually, the easiest way to calculate COCF is to begin with the operating income line on the income statement and add back depreciation, amortization, and other noncash charges. However, there are some cautions to observe for the particular measure used here.

Consolidated operating cash flow, or COCF, is similar to the better-known EBITDA, or earnings before interest, taxes, depreciation, and amortization. The main difference lies in the word "consolidated," which implies that we do not include earnings from nonconsolidated operations. (In fact, not all EBITDA definitions include these earnings, either.) As a general rule, nonconsolidated operations must be evaluated separately: Their reported "value" is often given on balance sheets as a historic cost, while the "earnings" from these operations are frequently not cash flow earnings at all, but instead the dividends paid to the parent company. COCF does, however, include minority interest, as this is also included in the asset values (fixed and otherwise) listed on the balance sheet.

Perhaps the biggest problem in the calculation of COCF arises with Germany's methods of pension-fund accounting. In Germany, corporations usually do not have an independent entity for managing funds set aside to meet future pension-fund obligations. Instead, the cash remains inside the company, although the cost of these future obligations is taken as an expense on the income statement and listed as a liability— generally labeled "pension provisions"— on the balance sheet. In American terms, it's as if DuPont calculated that the accrued pension liability was $500 million for the year, but after the company gave this money to its pension fund to manage, the fund immediately turned around and purchased a "Special DuPont Pension Bond" in an equal amount. The net effect would be that the cash would remain at the disposal of management, but a new liability would appear on the balance sheet.

This element is often missed in the analysis of German companies, and sometimes in the reporting of results by the companies themselves. Three points need to be made regarding this item: (1) The creation of the liability itself is an operating expense and should be treated as such (the company owes its employees future payments as a result of their employment for that particular year). (2) Despite it being a noncash event, the liability should not be included in operating cash flow and should instead be reported in the same manner as an increase in long-term debt (there is a temptation for companies to "cheat" on this one, as the inclusion of pension provisions makes the cash flows look better). (3) Interest accruals on existing pension liabilities are financing costs, not operating expenses, and should be excluded from operating income and added to interest expense. Most companies do not do this last adjustment, and thus for proper cross-border analysis, analysts must estimate this expense (by multiplying the reported pension provision by the accrual interest rate) and make their own calculations in order to arrive at a truly comparable cash-flow figure.[9]

---

[9]As an aside, some analysts and economists contend that the fact that pension funds remain within the corporations in Germany results in a source of cheap financing for these firms. Technically, this is not true. Even though the "interest expense" that is annually allocated to the liability tends to be calculated at a rather low rate, normally 5% to 6%, the counterbalance to this "cheap" financing is the fact that the current value of the future pension liability must also be calculated using this low interest rate as a discount rate; thus what the firm "saves" in a lower implied interest expense is actually "spent" on a higher charge in net present value terms for the future liability.

In addition to pension-fund reporting, several other, less significant accounting items technically require adjustments to be made in the calculation of COCF. The most important of these in respect to the chemical industry is the difference between last-in, first-out (LIFO) and first-in, first-out (FIFO) accounting. Our modeling generally assumes LIFO accounting, which sets values according to the most recently purchased items (those last in), and thus close to current market values. If a company is using FIFO accounting, the reported operating-income figures must be adjusted for inflation gains to be truly comparable.

APPENDIX 2: MEASURING CURRENT COST INVESTMENT (CCI)

The other number required for these return calculations is current cost investment (CCI). Any measure of capital employed is fraught with error. But while some people might not be comfortable with assumptions required to calculate CCI, these assumptions are *much* less onerous than the unseen, but nontheless inherent, assumptions imbedded in any accounting-system-dependent capital-employed figures. Relying on a depreciation formula, usually determined for tax purposes and unadjusted for inflation or other factors, to generate an economically meaningful measurement of capital is clearly questionable. For this chapter, we will restrict our discussions to a simple company with only two types of investment, working capital and fixed assets, although we acknowledge the importance of other investments such as R&D, advertising, and goodwill. CCI represents, in current dollars, the amount that would be required to replace each physical asset with a brand-new asset, plus net working capital.

Working capital often can be taken straight off the balance sheet, although it should be "reality checked" in a couple of ways. For example, if the firm uses the first-in, first-out (FIFO) accounting method, then its working capital may reflect the historical cost of that capital rather than the current cost. Also, it is useful to look at a few simple ratios (mainly relating to sales) to make sure that the working capital for the year in question isn't particularly abnormal.

The natural starting point with fixed assets is the level of gross fixed assets (before accumulated depreciation), as listed on the balance sheet. The next step is to convert the historic numbers to current dollar values. This can be done painfully by looking at the historical investment patterns of each company and the cumulative local inflation rate since that investment, or it can be approximated with a simple model that makes assumptions about asset lifetime, real investment growth, and average inflation. The main worry about such approximations is that errors in the assumed inflation rate can compound over long asset lifetimes. Fortunately, the resulting current-dollar investment can be reality checked. In general, two companies in the same business and at the same level of vertical integration should have the same ratio of current cost investment to sales. As a result, a meaningful discrepancy in this ratio can be treated as a signal of a likely error in the assumed historic inflation rate or some other inconsistency. In the end, common sense should prevail over blind application of the reported numbers.

For managers in a real hurry, a short cut can be taken to achieve a reasonably accurate estimate of the CCI for a large diversified chemical company. First, there is a relatively stable relationship between sales and current cost investment for many broad-based chemical companies: That is, the ratio of fixed assets (at cost) to sales is about 1.25, and the ratio of working capital to sales is around 0.17, making total CCI around 1.42 times sales. This relationship is clearly not perfect and may vary quite widely for smaller, more focused companies, but for the large diversified chemical organizations it works fairly well. Thus, dividing COCF by sales, which is a far more easily gathered figure, and then dividing the result by 1.42 will provide a reasonably accurate estimate of COCF/CCI.

APPENDIX 3: FINANCIAL CHARACTERISTICS OF THE CHEMICAL INDUSTRY

With the general framework for analyzing corporate internal rates of return in hand, the next task is to utilize this framework to measure performance. Such a measure requires specifying certain financial characteristics of the chemical industry, including asset lifetime and scrap value, capital intensivity, and so on. We discuss these items in turn.

Asset Lifetime and Scrap Value

As noted earlier, the chemical industry has an average asset lifetime of somewhere between 18 and 22 years. This figure can be estimated in several ways. First, there is the common-sense observation that most chemical plants operating today were built within the last two decades. However, there are a number of other approaches. One is to calculate the number of years of historic capital investment that would have been required to reach the current level of gross fixed assets, either by looking at actual historic investment data (problematic) or by assuming that the growth of capital expenditure has been constant over time. Another approach utilizes the fact that the ratio of gross fixed assets to annual investment is a function of both growth and asset lifetime, so that asset lifetime can be inferred from observations of the other two variables. In the end, all of these approaches have their shortcomings, and the volatility of most asset-lifetime estimates seems quite high. As a result, it would be unwise to place much confidence in any estimate of asset age based on evidence from a single company; the variations in measurement and accounting are just too wide. In our view, the best methodology is to examine a number of different firms in the same industry and take the average.

In theory, the scrap value of investments must be included in any model of discounted cash flow. After all, not all fixed assets wear out completely. For example, a building's purchase price includes the land on which the building sits, and even when the building becomes obsolete, the land retains value. For the chemical industry, however, land is generally a relatively small amount of the overall cost of constructing a chemical-production complex. In addition, significant decommissioning costs associated with shutting down a production site, often including environmental remediation, offset any remaining scrap value. As a result, it is reasonable to assume a scrap value of zero for the chemical industry.

Capital Intensivity and Capital Consumption

Although calculations of capital intensivity and capital consumption are not necessary for the calculation of IRR, they are worth discussion simply to avoid some common confusions. For starters, there are two potentially conflicting definitions of "capital intensivity," the first being the investment required per unit of sales and the second the amount of capital used per unit of sales per year. We will call the former "capital intensivity" and the latter "capital consumption rate." To clarify the distinction: Two companies might build different plants, each costing $100 million and each creating $100 million in sales. Both of these companies could be said to have a capital intensivity of 1.0, as each dollar invested is producing a dollar in sales, in both cases. However, if the first plant lasts 10 years while the second plant lasts 20 years, the annual capital consumption of each facility is meaningfully different. In our terms, the capital intensivity is the same, but the capital consumption is twice as high for the plant that only lasts 10 years.

In the chemical industry, roughly $1.25 must be invested in fixed assets to produce $1.00 in sales per year. Somewhat surprisingly, this level is relatively constant across a broad range of diversified chemical companies. This is demonstrated in Figure 14.14, in which we show the ratio of sales to gross fixed assets for a number of chemical companies (the average ratio is less than 1.25 because assets are reported in historic currencies). Although capital intensivity varies greatly between different branches of the chemical industry — e.g., paints are low and industrial gases are high — such differences often average out about the same for the large diversified firms. These chemical company capital intensivities also have not changed much in the last two decades, although the ratios do move with the business cycle, as sales rise and fall.[10]

If $1.25 is required to build a plant that produces a steady stream of annual sales for 20 years, then the capital consumption rate of that plant is 1.25/20 or 0.0625. Therefore, if a company is operating at steady state—meaning without growth—depreciation would be expected to average around 6.25% of sales. Introducing inflation into the calculations would tend to decrease this ratio, since depreciation is in historic dollars, while a positive growth rate would increase the ratio, since depreciation tends to be taken relatively early in an asset's lifespan and a higher growth rate would mean a higher proportion of younger assets. Figure 14.15 shows that the depreciation/sales ratio for our standard set of chemical companies is relatively close to this figure (this has also been true historically), although differences have, of course, occurred for certain companies at certain times.

Finally, capital intensivity should not be confused with cost structure or with value added. It's important to remember that, in its pure economic sense, value added is calculated by subtracting from revenue all nonlabor, noninterest expenses. Thus, value added does not mean the same thing as "profitability" or "high margins." Instead, it tends to describe whether labor expenses are high or low relative to nonlabor expenses, which often results in a high level of value added or vice versa.

These differences are best illustrated with some simple examples. Conventional wisdom is that BASF makes low value-added, highly capital-intensive products, while Bayer is in the opposite position. However, in Figure 14.14 we showed that BASF's and Bayer's ratios of gross fixed assets to sales are almost exactly the same, and historical data shows that this similarity has persisted for many years. But although the capital intensivity of these two producers is nearly identical, their cost structures are different: BASF pays more for raw materials and less for labor, while higher-value-added Bayer spends more on labor and less on raw materials. We'll revisit these points later.

[10]In addition, the sharp rise in the dollar in the mid-1980s did raise the *book* value of American-based firms such as Dow relative to its German counterparts, and thus also raised their ratio of gross fixed assets to sales. However, when the dollar returned to close to its earlier value in the later 1980s, the ratios again returned to their historic levels.

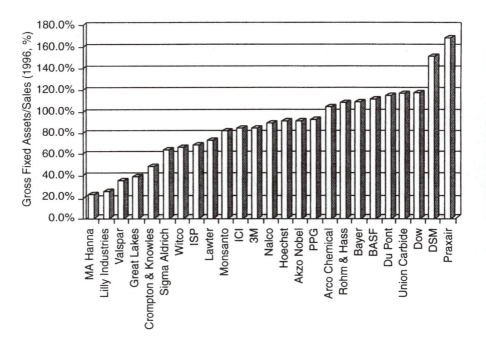

Figure 14-14. Gross Fixed Assets/Sales, 1996.

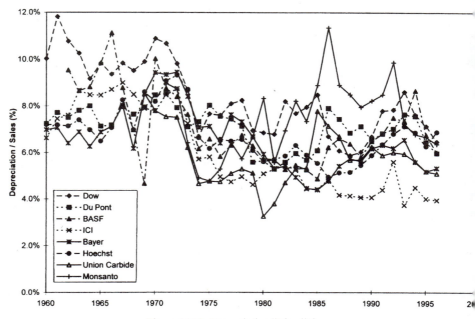

Figure 14.15. Depreciation/Sales (%).

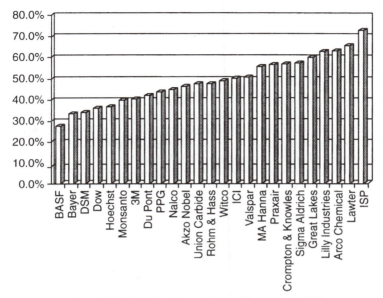

Figure 14.16. Net Fixed Assets/Gross Fixed Assets, 1996.

The confusion between capital intensivity and value added often comes up in an assumption that so-called specialty chemicals, often defined as those with high value-added margins, also have low capital intensivities compared with commodity-chemical products. This is not, however, always the case. Many basic paints, for example, are commodity products with relatively low value added and relatively low capital intensivity. In contrast, the amount of capital required per unit of sales for many specialty chemicals is often not much different than that required for commodity chemicals (industrial gases, arguably a specialty product, are considerably more capital-intensive than petrochemicals).

Depreciation Schedule and Tax Rate

The depreciation schedule and tax rate that a company faces will have a significant influence on its after-tax internal rate of return. There are several ways of finding what type of depreciation is used by a company. (The simplest is to just ask the management.) In general, we assume that most American chemical companies use a sum-of-the-year's-digits depreciation policy. Under common business conditions of moderate growth and inflation, this results in a net-fixed-assets/gross-fixed-assets ratio of around 0.45; almost exactly what is observed in the annual reports of the larger companies themselves, as shown in Figure 14-16. Some of the specialty companies tend to have higher ratios, reflecting the fact that these firms have had higher growth rates and correspondingly newer assets (which have been depreciated less). For the base model, we also tend to use American tax rates, with 35% seeming rather reasonable.[11] As noted earlier, countries such as Germany, with higher corporate taxes, also tend to have more

[11]There are differences between tax and book accounting, but delving into this distinction is well beyond the scope of this paper.

rapid depreciation. As a result, actual taxes paid, as a percentage of operating cash flow, are not that different between German and American firms — a result that surprises many analysts and managers alike.

Fixed Assets/Working Capital

The next financial characteristic needed in our IRR model is the amount of working capital required per unit of fixed investment. Working capital can be thought of as current assets (cash, receivables, and inventory) minus current liabilities (payables and short-term debt). It seems generally true that the cash and short-term debt levels of many companies — chemical companies included — while often quite variable with time, are on average in reasonable balance. Furthermore, excess cash should really be thought of as an investment in a separate business, while excess short-term debt is usually a financing decision (short rates look attractive, a long-dated liability is about to expire, etc.), rather than an ongoing business requirement. The asset/liability balance also extends to receivables and payables, although this latter case is more confined to chemicals and a number of similar industries than a general axiom. (The retail industry, for example, tends to combine very few receivables with a high degree of payables.) Thus, for chemical companies, the vast majority of net working capital is comprised of inventories. The ratio of inventories to fixed assets for chemical companies is typically 1 to 7.5. Thus, if the initial investment in fixed assets is 125% of sales, this implies a sales/inventory ratio of 7.5/1.25 or 6.0, a number that will be familiar to most chemical-company CFOs.

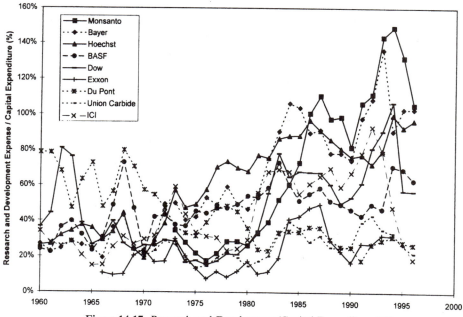

Figure 14-17. Research and Development/Capital Expenditure (%).

Research and Development

Now let's look at research and development expenditures for the major chemical companies over time. The level of R&D can be examined in a number of different ways, but our preferred approach is as a percentage of all capital expenditures, as shown in Figure 14.17. Sizable variations in research and development spending are obvious from a quick glance at this figure. Generally, the more health-care-focused companies have relatively high ratios of R&D investment to fixed investment: Monsanto, Bayer, and Hoechst. At the bottom of the chart are Union Carbide and Exxon, two fairly high-volume petrochemical companies that do not rely as heavily on technology input into their products. DuPont, too, occupies this latter position, due more to its large oil and gas operations (which require very little R&D investment) than to any deeper cause such as product characteristics. The plunge in R&D expenditures for ICI after 1993 occurred because of the de-merger of ICI and the highly R&D-intensive Zeneca operations. Note that different levels of R&D often reflect the competitive situation and product portfolio of the individual companies. It would be unwise to conclude from the chart that certain companies are spending "too much" or "too little" on R&D.

While we can define an intensity and consumption rate for research and development in much the same way as we calculated equivalent figures for fixed assets, the relevance of these figures is suspect. R&D intensity, defined as the amount of R&D investment per unit of sales per year, is particularly difficult to define because investment in research and development alone does not produce sales; some sort of fixed-asset investment is required, as well.

PART VI
Insights from this Study

15 Conclusions

ASHISH ARORA, RALPH LANDAU, and NATHAN
ROSENBERG

A study of the history and development of the chemical industry is important in its own right. Its close links with a wide range of industries across the economy—automobiles, textiles, oil refining, consumer products, pharmaceuticals, and many others—have meant that its fortunes in a country wax and wane with those of the economy as a whole. But the chemical industry is more than just a bellwether of the business cycle; rather, through feedbacks on both supply and demand its growth has stimulated the growth of other sectors, and therefore, of the economy as a whole. Accordingly, we undertook this study to understand what the history of the first science-based industry could tell us about the dynamics of long-term growth in modern industrialized societies.

In the long sweep of history, technological innovation has proved to be the most important factor for growth in this industry and, very likely, in other science-based industries. Science, however revolutionary, is not enough. It must be commercialized, and that is the process which creates wealth. This commercialization is an investment—physical, human, and intangible (R&D, etc.)—strongly affected by tax rates, monetary policy, exchange-rate variations, financial institutions, regulatory, and other economic policies. The commercialization process involves risks at every stage. Examples of such risk are found in Landau's chapter illustrating the diverse experience of Philips Petroleum and Montecatini in polypropylene polymers; the success of DuPont and Hoechst in polyester fiber (although the latter is now seeking to move away from concentrating on this area, while DuPont, by acquiring ICI's position, has increased its role in polyesters), contrasted with the reduction of the role of the original fiber developer ICI; and the success of DuPont in nylon, despite the earlier discoveries in Germany of polymer science by Hermann Staudinger.

Our study clearly shows that other forces affect the outcome of the commercialization process. History matters, as the very different post-World War II experiences of Japan and Germany show. Germany's long tradition in chemicals enabled her not simply to catch up but also to claim her place among the commercial and technological leaders in the industry. By contrast, the Japanese chemical industry, despite its size, is not generally reckoned among the technological leaders. Clearly, institutional and social and political differences in society matter, as do details of how firms and industry are organized. Many of these forces cannot be readily captured in the mathematical models of the growth of the economy that economists currently use, which leave an unexplained "residual" labeled by many as "technology", after accounting for all measurable inputs. Indeed, the notion of "technology" that is used in these models encompasses more than technology as understood by an engineer or scientist; it includes many of the other factors that we have tried to tease out in our study. And even if capital and technology move across national boundaries and markets become global, political and social institutions change only slowly, as does the mobility of people across national boundaries. They retain their national moorings for much longer and are an important part of what makes a country hospitable to innovation and the commercialization of inventions, and hence, to economic growth.

Although generalizations can sometimes be misleading, we believe that several broad themes emerge from our study that should be more generally applicable to other science-based industries, as well. The historical record shows very clearly that even in a science and technology intensive industry such as the chemical industry, commercial advantage arises from many sources whose importance varies over time and across sectors. For instance, in the United States, while initially the raw-material endowment was crucial to the industry's growth, over time, the size of the market and the technological advances, both in products and processes, came to be more important. No single source of advantage appears to have lasted for very long, although some have been very important for extended periods of time; the development of new, global markets and international competition has seen to that. Alternative sources of raw materials, or synthetic substitutes, were developed when raw materials became a major bottleneck. For example, the shortage of naturally occurring Chilean nitrate fertilizers first led to the Cyanamid process and then to the famous Haber-Bosch process in Germany for fixation of atmospheric nitrogen to produce synthetic ammonia for fertilizer. The United States' endowments of oil and natural gas, while they played a crucial role in getting the petrochemical industry off the ground, did not prove to be a source of overwhelming long-term advantage, as others sought and developed the vast oil and gas deposits in the Middle East. Mobility of personnel, wars, and other historical accidents have also proved to be disruptive of long-lived leadership by individual countries.

The American experience also underscores the complementarity among market size, resources endowment, and technology. The compound effect of a large market, abundant mineral resources, and technological change — exemplified, for instance, in the rise of petrochemical-based synthetic polymers — was much greater than simply the sum of the individual parts. Moreover, these complementarities were realized within a broad macroeconomic, financial, and sociopolitical legal context, which has often played a very important role. In other words, long-term growth does not just depend on getting one thing right; rather a constellation of factors must be in place. This realization is what lies behind the analytical framework of this book and the use of the matrix of the levels of the sources of comparative advantage as its organizing principle. Put differently, rather than one or two isolated factors, sustained international leadership depends on the effective working of a *growth system*.

It is not surprising, therefore, that very few direct government interventionist policies have been shown to have been successful in the sense of having had long-lasting effects on the success of the national industries. As Eichengreen's chapter shows, nearly every country attempted trade and industrial policies to promote domestic chemical industries at certain periods, and this did provide the latter with some breathing space. In some cases, this move enabled firms to develop new capabilities that proved valuable in the future. More often, protectionism, by insulating firms from competition, led to long-run weaknesses that markets later exposed. And even when the protected industry has prospered, it does not follow that the protectionism was worthwhile from the point of view of the economy as a whole. The creation of ICI in Britain is an example of successful direct government intervention, as an apparently failing set of chemical firms were transformed into a fairly innovative and progressive enterprise, capable for the first time in more than 50 years of competing with the successful large German chemical firms. But, as described further on, the merger was not an unalloyed long-term success.

The growth of Japanese petrochemicals immediately after the war has been nothing short of spectacular and the Japanese government, especially through its Ministry of International Trade and Industry (MITI), has had a hand in that success. Stimulated by such government planning, the primary source of growth nevertheless was a combination of imported technology and domestic investment by private firms, against a backdrop of high saving and investment rates in the economy as a whole. However, this combination appears to have been far less successful in the 1980s and 1990s. Indeed, as Hikino et al. clearly show, the industrial group structure, which was a source of strength during the earlier catch-up phase, now is a source of inflexibility, retarding adjustment to the changed international conditions. The Japanese industry, although the second largest in the world, is not generally regarded as a major player on the technological or international scene, and this is consistent with the idea that direct government intervention (often called industrial policy) is no longer feasible in complex, technology-based industries. The late 1997 financial meltdown of the Asian currencies where the governments generally followed the Japanese model, seems, at this early stage, to have further demonstrated a number of important points. It appears that cozy relations between politicians, the financial institutions (especially the banks, which are much more prominent in financing than in the American model), and private companies can lead to misallocation of capital and possibly corruption, which the more impersonal markets are more likely to avoid. One consequence of this misallocation seems to have been the creation of overcapacities and excessive real estate building. It will also affect adversely, at least for a time, the growth among other industries of the chemical industry and its plans to expand aggressively on a global basis. At this writing, the effects on the world economy of these events are still unclear. It is possible that a moral hazard problem will be created if bailouts at the International Monetary Fund (IMF) and various western countries could hold harmless lenders and speculators, who might continue such behavior in the future assuming there is a guarantee for their risk taking.

Another very important point is demonstrated by the experiences of the German chemical industry in the 1930s under National Socialism. The drive for self-sufficiency focused efforts on research in areas such as coal hydrogenation, and acetylene chemistry. These efforts were technically successful but did not make commercial sense, because by the 1940s it was highly likely that petroleum, rather than coal, would be the feedstock of the future. Not only did protection not help, in this instance it actually had harmful effects, because it prevented the industry from adapting to changed circumstances.

In more recent years, the German social market compact among government, labor, and business contributed to a spectacular recovery and growth, but it has resulted in a country with enormous labor costs and high unemployment, so that the industry is seeking to grow by expanding abroad rather than by more domestic investment. This is an illustration of how national institutions and socioeconomic conditions change slowly, even in the face of major global change. Despite the imminent creation of a common currency in Europe (the euro), there are still many national differences, for example the varying cost of energy, and tax policies which have tended to preserve national chemical industries rather than create one truly European in scope.

Public support of research universities has been perhaps one of the only industry-level policies with some measure of long-lasting success. A successful education system takes time to build and must be nurtured with care. As Rosenberg's chapter shows, what is crucial, although much harder to achieve in practice, is a way of linking universities with industry that makes universities responsive to the needs of industry while preserving the autonomy of the universities. Indeed, the much-vaunted German educational system, which was an important component of the German growth system based on organic chemistry, seems to have turned less responsive to industry needs, as Murmann and Landau note in their chapter. However, it would be a mistake to see universities solely as producers of new knowledge or trained engineers and scientists, although these functions are undoubtedly important. A first-rate university system may be important in other subtle ways as well. As the history of synthetic dyestuffs in Britain (their discoverer) shows, the willingness of firms to invest in promising, but risky, new technologies depends on their ability to understand the new technologies. Close links with leading researchers in the field, as well as the close involvement of researchers in the technological activities of firms, seem to have been essential for firms to be willing to make such investments. There were far fewer of these in Britain than in Germany, whose university system was superior.

Government policy can affect the willingness of firms to make such risky, but vital, investments by providing a stable macroeconomic environment with low inflation rates conducive to long-term investing, and low tax rates, as suggested by Eichengreen's chapter. But there are other ways in which government policy affects the ability and willingness of firms to invest in the generation and commercialization of new knowledge. Commercial success in a science-based industry also depends on a system of intellectual property rights; but, as the tort law controversies in the United States in the 1980s and 1990s and the patent litigation in Britain in the 1860s show, systems of property rights must be crafted with care. Similarly, labor-market flexibility turned out to be very important in the post-World War II success of Germany, whereas the opposite was the case in Britain until recently. The environment has been another arena in which government policy has potentially important effects on the fortunes of firms and the industry as a whole. Esteghamat's chapter confirms that environmental regulation has not been a decisive factor in the competition among developed countries (possibly because their environmental laws are broadly similar), and that in all the major countries, the chemical industry has dramatically reduced the environmental impact of its activities. However, some consequences of environmental regulation have led to diversion of investment in more polluting sectors (such as dyestuffs) to developing countries and away from Europe and the United States. Dyestuffs are no longer critical businesses in countries where they once provided dynamic growth possibilities such as in Germany. Arguably, economic growth and the energy use that such growth entails have much greater environmental impact, and although our study does not deal with it, environmental concerns will be important in determining the extent and pattern of economic growth in the future.

Competition is an important spur to innovation and efficiency, and government antitrust and other generic science and technological support policies, for example are generally helpful in this respect. To pick up a theme that runs through much of our study, however, openness to global markets can play an important role here. For example, the export orientation of the German chemical industry helped keep the leading companies innovative, and competitive, even after extended periods of cartelization. One must not view openness simply in terms of trade of products. As the chapter by Arora and Gambardella shows, the easy flow of capital and technology across national boundaries can be as important, as is illustrated by the influence of the specialized engineering firms, mostly American, in diffusing petrochemical technology around the world. For developing countries, the availability of technology was key to the establishment of a domestic chemical industry. A similar diffusion of advanced technology has taken place in licensing and overseas investments by firms. The lesson for both policymakers and managers is that global markets in technology and capital can provide followers and often rivals that have some competitive advantage, such as access to cheap raw materials or large markets with the opportunity to rapidly catch up with leaders. By the same token, the availability of many process technologies on a worldwide basis has meant that, in many cases, process technology has ceased to be the decisive source of competitive advantage for established firms, unless constantly improved upon, or replaced by even more modern technology, often involving new catalyst systems. Furthermore, the accumulated experience ("knowhow") of major firms in manufacturing, research and development, marketing, and strategic direction is the best competitive tool that such firms enjoy. It is not too easy to diffuse it to others in the sense described above, especially because much of this "knowhow" is in the minds of their personnel. In this sense, technology as understood by the engineer or scientist is by no means the same everywhere in the world, quite apart from the socioeconomic and policy differences that were described above as constituting a portion of the "technology" measured by growth economists.

In other words, as technology and markets change, so must firms, especially if they compete in the global market. One implication of our study is that even with favorable policies and institutions, firm behavior is in no way an automatic result. Some companies are simply more successful in building long-term capabilities that can sustain them over a long period of time, such as DuPont and IG Farben and its successors. Others, such as Dow, have risen rapidly from relative obscurity to achieve world-class status; and still others, such as Allied Chemical, which favored high dividends and relatively little research and development, have been less successful. As the chapter by Chandler, Hikino, and Mowery shows, commercial leadership at the level of the firm has often lasted for long periods, when the firm has made the required investments in technology, production, and marketing, as well as in research and commercialization of innovation. Thus, early-mover advantage has been an important component of firm success in the chemical industry; but so has the nature and quality of the firms' personnel and management. Strategies that made sense in some periods, however, may turn out to be ill-suited when times change.

During the first two decades of post-World War II, the United States, and its chemical industry, dominated the world chemical industry and was at its most innovative. This success led its firms to expand aggressively into new businesses. But with the two oil shocks and the unusually high-world inflation, these moves turned out to be mistakes. At this time, also, the United States chemical industry was maturing, followed by that in Europe, and with the increasing commodification of so many of its businesses, cost considerations became more important. Accordingly, profits became harder to sustain, and the rise of the global financial markets in the 1980s increased the pressure on managements to cut costs and identify business segments to be separated out to allow each company to focus on its core strengths.

An illustrative case for this view can be made, perhaps only in hindsight, that ICI's management, backed by creative scientists and technologists, chose after the war to be involved in many sectors of the chemical industry. In the more recent decades as new inventions became scarcer, producing commodity chemicals profitably required a large market share, the most modern technology, and judicious investment and restructuring — and ICI lacked these in many of its businesses. Hence, when its most successful activity — pharmaceuticals — was divested into Zeneca, its remaining, mostly commodity businesses (although it has a strong position in paints), were not truly competitive. This led to ICI's acquisition of the Unilever specialty chemicals business and the divestiture of many of its commodities. ICI is now a more focused company that bears little resemblance to the one pre-1990. By contrast, BP Chemicals focused in recent years on a few key commodities and seems to have a secure position in polyethylene, acrylonitrile, and acetyl chemistry.

Other companies are pursuing similar strategies: Hoechst, is planning to concentrate on life sciences (including pharmaceuticals), a path also sought by Monsanto, American Cyanamid, (now acquired by American Home Producers), and Rhône Poulenc, among others. Yet, there is a real question whether dividing up a chemical company into its constituent elements always makes good sense, or potentially limits future opportunities for taking important technological risks for future growth which only large companies can afford (as the polyester and polypropylene story in Landau's chapter demonstrates). The Bayer company seems to be following this pattern as an integrated company; on the contrary, Hoechst, as mentioned above, is in the process of divesting itself of its businesses so that it can remain a pharmaceutical and agro-chemical company. In this sense its strategy is different still from ICI's. Many companies are seeking to exit commodity chemicals and focus on specialties or performance chemicals, but these are unlikely to offer a way out for the large number of commodity manufacturers, and require different kinds of management skills.

As a result, many mergers, acquisitions restructurings, and divestitures have occurred over the last 15 years. One vehicle for this appeared in the form of entrepreneurial companies founded by Gordon Cain and Jon Huntsman, who relieved larger companies such as DuPont, Union Carbide, Monsanto, and Celanese, of their unwanted units and greatly improved their performance by reducing overheads and other means. The restructuring has also put R&D budgets under close scrutiny. In many companies, budgets for basic research have been severely reduced. No doubt some reductions were necessary, but the decline of basic research in a sector that has traditionally financed virtually all of its large R&D from within itself raises questions about the long-term prospects. It appears unlikely that growth can be sustained through foreign expansion alone! Similarly, in the 1990s, the European chemical industry has entered into major restructurings, mergers, alliances, and the like.

These restructurings of course result in job reductions, bu the industry is not labor intensive, so that this effect is relatively small in the country as a whole. The issue of jobs, equity, and long-term growth has not been a subject of this book, which is devoted to the creation of long-term wealth for society as a whole and is further touched on just below. Although the United States has done well in this regard it will be a major social and economic issue in many countries in the years ahead.

The restructuring has two important lessons. First, change happens at the level of the industry, not just within firms. The rise of new firms and the disappearance of old ones can sometimes be an important, albeit painful, form of renewal for an industry. Small, innovative firms that have played an important role earlier in the industry are now confined to a few subsectors. Thus, while bigness does bring longevity and durability, only ceaseless change can sustain firms. Similarly, as the industry has matured, process improvements and efficiency have increased in importance, and along with them, the importance of the engineer. As Landau's chapter shows, this is an important difference between the rise to leadership of the United States in the 20th century, thanks to large-scale production of petrochemicals based on chemical engineering, and the German ascendancy in the 19th century, based on organic chemistry. This pattern is likely to apply to other science-based industries, as well, as they mature.

The second lesson of restructuring is that the speed with which an industry responds to change depends in important ways on the existing institutions, particularly the financial institutions. Da Rin's chapter details how the different financial institutions of Britain and the United States on the one hand, and Germany and Japan on the other, have had a profound long-term impact on the chemical industry in these countries. The close relationships of the banking systems in Germany and Japan clearly fostered the rapid rise of this industry and insulated managements from short-term pressures to perform — clearly advantageous to a research-intensive industry where time horizons are long, risks great and managements wise. The U.S. financial system evolved very flexibly, thus permitting well-managed companies to grow. In Britain, however, the financial system was less friendly and contributed to the difficulties that entrepreneurs in this industry faced. As the financial markets in different countries became more internationally integrated over time, however, they will partially converge. The recent financial crises in the Asian countries, following the difficult period triggered by the collapse of the Japanese "bubble" economy in the 1990s, suggest that their financial institutions are immature and lack adequate government oversight and appropriate regulation, particularly in an era when large international flows of capital are invited in. Speculative activity has been rife in such situations, and imprudence by both borrowers and lenders is apparent. These events seem to confirm, on a much larger canvas, that even fine technology and hard working people (which was certainly the case for the Asian countries) are insufficient in the face of unfavourable and politicized macroeconomics, institutions, and policies, as we have pointed out earlier in this volume.

The chapter by Richards offers many illuminating insights into various matters taken up in the rest of this volume, and has argued convincingly that while equity markets do a good job of evaluating past performance and provide a good measure of future expected performance, they are not by themselves capable of forcing managers into a particular course of action; e.g. maximizing investment returns. In other words, markets themselves do not provide a corporate-governance function, and this must instead be provided by active owners. Already, over the last 10 years, firms have felt the pressure to increase shareholder value by focusing on their core competencies and cutting costs, and this focus has certainly benefited both the economic growth rate and the unemployment rate in the United States. Given the large weight of the United States in the world financial system and the increasing attention by owners of capital (or their agents, financial institutions) to seek the best return globally on their investment, it is likely that even in the bank-centred systems of Germany and Japan, managers will have to pay more attention to the interests of their shareholders. Richards presents a large array of data to show how successful some firms have been in creating walth for their owners, while others, even though large, have operated rather more in the interest of their managements than of their shareholders. This seems also to be largely true for Japanese companies, whose low profitability, as depicted by Hikino et al., suggests the dominance of managerial over shareholder control. Richards argues further that, while the short-term well-being of a specific corporate workforce may be improved by a benevolent management, the longer-term wealth of society overall (and hence the entire workforce) is best served when individual managers pursue maximum investment returns thus allocating capital to its best use. Of course, to be effective, managements while maximizing shareholder value are constrained by the interests of other stakeholders, such as customers, employees, and the public at large. His proposed "circle of governance" is a suggested institutional and policy environment that he believes is best suited to increasing nation wealth, while dealing with short-term unemployment volatility and income inequities is the proper function of society and its governments and institutions as a whole, not of individual corporations. There is still some controversy among economists on this issue, but Richards supports his case as being increasingly important to a rapidly globalizing economy, where competitive pressures on firms are rising. The increasing attention of firms to shareholder value seems to support his case as well, although it is true that the changing technology of this industry is difficult for analysts to comprehend fully.

As this book is completed, it also turns out that continuing inquiries into the subject of longer-term growth have become the focal point of much debate among politicians, economists, journalists, and scholars. With the growing globalization of the economy and the increasing world population, as well as puzzlingly high unemployment rates in many countries, substantial dissatisfaction is being expressed with the apparently sluggish rate of growth. These strains are being felt not only in the United States and other industrial countries, but also by leaders in many developing countries, who see the East-Asian success stories (although this has recently faltered) as challenging their own slower growth rates.

It is in this context that a renewed drive for economic nationalism and more protectionism flourishes. In the past, as between 1890 and 1930, the United States was indeed a protectionist country. Yet, after the Second World War, productivity growth was even greater, when trade barriers fell, and technology flows became much more global. Other countries grew even more rapidly than the United States, but this did not harm it. On the contrary, the country benefited from the greater demand for its goods and services abroad and the imports of valuable products (such as consumer electronic devices and automobiles) that were much more desirable to American consumers and stimulated competition among firms. Today, the economies of the world are so closely linked by currency and trade flows that protectionism, despite its populist appeal, would be both difficult and harmful. Most economists would agree that whatever problem the United States has had in recent years, it has not basically been due to the successes or excesses of other countries.

If sustained innovation and productivity growth are the key to relieving these political stresses, one cannot achieve this simply by manipulating fiscal and monetary policy. The policies for economic growth have a time span considerably longer than that of the stablization policies that occupy much of the political center stage. The firms that actually create wealth require an environment that is as favorable to wealth creation as possible, including the right incentives.

The widely applicable principles underlying economic growth, described in this chapter, hold both a promise and a warning. The promise is that investment in the three-legged stool of physical, intangible (including research and development), and human capital can stimulate other investment, which can lead — at least in the theoretical models — to a sort of virtuous circle of feedbacks to yield increased productivity growth. The corresponding warning is that the policies for achieving this continuing increase in the standard of living are not obvious. Some advocates of "industrial policy" speak as if government can guarantee future prosperity by showering favors on some private industries, choosing to ignore or minimize the critical import- ance of a propitious macroeconomic and policy climate within which all business must work. Frequently, the negative consequences of such intervention lie elsewhere in the economy, such as in reduced competition or flagging innovation in other industries, or a greater risk of inflation. Without an appropriate blend of macroeconomic and microeconomic insight, there is no guarantee that a nation can sustain a high standard of living in the world.

ABOUT THE CONTRIBUTORS

ARORA, ASHISH,   Ashish Arora received his Ph.D. in Economics from Stanford University in 1992 and at present is Associate Professor of Economics and Public Policy at the Heinz School of Public Policy and Management, Carnegie Mellon University, Pittsburgh. His research focuses on the economics of technological change, management of technology, and technology policy. In the past, he has carried out research on biotechnology, technology licensing and technology transfer, the role of intellectual property rights and science policy.

CAMPBELL, TOM,   Tom Campbell holds a Ph.D. in Economics from the University of Chicago and a J.D. from Harvard. He is a tenured professor at Stanford Law School, where he has taught microeconomics, international trade law, and advanced antitrust, among other subjects. He has been a White House Fellow, a Supreme Court law clerk, a California State Senator, and is currently a United States Congressman.

CHANDLER, ALFRED D., JR.,   Alfred D. Chandler, Jr., Straus Professor of Business History Emeritus, is an industrial historian whose major works include *Strategy and Structure: Chapters in the History of the Industrial Enterprise* (1962), *The Visible Hand: The Managerial Revolution in American Business* (1977), and *Scale and Scope: The Dynamics of Industrial Capitalism* (1990). Professor Chandler got his Ph.D. in History. He became Straus Professor of Business History, emeritus, in July 1989.

DA RIN, MARCO,   Marco Da Rin received a B.A. in Economics from Università Bocconi (Milano) in 1988 and a Ph.D. in Economics from Stanford University in June 1995. He taught at the Universitat Autonoma of Barcelona before joining the Innocenzo Gasparini Institute for Economic Research (IGIER) of Università Bocconi (Milano), where he has been a Fellow since 1995. His research interests cover financial economics, financial history, and comparative institutional analysis, themes on which he has published in professional journals and collective volumes.

EICHENGREEN, BARRY,   Barry Eichengreen is the John L. Simpson Professor of Economics and Political Science at the University of California, Berkeley; Research Associate of the National Bureau of Economic Research; and Research Fellow of the Centre for Economic Policy Research (London). He is author of *Globalizing Capital: A History of the International Monetary System* and has written widely on the histories of the U.S. and European economies in the 20th century.

ESTEGHAMAT, KIAN,   Kian Esteghamat has been an engineer, consultant, and project manager in the fiber optics, energy, and computer-software industries. His research interests include financial economics, global investments, and impact of risks and regulations on international investments. He is a doctoral candidate in the Department of Engineering-Economic Systems and Operations Research at Stanford University and holds a Master of Science degree in Electrical and Computer Engineering from the University of California, Irvine.

GAMBARDELLA, ALFONSO,   Alfonso Gambardella is Professor of Economics and Management at the Faculty of Economics of the University of Urbino, Italy. He obtained his Ph.D. from Stanford University, Department of Economics. His research interests are in the economics of technical change, applied industrial organization, and management economics. In 1995, he published a book on the effects of advances in the life sciences on the innovation process, competition, and organization of the U.S. pharmaceutical industry.

HARADA, TSUTOMU,   Tsutomu Harada is assistant professor of economics at the Graduate School of Business, Kobe University, where he focuses on industrial organization and the economics of technological change. He received his B.A. and M.A. from Hitotsubashi University and his Ph.D. in Economics from Stanford University. His current research interest is in technological change in the Japanese chemical and machine-tool industries.

HIKINO, TAKASHI,   Takashi Hikino is Senior Research Associate of Business History at the Harvard Business School and Research Fellow at the Center for International Studies at MIT. He has published extensively on comparative business history and international management. He is currently editing two volumes on international business and economic development: with Alfred Chandler and Franco Amatory, *Big Business and the Wealth of Nations* and with Hideaki Mijajim and Takeo Kikkawa, *Policies for Competitiveness: Industrial Economies during the Golden Age of Capitalism.*

HORSTMEYER, MICHELINE,   Micheline Horstmeyer has a B.A. in American History from Middlebury College, 1971, and a Ph.D. in American History and Education from Columbia University, 1979. She has taught history at the seconday level; developed curriculum materials for secondary-level history; and been a teaching/research assistant at the college level (Columbia University); as well as an independent consultant doing research and writing on a wide range of topics for various clients including Stanford University.

LANDAU, RALPH,   Ralph Landau was the cofounder, in 1946, and CEO of Halon Scientific-Design Group, which was a prominent innovator in the chemical industry. He is a Consulting Professor of Economics at Stanford University and a Research Fellow at Harvard's Kennedy School. Dr. Landau earned an Sc.D. in Chemical Engineering at MIT. In 1985, he received the National Medal of Technology. He has also received the Perkin Medal (1981) and the Othmer Medal (1997). He has been named one of Chemical & Engineering News' Top 75 Contributors to the Chemical Enterprise. He has published numerous works dealing with technology, economic growth, and innovation.

MOWERY, DAVID C.,   David Mowery is Professor of Business and Public Policy in the Walter A. Haas School of Business at the University of California, Berkeley. He received his undergraduate and Ph.D. degrees in economics from Stanford University and was a postdoctoral fellow at the Harvard Business School. His research deals with the economics of technological innovation and the effects of public policies on innovation. Dr. Mowery has published numerous academic papers and has written or edited a number of books.

MURMANN, JOHANN PETER,   Johann Peter Murmann is an Assistant Professor of Organizational Behavior at the J. L. Kellogg Graduate School of Management of

Northwestern University, where he focuses on strategic management and the management of innovation. He holds a B.A. in Philosophy from the University of California, Berkeley, and a Ph.D. in the Management of Organization from Columbia University. He currently investigates how institutional differences across countries — in particular, Great Britain, the United States, and Germany — influenced the competitive position of firms in the synthetic-dye industry since 1850.

RICHARDS, ALBERT D.,   Albert Richards is a Managing Director and Head of European Equity Research at Salomon Smith Barney in London. Prior to his current position, Dr. Richards spent eight years with CS First Boston, most recently both as the Director of Sector Research for European Equities and the European Chemical Analyst, based in London. In the latter position, he was responsible for the study of world chemical markets and the major European companies that operate in those markets. Dr. Richards holds a Ph.D. in Chemical Engineering from the Massachusetts Institute of Technology, an M.B.A. from the Sloan School of Management, MIT, and is a Chartered Financial Analyst.

ROSENBERG, NATHAN,   Nathan Rosenberg is the Fairleigh S. Dickinson, Jr., Professor of Public Policy in the Department of Economics at Stanford University. His research has dealt primarily with the economics of technological change and the relationships between scientific research and changes in technology. Among his books are: *Perspectives on Technology, Inside the Black Box, Technology and the Pursuit of Economic Growth* (with David Mowery), *Exploring the Black Box*, and *How the West Grew Rich* (with L. E. Birdzell). He has received honorary degrees from the University of Bologna and the University of Lund. In 1996, he was awarded the Leonardo da Vinci Prize by the Society for the History of Technology for his extensive contributions to the history of technology.

TOKUHISA, YOSHIO,   Yoshio Tokuhisa received a bachelor's degree from Tohoku University (Japan). While at Mitsubishi Petrochemicals (1960–1994), he held various positions, most recently being Managing Director in charge of Corporate Planning. He is also associated with chemical industrial associations (and government councils), often involving coordination and negotiation with MITI. His recent publications on the subject of the Japanese chemical industry include: *Is There Future for the (Japanese) Chemical Industry?* (1995) and *Peculiarity of the Chemical Industry* (1993).

YOSHIDA, JAMES A.,   James Atsushi Yoshida is President of Toho Carbon Fibers in Palo Alto, California. He holds a B.A. from Pepperdine University and an M.A. in International Relations from the University of Southern California. Prior to coming to the United States, he worked in Japan for 30 years in the fields of technology transfer and market development of basic chemicals and petrochemicals. He held executive positions at BASF Japan and Halcon/Scientific Design Co., and in joint ventures copartnering with Arco Chemical, Mitsui, Showa Denko, Sumitomo, and Takeda Chemical, etc.

Index

In this index, *italic* page numbers refer to figures; page numbers followed by "t" and "n"; refer to tables and footnotes; See also refers to related topics.